PHOTOCHEMICALLY-GENERATED INTERMEDIATES IN SYNTHESIS

PHOTOCHEMICALLY-GENERATED INTERMEDIATES IN SYNTHESIS

ANGELO ALBINI
MAURIZIO FAGNONI
PhotoGreen Lab, Department of Chemistry
University of Pavia
Pavia, Italy

WILEY

Published by John Wiley & Sons, Inc., Hoboken, New Jersey
Published simultaneously in Canada

Library of Congress Cataloging-in-Publication Data:

Albini, Angelo.
Photochemically-generated intermediates in synthesis / Angelo Albini, Maurizio Fagnoni, PhotoGreen Lab, Department of Chemistry, University of Pavia, Pavia, Italy.
pages cm
Includes bibliographical references and index.
ISBN 978-0-470-91534-9 (cloth)
1. Carbocations. 2. Carbanions. 3. Intermediates (Chemistry) 4. Photochemistry. I. Fagnoni, Maurizio. II. Title.
QD305.C3A43 2013
547′.1372–dc23

2013007093

MIX
Paper from responsible sources
FSC® C013604

CONTENTS

PREFACE

The concept of intermediates has had a particular role in photochemistry. The distinction between electronically excited states and (highly reactive, but ground state) intermediates has taken some time to become generally accepted. At the beginning of the twentieth century, the founder of photochemistry, G. Ciamician, felt that there may be a relation between the chemical effect of light and the modification of the electronic structure of molecules photons caused, as physics was beginning to explain at that time. The fact that an excited state was an independent species with a thermodynamics of its own was clearly recognized shortly afterwards. However, for a long time, chemists continued to use terms such as photoaccelerated or photocatalytic for any chemical effect of light, with no clear idea about how the impinging photons interacted with the molecules. With the renaissance of photochemistry after World War II, many more reactions were discovered and the physical characterization of electronic excited states was well established. Some scholars felt that excited states could be considered as intermediates, whereas others stressed the difference, in particular because electronic energy could be transferred between molecules without any chemical interaction. Extensive experimental work in the following decades and the advancement of computational chemistry have laid the solid foundation of modern photochemistry and distinguished the role of excited states as intermediates.

An excited state lies on a potential energy surface much higher than the lowest one for that molecular configuration. Such a state has a different electron distribution with respect to the ground state (of which it is an electronic isomer) and may have a more or less polar or radical nature. These states convert back to the ground state surface either in the starting configuration or in a different one by some processes that at a certain point arrive at the lowest possible surface for a certain configuration. This is, however, still much higher in energy than either the reagent or the products. This is an intermediate that can—at least in principle—be arrived at also from a ground state by thermal excitation. Arriving at high-energy intermediates starting from excited states is, however, generally an exothermic process through a small energy barrier.

Thus, photochemistry allows us to arrive also at intermediates that are not attainable by (known) thermal reactions and, as least as importantly, generates intermediates that are reached also thermally under *mild* conditions, which allows us to better direct the ensuing chemistry of such species. Realizing this fact may make synthesis practitioners better aware of the potential of photochemistry and encourage them to include it among their favorite tools. Actually, the photon has to be considered a reagent that is able to generate intermediates and cause in-depth transformation but leaves no residue (the ideal "green" reagent).

It was thus deemed appropriate to highlight the classes of intermediates formed photochemically and which advantages this method offers, maintaining a synthetic perspective. It is often stressed that photochemistry is much less frequently applied in synthesis than it would deserve in view of the many selective transformations it allows. This is not completely true. The examples chosen in the following chapters have been mostly taken from papers published in the third millennium, and many of them embody a significant advancement and would be difficult to predict ten years before. This gives the idea of a living science, ready to acquire a more important role in synthesis. It is hoped that this presentation may be of interest to the synthetic and the photochemical communities alike. As mentioned above, a remarkable amount of photoreactions of synthetic value have been developed in recent years; however, the advancement in the field may further improve by a closer contact between the two communities that at present remain somewhat separated.

What is presented here is based on a long-standing activity in the field of organic photochemistry. Inevitably, personal points of view or even biases may have escaped detection and be present in the text. We apologize for any such inappropriateness or insufficient quotation of work from other laboratories.

Last but not least, we express our warmest thanks to Drs. Stefano Protti and Davide Ravelli, who have had a prominent role in the recent research in this laboratory and have given invaluable help in the preparation of this text, as well as to the other co-workers, students, and colleagues that have participated in this effort.

Angelo Albini
Maurizio Fagnoni

University of Pavia

1

PHOTOGENERATED INTERMEDIATES: PRINCIPLES AND PRACTICE

1.1 INTRODUCTION

With a few exceptions, notably pericyclic processes (and even in this case not everybody concurs with this distinction), organic reactions involve intermediates. According to the IUPAC Gold book, *an intermediate is a molecular entity with a lifetime appreciably longer than a molecular vibration (corresponding to a local potential energy minimum of depth greater than RT) that is formed (directly or indirectly) from the reactants and reacts further to give (either directly or indirectly) the products of a chemical reaction* [1]. This definition excludes other species that intervene on the reaction path, such as vibrationally excited states and transition states, both of which have by definition a lifetime shorter than a molecular vibration. Actually, rationalizing a reaction—that is, recognizing the mechanism—essentially involves determining the intermediate involved and describing how this is formed and reacts. This is a key issue in all aspects of chemistry, a target for scientific studies, a means for making learning chemistry different from memorizing a long list of reactions, and a way for finding and optimizing practical applications. In fact, when the nature of the intermediate has been clarified, one is able to control the course of the reaction. Computational studies are nowadays much more commonly available and powerful, and they help in giving confidence in depicting the shape of the potential energy surface involved, the processes competing, and the location of minima corresponding to the intermediates (**I**, Figure 1.1).

Photochemically-Generated Intermediates in Synthesis, First Edition.
Angelo Albini and Maurizio Fagnoni.

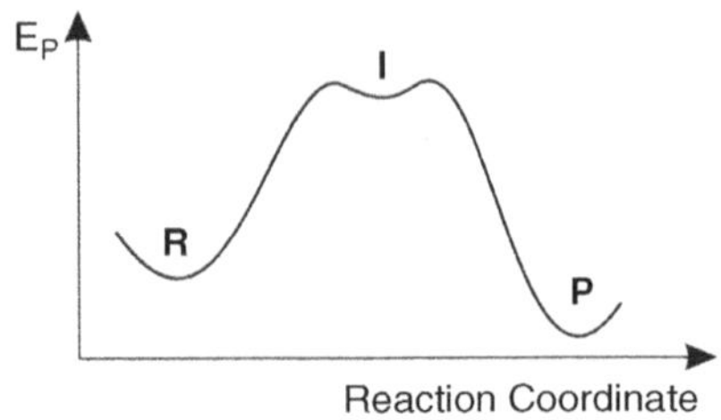

Figure 1.1. Energy profile for a thermal reaction from reagent **R** to product **P** through intermediate **I**.

The theoretical support gives firm ground to the qualitative picture of the structure of intermediates chemistry practitioners use and is an obvious improvement with respect to pen drawings commonly used for discussing the mechanism of reactions [2, 3]. Certainly, it becomes more and more customary to supplement experimental studies with theoretical investigations, but the importance of the first ones will not decline and both aspects contribute to clarify the course of a reaction over the intermediate and suggest how to improve yields, scope, and (environmental, economic, etc.) performance of chemical processes (Figure 1.2).

Intermediates are formed by heating, or, more often, by treatment with a chemical reagent. It is not always easy to identify the structure and the role of such species, because these may be present at such a low concentration to make spectroscopic investigation difficult. Likewise, it may be easy to establish that the kinetics of the overall process is compatible with the involvement of a given intermediate, but it is often difficult to achieve positive evidence for a specific mechanism.

Photochemistry enjoys a peculiar state in this connection. Absorption of a photon leads to the electronically excited states of the reagent, the chemical reactions of which basically differ from those of ground state molecules. In fact, in thermal reactions the system may encounter high barriers and surmount them, but at each configuration will remain at the lowest possible energy level (Figure 1.1). On the contrary, photochemical reactions by definition start from a high energy level. An electronically excited state, as the name indicates, lies much above the lowest energy possible for a given atomic configuration and will remain above the lowest potential

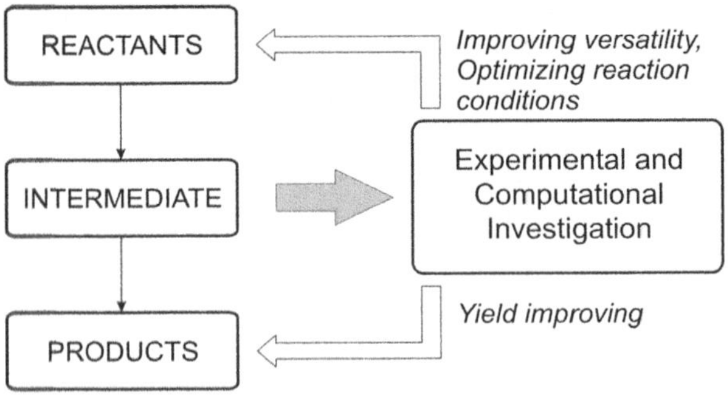

Figure 1.2. Knowledge of intermediate helps improving the reaction.

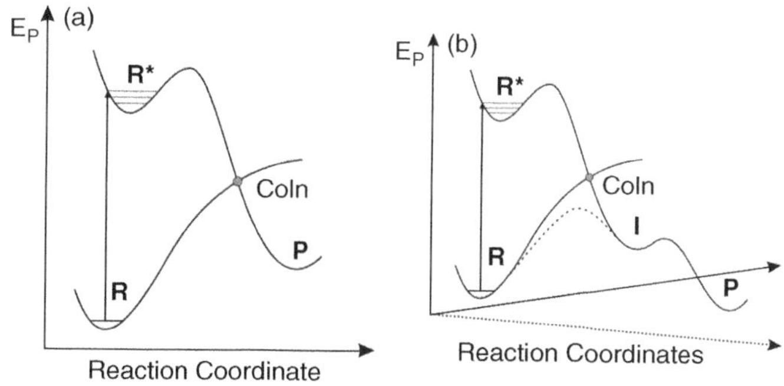

Figure 1.3. A photochemical reaction starts from an excited state of the reagent **R*** and returns to the ground state through a conical intersection (CoIn), finally reaching product **P**. (a) The general case. (b) The particular case where the photochemical path reaches the configuration of an intermediate (**I**) formed also in a similar thermal reaction.

energy surface (PES) while the system evolves along a certain coordinate until a conical intersection (CoIn) is encountered. This funnels back the system to what is the ground state for that specific configuration (Figure 1.3a). This is a very important characteristic and is what actually makes unique the contribution that photochemical reactions give to chemistry—that is, adding a new dimension to that defined by thermal reactions, based on the fact that *the PES implied is different* from that of ground state reaction. This, as remarked by Noyori [4], "enhances the power of chemical synthesis by removing current thermodynamic restrictions," making photochemistry an appealing tool for organic synthesis. The Nobel laureate recommends this as one of the two main targets "that our young generation (should) develop" in order to "facilitate a thermally unachievable, energetically uphill reaction."

Certainly, also in photochemical reactions small barriers have to be overcome, and situations corresponding to an intermediate are encountered. This fact is quite important, primarily for photochemical studies, in order to understand how such reactions happen. Proposing a photochemical mechanism is not necessarily a question of "chemical good sense." Quite often these processes involve in-depth transformations of the molecule backbone that at first sight seem to have little in common with the general mechanistic tenets that have been learned for "normal" (thermal) reactions. Things have improved, though. Actually, organic photochemistry grew hefty in the period from 1955 to 1970, with the introduction of many new processes that indeed were ill reconciled to currently recognized mechanisms. Take the case of George Büchi, one of the main contributors to this area (who rediscovered the [2 + 2] carbonyl–olefin cycloaddition to which now his name is associated along with that of the original discoverer, Paternò, as well as the intramolecular [2 + 2] cycloaddition originally discovered by Ciamician). This scientist became disenchanted with photochemistry because, to quote from an interview, he thought that "useful applications were not forthcoming, and because the course of the transformations could rarely be predicted, thus robbing the investigator of the pleasure derived from designing new reactions" [5].

Photochemistry still seems to retain somewhat "magic" that makes it sometimes difficult to teach this topic in an organic chemistry course. As has been recently expressed, at times photochemistry seems to involve a "somewhat mysterious effect" so that, "by simply absorbing a UV–Vis photon, a molecule becomes something different, and in many cases behaves in a way completely opposite to that of its ground state" [6]. As mentioned above, it is certainly true that photochemical reactions are basically different from thermal ones, since they proceed from a different electronic state, in a sense from a different isomer of the molecule. However, the situation has fundamentally changed in the last decades, mainly because the nature of the intermediates involved has been documented also in photochemistry. Indeed, advanced techniques of detection and characterization as well as modern computational methods now allow us to recognize the structure of excited states and the course of their reactions in detail and to identify intermediates. Thus, although the fundamental difference remains that in a photochemical reaction a change of PES is involved and the key point is to understand how the system converts to the ground state (see Figure 1.3), it is possible to discuss such reactions with the same vocabulary and the same attitude that is used for thermal reactions. This is done as usual by evaluating the effects that changing the molecular structure of the reagents and/or changing the environment have on the generation and the chemistry of intermediates in terms of steric/electronic effects, acid/base, electrophile/nucleophile character, and so on. As a matter of fact, photochemical reactions should be more often included in textbooks and courses where the reactivity of organic molecules is introduced, exactly because of the variety of the intermediates involved and the large structural changes occurring. Thus, mechanistic photochemistry may help mechanistic chemistry in general, both in teaching and in research.

The course of photochemical reactions is, as mentioned, intrinsically complex, due to the change of surface occurring during the chemical transformation. In particular, it may well happen that the return to the lowest-energy configuration occurs in correspondence to an atomic configuration identical (or very close) to one that corresponds to that of an intermediate formed under nonphotochemical conditions from the same or from a different system. In other words, it may well happen that a "thermal" intermediate is actually formed also by irradiation (Figure 1.3b). The conditions for the generation of the same intermediate in the two cases are quite different, though. Generating an intermediate thermally is generally a slow step, because this implies overcoming a high barrier. In other words, molecules are quite stable and correspond to a precise atom configuration. Moving atoms somewhat away from the assigned location—that is, stretching some bonds—is an energetically expensive step. Thus, drastic conditions are required for making the reaction to occur at a significant rate.

On the contrary, a photochemical reaction is, by necessity, fast. Electronically excited states undergo physical decay down to the ground state in a very short time (typically in the 10^{-9} to 10^{-6} s range depending on the multiplicity, or even below); and in order to compete with such fast decay, chemical reactions necessarily have to confront only a low energy barrier, typically a few kilocalories per mole. No problem with that, since the starting point, the electronically excited state, is much higher in energy than the ground state, indeed by an amount comparable to the energy of covalent bonds, and thus it will often lay above high-energy intermediates. Thus, it is to be expected that the photochemical generation of an intermediate is thermodynamically viable, since the starting point is itself very high. This does not mean that every excited state reacts or that any bond may cleave unselectively for the mere fact that so much energy has been injected in the molecule. One needs also to have a viable path to reach the new chemical structure, and it may well happen that no path having a low enough barrier to be surmounted during the fleeting existence of the excited state is available. In that case, return to the ground state and degradation of light to heat (internal conversion) is the only possibility. Or it may be that only a single low barrier path is available and a very selective process occurs. In practice, many photochemical reactions are quite clean. To summarize, photochemical reactions—by definition—have a low energy barrier. A noteworthy consequence of this fact is that they are weakly dependent on temperature and generally on experimental conditions, unless these cause a change in the nature of the excited state involved.

These characteristics make photochemistry the ultimate choice for the *identification* of intermediates. Indeed, this method allows to overcome the main limitation to their detection—that is, the low concentration of such species under steady-state concentrations. For example, intermediates can be generated at a relatively high concentration by means of a high-energy, short light pulse delivered on a small volume of a solution of the precursor. Photochemical reactions are fast and a transient signal will be detected by time-resolved spectroscopy, giving information about the nature of intermediates and allowing to directly follow their decay (this method is known as flash photolysis) Alternatively, they can be generated under steady-state conditions, but their decay can be blocked. This can be done by irradiating the precursor in a cryogenic matrix (at very low temperature, e.g. at 5 K in the matrix obtained by co-deposition of the precursor and argon), by taking advantage from the above-mentioned scarce temperature dependence of photoprocesses. In this way the intermediates accumulate and are available for spectroscopic investigation. For further details, see the next section.

When possible, the photochemical generation of intermediates offers a convenient point of view for documenting the course of a reaction and for any mechanistic study. The advantage photochemistry has, in comparison with other methods, been extensively exploited and can be easily appreciated, for example, by perusing the monumental treatise on intermediates edited by Moss, Platz, and Jones [7]. Indeed, everybody would concur that much of what is known about the structure and reactivity of intermediates such as radicals and carbenes arise from photochemical investigations. The knowledge about the chemistry of intermediates resulting from such studies can be used for predicting and recognizing their role in preparative experiments under steady-state conditions, whether thermal or photochemical.

There is a second way in which the characteristics of photochemical reactions become useful, however, and this is the generation of an intermediate by photochemical rather than thermal methods for a *synthetic* purpose. In fact, arriving at an intermediate through electronic excitation and reaction of an excited state is potentially convenient and may become the method of choice with the aim of:

- generating intermediates that are ill reached by nonphotochemical means;
- controlling the chemistry of such species, in particular by limiting the decomposition processes, taking advantage of the fact that the generation of the intermediates does not require heating or the treatment with aggressive chemicals;
- increasing the versatility of the reactions by generating the intermediates in an environment that could not be chosen for a thermal reaction, because some interaction with the precursor or, at any rate, an interference with the generation of the desired intermediate would result.

The advantage that generating intermediates by irradiation gives in synthetic planning is less well known by synthesis practitioners than is the physical characterization of such species by scientists investigating mechanistic issues. This is unfortunate, because activation by a photon, an unusually powerful chemical reagent that leaves no residue behind, offers new possibilities. The limited effect on the *excited state* reaction mentioned above allows to choose the appropriate conditions for *directing* the reaction toward the chosen target by governing the chemistry of the *intermediate*, not having to worry about its *generation*, most often the key issue in nonphotochemical processes. In fact, when planning a synthesis, attention is most often concentrated on ensuring that enough of the intermediate is formed for giving the desired reaction efficiently, through the judicious choice of appropriate reagent/solvent/temperature parameters. Furthermore, since an intermediate is highly reactive by definition, any potential chemical interference with the system must be carefully eliminated and any component present in the system must not play an unwanted role; for example, a trapping agent must not interfere with the precursor but only with the intermediate. If a photochemical reaction is found that gives the intermediate efficiently, then there is no need to worry about any interference by reagent, solvent, or impurities at the *generation* step, while all of these factors play their role at the *trapping* step. This is often but not always true. As an example, with some compounds a protic solvent changes the very nature of the excited state and thus its chemistry (see further below). As indicated in Figure 1.4, most of the effects on a reaction caused by varying the conditions is exerted primarily at the intermediate generation step in thermal reactions, but primarily at the trapping step in photochemical reactions.

A number of processes are now known that are sufficient for a detailed discussion of the synthetic role of photogenerated intermediates (anions, cations, carbon- and heteroatom-based radicals, radical ions, carbenes, nitrenes, etc.), and the perspectives of the development of this method are, in the opinion of the present authors, really important. This is the subject treated in the following. The idea is not to discuss the chemistry of all of the intermediates, but only of those that have been, or reasonably may be, generated *both* thermally and photochemically. Analogously, not all of the photoreactions will be examined, but only those that have been shown, or

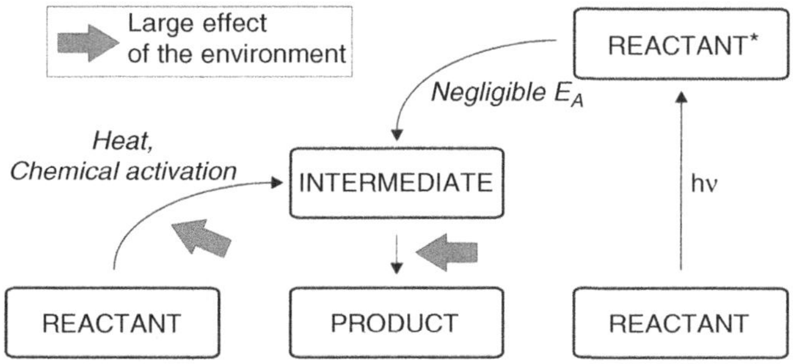

Figure 1.4. The most important point of a thermal reaction is generating the intermediate, and environmental effects mostly act on that step. In photochemical reactions, however, the generation step is relatively independent on conditions, and attention can be given to the conversion to products almost exclusively.

are expected, to involve an intermediate accessible also through a thermal path (see Figure 1.3b). It is hoped that focusing on known intermediates help synthetic chemists to enjoy and actually use the large repertoire of preparative alternatives that photochemistry offers. As will be seen, in some cases, going through excited states involves a straightforward change of conditions that leads to improved results, whereas in other cases a completely new path through different intermediates has been discovered. The choice of topics and examples is, by necessity, limited; the interested reader is referred to general presentations of photochemistry in general [8, 9] and of the preparative aspects in particular that are easily available [10–19].

The presentation is organized into three parts, beginning with an overview on how to carry out a preparative photochemical reaction and how to use for preparative purposes the (mainly mechanistic) photochemical literature (Chapter 1). An overview of photochemical reactions that have been demonstrated to (or are reasonably expected to) involve an intermediate follows (Chapters 2 to 7), and finally a comparison of the merit of different methods is made (Chapter 8). While the identification and characterization of an intermediate may be *per se* the highest point of a mechanistic investigation, the motivation here is putting such knowledge at the service of synthesis. The target of this book is not to summarize the state of the art on intermediates (which has already been well done in particular in the Moss–Platz–Jones book [7]), but to show (potential) applications of what we are learning about intermediates and stimulate further (synthetic) work that the knowledge on the photochemical generation of such species makes possible.

1.2 STUDYING THE INTERMEDIATES: THE PHOTOCHEMICAL WAY

As mentioned in the introductory section, the role of short-lived intermediates in chemical reactions is hardly overemphasized. The detection and identification of such species is mainly based on spectroscopic techniques that must be appropriate for the specific case, and thus primarily with respect to the lifetime of the species investigated. This obviously depends on the definition of intermediates adopted. If this applies to a "normal" molecule with all the required bonds at their place, but that under the experimental conditions undergoes a rapid transformation, as is the case for the sequence of intermediates that usually intervenes in an enzymatic reaction (lifetime from 1 to 10^{-3} s), then a stopped flow apparatus is appropriate. This instrument provides rapid mixing of the reagents solutions and allows the measurement of some optic parameter. By using an appropriate deconvolution analysis, this technique may reveal the role of several intermediates that participate at the overall process and reach a sufficient steady-state concentration.

If, however, the definition of intermediates is limited to high-energy species that do not obey the Lewis octet rule or the tetravalence of carbon atoms, then the lifetime is much shorter (typically from 10^{-3} down to 10^{-10} s) and the upper limit of the steady-state concentration attainable is very low. In a typical radical reaction, as an example, the process is put in action by the decomposition of the initiator that typically occurs at a rate around $k_D \sim 1 \times 10^{-5}\,\mathrm{s}^{-1}$. These radicals are rapidly trapped, and at any rate radical–radical coupling occurs at diffusion controlled rate ($k_R \sim 2 \times 10^{10}\,\mathrm{L\,mol^{-1}\,s^{-1}}$ in a nonviscous solvent) or close to that. This puts the limit of the radical concentration attainable under steady-state conditions at the micromolar level or below. Mixing the reagents sufficiently fast for studying such reactions in real time is impossible. The reaction can be studied under steady-state conditions, but the very low amount of intermediates that accumulate will limit the techniques that may be used for detection; thus in practice we use only electron paramagnetic resonance (EPR) for spin-unpaired species.

In this case, photochemistry demonstrates its potential. Apart from increasing the number of cases where techniques suited for studying steady-state reactions, such as EPR and chemically induced dynamic nuclear polarization (CIDNP), the fact that photochemical reactions are fast and independent on conditions makes two new approaches possible. In the first one, the decay of the intermediate, whether it is a thermally activated unimolecular process or a bimolecular reaction, is slowed down. This is done by lowering the temperature and thus the mobility and reactivity of any species in a rigid medium. Typically a clear glass must be obtained by cooling at 77 K a solution in a cuvette or the reagent is co-distilled with a rare gas on a cold finger forming a matrix at a temperature of a few Kelvin degrees. The former method is suited for UV–visible detection (see Figure 1.5), the latter one also for the more informative IR and Raman spectroscopy. The photochemical reaction occurs even under these conditions, but the intermediate is frozen and does not evolve further. Matrix isolation allows to obtain a plethora of information about the structure of the species that can be analyzed by various spectroscopic methods, and thermolysis cannot rival photochemistry in this field. Most usually, the rich pattern of lines of an IR spectrum is compared with the calculated values affording a strong support for structure identification.

On the other hand, the very nature of the experiment demands that the intermediate generated does not evolve significantly under the conditions used. The occurrence of further photochemical reactions of the matrix isolated species can

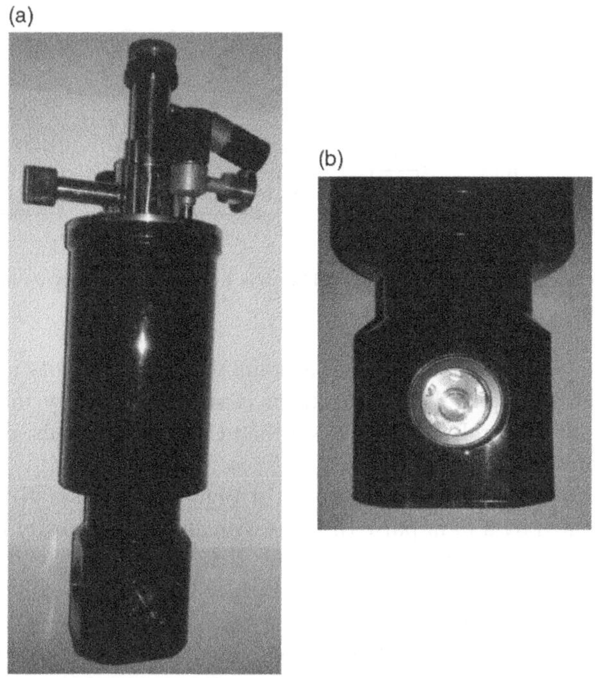

Figure 1.5. A cryostat for studying photochemical reactions at low temperature. (a) General view. (b) Enlargement of the cell compartment; the UV spectrum of the solution is measured perpendicularly to the irradiation path.

be studied by continuing the irradiation, but any evidence refers to the actual situation of both reagents and products in the matrix. What actually is of interest for mechanistic studies, however, is the behavior (thermal or photochemical) of such intermediates *in solution*. This may make the information from matrix isolation experiment incomplete or sometimes misleading if applied to reactions in solution.

This limitation is lifted by the second approach that is based on time-resolved spectroscopy in order to actually measure the kinetics of the processes undergone by such species. This in turn requires that a high enough concentration of the intermediate is formed at the beginning of the experiment and thus that a large amount of energy (of photons) is delivered to the sample essentially instantaneously on the scale of the intermediate lifetime. As is well known, this technique was originally developed by Norrish and Porter in the flash photolysis apparatus, which delivered a short (<1 ms) pulse of light from a discharge lamp. This technique demonstrated to be quite useful for the detection of radicals. Traditional flash photolysis has now long been substituted by instruments where a pulsed laser is the source and an appropriately sensitive detector makes possible the characterization of much shorter-lived species down to the picosecond region and below, where electronically excited states and intermediates such as carbocations, carbanions, carbenes, nitrenes, and so on, are revealed. Both optical (UV, IR, and magnetic

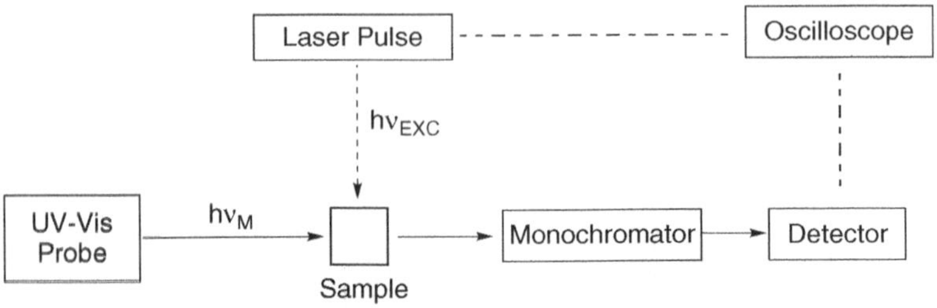

Figure 1.6. Schematic outline of the UV–Vis laser flash photolysis apparatus.

detections [20–22]) are used, taking into account that the sensitivity issue is here all important. As an example, the IR spectrum is structurally very informative, but detecting IR bands, generally much weaker than UV–visible bands, is technically a more difficult problem.

The scheme of the instrument is shown in Figure 1.6. A small volume of the solution is irradiated by a light pulse ($h\nu_{EXC}$) with an energy density in the order of 1 mJ cm^{-2} over a short interval—for example, 10 ns down to <1 ps. One takes care that a significant fraction of this energy (typically ~50%) is absorbed by the reagent, so that a relatively high amount of excited states is formed "instantaneously." If a photochemical reaction occurs with a reasonable quantum yield and gives products with a high extinction coefficient in an accessible part of the spectrum, well-detectable transient signals will appear, and the monitoring light ($h\nu_M$) reveals the role of intermediates. Information about the structure, as far as the often broad UV–Vis bands allow, and on the kinetics of the intermediates can thus be obtained, in less simple cases, by applying an appropriate mathematical analysis for the deconvolution of the individual spectra and the decay rates.

An example may be appropriate for illustrating the time scale of the experiment. A viable method for the generation of aliphatic radicals is photocatalysis starting from alkanes or simple derivatives by strong oxidants, such as aromatic nitriles in the excited state (see Chapter 5). A solution of 1,2,4,5-tetracyanobenzene (TCB) and aliphatic compounds in acetonitrile was flashed at 266 nm under conditions where the overall time resolution was about 300 fs. The flash length was around 100 fs, delivering 10^{12} exciting photons in a volume of 0.255 mm^3—that is, a 7×10^{-6} M concentration. The energy of the photons corresponded to the excitation to a higher singlet state ^{1n}TCB, and the examination of the time domain up to 2 ns allowed to follow the intramolecular vibrational energy redistribution and vibrational cooling that finally led to the thermally equilibrated lowest excited singlet, ^{1}TCB, with a lifetime of ~4 ns (Figure 1.7a) [23].

The following evolution of the system was monitored in a further experiment with a laser delivering a flash in the nanosecond domain, since the femtosecond apparatus is built on the basis of a different physical principle and the time-range explored cannot be further extended. Under these conditions, electron transfer from the aliphatic donor occurs and the radical anion $TCB^{\bullet-}$ is revealed (Scheme 1.1). The radical anion is a stable, nonbasic species and whether the initial step leads to

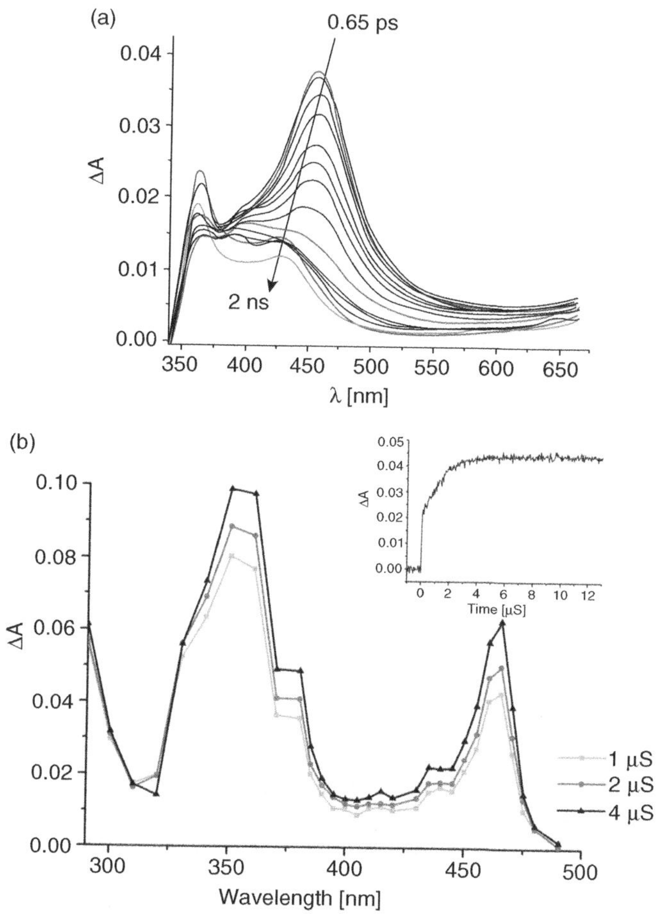

Figure 1.7. (a) Transients formed upon the excitation of a 7×10^{-4} M solution of TCB in MeCN by means of a 70-fs FWHM laser pulse at 266 nm up to 2 ns. The thermally equilibrated singlet excited state of TCB is formed. Reprinted with permission from Protti, S., Fagnoni, M., Monti, S., Réhault, J., Poizat, O., Albini, A. (2012). Activation of aliphatic C–H bonds by tetracyanobenzene photosensitization. A time-resolved and steady-state investigation. *RSC Advances*, **2**, 1897–1904. (b) Transients formed upon excitation of the same system followed on a longer time scale, up to 4 ns. The radical anion of TCB is formed (see inset).

a chemical reaction or to back electron transfer depends on whether the radical cation has a path available. Deprotonation is the obvious mechanism, but cleavage of the nonpolar C–H bond is slow—for example, when cyclohexane is the donor. In that case, no change of the $TCB^{\bullet -}$ signal occurs on a microsecond time scale, while later back electron transfer between the free radical ions predominates (path *a*). On the other hand, electron transfer (ET) quenching by amines (path *b*) is followed by fast deprotonation of the amine radical cation (<0.1 μs) and the resulting α-aminoalkyl radical reduces a further molecule of TCB (path *b*′, in some

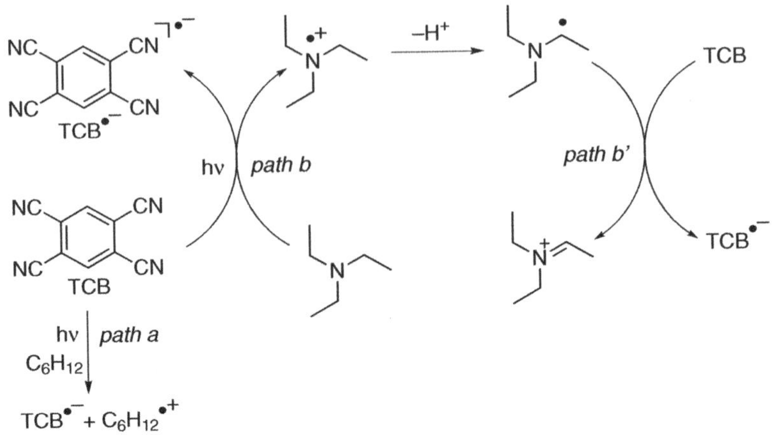

Scheme 1.1. ET to excited TCB from cyclohexane (path a) and from triethylamine (path b).

microseconds), causing the accumulation of the radical anion which is indefinitely stable in the absence of oxygen. [23]

Another example of a detailed description of a stepwise process was given by the photocatalytic oxygenation of aliphatic donors by using a ruthenium tetraphenylporphirinato complex **1**, where water had the role of both an electron and an oxygen donor. Laser flash photolysis has been instrumental in optimizing the process and suggesting a viable mechanism. This was a one-photon process involving two electron transfer steps, where alkenes were converted to the corresponding epoxides according to Scheme 1.2. Laser flash photolysis revealed the formation of the triplet ruthenium complex (^{3}**1**, bands at 450 and 510 nm) that was quenched by hexachloroplatinate within 20 μs under optimized conditions. The oxidized photocalyst equilibrated with water and deprotonated to give first an intermediate bearing a OH group within 50 μs and then, upon reaction with cyclohexene, a second intermediate within 200 μs. The rate depended on the chloroplatinate concentration (involved in further electron transfer step). On a slower scale the starting complex was regenerated [24].

The examples illustrate the large time span that can be explored by modern time-resolved methods. The region at short time intervals ($<$1 ns) is of interest mainly for "professional" photochemical practitioners, since on this time scale (mainly physical) processes involving excited states occur, but the part from 1 ns up allows to follow the *chemical* evolution of the system, with the formation and reaction of intermediates. Clearly, not every intermediate is detected; for example, in the case presented in Scheme 1.1 that is based on optical detection, aliphatic radicals absorb too weakly to be detected, and their role can only be indirectly deduced from the formation and evolution of the strongly absorbing TCB radical anion. The conditions used for time-resolved studies are basically different from those of preparative applications. In the first case, a packet of photons is administered on a small spot in a short time—for example, an amount of 10^{14}–10^{15} photons cm^{-2} or more in a time interval from 100 fs to 10 ns. The same experiment by using a focalized steady-state

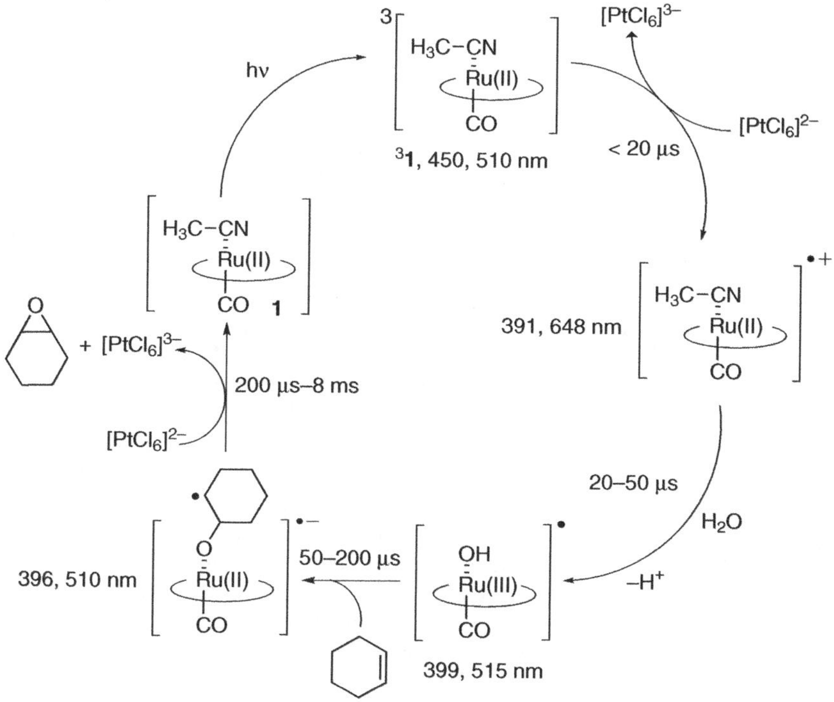

Scheme 1.2. Photocatalyzed cyclohexene epoxidation. Four subsequent intermediates are detected by flash photolysis.

lamp would deliver 10^{16}–10^{17} photons cm^{-2} per minute, which is a flux lower by a factor of 10^8 to 10^{10}. In the latter case the steady-state regime is "immediately" entered and the concentration of any intermediate is too low for detection. On the contrary, in the first one the starting concentration is sufficient for detecting the presence of the intermediates and their evolution, provided that the analytic technique used is sufficiently sensitive for that case. Obviously, such high energies are obtained only by using a focalized laser as a source, which is appropriate for the irradiation of a small volume of solution and is not generally suited for preparative purposes. The flux delivered by a lamp may be much more diluted in the time dimension but irradiates larger surfaces, thus leading to the excitation and reaction of large amounts of reagent.

Also, some care must be used for extracting information about the chemical behavior of intermediates, because the high energy of the pulse may cause processes that have no role under "normal" preparative conditions. In particular, two-photon excitation of the reagent to an upper state becomes a possibility [25]; this may lead to processes, such as photoionization, that are uncommon in steady-state irradiation. Furthermore, secondary excitation of the primarily formed intermediates may occur. Thus, it is advisable to check for any dependence of the intermediate trace on the laser pulse intensity, since this may indicate a double excitation, as well as to check whether photoionization occurs as manifested by an intense transient absorption in

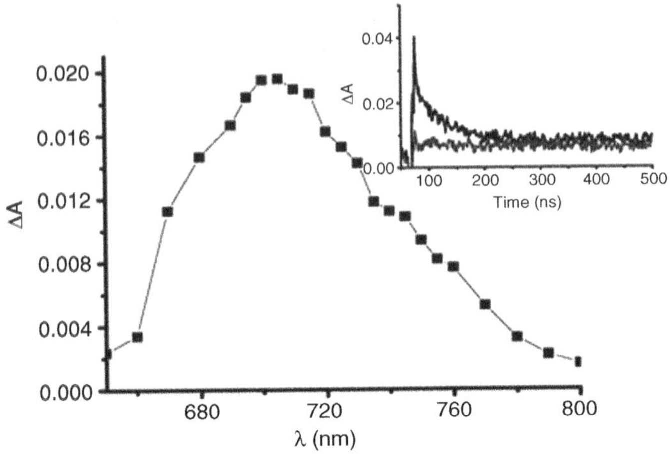

Figure 1.8. Difference absorption spectrum of a 1×10^{-4} M solution of Linezolid in water 0.1 μs after flashing at 266 nm, spectrum in the 650- to 800-nm region. *Inset*: decay of the 675-nm absorption attributed to the solvated electron in the nitrogen flushed (*upper trace*) and nitrogen oxide flushed (*lower trace*) solution. Reprinted with permission from Fasani, E., Tilocca, F., Protti, S., Merli, M., Albini, A. (2008). An exploratory and mechanistic study of the defluorination of an (aminofluorophenyl)oxazolidinone: $S_N1(Ar^*)$ versus $S_{R+N}1(Ar^*)$ mechanism. *Organic & Biomolecular Chemistry*, **6**, 4634–4642.

the visible attributable to the solvated electron (further note that this can be eliminated by saturating the solution with N_2O; see Figure 1.8).

1.3 PHOTOGENERATED INTERMEDIATES. FROM MECHANISTIC INVESTIGATION TO ORGANIC SYNTHESIS

1.3.1 Light Sources for Preparative Photochemistry

A look at the literature certainly shows that photochemistry is not exploited to full advantage in organic synthesis. Actually, it appears that photochemical reactions and the way they are carried out are not as familiar to synthesis practitioners as they should be. This situation has several causes, the most serious one probably being the fact that photochemical reactions on a significant industrial scale requires to set up a devoted reactor with a suitable light source. This involves a large technical and financial investment and makes photochemistry a convenient choice only when a high value product, typically a drug, can be arrived at only by this way or when the method offered is by far a better-performing one. Otherwise, a thermal method that uses a versatile all-purpose thermal reactor will doubtless be preferred. As a consequence, photochemistry is considered less often than it may be when planning a synthesis, even at a small scale, where there is no problem about the apparatus, as will be discussed below. This is probably a problem of chasing one's tail. One needs more work in photochemistry in order to have a larger palette of technically

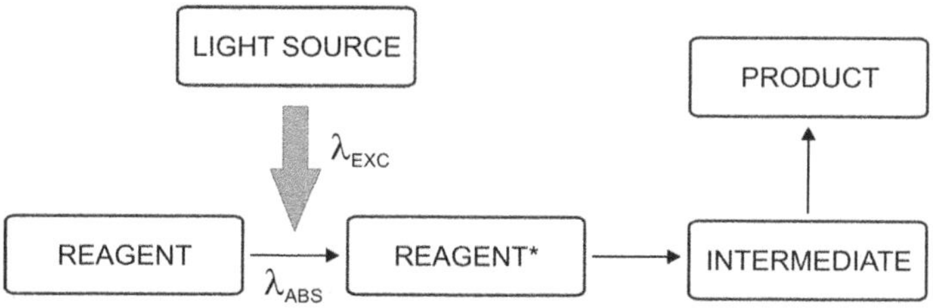

Figure 1.9. Absorption of light is required for a photochemical reaction.

well-developed photochemical reactions that can be chosen for industrial application, but until photochemistry is not largely applied in the industry, attention to (and funding of) research in photochemistry will be limited. Perhaps again because less familiar to many scientists, photochemical reactions are sometimes considered less dependable and easily exposed to experimental errors. This is certainly not true, and unexpected failures can generally be avoided by paying attention to a few directions that are typical of photochemical reactions and that are certainly not more complicated than those commonly applied for thermal reactions. Therefore, it appeared worthwhile to briefly review the photochemical preparative methods available and discuss the appropriate indication for correct use.

The so-called first law of photochemistry asserts that photochemical reactions necessarily involve the absorption of light (Figure 1.9). This is certainly obvious, but it must be checked that light is absorbed by the compound which is meant to react. Actually, oversight of this precaution is a prime cause of failure. Thus, it should be checked that:

- The lamp chosen emits in a convenient wavelength range (emission spectrum available from the manufacturer), which matches at least in part the absorption spectrum of the reagent (taken from the literature, but advisably also measured on the actual sample used, which gives an idea of possible competitively absorbing impurities).
- The light flux emitted, or a sizeable fraction of it, does reach the solution to be irradiated. Light is emitted toward all of the directions by the lamp (except when these are focalized; see below), and it must be checked that the solid angle hitting the sample is not too small.
- Nothing absorbs the photons before they reach the target molecule. In particular, the walls of the vessel and the solvent must be transparent to the irradiation wavelength.
- No efficient quencher of the excited state involved is present that may de-excite the reagent.

The above applies when it is the reagent that absorbs light. A number of reactions are, however, carried out under photosensitized or photocatalyzed conditions. In this case, a molecule (photosensitizer or photocatalyst) absorbs light and in a physical or chemical way activates the nonabsorbing reagent (by energy transfer, generating indirectly an excited state; by electron transfer, generating radical ions; by atom transfer, generating radicals). In such cases the same requirement applies, but the absorption step involves the sensitizer or the catalyst, not the reagent. The solvent itself may act in this role; as an example, acetone absorbs up to 300 nm and has been largely used as triplet sensitizer.

As for the practical way of carrying out a photochemical reaction, one choice is buying a dedicated apparatus. There are several firms supplying lamps as well as "complete" photochemical reactors which are also lamps (most often a medium pressure mercury arc; see below)—along with the appropriate power supply and the reaction flask with accessories (e.g., for gas inlet). The complete set may be quite expensive and is scarcely versatile. In the following, a brief review of the available alternatives is presented, along with indications on how to use home-made devices that operate almost as well as the commercial ones and come for a fraction of the price. A caveat that must be clearly stated is that the wattage of (the electric power required by) a lamp has no general relation to the flux emitted by the lamp, and in particular to the intensity of the flux of the desired wavelength that impinges on the vessel, so that it can be absorbed by the reagent. The efficiency of the conversion of electricity into light, the wavelength distribution of the emission, the dimension and the shape of the lamp, and the geometrical arrangement with respect to the reactor vessel must all be taken into account for arriving at this value. In practice, it is more expedient to measure the light flux by chemical actinometry—that is, by carrying out a reaction of known quantum yield and calculating the light flux from the actinometer converted.

Perhaps, from the point of view of the experimental arrangement the main distinction is between focalized sources, where the beam is directed on the object to be irradiated—in this case the vessel containing the solution of the starting material—and nonfocalized sources, where light is emitted over the entire sphere and the first factor that must be considered is which fraction of the total emission hits the target.

1.3.1.1 Nonfocalized Sources Nonfocalized sources are lamps that convert electric energy into light more or less efficiently [26]. Incandescent lamps (Figure 1.10a) that have long been universally used for domestic illumination are based on the emission from a tungsten wire heated by conduction and protected from oxidation by air by a glass bulb filled with inert gas or evacuated. These exhibit a continuous emission with no UV component and peaking in the 550- to 650-nm region, where most organic molecules do not absorb (except, of course, dyes) and also are rather inefficient (generally the conversion of electric power to light is $<5\%$) and produce much heat. Tungsten–halogen lamps (Figure 1.10b) can be devised to have the centre of emission shifted toward the blue, but still produce practically no UV and much heat. Mounting in a projector gives a uniform illumination on a certain area (floodlight lamps, used for example in shop windows, Figure 1.10c); but these remain a poor source for photochemical application, except when both light and heat are required or when light intensity is not important—that is, typically in chain

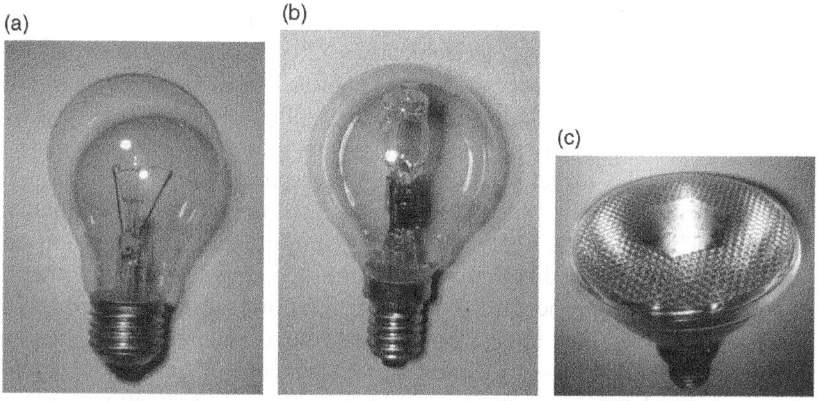

Figure 1.10. (a) A traditional tungsten bulb incandescent lamp. (b) A tungsten–halogen lamp. (c) A 150-W floodlight lamp, 10-cm diameter.

reactions, such as halogenation reactions, aromatic substitutions via S_NAr1 mechanism, and so on.

At any rate, these have been superseded also for general illumination application by arcs. "Arc lamps" or "arc lights" produce light by an electric arc. The lamp is an ampoule containing two electrodes made from carbon, but more often by tungsten and a substance that is gaseous under the conditions of use. Electrodeless discharge lamps that are energized by an external field are also available. The filling gases include noble gases (neon, argon, xenon, krypton) or metals and their compounds (sodium, mercury, or metal halides; see Figures 1.11 and 1.13). Light is generated by sending an electrical discharge through the ionized gas or plasma, generated by the

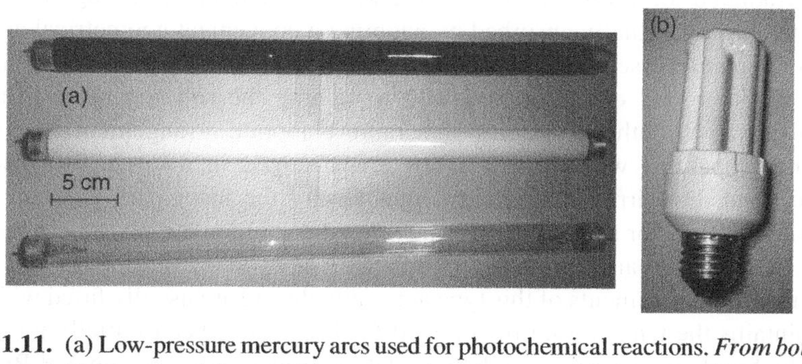

Figure 1.11. (a) Low-pressure mercury arcs used for photochemical reactions. *From bottom to top*: A germicidal lamp emitting at 254 nm; two phosphor-coated lamps: the short-wavelength radiation emitted is absorbed by the phosphor that emits over a range of longer wavelengths, centered at 305–310 nm in the first case, at 360 nm—a "black light"—in the latter one. Light is emitted over the whole length of the tube (40 cm for these 15-W lamps; different dimensions and wattages are available). Care must be taken that a sizeable part of the flux impinges on the vessel with the solution to be irradiated, or several lamps can be arranged in a multilamp apparatus as in Figure 1.12. (b) A 12-W fluorescent lamp for household lighting, 5 cm long.

appropriate frequency or modulation of the current. Under these conditions, the atoms are excited by collision to electronically excited states. Return to the ground state is then accompanied by the emission of a photon. The most largely used are the highly efficient mercury arcs. At low concentrations, these emit lines corresponding to the electronic transitions from the singlet (185 nm, filtered off by many materials, however) and from the triplet (254 nm). At larger concentration, emission from bimolecular species (e.g., Hg_2^*) becomes increasingly important and additional lines appear at longer wavelength. Gradually, the 254-nm monoatomic emission disappears because it is reabsorbed before getting out of the ampoule and the other lines gain importance. A further increase of the concentration leads to the appearance of the blackbody emission and, thus, a continuous ground underlying the lines.

The low-pressure mercury arcs known as fluorescent or germicidal lamps or mercury resonance lamps are largely used for a variety of applications, including sterilization. These are supplied as tubes of various lengths, typically 20–60 cm (but lamps >1 m long are available), 1.6–2.4 cm in diameter, made of quartz (or rather of the so-called fused silica, a synthetic amorphous SiO_2) with an electric power of 6–16 W. In these lamps, the gas pressure under operating conditions is *ca.* 10^{-5} atm and the emission is essentially monochromatic. The normally used "fused silica" filters most of the 185-nm emission, and thus the lamp is essentially a monochromatic source at 254 nm. The low wavelength can be conserved if a high-purity (UV grade) quartz is used. The 254-nm radiation is completely absorbed by Pyrex and other glasses used for laboratory glasswares, and thus only quartz vessels may be used for irradiation. Due to the many applications in various fields, these easy-to-use lamps are quite cheap.

The negative side for application in preparative experiments is the weak density of the flux that is emitted over the large surface of these lamps. Therefore, these lamps are most useful for external irradiation by using tubes (obviously made from quartz) for the irradiated solutions, and it must be taken care that a large enough fraction of the emitted flux is actually absorbed by adopting a convenient geometrical arrangement of lamp and vessel to be irradiated. A simple, but not very efficient, solution is having the lamp and one or more tubes containing the solution to be irradiated parallel one to the others. A somewhat more elaborate arrangement is having a multilamp apparatus where 6–12 lamps are arranged in a circle (40–60 cm in diameter) and the mirror walls reflect light toward the inner part, where a single vessel (80–400 mL) or several tubes containing the solution are placed. This choice limits air circulation and some heating results.

Commercial instruments of this type are available and are usually fitted with a fan that maintains the temperature below ~40°C. These have been initially put on the market by Southern New England Ultraviolet Company under the name of Rayonet, now often used for similar devices by other firms (see Figure 1.12a). A cheap implementation of the same scheme can be assembled by putting one to three pairs of lamps (each pair mounted on a normal lamp holder for household fluorescent lamps) around a small space where two to four test tubes or a single cylindrical vessel of larger diameter can be placed (see Figure 1.12b). Furthermore, coiled and variously shaped lamps are also available. These lamps facilitate devising alternative

Figure 1.12. Rayonet-type irradiation apparatuses. (a) At the center, a series of test tubes are fitted in a rotating merry-go-round. This ensures that all of the samples are equally illuminated and are used for exploratory studies or for the optimization of a process. (b) An inexpensive homemade variation.

arrangements, such as the use in an immersion well apparatus with internal irradiation (see below).

Irradiation at 254 nm is effective for the excitation of most classes of organic compounds that have a significant photochemistry. The problem may be that many common solvents absorb that wavelength too, thus precluding excitation of the desired reagent and that the reactor walls traversed by the light beam must be made of quartz, since, as mentioned, any glass absorbs completely this wavelength. Low-pressure mercury arcs (Figure 1.11) are by far the cheapest light source (obviously when bought as such, not incorporated in an instrument for sterilization or, worse, for photochemistry). Furthermore, these lamps are long-lived (>10,000 h, although this time may be shortened by excessive on/off switching), have a low energy consumption, and require only an inexpensive transformer and a starter for operation.

Lamps with a variety of emission ranges are manufactured on the same basis by applying onto the lamp walls a coating made of a phosphor (or a combination of phosphors) that absorbs the almost monochromatic Hg radiation and emits a more or less wide range (or the sum of some ranges) of longer wavelengths. Phosphor-coated lamps maintain the same advantages of quartz lamps (for the most common types including the price, because of the large-scale manufacture for different uses, including household illumination). Among the types that are often utilized, some are useful for photochemical synthesis. These include UV-B lamps with emission centered at 305–310 nm (a white phosphor), the "black light" (also called "Wood light") lamps (see again Figure 1.11), painted with a dark phosphor that emits at 350–370 nm, as well as lamps with the emission centered at various wavelengths in the visible—for example, in the blue around 450 nm or over the whole spectrum (fluorescent lamps for household illumination) giving a white light with various hues (see Figure 1.11b). The 305- to 310-nm emitting lamps (also indicated as 313 nm because the Hg emission has a line at that wavelength) are produced for the treatment of skin diseases. The emission of these lamps is still in part blocked by Pyrex, but become quite useful for exciting many moderately conjugated chromophores. The black light lamps, often indicated as emitting 366 nm, again due to the mercury line at that wavelength, but actually centered at a λ varying from ~350–355 nm ($BaSi_2O_5:Pb^+$ phosphor) to ~368–370 ($SrB_4O_7F:Eu^{2+}$ phosphor) are popular in night clubs and can be used for illuminating vessels made of glass that is obviously less expensive.

Most conjugated chromophores are colorless, but absorb in the UV-B and A regions, thus making the 313- and the 366-nm lamps quite useful in photochemistry, as are the visible emitting ones for colored reagents and dyes. The phosphor coating does not change the electrical characteristics, and these lamps can be interchanged with germicidal lamps in all of the settings mentioned above. A laboratory having available two to three pairs each of 254-, 313-, and 366-nm lamps, as well as of lamps emitting in the visible, has the possibility to test any type of small-scale photochemical reaction with a minimal initial investment and virtually no expenditure for maintenance. Lamps doped with different metals are also available and yield an emission enriched in some region of the spectrum. Thus, these may be better suited for a particular photoreaction (but are generally more expensive and less versatile).

The external irradiation arrangement, and better still a multilamp arrangement, is often the most convenient choice for small-scale or explorative studies, by having the solution distributed in a number of quartz test tubes (20 mL) or in a single cylindrical vessel (100 mL or more). Furthermore, this setup is convenient for optimizing the reactions, since one can easily compare the results under different conditions but under consistent irradiation conditions by putting different solutions in the tubes; a rotating merry-go-round holding several tubes that ensures equivalent irradiation in all of the positions is available as an accessory for multilamp apparatuses (see again Figure 1.12a).

As hinted above, the emission of mercury arcs changes from monochromatic to polychromatic with growing pressure. Medium-pressure lamps (sometimes called high-pressure arcs) operate at a pressure in the range 1–10 atm (Figure 1.13).

The emission consists of several lines, the most intense ones are those at 313, 366, 405, and 550 nm superimposed to a continuum, while the 254-nm line is strongly diminished. These are available by various suppliers as small quartz ampoules, from 3 to 15 cm long depending on the power. These mercury arcs are available in different sizes, typically with an absorbed power of 125, 150, 450, and 1000 W (the last one are also used as the inner part of phosphor-coated street lamps). The overall emission from these lamps is ≥10 times stronger than that of low-pressure arcs and occurs over a much smaller surface. Although the conversion to light is still reasonable, much heat is developed that, given the small dimensions of the ampoule, cannot be dissipated in the atmosphere. These lamps are therefore used within a double-welled cooler where water is circulated, a small stream of tap water being sufficient. Furthermore, in contrast to the low-pressure lamps that reach the maximal output

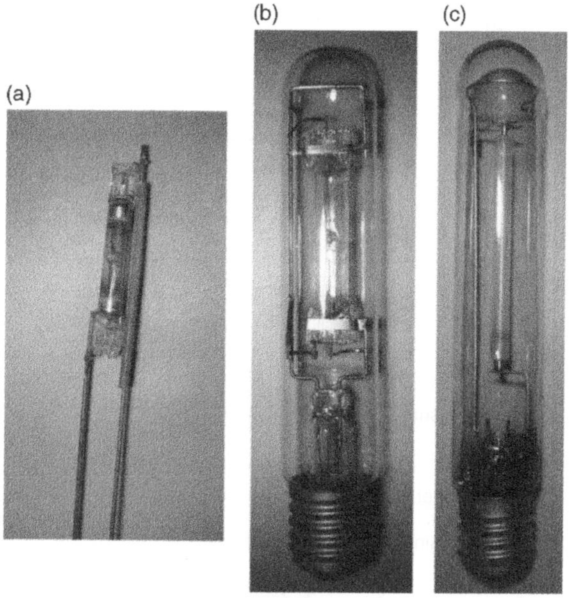

Figure 1.13. Lamps used for photochemical reactions. (a) A 125-W medium pressure mercury arc; the ampoule, 3 cm long, is mounted for fitting in an immersion well apparatus (see Figure 1.14). (b) An analogous 250-W arc; the ampoule is 6 cm long. (c) A 450-W sodium arc (emission at 589 nm).

level practically instantaneously, in this case 1 min or more is required for reaching the maximal intensity. These lamps are typically used in an immersion well apparatus with circulating water as shown in Figure 1.14.

The cooling well may be made either of quartz, in which case the small fraction of emission below 310 nm is conserved, which may be useful if the compound to be excited absorbs in this region, or of Pyrex, which is less expensive and may be useful to avoid the excitation of compounds absorbing below 300 nm—for example, the solvent. In this arrangement these compact sources give the best results, with the emission in all directions being intercepted. With these arcs, alkenes and simple benzenes do not absorb (or do it quite poorly) even when a quartz well is used, simple carbonyls are borderline, but practically all of the conjugated molecules that are so often photochemically reactive absorb the Pyrex filtered light. These lamps are more expensive, require a dedicated power supply for operation, have a considerably shorter lifetime, and require more attention to the safety issue (see below), but the compact design is a large advantage that makes this lamp the optimal choice for a laboratory that has recourse regularly to photochemical reactions, because grams of products are easily transformed. Furthermore, this arrangement is particularly convenient for controlling the conditions. Thus, one can easily stir the solution (efficient only if the design is appropriate, however, see below), circulate an inert gas or a reactive gas such as oxygen or ethylene, and gradually add reagents from a dropping funnel, as well

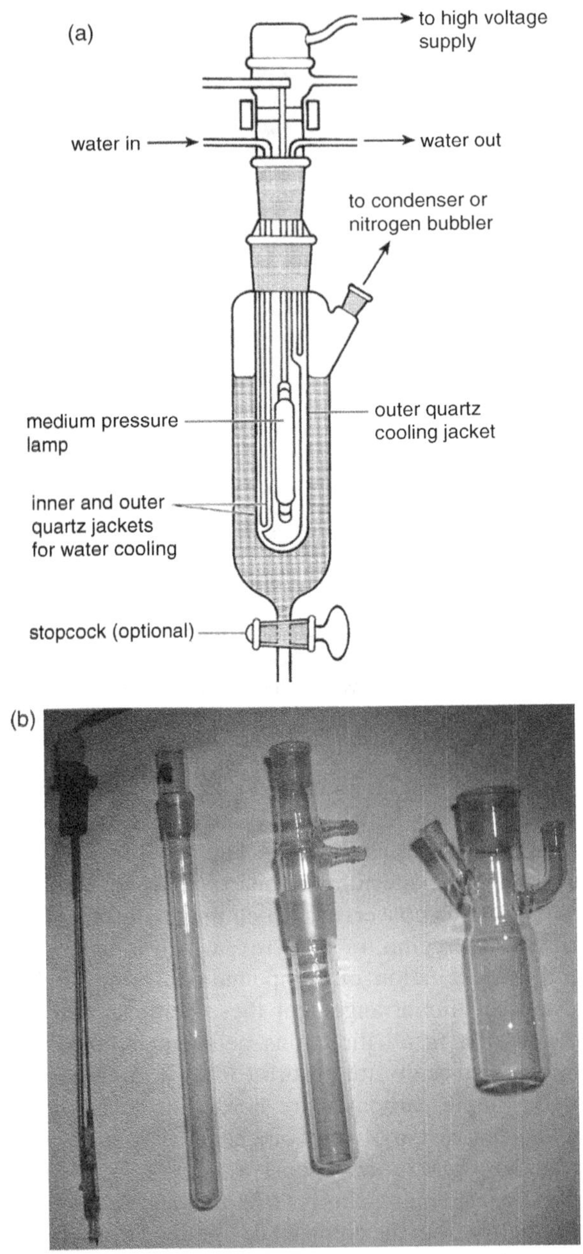

Figure 1.14. (a) An immersion well apparatus for internal irradiation by a lamp fitted in an immersion well. Reprinted with permission from Hook, B. D. A., Dohle, W., P. R. Hirst, P. R., Pickworth, M., Berry, M. B., Booker-Milburn, K. I. (2005) A practical flow reactor for continuous organic photochemistry. *The Journal of Organic Chemistry*, **70**, 7558–7564. (b) The parts of the apparatus (from left to right): medium pressure mercury arc, internal tube (made of quartz for resisting heating), cooling vessel (made of quartz if emission below 310 nm is required, otherwise of Pyrex), irradiation vessel made of Pyrex.

as control the temperature. Thus, low-temperature experiments (e.g., at −78°C), are easily carried out by circulating a refrigerating liquid from a bath in the lamp jacket and immersing the vessel in an external CO_2/acetone cooling bath. In this case the lamp has to be ignited outside the well and put in its place when lit, because otherwise it won't work.

1.3.1.2 Focalized Sources Focalized sources may be obtained from lamps by physical means. The intensity of the previously mentioned lamps is too low to make this convenient, but mercury arcs operating at a still higher pression, around 200 atm under working conditions, are indicated as "high" or "very high" pressure mercury arcs and do fit with this approach. These are commercially available with an electrical power from 150 to 1000 W and more. These are the smallest mercury arcs, made from thick-walled quartz ampoules, with a distance between the electrodes of a few millimeters. In this case the emission is primarily a continuum, over which the strongest mercury lines are still distinguishable. These arcs operate at a higher temperature than the previous ones and actually several minutes are required for reaching the optimal working temperature (Figure 1.15a).

Safety precautions (see below) prohibit that these arcs are used in glass apparatuses. Actually, these are used in explosion-proof cases fitted with fins for dispersing the heat and often also with a fan or, for the case of high-power lamps, cooled by forced water circulation. These arcs are rather expensive both when buying and for maintenance and are much shorter-lived. For these reasons, these arcs are not used for preparative irradiations. The typical application in photochemistry is for quantitative measurements, where these arcs are a polychromatic source from which a thin wavelength range is extracted. Thus, the arcs with their cases are mounted on an optical bench and a collimated emission is obtained by introducing a mirror behind the arc and a system of lenses (see Figure 1.15b). The beam, typically 4–5 cm in diameter, is used for the illumination of a (or a pair of) spectrophotometric cuvette(s) or of a horizontal cylindrical cuvette. The beam is filtered as desired, by using as appropriate a cut-off filter, a large-band colored filter, or an interference filter (these select wavelength range within

(a) (b)

Figure 1.15. (a) A 150-W high-pressure mercury arc; the electrodes are a couple of millimeters apart; it makes sense to use this lamp only in a protective case with mirror and lamp for focalizing the emitted beam. (b) Optical bench; the flux from the lamp contained in a protective case is focalized by a mirror and lenses, and a range of wavelengths is selected by a filter and reaches the sample.

10 nm). The light intensity is conveniently monitored for invariance by means of a radiometer placed after the cuvette and is measured by means of an actinometer. As mentioned, these lamps are reserved for kinetic studies and quantum yield measurements or for mechanistic studies where a specific wavelength or intensity is required (the latter is varied by inserting a gray filter). In other studies where a geometrical control is required, this powerful lamp may be useful for the uniform illumination of a small surface as well as for directing a strong beam where desired by adding an optical fiber.

Several other high-pressure arcs are commercially available and share the using characteristics of the above mercury arc—that is, the external usage and scarce application in photochemistry. The xenon arc has a continuous emission from which a monochromatic component can be extracted, but has a weak flux per nanometer. More reasonably, this arc is used because it mimics the solar emission and thus is used for testing the light fastness of various materials, such as polymers, dyes, drugs, and many others. These are adopted in solar boxes, where a certain area is illuminated evenly (and the flux is easily measured) and the choice of an appropriate filter allows to reproduce indoor/outdoor irradiation (see Figure 1.17a below). The mercury–xenon arc adds the lines of the metal, while the antimony–xenon arc is rich in the UV.

The sodium arcs (see Figure 1.13c) are most efficient. The low-pressure version emits only the two close-lying lines at 589 nm, the well-known yellow light. The high-pressure version has a more extended emission and is available also in a version for the use in immersion well apparatuses. These are useful for dye-sensitized reactions, in particular for oxygenations. Being free of a UV component, these lamps have the advantage of avoiding competitive excitation of other components of the solution, but are rather expensive and short-lived. Excimer lamps (an excimer is a dimer that is stable only in the excited state) are a convenient source that emits quasi monochromatically and does not emit heat. These are usually manufactured in the coaxial tube arrangement. In particular, the 308-nm emission of XeCl is well suited for photochemical synthesis. A commercially available such lamp (60 cm in length) has been used in a falling film reactor (see below) with remarkable success for the conversion of hundredth of grams of chemicals [27]. However, these light sources are rather expensive, so that they cannot be considered for adoption by an organic laboratory requiring a versatile tool for preparative applications.

Intrinsically focalized sources are lasers and LEDs. In view of the high intensity, lasers (Figure 1.16a) are used for preparing and modifying nanoparticles or nanocomposites and for the functionalization or ablation of surfaces, rather than for molecular synthesis.

The future for focalized sources is certainly based on light-emitting diodes (LEDs). These are semiconductors that emit an incoherent electroluminescence over a short wavelength range (see Figure 1.16b and c). These small devices are mounted in an optical setup that gives to the emitted beam the desired form, either parallel or with a certain divergence angle. The system of fabrication makes possible obtaining almost any wavelength or, at least, wavelength range (typically a 20-nm

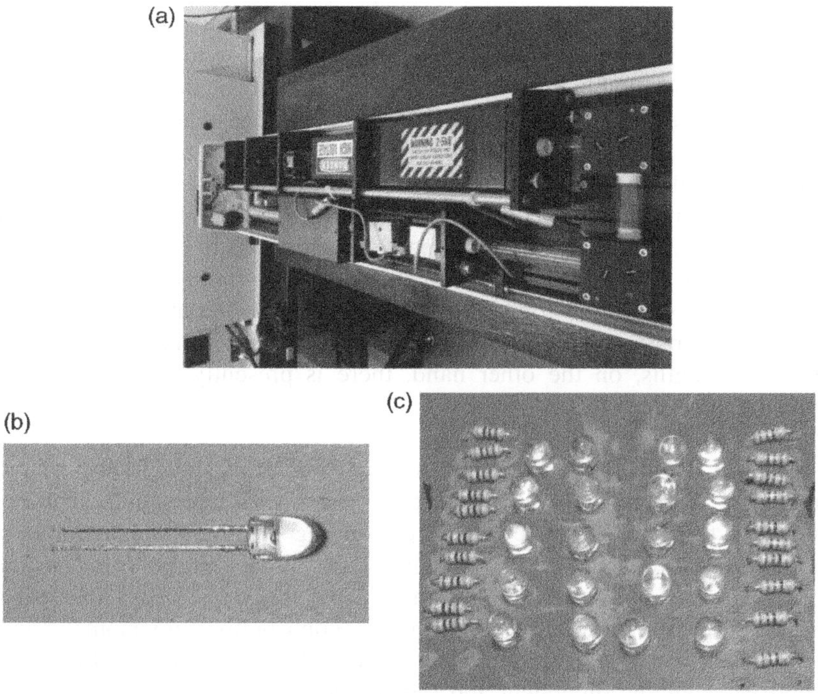

Figure 1.16. (a) A Nd-YAG laser. (b) An LED emitting a close to focalized beam of 4-mm diameter (models with a divergence between 10° and 50° are commercially available). (c) Twenty such LEDs mounted on a surface for parallel irradiation.

range) desired. Actually, LEDs emitting in the visible are commercially available for any 10-nm interval over all of the visible range. These are characterized by a low absorbed power, typically in the hundreds of milliwatts range, allow for a good conversion into light, and are cheap and long-lived ($>10^4$ h). On the other hand, the performance drops rapidly with decreasing wavelength and, although LEDs emitting in the UV-A and down to the UV-B limit (320–310 nm) are also available, their emission is much less intense (less than one-tenth with respect to those emitting in the visible) and their price is higher. Apart from this limitation that may be at least in part lifted by future research, these are an appealing solution for photochemistry. The small LEDs presently available may be conveniently used by assembling a set of them (e.g., 20) and mounting them on an appropriate surface (see Figure 1.16c).

This may be a cylindrical surface at the center of which test tubes containing the solution are accommodated, a sensible choice for carrying out photochemical reactions on small volumes (2–30 mL), or a flat surface for the irradiation of spectroscopic cuvettes or of flat microreactors (see below).

The versatility in the arrangement gives to LEDs a large advantage, and the narrow range of emission gives the possibility of a better control of the reaction. The last point may be a key advantage in specific classes of reactions where the fine-tuning of the wavelength is important. Paradigmatic cases are that of a reactive primary product absorbing in a region quite close to that of the starting material (where being able to choose λ_{EXC} precisely may be the way for obtaining a clean reaction) and that of a dye-sensitized reaction where using a source exactly matching the absorption of a specific dye may give an advantage in terms of selectivity and lower energy consumption (see a caveat below, however). At present, having available a couple of sets of LEDs, one emitting in the UV and one in the blue, appears as a reasonable alternative to having two sets of phosphor-coated low-pressure mercury arcs. For the reasons mentioned above, LEDs are more expensive but more versatile than lamps and might be chosen if a narrow band is important for the studies usually carried out in a laboratory. For shorter wavelengths, on the other hand, there is presently no alternative to a "germicidal" arc.

1.3.1.3 Solar Light Finally, a time-honored but often disregarded source is sunlight [28]. This is a perennial source, is freely available, and can be used without affecting the carbon cycle by consuming electricity. In a time where attention to the ecological factor is growing, this is certainly a convenient alternative in this sense. The limitations are the low density and the fact that the emission spectrum is continuous, with a low contribution per nanometer and, in particular, with a scarce UV component. Obviously, glass windows further filter the short-wavelength component and thus outdoor irradiation has to be chosen. The flux is partly perpendicular and partly tangential due to the reflection by objects or the atmospheric particulate, and the latter component becomes important with a clouded sky. Despite these limitations, simply putting a solution on the window ledge on a sunny day is a convenient way for carrying out photochemical reactions in the 100-mg to 1-g scale, provided that the compound considered absorbs at least a part of the UV-A component (>330 nm). In this case, exposure to sunlight is a convenient method for carrying out an exploratory study for product determination, although it requires a longer time than when using artificial sources. The flux impinging and thus the course of the reaction obviously depend on the geographical location, the time of the year, and the weather conditions, which makes it difficult to reproduce the data and identify the most convenient irradiation time in the same laboratory, not to mention comparing results from different laboratories. On the other hand, something of this type happens also when artificial lamps are used, because even if the same type of arcs are used, different makes, mounting, and age of the lamps lead to different yields at the certain time. Periodic soaring of the oil price have induced attempts to exploit sunlight for organic synthesis on a (relatively) large scale.

Various apparatuses have been built, generally circulating the solution through tubes exposed to the sun by means of a pump, with the addition of a heat exchanger in order to avoid overheating. Tubes have been either simply exposed in a flatbed arrangement or, when it was desired to concentrate the solar emission by using a parabolic mirror, placed in the axis of the mirror itself. Advanced models had a motor-driven system that optimized the arrangement depending on the

Figure 1.17. Using solar light or an artificial model. (a) A SolarBox where light from a (filtered) xenon arc is used for the even illumination of samples lying on the bottom. (b) A Pyrex cylindrical vessel is irradiated an a window ledge. (c) Tubes exposed to solar irradiation through which the solution is circulated; in the upper part, the tubes are mounted in the axis of parabolic mirrors that can be oriented for the best use of solar light (courtesy of Professor M. Oelgemöller, JCU, Australia).

atmospheric situation and the position of the sun. In this way, some reactions have been scaled up to 500 g (Figure 1.17).

1.3.2 Further Experimental Parameters

Obtaining a homogeneous transmission of heat is a major problem for the engineering of industrial reactors, but is usually easy in the laboratory simply by stirring. On the other hand, achieving a homogeneous excitation of molecules in solution by light is virtually impossible. This is obviously an all important issue that has a central role at the moment of scaling up and actual industrialization. As such, it will not be discussed here, since exploratory, small-scale studies are considered at the moment,

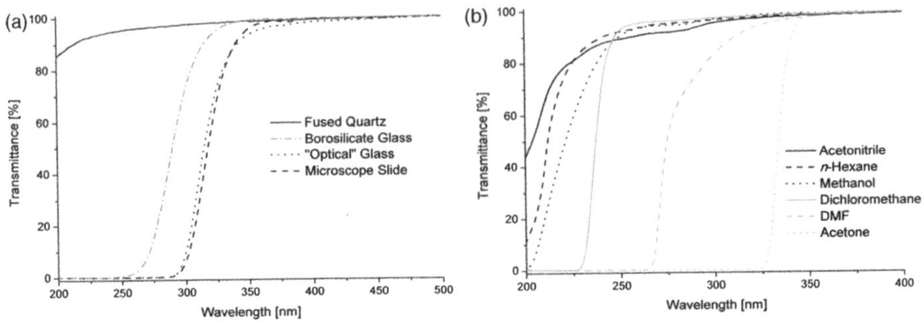

Figure 1.18. Checking that the conditions for a successful photochemical reaction are met. (a) Cut-off by materials used for the reactors (1 mm thick). (b) Cut-off by some commonly used solvents (1-cm optical path).

but the reader will simply be referred to some key works in chemical engineering [29].

However, even if a nonoptimal situation for a small-scale exploratory study is tolerated, it is at least important to check that the planned photochemical reaction is not prevented by some trivial factor, in the sense that light does reach the desired reagent and in the amount sufficient for converting it (see below). Thus, after the choice of the lamp on the basis of the match with the absorption by the reagent, it must be checked that any object placed between the lamp and the solution to be irradiated (reactor vessel, cooling well) is made of a material transparent to the wavelength range concerned (e.g., only quartz for germicidal lamps emitting at 254 nm; see Figure 1.18a). Furthermore, the solvent must not absorb competitively, unless it acts also as the sensitizer. Some relevant cut-off wavelengths for common solvents are reported in Figure 1.18b. It is always convenient to measure the absorption of the actual samples of reagent and solvent used, for a comparison with those taken from the literature, in order to check for absorption by impurities.

The next question is, Which molecules absorb (see Table 1.1 for some selected molecules)? The Lambert–Beer law states that the absorbance of a solution is proportional to the concentration of the reagent, according to the expression $A = \varepsilon bc$, where c is the molar concentration and b is the optical path. The value of ε_{MAX} ranges from $\sim 10\,cm^{-1}\,M^{-1}$ for a partially forbidden band (e.g., the $n\pi^*$ band of aliphatic ketones at 280 nm) to 10^4 and more for allowed transitions (e.g., polyenes, some bands of aromatics, dyes). Given these high values and considering that common spectrophotometers give dependable readings for $A < 3$, spectra are usually measured in dilute solutions. However, carrying out a preparative irradiation is meaningful only if a significant concentration is used (e.g., 0.1 M).

A light beam that enters the vessel will be absorbed in subsequent shells in a proportion depending on the absorbance. As an example, a reagent with molar absorptivity $\varepsilon = 10\,cm^{-1}\,M^{-1}$ at the wavelength considered, at a concentration 0.1 M, has $A = 1$ over a 1-cm depth; in other words, it absorbs 90% of the light flux over a 1-cm layer (and 21% over the first millimeter). With a reagent absorbing more strongly, however, the absorption is more localized. Thus, in the not uncommon

Table 1.1. Choosing a Wavelength with Reference to the Reagent Irradiated

Reagent	λ^a (nm)	Reagent	λ^a (nm)
Allyl alcohol	205	Acetanilide	288
cis-1,2-Dichloroethylene	230	Resorcinol	290
1,1,2,2-Tetrachloroethylene	260	Indene	291
Allyl amine	232	Benzaldehyde	298
Cyclohexanone-oxime	238	Acetophenone	304
Pyrrole	238	Aniline	308
Allyl bromide	253	2-Cyclohexenone	310
Fluorobenzene	270	Stilbene	333
2,5-Dimethyl-2,4-hexadiene	275	Phenanthrene	345
Chlorobenzene	275	Carbazole	350
Toluene	275	Nitrobenzene	350
Bromobenzene	276	Benzophenone	360
Iodobenzene	280	Quinine	370
Thiophenol	280	Anthracene	378
Uracil	285	Acridine	395
Phenol	285	Retinol (vitamin A)	395

[a]The longest wavelength at which the reagent has absorbance $A = 1$ at a 0.01 M concentration.

case where a solution likewise 0.1 M has an $\varepsilon = 10^4\,\text{cm}^{-1}\,\text{M}^{-1}$ at the irradiation wavelength, then the value $A = 1$, where 90% of the light flux is absorbed, is reached over the first 0.1 mm. The optical path in a tube with external irradiation or the thickness of the solution in an immersion well apparatus is typically 1–2 cm. This means that in both cases the flux is absorbed in good part or completely, up to 90% and 99.9% respectively, but with a difference. In the low absorbing solution, excitation by light takes place—if not uniformly over the entire solution, not that far from it—and convection of the liquid or stirring ensures that the molecules exposed to irradiation are regularly changed. In fact, when using low-flux sources where the emission is distributed over a large surface—for example, a low-pressure mercury arc and a tube in front of it, or a vessel exposed to sunlight—spontaneous mixing is often sufficient.

In the latter case, however, with a high absorbance, practically all of the solution is unaffected except for a thin layer adjacent to the surface exposed to light, and effective mixing is a must (this is the "mass transfer" problem—that is, one of the key issues for the design of a large-scale photoreactor). This is the typical case when high–pressure mercury arcs, compact lamps that emanate a high-density flux, are used with high absorbing solutions. Under such conditions, stirring is often nonsatisfactory in the cylindrical "photochemical reactors" most often used (and commercially available), and the likelihood that secondary photoreactions take place before the starting material is consumed increases. High-molecular-weight products or polymers may be formed and a film of tar sometimes covers the surface of the well, making the reaction stop altogether because the remaining reagent in solution receives little, if any, light.

A further common trouble occurs when the photoproducts absorb in the same wavelength range as the starting reagents. This may have two negative implications. If the primary product is not reactive, a progressive slowing down of the reaction is observed (this is often referred to as the "inner filter" effect). On the contrary, if it reacts with the same or larger efficiency, a second photochemical step takes place and a mixture is obtained (in lucky cases, one may arrive in a clean way to a secondary product; see Chapter 8). Secondary reactions are a problem also in thermal chemistry, obviously, and photochemical reactions have the advantage that when the product absorbs at a shorter wavelength than the reagent, the reaction remains clean even when the primary product is photoreactive, provided that the irradiation is limited to the long-wavelength part of the reagent absorption (see Chapter 8).

The dilution problem remains a difficult point in devising the conditions of a photochemical reaction. For the reasons presented above, clean reactions are typically obtained by using dilute solutions (optimal results are rarely obtained above a 0.01 to 0.1 M concentration) and by avoiding an intense light flux. This choice is certainly advisable for mechanistic investigations (and, at any rate, dilution simplifies the mechanism), but obviously not for preparative purposes.

The general answer to this problem is shortening the optical path. This can be done on a large volume by circulating the solution stored in a reservoir by means of a peristaltic pump, in such a way that a thin layer is exposed to the lamp at any time, rather than irradiating and stirring the whole batch. Several approaches have been used and in some cases have arrived at commercial development, although homemade devices can be effective as well (see Figure 1.19).

A simple way is coiling a plastic tubing (transparent to the light used and preferably made of fluorinated polyolefins for increased chemical resistance) around the cooling well of a medium-pressure lamp and circulating the solution in it. In this way the portion of the solution irradiated is continuously changed, and the limited

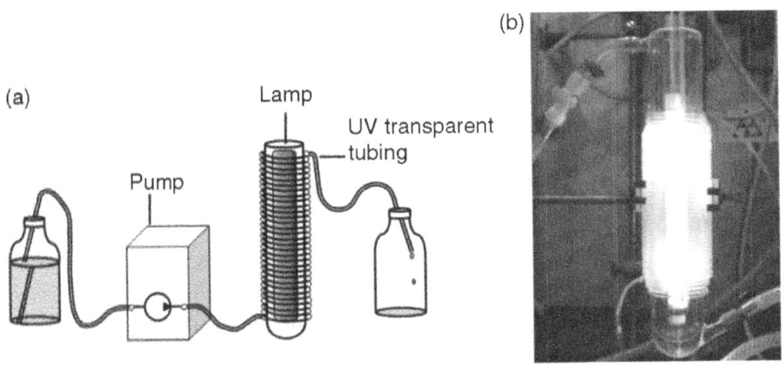

Figure 1.19. (a) Typical scheme of a flow photoreactor. (b) A circulating apparatus built by coiling a fluorinated ethylene propylene (FEP) tube (internal diameter: 0.7 mm) around an immersion well apparatus. Reprinted with permission from Knowles, J. P., Elliott, L. D., Booker-Milburn, K. I. (2012). Flow photochemistry: Old light through new windows, *Beilstein Journal of Organic Chemistry*, **8**, 2025–2052.

depth diminishes somewhat the lack of homogeneity. Experiments based on this concept have been reported, highlighting that considerable volumes (up to several liters) can be irradiated in a limited overall time, and a better conversion and yield have been obtained than when irradiating in a bulk reactor [30]. A tube has been used that has an internal diameter wide enough that no significant increase of the pressure occurs over the path (e.g., >1 mm). A further increase in the scale can be achieved by coiling a new portion of the tubing around another lamp and then around another one. Thus, several lamps may be connected in series by the tubing so that the irradiated volume and the amount of reagent transformed reach an industrially significant amount. Experiments with this type of setup and a single or several lamps have shown that a production in the order of hectograms per day can be obtained with a little change in the normally used setup, provided of course that the lamps are available, with a better use of the light energy despite the loss by refraction.

A way to overcome the last limitation is eliminating the tubing and having a thin layer of the solution circulating around the lamp. In such a "falling film" arrangement (see Figure 1.20) a solution from a reservoir is sprayed at the top of the cooling well where the lamp is placed by means of a peristaltic pump and collected at the bottom. An external cylinder may be used for maintaining the chosen atmosphere, and in fact this type of reactor is best used for gas–liquid reactions. Apparatuses of this type have been commercialized. As a matter of fact, however, the coverage of the surface is often not uniform due to the formation of preferential flowing channels. An improvement has been obtained by placing the lamp horizontally rather than vertically in a rotating tube. In this way the path is shorter and the inhomogeneity diminished [31].

The most radical advancement, however, is miniaturization, which fits into the general move in this direction that has been developed in various fields of synthesis. This is based on the introduction of microstructured reactors (Figure 1.20b), consisting of a solid support where thin channels are engraved (1 mm or down to 10 μm) that form a serpentine channel [32–34]. Here, a multitude of thin falling films move by gravity in parallel microstructured channels. The reactors are commercially available or may be prepared in house, according to the specific needs. Thus, thin grooves are engraved and then covered by a transparent plate so that the solution is circulated through the resulting channels and is illuminated. Contrary to the previous cases, here it may be that absorption is partial. However, excitation is homogeneous. As mentioned, this condition is difficult to reach otherwise and, if obtained, may lead to an impressive simplification of the product mixture, since secondary reactions are minimized [32–34]. At any rate, the process can be optimized by changing the time of exposure of the solution to the light source, simply by changing the flow rate. Further useful characteristics are the control of the temperature, easily achieved given the high surface of the channels, and the easy coupling with a system for the on-line monitoring of the reaction.

This "lab on the chip" concept is best applied when the light source is likewise miniaturized; indeed this is the best field of application of LEDs, which add the further advantage of the narrow emission wavelength and thus another way of avoiding secondary reactions. A significant number of photochemical reactions have recently been tested under these conditions and compared with batch experiments,

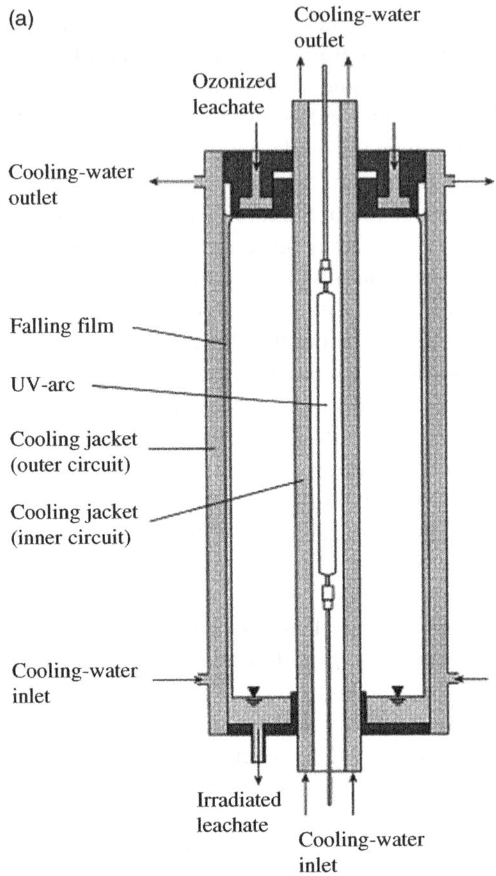

Figure 1.20. Falling film reactors. (a) By using a conventional arc lamp. Reprinted with permission from Wenzel, A., Gahr, A., Niessner, R. (1999). TOC-removal and degradation of pollutants in leachate using a thin-film photoreactor. *Water Research*, **33**, 937–946. (b) By using a microreactor. Reprinted from Knowles, J. P., Elliott, L. D., Booker-Milburn, K. I. (2012). Flow photochemistry: Old light through new windows. *Beilstein Journal of Organic Chemistry*, **8**, 2025–2052.

evidencing large improvements. As an example, the photocycloaddition of cyclohexenone to vinyl acetate was found to give 88% of the product in 2 h in a microchannel reactor and—all other conditions being unchanged—only 8% in a batch reactor [35].

The reasons that make difficult a photochemical reaction on a concentrated solution apply a fortiori to the solid phase. In principle, a molecule that is photoreactive may well be reactive in the solid state and in particular in the crystal state, but this applies only when it is actually excited—that is, when the light flux is able to penetrate into the crystal and makes subsequent layers of molecules react. This means that the product formed must not absorb at the wavelength used, so that the

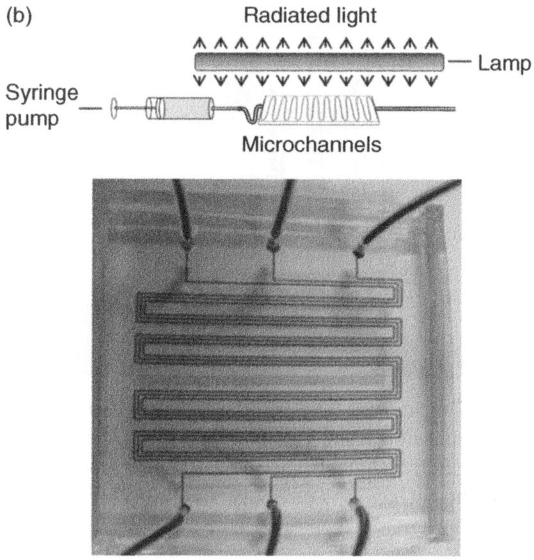

Figure 1.20. (*Continued*)

photoreaction makes that layer transparent and the subsequent photons can reach the following layer. At any rate, it is generally convenient to irradiate at the border of the absorption band, where the extinction coefficient is smaller.

Furthermore, changes in the crystal habit are to be expected when the molecular structure changes; this may also lead to the loss of most of the light by refraction, once again stopping the reaction. When this is the case, crystals exposed to light may undergo a conspicuous change in their appearance that, however, corresponds to a minimal change in the chemical composition, as it may be checked by analyzing the dissolved crystal, since only the very first layers of molecules have reacted. This is not to say that solid-state photochemistry is no interesting field. In fact, although investigations in the field have proceeded somewhat intermittently, a remarkable number of crystal-phase photochemical reactions have been reported, and indeed these are rather common in some classes of processes (e.g., cycloadditions), where it is almost the rule because the less conjugated product absorbs drastically to the blue with respect to the starting material [36, 37]. Many cases of high selectivity have been reported; and, although the rationalization is not always unchallenged, the clean course has been related with the location of the molecules in the crystal lattice. Notably, a number of enantioselective processes have been obtained by using chiral auxiliaries. From the practical point of view, these experiments are usually carried out either on a thin layer of small dimension crystals on a glass plate or with the material deposited on the internal walls of a rotating tube (from a rotating evaporator). In the latter case, a solution can be slowly evaporated in a rotating tube taken horizontally, and the thin coating obtained is then irradiated [38]. Another approach for obtaining small crystals is precipitating the reagent by adding a solution

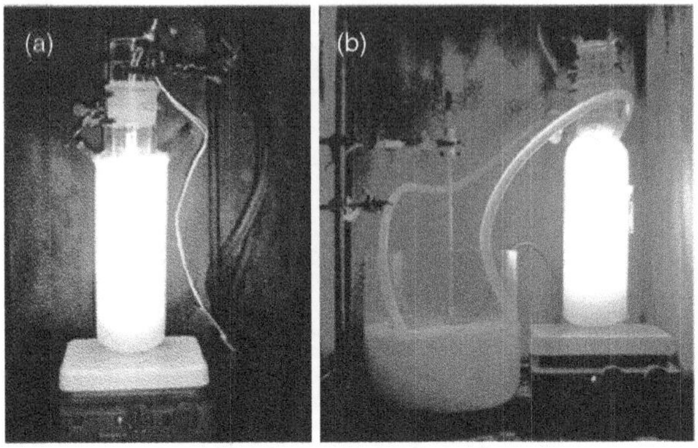

Figure 1.21. Irradiation of nanocrystalline suspensions. (a) In a bulk reactor. (b) In a flow reactor. Reprinted with permission from Veerman, M., Resendiz, M. J. E., Garcia-Garibay, M. A. (2006). Large-scale photochemical reactions of nanocrystalline suspensions: A promising green chemistry method. *Organic Letters*, **8**, 2615–2617.

in a water-miscible organic solvent to a water layer while stirring. The microcrystals suspension obtained under these conditions has been shown in some cases to be effectively transformed photochemically (see Figure 1.21) [39].

1.3.3 Practical Hints

In the paragraphs above, attention to the key fact that light has to arrive at the molecule intended to react has been recommended. The next move goes to the molecule and is a fast one. In fact, the lifetime of excited states is short, with typical values $\tau < 1$ μs for triplet and $\tau < 1$ ns for singlet states. As mentioned in Section 1.1, only a very fast chemical reaction may compete with mere physical decay to the ground state—in practice, only those that have activation energy of a few kilocalories per mole. This fact gives to photochemical reactions their characteristic independence with regard to the temperature, and indeed it makes them relatively little affected by impurities as well—or at least by some types of impurities. The fastidious attention to the purification of reagents and solvent, the flaming and protecting the vessel from contact with the atmosphere, and most of the precautions that are so typical of advanced organic synthesis are rarely used in photochemistry. Certainly, a change in conditions (changing the solvent, adding an acid, etc.) may cause a large change in a photoreaction when it implies shifting to an excited state of different nature. The usually quoted example is that of ketones having a $n\pi^*$ state as the lowest (and thus the reactive) triplet and a $\pi\pi^*$ triplet slightly higher in energy. In a protic solvent, hydrogen bond stabilizes the n orbital and thus increases the energy of the $n\pi^*$ state that overreaches the $\pi\pi^*$ state, thus making the latter the reactive state. Furthermore, a large change in the solvent characteristics may change the nature of the photochemical process. The typical example is that of aryl halides that decompose homolytically in apolar or moderately polar solvents, but heterolytically in

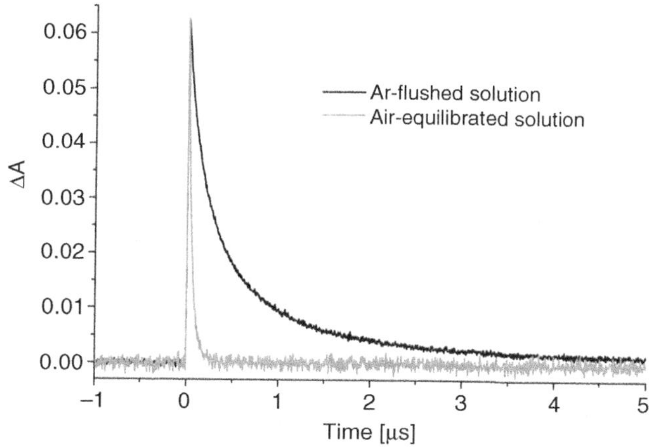

Figure 1.22. Decrease of the lifetime of benzophenone triplet in air equilibrated versus argon flushed solution.

polar (protic) ones [40, 41]. Apart from this, extensive purification of reagents and solvents is not generally required, provided of course that neither strongly absorbing impurities (the presence of which should always be checked by running a UV spectrum of the compound used and comparing it with the literature) nor high enough amounts of reactive compounds (that introduce an undesired bimolecular reaction) are present. This makes alkanes generally a nonsatisfactory solvent because these contain significant amount of alkenes, difficult to detect spectrophotometrically but highly reactive (e.g., with ketones).

In this category an important case is that of oxygen, which quenches most excited states at a rate within one order of magnitude from diffusion-controlled. In many air-equilibrated organic solvents the amount of oxygen present in solution is in the range 0.002–0.003 M (0.0005 in water), high enough for quenching 80–98% of a long-lived triplet ($\tau = 1\ \mu s$). An example is shown in Figure 1.22 for the case of benzophenone. The effect is much smaller, but not negligible, with singlets or with short-lived triplets (<30% for a lifetime of 10 ns). A slowing down of the reaction is often observed in the presence of oxygen, and impurities may result from some chemical reaction of oxygen with the excited state or, even more commonly, with some ensuing intermediate, particularly when odd electron species such as radicals or radical ions are formed. In that case the course of the reaction can be actually diverted or stopped.

The effect is easily eliminated by flushing the solution with an inert gas for some minutes, so that the amount of oxygen dissolved drops by approximately three orders of magnitude, in the same way as is commonly done with radical reactions. In exploratory studies, it is advisable to carry out the experiments on nitrogen- or argon-flushed capped test tubes or in flushed reactors. In the continuation of the work, one may check whether oxygen causes a mere lengthening of the irradiation time required, as is often the case, and skip this precaution, thus further simplifying the experimental procedure.

It may be mentioned in this connection that the above characteristics of the excited states likewise affect a desired bimolecular reaction. Thus, even if, as mentioned above, photochemical reaction are by definition quite fast, and the rate constant is large (typically k_r 10^6 to $>10^9\,M^{-1}\,s^{-1}$), the concentration of excited states is low. Trapping by a reagent T must overcome decay to the ground state (k_d, which as mentioned above is 10^6 to $>10^9\,s^{-1}$). Traps must be added in a high enough concentration that reaction with the excited state A* is nonnegligible (quantum yield $= k_r\,[T]/(k_r\,[T] + k_d)$. In practice, with singlets, traps have to be used at a >0.1 M concentration, and thus in a large stoichiometric excess. Lower amounts are required for short-lived triplets or for intercepting intermediates formed from the excited states.

Photochemical reactions raise relatively small safety issues, but there are some important requirements that must be adhered to. Intense light sources (including solar light) may be dangerous for the skin and the eyes—in particular, due to the UV component. The UV-B wavelength range, 280–320 nm, causes erythema while UV-C, below 280 nm, is genotoxic. Even exposure to a strong UV-A (up to 400 nm) or visible source may cause damage. The UV component is filtered off by the conjunctiva, and therefore does not penetrate in the inner part of the eye, but for this reason causes a quite painful *conjunctivitis*. Importantly, the pain begins to be remarked several hours after the exposure; thus there is no warning with regard to the exposure to short-wavelength light. Thus, photochemical reactions must be carried out in a protected area, either in a devoted room or in a separated part of the laboratory protected by a curtain. Several commercial instruments (e.g., some multilamp apparatuses) are fitted with a switch that automatically shuts down the lamps when the cabinet is opened. Otherwise, home-mounted setups should be protected by a suitable opaque screen. Taking into account that it is very difficult to avoid that indirect or reflected radiations arrive at the operator, appropriate precautions must be taken when coming close to an operating irradiation apparatus, unless one prefers putting off the lamps, which causes no negative effect with low-pressure arcs, while high-pressure arcs require a starting time when reignited and age faster when frequently switched. Thus, besides a laboratory coat, protective glasses of the type purposely designed for UV light must be worn. Another issue is the formation of toxic ozone. The amount generated by low-pressure mercury arcs is limited, but with high-pressure arcs (which, differently from the previous ones, are usually mounted in a confined space) a high concentration may build up locally. Therefore, photochemical reactors of this type must be operated under a fume hood. Better still, high-power lamps should be operated while maintaining nitrogen flushing in the close vicinity of the arc, which avoids the inconvenience.

Summing up, one may conclude that having available some lamps, the most convenient choice being one to two pairs each of 15-W low-pressure arcs with emission centered at 254, 313, and 366 nm (or as an alternative to become more common in the future, some sets of LEDs) and some quartz tubes, is a sensible move for every synthetic laboratory and allows exploring the viability of photochemical steps. The limited precautions and safety requirements involved make photochemistry a method much more easily used than many reactions involving thermal/oxygen/moisture labile and/or toxic/flammable reagents or catalysts that are routinely considered and carried out in synthetic laboratories. Therefore an experiment can be easily done; failure by a beginner is more easily due to some naive oversight (see below) than to inadequate experimental capability. Furthermore, an experiment with a time-honored method such as exposure to solar light may be appropriate, at least as a first indication, for reagents absorbing in the near UV and has the nonnegligible advantage of using a freely available source and of requiring no investment at all.

As indicated above, the choice of the lamp must be based on the match of the source emission with the absorption by the putative reagent, while the material of the irradiation apparatus must be transparent to that wavelength. The same obviously holds for the solvent. In Figure 1.18b the transmission by a 1-cm layer of (spectrophotometric grade) common solvents is shown. From these data it can be concluded that when using a low-pressure mercury arc (254 nm) the apparatus must be made of quartz and the solvent chosen among alkanes, alcohols, ethers, acetonitrile, or simple haloalkanes (e.g., dichloromethane), besides water. Lamps emitting in the UV-B and A allow for a larger choice, including benzene (obviously not a choice on other grounds), *N,N*-dimethylformamide, dimethylsufoxide, and others. Acetone has a cut-off over 300 nm. All of the common solvents are transparent in the visible.

The best setup will be chosen also with reference to the scale desired. Small-scale photochemical experiments—in particular, comparative studies—are conveniently carried out by using either multilamp apparatuses, or lamps centrally mounted immersed in a thermostating bath or focalized lamps on a optical bench, choosing the vessel according to the analytical technique used (absorption/fluorescence cuvette, NMR tube, etc.) and finally measuring the conversion of an appropriate actinometer when the determination of the quantum yield is desired. For more synthetically addressed studies, the desired scale must be taken into account. An indication is given in Table 1.2, where we reported the time required for the complete conversion

Table 1.2. Some Key Indications for Planning a Photochemical Reaction

Lamp	Irradiation (nm)	Einstein (min^{-1} cm^{-2})	Volume irradiated (mL)	Converting a 5×10^{-2} M solution ($\Phi = 1$) (h)
Low-pressure Hg	ext, 254[a]	$\approx 4 \times 10^{-6}$	10–100	0.25
Phosphor-coated	ext, 305[a]	$\approx 1 \times 10^{-6}$	10–100	1
Medium-pressure	ext, 350[a]	$\approx 8 \times 10^{-7}$	10–100	1.2
Hg immersion	300–400[b]	$\approx 10^{-5}$	80–1000	0.1
LED	ext, 310[c]	$\approx 2 \times 10^{-8}$	1–10	40
LED	ext, 400[c]	$\approx 1 \times 10^{-6}$	1–10	1
Solar light	ext, 330–400	$\approx 10^{-7}$	any	10

[a]In a multilamp apparatus fitted with 6 × 15-W lamps.
[b]Referred to the wavelegth range indicated, neglecting emission in the visible.
[c]Six LEDs circularly placed around the test tube containing the solution.
Source: Reproduced with permission from ref. 11, p. 18; Copyright Wiley-VCH Verlag GmbH & Co. KGaA.

of 100 mL of a 5×10^{-2} M solution of a reagent that absorbs completely the flux impinging (up to 400 nm for the case of solar light), on the assumption that the quantum yield is unitary and there is no inner filter effect. In other words, this is the minimal time required for converting the above amount.

As it appears from these data, an amount up to a few grams requires an irradiation time of 5–10 h. Longer irradiation times are required under less favorable conditions, such as a lower quantum yield or an incomplete light absorption. This can be summarized by stating that a new product can be photochemically prepared in the scale between 100 mg and some grams in 5–24 h, provided that the quantum yield is $\Phi \geq 0.05$. Reactions with a quantum yield of 0.01 may still be of preparative interest, if they are clean.

A number of excellent textbooks of organic photochemistry are available, and reference to such texts is recommended for an appropriate introduction to the field [8, 9]. Presentations of the synthetic aspects of organic photochemistry are likewise available in books and reviews [10–19], arranged according to the chemical function reacting or to the type (mechanism) of the reaction involved.

REFERENCES

1. Muller, P. (1994). Glossary of terms used in physical organic chemistry (IUPAC 2 Recommendations 1994). *Pure and Applied Chemistry*, **66**, 1077–1184.
2. Liu, Y.-J., Roca-Sanjuan, D., Lindh, R. (2012). Computational photochemistry and photophysics: The state of the art. *Photochemistry*, **40**, 42–72.
3. Griesbeck, A., Oelgemoller, M., Ghetti, F., Eds. *CRC Handbook of Organic Photochemistry and Photobiology*, third edition, CRC Press, Boca Raton, FL, 2012, pp. 1029–1056.
4. Noyori, P. (2010). Insight: Green chemistry: The key to our future. *Tetrahedron*, **66**, 1028.
5. White, J. D. (1998). *George Buchi. Obituary. Organic Syntheses*, **77**, xxiii–xxvi.
6. Griesbeck, A. (2009). Photochemistry of organic compounds. *Angewandte Chemie International Edition*, **48**, 4671–4672.
7. Moss, R. A., Platz, M. S., Jones, M., Jr., Eds. *Reactive Intermediates Chemistry*, John Wiley & Sons, Hoboken, NJ, 2004.
8. Turro, N. J., Ramamurthy V., Scaiano J. C. *Modern Molecular Photochemistry of Organic Molecules*, University Science Books, Sausalito, CA, 2010.
9. Klàn, P., Wirz, J. *Photochemistry of Organic Compounds. From Principles to Practice*, John Wiley & Sons, Chichester, 2009.
10. Griesbeck, A., Oelgemoller, M., Ghetti, F., Eds. *CRC Handbook of Organic Photochemistry and Photobiology*, third edition, CRC Press, Boca Raton, FL, 2012.
11. Albini, A., Fagnoni, M., Eds. *Handbook of Synthetic Photochemistry*, Wiley-VCH, Weinheim, 2009.
12. Griesbeck, A. G., Mattay, J., Eds. *Synthetic Organic Photochemistry*, Marcel Dekker, New York, 2005.
13. Ramamurthy, V., Schanze, K. S., Eds. *Organic Photochemistry and Photophysics*, Marcel Dekker, New York, 2006.

14. Carreira, E. M., Griesbeck, A., Eds. (2001). Organic photochemistry. *Synthesis*, 1111–1280.
15. Horspool, W. H., Armesto, D. *Organic Photochemistry: A comprehensive treatment*, Ellis Horwood, London, 1992.
16. Müller, E., Ed. *Methoden der organische Chemie (Houben.Weeil) Photochemie*, Vols. 1 and 2, Georg Thieme V, Stuttgart, 1975.
17. Mattay, J., Griesbeck A., Eds. *Photochemical Key Steps in Organic Synthesis*, Wiley-VCH, Weinheim, 1994.
18. Horspool, W. M., Ed. *Synthetic Organic Photochemistry*, Plenum, New York, 1984.
19. Hoffmann, N. (2008). Photochemical reactions as key steps in organic synthesis. *Chemical Reviews*, **108**, 1052–1103.
20. Bakac, A., Ed. *Physical Inorganic Chemistry: Principles, Methods, and Models*, John Wiley & Sons, Hoboken, NJ, 2010, pp. 199–267.
21. Khudyakov, I. V., Turro, N. J. Laser flash photolysis of photooinitiators: ESR, optical, and IR spectroscopy detection of transients. In Forbes, M. E. D., Ed. *Carbon-Centered Free Radicals and Radical Cations*, John Wiley & Sons, Hoboken, NJ, 2010, pp. 249–279.
22. Lackner, M., Ed. *Lasers in Chemistry*, Wiley-VCH, Weinheim, 2008, pp. 1073–1091.
23. Protti, S., Fagnoni, M., Monti, S., Rehault, J., Poizat, O., Albini, A. (2012). Activation of aliphatic C–H bonds by tetracyanobenzene photosensitization. A time-resolved and steady-state investigation. *RSC Advances*, **2**, 1897–1904.
24. Shimada, T., Kumagai, A., Funyu, S., Takagi, S., Masui, D., Nabetani, Y., Tachibana, H., Tryk, D. A., Inoue, H. (2012). How is the water molecule activated on metalloporphyrins? Oxygenation of substrates induced through one-photon/two-electron conversion in artificial photosynthesis by visible light. *Faraday Discussions*, **155**, 145–163.
25. Scaiano, J. C., Johnston, L. J., McGimpsey, W. G., Weir, D. (1988). Photochemistry of organic reaction intermediates: Novel reaction paths induced by two-photon laser excitation. *Accounts of Chemical Research*, **21**, 22–29.
26. Heering, W. (2004). UV sources—Basic, properties and applications. *IUVA News*, **6**, 7–13.
27. Griesbeck, A., Maptue, N., Bondock, S., Oelgemöller, M. (2003). The excimer radiation system: A powerful tool for preparative organic photochemistry. A technical note. *Photochemical & Photobiological Sciences*, **2**, 450–451.
28. Protti, S., Fagnoni, M. (2009). The sunny side of chemistry: Green synthesis by solar light. *Photochemistry & Photobiological Sciences*, **8**, 1499–1516.
29. Romero, R. L., Alfano, O. M., Cassano, A. E. (2003). Radiation field in an annular, slurry photocatalytic reactor. 2. Model and experiments. *Industrial & Engineering Chemistry Research*, **42**, 2479–2488.
30. Hook, B. D. A., Dohle, W., Hirst, P. R., Pickworty, M., Barry, M. B., Booker-Milburn, K. I. (2005). A practical flow reactor for continuous organic photochemistry. *The Journal of Organic Chemistry*, **70**, 7558–7564. Knowles, J. P., Elliott, L. D., Booker-Milburn, K. I. (2012). Flow photochemistry: Old light through new windows. *Beilstein Journal of Organic Chemistry*, **8**, 2025–2052.
31. Saien, J., Soleymani, A. R. (2012). Feasibility of using a slurry falling film photo-reactor for individual and hybridized AOPs. *Journal of Industrial and Engineering Chemistry*, **18**, 1683–1688.

32. Coyle, E. E., Oelgemöller, M. (2008). Micro-photochemistry: Photochemistry in microstructured reactors. The new photochemistry of the future. *Photochemistry & Photobiological Sciences*, **7**, 1313–1322.
33. Oelgemöller, M., Shvydkiv, O. (2011). Recent advances in microflow photochemistry. *Molecules*, **16**, 7522–7550.
34. Matsushita, Y., Ichimura, T., Ohba, N. Kumada, S., Sakeda, K., Suzuki, T., Tanibata, H., Murata, T. (2007). Recent progress on photoreactions in microreactors. *Pure and Applied Chemistry*, **79**, 1959–1968.
35. Fukuyama, T., Hino, Y., Kamata, N., Ryu, I. (2004). Quick execution of [2 + 2] type photochemical cyclo-addition reaction by continuous flow system using a glass-made microreactor. *Chemistry Letters*, **33**, 1430–1431.
36. Natarajan, A., Bhogala, B. R. Bimolecular photoreactions in the crystalline state. In Ramamurthy, V., Inoue, Y., Eds. *Supramolecular Photochemistry: Controlling Photochemical Process*, John Wiley & Sons, Hoboken, NJ, 2011, pp. 175–228.
37. Sonoda, Y. (2011). Solid-state [2 + 2] photodimerization and photopolymerization of α,ω-diarylpolyene monomers: Effective utilization of noncovalent intermolecular interactions in crystals. *Molecules*, **16**, 119–148.
38. Cohen, M. C. (1987). Solid phase photochemical reactions. *Tetrahedron*, **43**, 1211–1224. Williams, J. R., Abdel-Mayed, A. (1981). Photolysis of 3-oxo-$\Delta^{5(10)}$-steroids in alcoholic solvents and in the solid phase. *Tetrahedron*, **27**, 1675–1677.
39. Kuzmanich, G., Natarajan, A., Chin, K. K., Veerman, M., Mortko, C. J., Garcia-Garibay, M. A. (2008). Solid state photodecarbonylation of diphenylcyclopropenone: A quantum chain process made possible by ultrafast energy transfer. *Journal of the American Chemical Society*, **130**, 1140–1141.
40. Lazzaroni, S., Protti, S., Fagnoni, M., Albini, A. (2010). Participation of a heterolytic path in the photochemistry of chlorobenzene. *Journal of Photochemistry and Photobiology A: Chemistry*, **210**, 140–144.
41. Dichiarante, V., Fagnoni, M. (2008). Aryl cation chemistry as an emerging versatile tool for metal-free arylations. *Synlett*, 787–800.

2

PHOTOGENERATION OF CARBON-CENTERED RADICALS

2.1 INTRODUCTION

Among reactive intermediates, the use of radicals has become an increasingly important and attractive tool for organic practitioners, as clearly witnessed by the recent publication of books and monographies on this topic [1–7]. Many approaches are available for the *thermal* generation of carbon-centered radicals [1–7], but most of them are amenable to either of the two pathways depicted in Scheme 2.1. In the first one, a (relatively) weak C–X bond is present and undergoes homolysis (path *a*). Accordingly, alkyl (or aryl) bromides and iodides are the elective substrates for radical generation, and to some extent further groups such as SePh, SPh, and others. The key feature of these reactions and the one that has made this chemistry popular is that these are chain processes. Thus, the applied protocol requires an initiation step in which the thermal fragmentation of an initiator occurs. Peroxides, an azo compound such as azobisisobutyronitrile (AIBN) or its derivatives, and a few other molecules are used. These initiators fragment at a convenient rate at a temperature in the range 50–60°C and yield a pair of radicals that in turn abstract a hydrogen atom from a donor such as a metal hydride, most often Bu_3SnH [8]. As is well known, the thus formed stannyl radical operates as a chain carrier that abstracts a halide atom from an organic derivative forming a carbon-centered radical. Then, this nucleophilic species adds to an electrophilic multiple bond forming the planned C–C bond [9, 10]. The cyclic process is completed by hydrogen abstraction from the stannane by the adduct radical. Different chain carriers can be used; for example, hydrostannanes can be replaced by

Photochemically-Generated Intermediates in Synthesis, First Edition.
Angelo Albini and Maurizio Fagnoni.

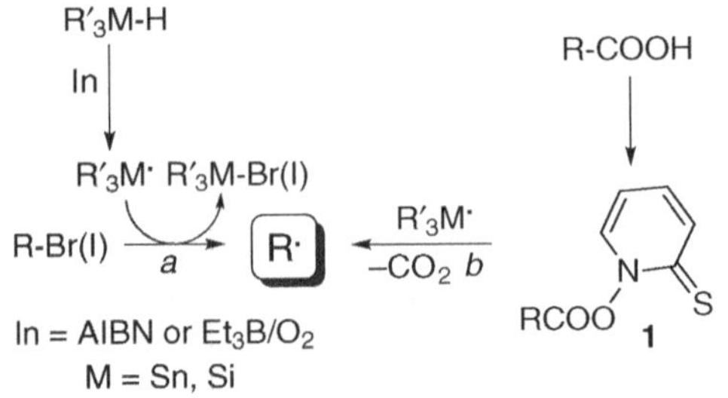

Scheme 2.1.

the less reactive, but also less toxic, silicon hydrides, such as tris-(trimethylsilyl)silane $(TMS)_3SiH$ [11]. These compounds are active only at a higher temperature (at least 80°C). Alternative initiation methods that work efficently below room temperature, such as that based on Et_3B/O_2 [12] (active at −78°C), have been also developed and applied to the cases where the selectivity of the radical addition is crucial.

In the second pathway, a highly stabilized leaving group is introduced and makes fragmentation viable. A typical case is the decarboxylation of acids that involves prior conversion into thiohydroxamate esters, typically *N*-acyloxythiopyridones (Barton esters, **1**). When the C–S double bond of such a compound reacts with an R'_3M radical (M = Sn, Si), a carboxyl radical R-COO$^•$ is liberated and gives the desired radical upon carbon dioxide loss. An advantage of this method resides on the large choice of the carboxylic acids used for the synthesis of the Barton esters [13].

The role of *photochemistry* is twofold. The first one is its application for the initiation of the process, in that irradiation causes fragmentation either of the initiator or directly of a bond in the precursor and from that point on the previously presented chain process takes place. In this way there is no temperature requirement for the initiation step, although it may remain for the chain process. The photochemical option makes sense when the initiator or the precursor absorb conveniently and electronic excitation leads to efficient homolysis.

Exemplificative pathways for the photochemical generation of carbon-centered radicals are illustrated in Scheme 2.2 [14–20]. A seemingly obvious source of alkyl radicals would be the photodissociation of a weak C–halide bond—in particular,

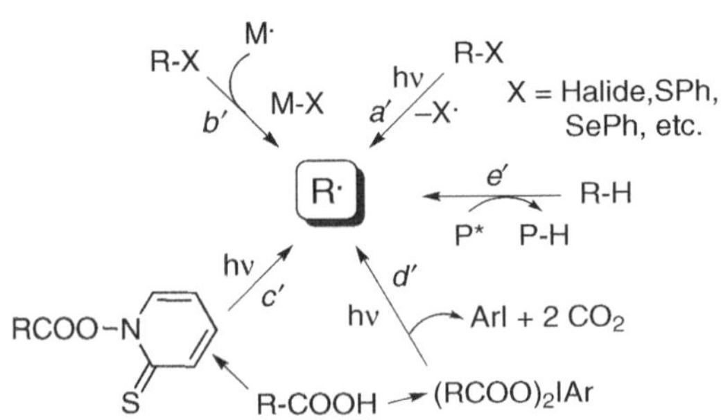

Scheme 2.2.

C–Br and C–I, the same precursors as in thermal reactions (Scheme 2.1, path *a*). However, this is generally a poor choice, both because of the weak absorptivity of such compounds in accessible (that is not covered by other reagents or the solvent) wavelength regions and because of the occurrence of competitive processes leading to ions and carbenes. Despite these limitations, a number of radical syntheses starting from alkyl and aryl halides (mainly iodides and bromides) and to a lesser extent phenyl sulfides and selenides have been reported (Scheme 2.2, path *a′*). Alternatively, it is the chain carrier radical $M^{\bullet}$ that is generated in some other way and cleaves a C–halogen bond (path *b′*). Well-characterized photoinitiated processes involve the photocleavage of Barton esters (path *c′*) that exhibit in fact a strong UV absorption [13, 21]. Furthermore, a considerable part of the reported radical syntheses involves another family of precursors based on carboxylic acids via hypervalent iodonium species (path *d′*) [22].

In contrast to the above applications, the actual breakthrough that photochemistry brings in synthesis is based on non-chain methods. In this case, one gives up the convenient aspects of chain methods, but obtains two important advantages. First, the use of a suitable excited photocatalyst (or a photomediator) P^* (path *e′*) allows to cleave a strong bond, such as a C–H bond [23–26]. The process is energetically expensive, but results in a shortening in the path for the functionalization of hydrocarbons, which is obtained in a single step and not via halogenation and following halogen substitution. Second, no chain carrier is required, which involves a large decrease of the waste and a simplification of the workup and thus results in an advancement in terms of atom economy and green chemistry.

Since many reports and reviews deal with the application of photogenerated radicals in organic synthesis [14–20], we selected the most significant examples (in our opinion) belonging to each class, with particular regard to examples reported after the beginning of the new millennium.

2.2 ALKYL RADICALS

2.2.1 Alkyl Radicals from Alkyl Halides

The generation of alkyl radicals by fragmentation has been largely documented, or at least there are a large number of reports where the photochemical dehalogenation of organic compounds is suggested to operate [27–29]. This may be the case, and indeed the process may be rather efficient. However, it is not always trivial to identify the actual mechanism operating under the conditions reported—and at any rate has been only rarely investigated—and whether the cleavage of the carbon–halogen bond actually involves a reaction of the excited halide (which, as mentioned in the introduction, often is not selective) [27]. Moreover, other mechanisms can operate—for example, some where the solvent has an active role or the halide release occurs from an intermediate, such as the radical anion formed in a photoinduced electron transfer process (see Section 5.2).

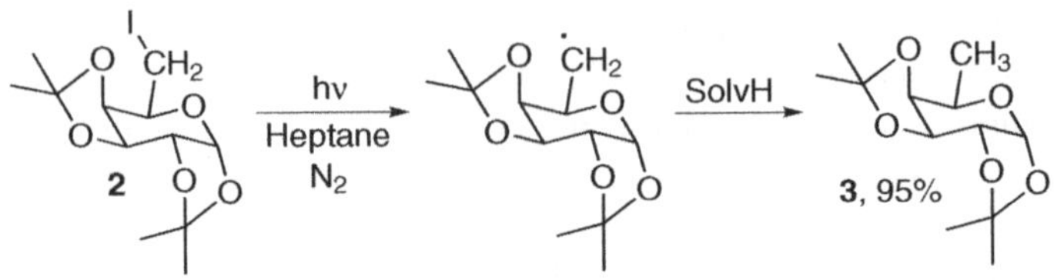

Scheme 2.3.

On the other hand, some photoinitiated radical reactions have been reported where Bu_3SnH or related chain carriers were activated by irradiation of the reaction mixture, but do not otherwise differ from thermal reactions (path b′); such cases will be not discussed here. Note, however, that the choice of the stannane and the conditions may dramatically change the result. As an example, allyl α-bromophenylacetate underwent reductive dehalogenation in the presence of tributylstannane under the usual thermal conditions (AIBN, 80°C), but photoinitiation in the presence of hexamethyldistannane under triplet sensitization by a ketone led to a different evolution of the benzyl radical that was better controlled at a lower temperature. Thus, in a non-hydrogen-donating solvent, the dimer was formed, while in a moderately hydrogen-donating medium (*i*-PrOH) 5-endo cyclization preceded reduction [30].

Taking into account these limitations, a selection of examples will be reported where photohomolysis of the radical precursor was established and clearly stated in the original paper. A representative example is the photoreaction of deoxyiodo sugars such as 6-deoxy-6-iodo-1,2,3,4-di-*O*-isopropylidene-α-D-galactopyranose (**2**, Scheme 2.3). This compound suffered iodine atom loss upon photolysis in nitrogen-flushed heptane, and the resulting radical abstracted hydrogen to give compound **3** in 95% yield (quantum yield = 0.31). Under aerated conditions a complex mixture resulted. On the contrary, irradiation of 3-deoxy-3-iodo-1,2:5,6-di-*O*-isopropylidene-α-D-allofuranose and 3-deoxy-3-iodo-1,2:5,6-di-*O*-isopropylidene-α-D-glucofuranose caused epimerization at position 3 via carbon–iodine bond homolysis and radical recombination, apparently a chain process (quantum yield >1) [31].

In particular cases a selective partial dehalogenation of polyhalides was achieved. In fact, *gem*-dibromocyclopropanes were smoothly converted to the corresponding monobromo derivatives in fair to good yields by irradiation in ether solutions [32].

An indirect way to generate alkyl radicals from alkyl halides by photochemical means makes use of dimeric metal complexes [33–36]. As is well known, dicarbonyl (η5-cyclopentadienyl)iron dimer or decacarbonyldimanganese undergoes metal–metal bond cleavage upon light absorption, forming monomeric metal-centered radicals. The last species was able to abstract a halogen atom from an alkyl halide (a primary alkyl iodide was preferred) and generated an alkyl radical that was trapped by conjugate addition onto electron-poor olefins [33–35]. This approach was successfully applied to the synthesis of (*R*)-coniine (**7**, Scheme 2.4) [36]. The synthetic plan was based on the addition of the 3-chloropropyl radical onto the C=N bond in a chiral *N*-acylhydrazone **5** activated by a Lewis acid such as $InCl_3$. Under

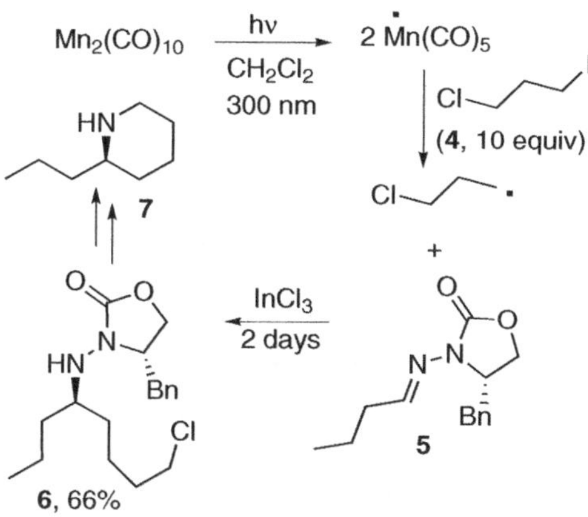

Scheme 2.4.

these conditions, the coniine precursor **6** was prepared in 66% yield, with the limitation that both a large excess of **4** and a long irradiation time (2 days) were required (Scheme 2.4) [36].

2.2.2 Alkyl Radicals from Carboxylic Acids

As mentioned in the introduction, carboxylic acids can be easily converted into the corresponding Barton esters by reaction with thiohydroxamic acid [13, 21]. The photolability of the N–O bond in these esters can then be exploited for the generation of alkyl radicals as illustrated in Scheme 2.5, where an adamantyl radical was generated and used for the alkylation of heteroaromatic compounds. Thus, lepidine gave the corresponding 2-substituted derivative **9** in a high yield [37].

The reaction between radicals photogenerated from Barton esters with methanesulfonyl **10** or *p*-toluenesulfonyl cyanide represents a smooth entry for the preparation of the corresponding nitriles **11** in high yields [38]. Interestingly, when the reaction was

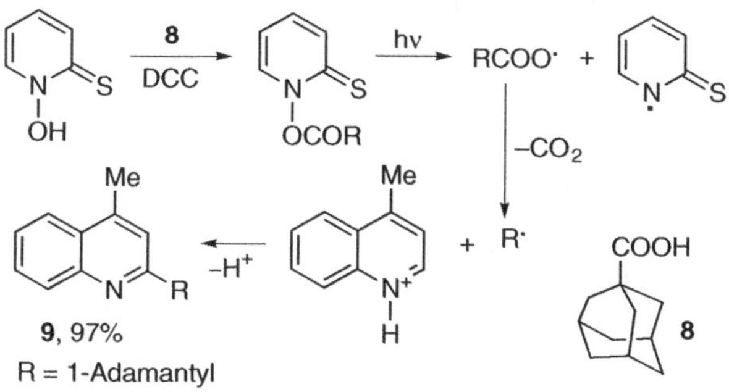

Scheme 2.5.

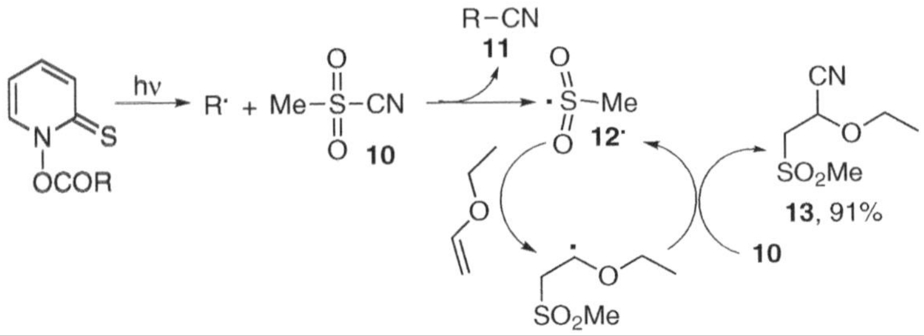

Scheme 2.6.

performed in the presence of an (electron-rich) alkene, a β-cyanosulfone (e.g., **13**) could be prepared in a satisfactory yield by a process initiated by addition of the sulfonyl radical **12**˙ onto the double bond (Scheme 2.6) [39].

A Barton ester was found to be a convenient starting material for the radical functionalization of protected D-gluconic acid **14** by tungsten lamp irradiation in the presence of ethyl α-(trifluoroacetoxy)acrylate (**15**) to afford adduct **16** in 53% yield (Scheme 2.7) [40]. A partial deprotection of the acetal groups appeared to justify the modest yield obtained in this case.

Thiohydroxamic esters have been likewise used for reductive decarboxylation, by adding a suitable hydrogen donor for the reduction of the intermediate alkyl radical. This approach was applied in the final step of the preparation of 1-methylcarbapenem antibiotic precursors starting from D-glucosamine; *t*-butylthiol was chosen in this case as the reducing agent [41]. This decarboxylation protocol was likewise used as the key step in the synthesis of *Kopsia* alkaloids, such as (±)-kopsidasine and (±)-kopsidasine-*N*-oxide [42]. Recently, the use of chloroform as an inexpensive hydrogen atom source for Barton reductive decarboxylation reactions has been reported [43].

The functionalization of the side chain of carbohydrates with a Barton ester (e.g., compound **17**) allowed the preparation of 2-C-methyl (**18**) and 2-C-bromomethyl (**19**) hexoses, pentoses, and disaccharides in yields up to 80% by irradiation in the presence of a reducing agent (*t*-BuSH) or a brominating agent ($CBrCl_3$), respectively (Scheme 2.8) [44]. The brominated carbohydrates (e.g., **19**) formed were susceptible of further interesting transformations.

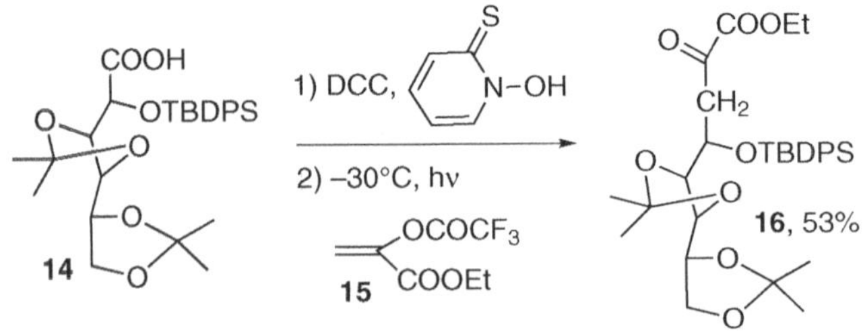

Scheme 2.7.

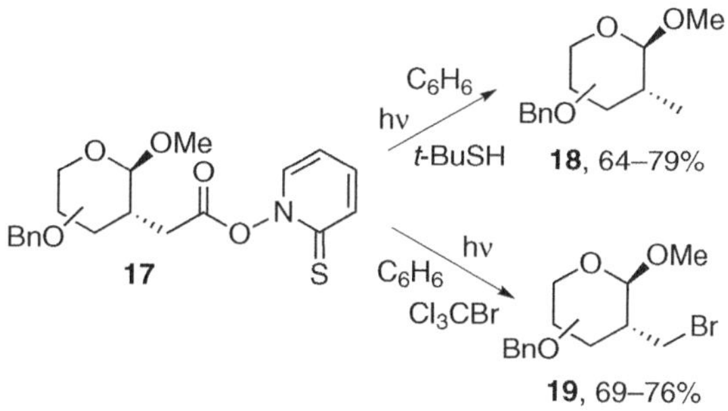

Scheme 2.8.

The formation of C-heteroatom bonds via alkyl radicals has recently been described. Thus, chiral β-silyl Barton esters were used for both free radical halogenation (by reaction with $CBrCl_3$) and azidation (by reaction with $EtSO_2N_3$). These reactions occurred with moderate to excellent diastereocontrol and yield [45].

Thiohydroxamic esters different from the classical thiopyridone derivatives have recently been developed for radical generation, in particular under photochemical conditions. As an example, photolysis of ester **20** with visible light under aerated (or oxygenated) conditions in the presence of a thiol gave hydroperoxide **21** that was easily converted to alcohol **22** in 70% overall yield (Scheme 2.9). This reaction sequence represents a nice example of decarboxylative hydroxylation of carboxylic acids [46].

Open-chain thiohydroxamate esters (e.g., **23**) are likewise excellent sources of alkyl radicals, as demonstrated in the preparation of oxime ethers **25** by radical addition onto the sulfonyl derivative **24** (Scheme 2.10) [47]. The presence of a -COOMe group in the place of a hydrogen directly linked to the oxime carbon decreased to some extent the radical addition yield.

A modified Barton ester (**26**) was likewise adopted for the synthesis of pseudomonic acid C from L-lyxose (Scheme 2.11) [48]. The α-oxy radical generated by irradiation of thionocarbonate **26** in toluene was trapped by allyltributyl stannane

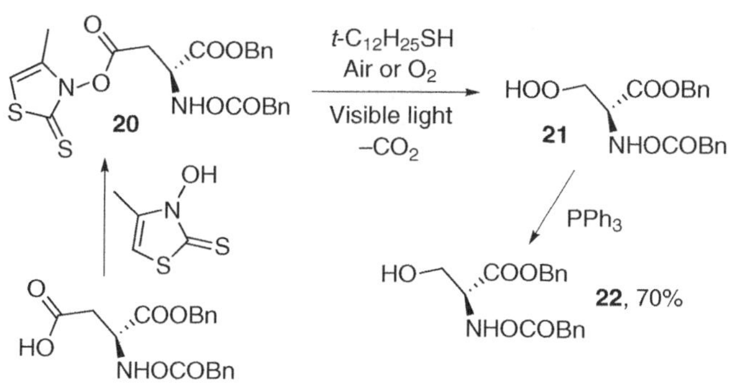

Scheme 2.9.

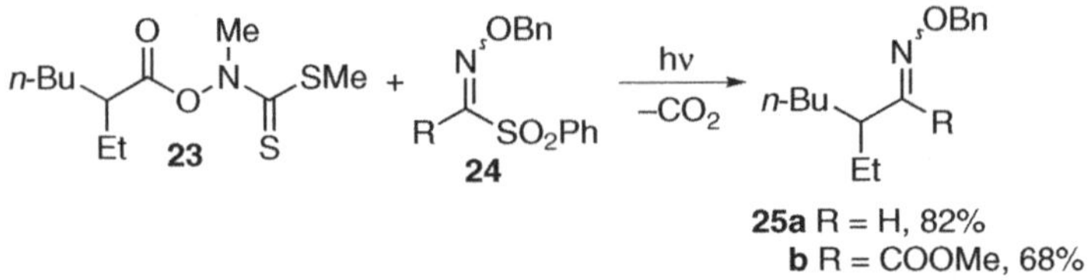

Scheme 2.10.

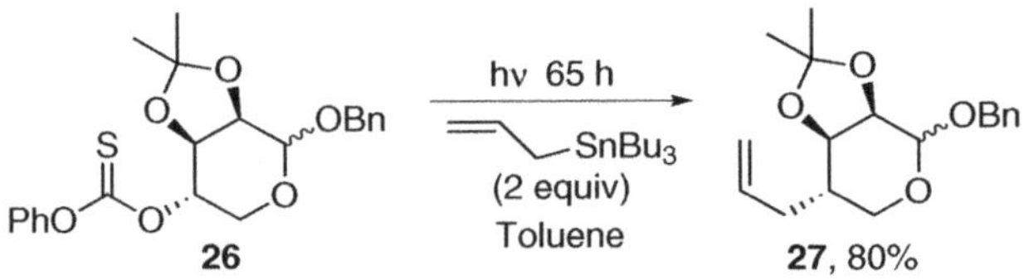

Scheme 2.11.

(2 equiv) to give allylated derivative **27** in 80% yield; a limitation in this case was the long irradiation time (65 h) required.

Another possibility for using carboxylic acids as radical sources involved conversion into (diacyloxyiodo)arenes as depicted in Scheme 2.12. In this case, the radical precursor is easily prepared by mixing (diacetoxyiodo)benzene and the acid. Then, the radical is smoothly liberated in solution either photochemically or thermally by decarboxylation [22].

In favorable cases the stereochemistry of the carbon bearing the COOH group was maintained in the reaction. This turned out to be the case for the preparation of new *C*-nucleosides by radical addition onto a heteroaromatic base such as lepidine (Scheme 2.13a). A new C–C bond was then obtained in a stereoselective fashion starting from sugar-derived acid **28** by treatment with $PhI(OAc)_2$ in the presence of lepidine (**29**) and under irradiation [49]. Under these conditions, compound **30** was obtained as a single isomer in 90% yield (Scheme 2.13a).

The cyclohexylation of lepidine was chosen as a test for comparing the photochemical and the thermal generation of alkyl radicals from trivalent iodine compounds (Scheme 2.13b). The radical precursor was prepared starting from cyclohexanecarboxylic acid and [bis(trifluoroacetoxy)iodo]benzene and then either refluxed in benzene or irradiated in DCM for the same time in the presence of **29**. As a result, in both cases the alkylated heteroaromatic base **31** was isolated in roughly the same yield (~50%, Scheme 2.13b) [50].

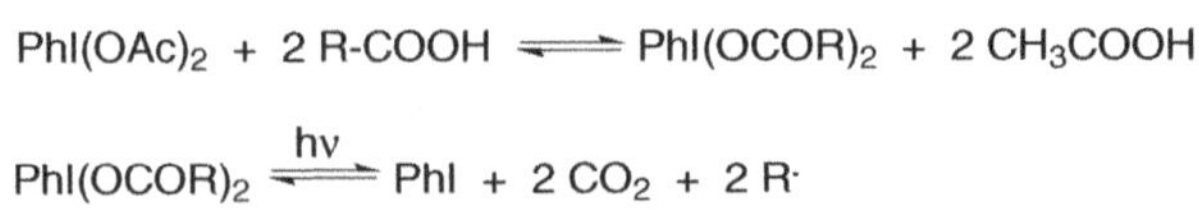

Scheme 2.12.

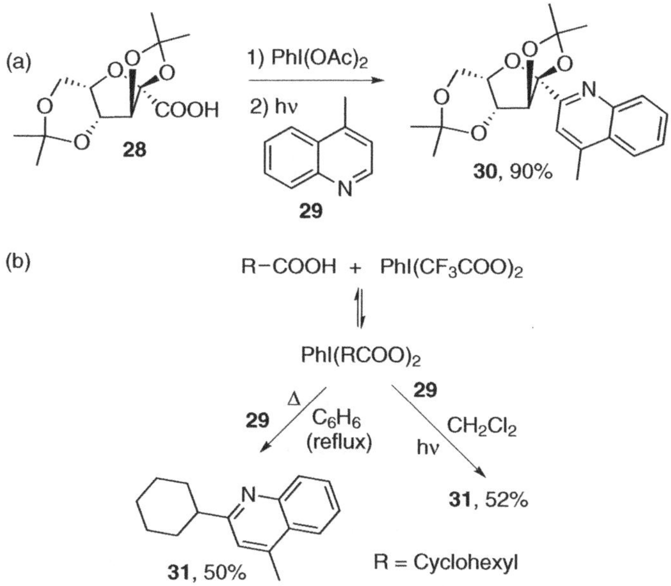

Scheme 2.13.

Photogenerated radicals from hypervalent iodine compounds have been likewise employed in the conjugate addition onto vinyl sulfones, vinyl phosphates, and acrylate esters in good to excellent yields [51].

2.2.3 Alkyl Radicals from Alkanes

In the realm of paper chemistry, an obvious breakthrough for the generation of alkyl radicals would seem to be using hydrocarbons in the place of alkyl halides or carboxylic acids and obtaining the direct activation of C–H bonds. This goal has often been considered as one of the "Holy Grails" in chemistry, and the strength of such bonds makes this a difficult task to be fulfilled thermally [1, 52]. An exception is the thermal cleavage of the labile oxygen–oxygen bond in peroxides or peresters where an electrophilic, oxygen-centered radical R–O$^{\bullet}$ is formed (see also Chapter 3) and is able to abstract homolytically a hydrogen from a C–H bond.

The driving force of the reaction is the formation of a strong O–H bond. In this case, peroxides should be used stoichiometrically, rather than in a small amount as when they act only as initiators in the alkylation in the presence of chain-carriers discussed above. The use of large amounts of peroxides involves practical problems, and this fact along with the scarce selectivity of the process limits their use for this purpose [1].

On the other hand, direct photochemistry offers no way of activation. Alkanes absorb only in the far UV (<200 nm), so that they can be considered transparent to all practical purpose. Also, there are no virtual (triplet) low-lying states that can be populated by sensitization.

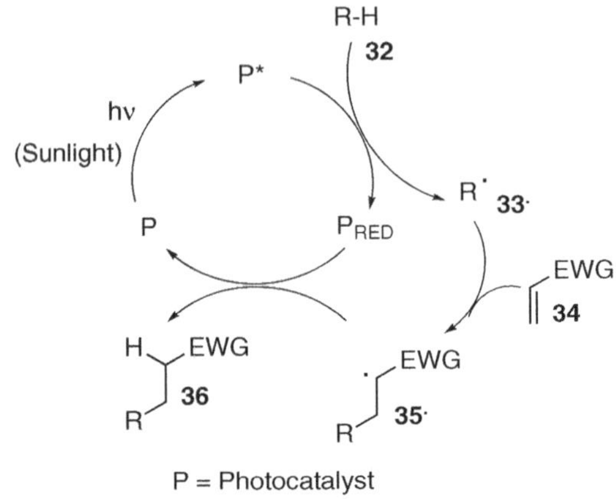

Scheme 2.14.

A new scenario can be devised by having recourse to photocatalysis [23–26]. Scheme 2.14 illustrates a typical photocatalytic cycle. In this case an excited photocatalyst (P*) behaves just as an alkoxy radical does; that is, it abstracts homolytically a hydrogen atom from a C–H bond in a donor R–H (**32**). The process generates the reduced form of the photocatalyst (P_{RED}) along with an alkyl radical (**33˙**) prone to give conjugate radical addition onto electron-poor olefins (**34**). This option is greatly advantageous in view of the really mild conditions required (no heating, no base or nucleophile required). Even more important is the fact that P_{RED} is able to donate a hydrogen atom back to the adduct radical (**35˙**), thus forming the alkylated olefin (**36**) with the concomitant regeneration of P (Scheme 2.14). As a result, the compound P can reasonably be defined a photo*catalyst* and is used in a substoichiometric amount, while the alkylation process requires the absorption of a stoichiometric amount of photons. This at least in the ideal case, since the back hydrogen transfer step may not be quantitative and the photocatalyst is consumed to some degree. As will be shown in more detail below, semiconductor oxides and inorganic photocatalysts approach the ideal case and are used in a small amount (e.g., 1–2%), while an organic material in the same function generally is consumed to a large extent, which forces to use it in a much larger starting amount (10–100%) [26]. This is less convenient also because the photocatalyst-derived side products make workup less simple [23–26].

The reaction is most often carried out by using cycloalkanes as the radical precursors in order to avoid regiochemical selectivity problems, as summarized in Scheme 2.15. Cyclohexane and cyclopentane have been the usual choice, also because a large excess of the hydrocarbon is required for an efficient activation, and low-boiling derivatives are expediently distilled away at the end of the reaction.

As for the photocatalyst, different choices are possible. Aromatic ketones (e.g., benzophenone) have been largely used, although these are effective only when used in a high, often equimolar amount. Accordingly, *β*-cycloalkyl (di)nitriles were smoothly synthesized starting from α,β-unsaturated (di)nitriles and cycloalkanes [54].

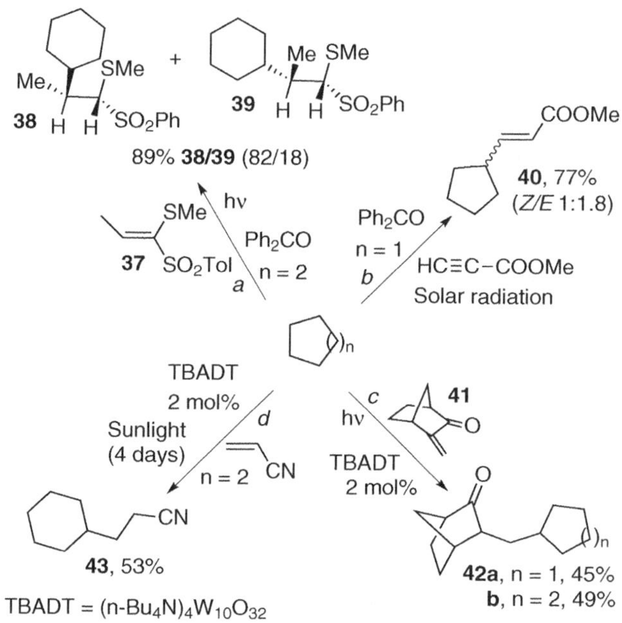

Scheme 2.15.

Captodative olefins were likewise convenient substrates, as in the photocatalyzed addition of cyclohexane onto ketene thioacetal *S,S*-dioxide **37**. The alkylation took place in a high yield (89%) and with some diastereoselectivity (**38/39** 4.5/1), apparently induced by the steric repulsion between the entering radical and the *p*-tosyl group (Scheme 2.15, path *a*) [55]. It is noteworthy that attempted alkylation of olefin **37** by using an alkyl iodide under the usual thermal radical chain conditions with a tin hydride as the transfer agent failed. This was reasonably due to the great stability of the radical adduct due to the captodative effect that made hydrogen abstraction from Bu_3SnH too slow, thus precluding the propagation of the radical-chain process [55].

Electron-poor alkynes as well were functionalized under these conditions [56–58]. The absorption of aromatic ketones extends to the UV-A region, with some superimposition with the solar spectrum. Indeed, solar irradiation turned out to be a viable choice for carrying out the photochemical addition of cyclopentane onto methyl propiolate (Scheme 2.15, path *b*) [57]. The cyclopentyl derivative **40** was isolated in 77% yield as a diastereoisomeric mixture. Despite the fact that compound **40** would appear liable to a further radical attack, no polyalkylated compound was detected in the irradiation mixture. In this case, an equimolar amount of Ph_2CO had to be employed again, despite the fact that the vinyl radical intermediate is known to be able to abstract hydrogen from cyclopentane and form a cyclopentyl radical. Alkynes were successfully alkylated also by using recyclable polymer-bound photocatalysts, such as benzophenone bound to Merrifield resin or to silica gel surface through an aminopropyl tether [58].

A second important class of photocatalysts involves inorganic compounds—in particular, polyoxometalates. Such ions were introduced in this role about 25 years ago [59] and have recently found increasing application in radical generation. Tetrabutylammonium decatungstate $(n\text{-Bu}_4\text{N})_4[\text{W}_{10}\text{O}_{32}]$ (TBADT) is soluble in acetonitrile and appears to be the most active salt in the class [60, 61]. In analogy with aromatic ketones, the excited state of TBADT is able to abstract hydrogen from alkanes. Two great advantages with respect to organic materials, however, are that it can be used in catalytic amounts and that it can be easily removed from the irradiated mixture—for example, by filtering over a silica gel layer [62, 63].

Accordingly, the photochemical alkylation of a range of conjugated olefins by both cyclopentane and cyclohexane was obtained by using a 2 mol% of TBADT. In the case of methylenenorbornanone (**41**), adducts **42a** and **42b** were formed in a comparable decent yield (~50%, Scheme 2.15, path *c*) via a process characterized by an excellent atom economy [64].

The absorption of TBADT covers a large fraction of the UV-A region, and this photocatalyst is again quite effective also under solar irradiation. Accordingly, the alkylation was expediently carried out by simply exposing the reagents solution on a window ledge. No artificial energy was used for heating, cooling, or stirring; thus this could be considered a model eco-sustainable synthesis. For instance, a mixture of cyclohexane (5 equiv), acrylonitrile, and TBADT was exposed to sunlight for four days, and 3-cyclohexylpropionitrile (**43**) was then easily isolated by evaporation of the solvent followed by bulb-to-bulb distillation (Scheme 2.15, path *d*). The irradiation was carried out in 100-mL cylindrical vessels and afforded about 10 g of the desired product [65].

Recently, the C–H activation in some alkylcyclohexanes, chosen as models for nonsymmetric alkanes, has been likewise investigated. Interestingly, hydrogen abstraction occurred with some regioselectivity, judging from the results with *tert*-butylcyclohexane, where five different kinds of hydrogens are present (Scheme 2.16). As a matter of fact, virtually exclusive activation of the hydrogens in position 3 and 4 resulted [66]. Moreover, a high degree of stereoselection was observed in the alkylation of β,β-disubstituted alkenes, such as compound **44**. In this case, a 1,3-cyclohexane derivative with *cis* stereochemistry (**45**) and a *trans*-1,4 isomer (**46**) were exclusively isolated in a good overall yield (70%). Trapping experiments with α-phenyl-*N*-*tert*-butylnitrone were carried out and demonstrated that the initial hydrogen abstraction step was responsible for the regioselectivity observed, not a following preferential trapping [66], supporting the notion that the use of a rather bulky photocatalyst can result in a selective activation.

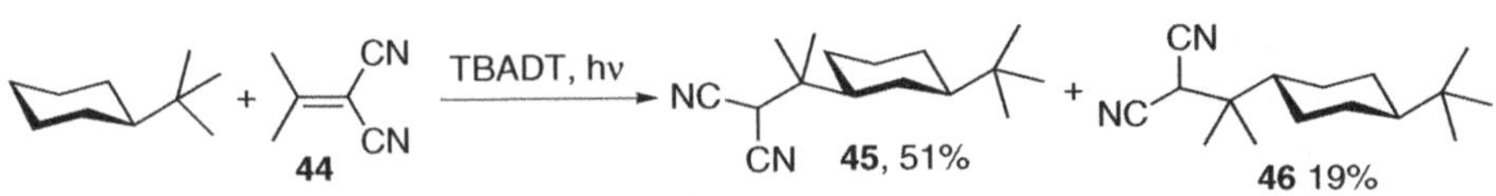

Scheme 2.16.

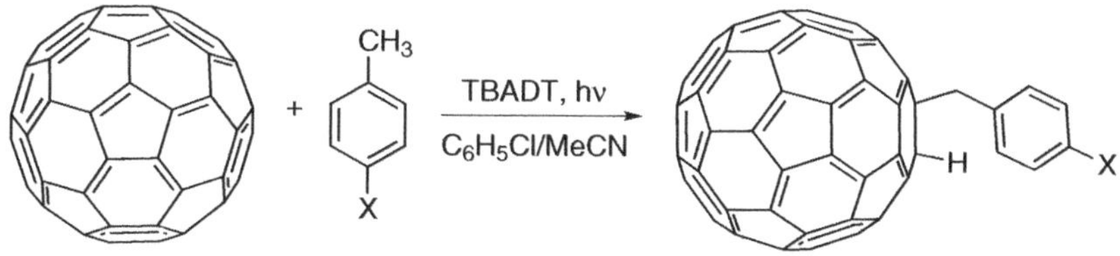

47a, X = H, 35%
b, X = CH_3, 45%
c, X = Br, 50%

Scheme 2.17.

Benzylic hydrogens in toluene derivatives were likewise liable to TBADT-photocatalyzed activation. The resulting benzyl radicals were efficiently trapped by [60]fullerene and yielded the functionalized 1,2-dihydro[60]fullerenes (**47**) in moderate to good yields as shown in Scheme 2.17. In this case the end products arose from the reaction of the carbon-centered radical with the radical anion $C_{60}^{\bullet -}$ formed upon photoinduced electron transfer between $^3C_{60}^*$ and the P_{RED} form of TBADT ($[W_{10}O_{32}]^{5-}$) [67].

2.2.4 Alkyl Radicals from other Precursors

Azoalkanes (e.g., AIBN) are commonly used as radical initiators due to their instability upon heating that causes nitrogen loss with the concomitant release of alkyl radicals [2]. These derivatives are likewise easily cleaved upon irradiation, and in some cases the resulting radicals have been used in alkylation reactions. As an example, perfluoroazooctane (**48**) was used for the preparation of perfluoroalkyl-substituted aromatics (Scheme 2.18). In order to have a good yield, the reaction had to be carried out in a two-phase system consisting of perfluorohexane (the bottom layer) and the aromatic to be alkylated (e.g., benzene, the upper layer). When the perfluorohexane phase (containing most of **48**) was exposed to far-UV light (185 nm), perfluorooctyl radicals were smoothly formed and reacted with the benzene dissolved in the fluorous phase or present at the interface. This radical process allowed to obtain the functionalized benzene **49** in a 70% yield, along with a small amount of the fluoroalkane $C_8F_{17}H$ (Scheme 2.18) [68]. On the other hand, when the same reaction was carried out in benzene solution, the yield of product **49** was reduced to 13% [68].

48

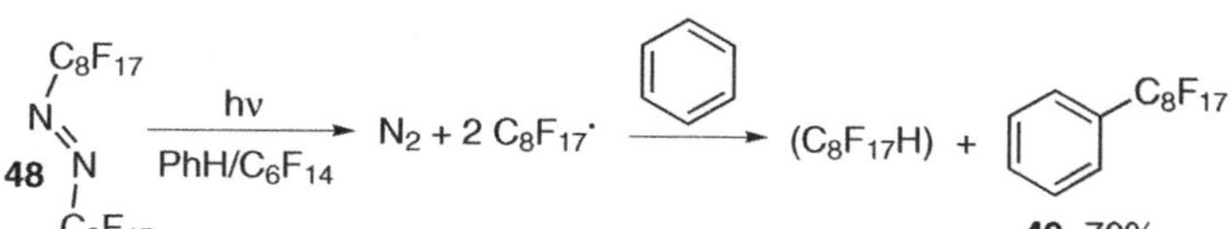

49, 70%

Scheme 2.18.

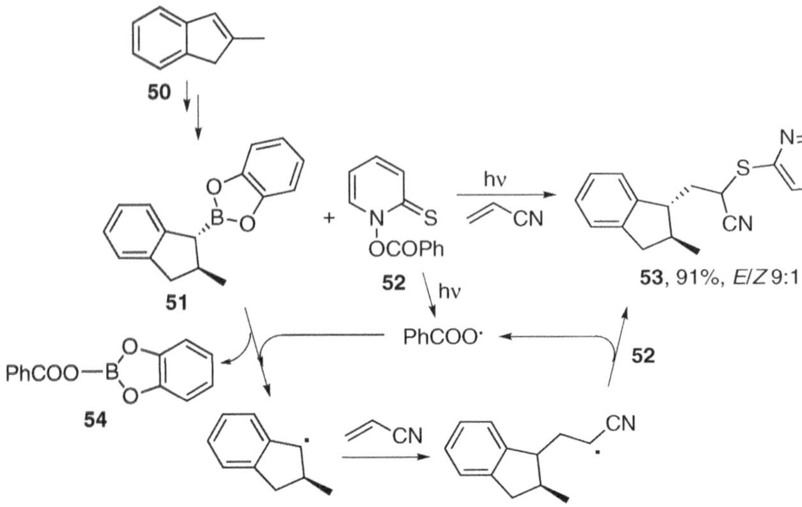

Scheme 2.19.

Following the same approach, a suspension of single-walled carbon nanotubes (SWCNTs) in perfluorohexane containing perfluoroazooctane was photolyzed for 8 h at room temperature. The desired side-wall modification was indeed obtained and, noteworthy, the mild conditions of the functionalization preserved the SWCNTs electronic behavior [69].

Moreover, organoboranes have been reported as an effective radical source when irradiated in the presence of a Barton ester (e.g., **52**) as illustrated in Scheme 2.19. The reaction was initiated by photolysis of phenylcarbonyloxy (pyridine-2-thione) (**52**) and the resulting acyloxy radical PhCOO$^\bullet$ added to catechol boronate **51**, prepared in a few steps from **50**. Borate **54** was then released with the concomitant formation of an alkyl radical that upon addition onto acrylonitrile and reaction with the starting ester **52** gave the indane derivative **53** in a very good yield [70].

Sparse examples deal with the generation of alkyl radicals by irradiation of organometallic compounds—for example, alkylcobalt complexes [71–73] or alkylmercury(II) compounds [74]. As an example, the irradiation of 1-substituted-2-oxocyclopentylmethyl cobaloxime **55** gave the cyclohexenone derivatives **57** in variable yields by rearrangement of the intermediate alkyl radical **56**$^\bullet$ (Scheme 2.20) [75].

A related photohomolysis of alkyl-cobalt(III) species was applied to amino acid derivatives for the synthesis of (−)-α-kainic and (+)-allokainic acid [76]. The use of air-stable alkyl or stannyl cobaloximes (bis(dimethylglyoximato)-cobalt) (e.g., **58**, Scheme 2.21) has recently emerged as a convenient method for the generation of carbon-centered radicals from the corresponding alkyl iodides [77]. Thus, the irradiation of a mixture containing an alkene, an alkyl iodide, and a trialkyl amine

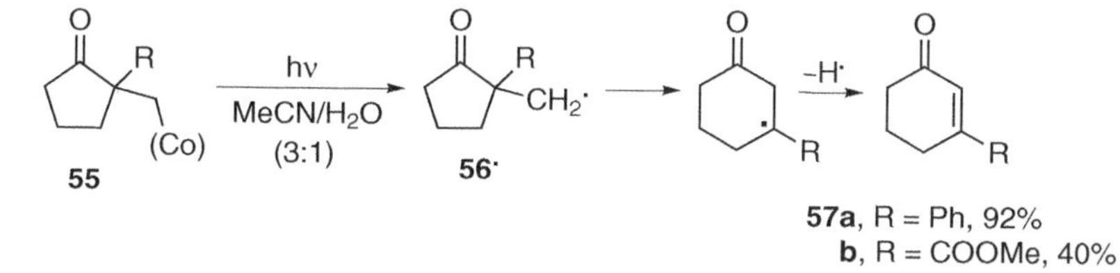

Scheme 2.20.

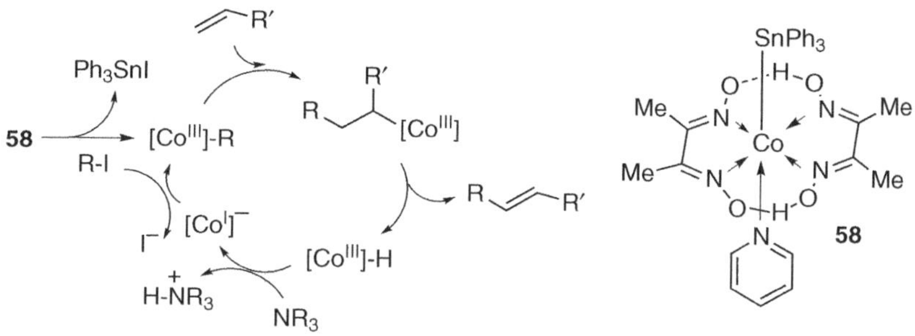

Scheme 2.21.

in the presence of a catalytic amount of **58** allowed for base-mediated alkyl Heck-type couplings as illustrated in Scheme 2.21.

A recent application of these alkylcobalt complexes is shown in Scheme 2.22, where the octahydroazulene moiety in the vasorelaxant daphmanidin E (**61**) was formed by a photoinduced Heck-type cyclization of iodide **59** in the presence of a catalytic amount of **58** and of a stoichiometric amount of the Hünig's base [78].

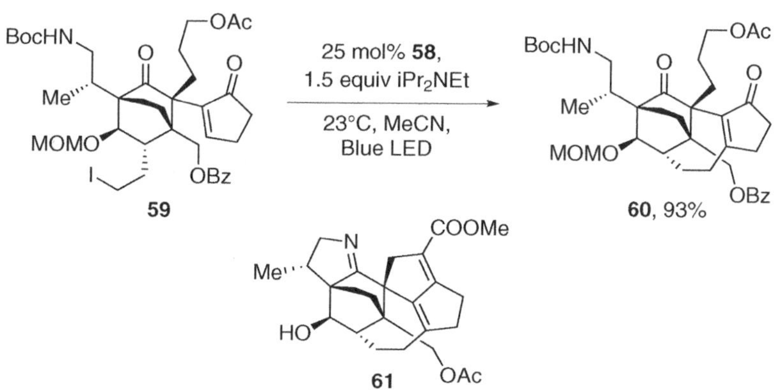

Scheme 2.22.

2.3 VINYL RADICALS

The photochemical generation of vinyl radicals is actually restricted to the photohomolysis of vinyl halides. An intrinsic drawback of this approach is that with these compounds the heterolytic cleavage of the C–X bond to yield vinyl cations competes. The ratio between radical and cationic pathways depends both on the substituents present on the double bond and on the solvent polarity [79, 80]. As a matter of fact, the clean generation of vinyl radical is uncommon, and only a couple of application in organic synthesis will be reported below.

The first example deals with the photochemical behaviour of β-iodoacrylamides. Thus, *N*-(*trans*-β-iodoacrylyl)-2,6-dimethylpiperidine (**62**) gave the open chain keto amide (**63**) in 77% yield when irradiated in MeOH [81]. The mechanism of the reaction is illustrated in Scheme 2.23. UV light absorption caused the homolysis of the vinyl–iodine bond and generated a reactive vinyl radical prone to hydrogen abstraction from the methine group adjacent to the nitrogen atom. An enamine was thus formed and the keto function was liberated upon hydrolysis (by adventitious water present in the solvent or during the purification step) [81].

In addition, vinyl radicals were exploited for the preparation of benzo[*b*]thiophenes (**65a–c**) by photolysis of β-(*o*-thiomethylphenyl)vinyl bromides (**64a–c**). The process relied upon the radical cyclization onto a sulfur atom; the target heterocycles were formed in a quantitative yield as determined by NMR analysis (Scheme 2.24) [82].

Phenyl iodostyrenes (**66a–c**) were suitable substrates for the synthesis of substituted phenanthrenes (**67a–c**). Here a C–C bond was formed by photolysis of (**66a–c**) followed by an intramolecular vinyl radical addition onto a phenyl ring (Scheme 2.25) [83]. The protocol was tolerant of other substituents present on the double bond (OAc, Cl, etc.).

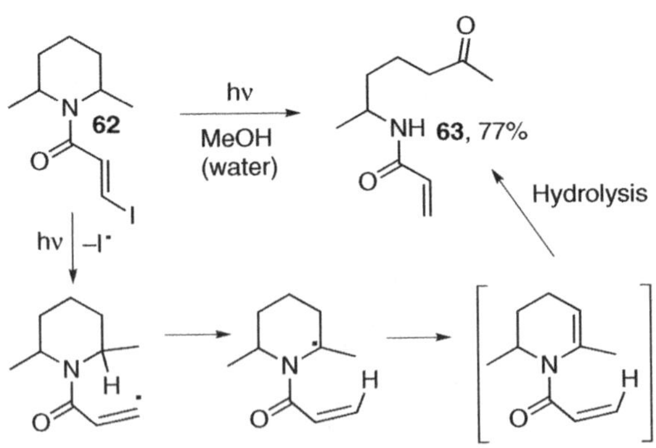

Scheme 2.23.

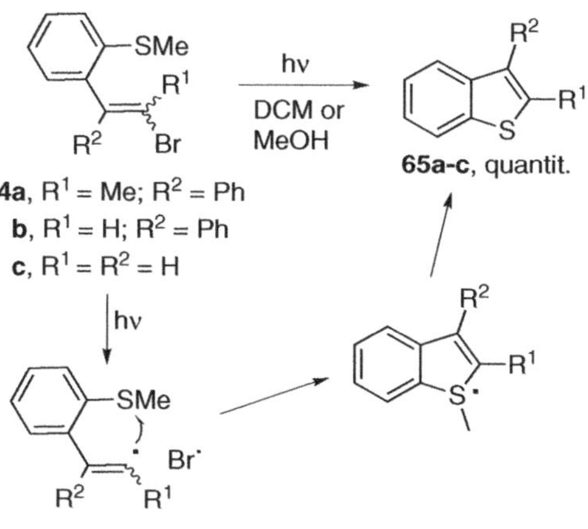

Scheme 2.24.

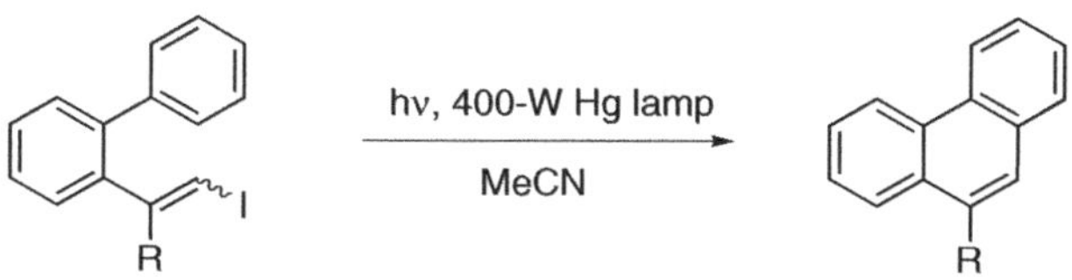

66a, R = H
b, R = OAc
c, R = Cl

67a, R = H, 95%
b, R = OAc, 80%
c, R = Cl, 64%

Scheme 2.25.

2.4 ARYL RADICALS

In analogy with vinyl radicals, aryl radicals can in principle be obtained by photohomolysis of Ar–X bonds (mainly an Ar–halogen bond) [84]. As a matter of fact, photodehalogenation is a rather ubiquitous process among aromatic derivatives [85]; but whether an aryl radical has been formed has neither been often established, nor necessarily an easy job to demonstrate. The mechanism may be different from straightforward homolysis (e.g., assistance by radical complexation was postulated in some cases), or a two-step mechanism may be involved—for example, monoelectronic reduction of the aromatic followed by halide anion loss from the radical anion (see Chapter 5). Moreover, the follow-up chemistry may involve competitive processes; in particular, reduction is generally efficient, thus limiting the potentiality of the more appealing arylation reaction [86].

About 50 years ago, one of the first studies based on the intermediacy of photogenerated aryl radicals in organic synthesis was the irradiation of aryl halides in aromatic solvents (Scheme 2.26a) [87–89]. A substituted biphenyl (or a mixture of biphenyls when R ≠ H) was isolated in variable yields. The reaction was effective

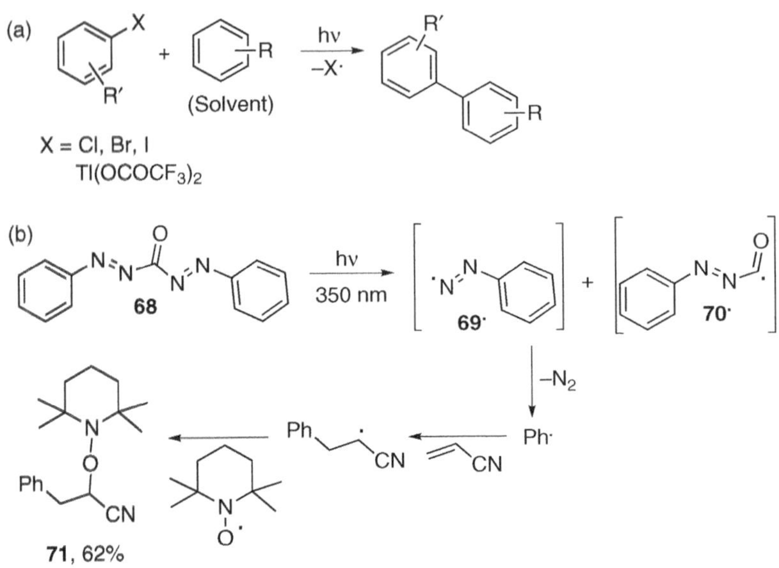

Scheme 2.26.

starting from aryl iodides [87], bromides [88], and chlorobenzene [89]. Arylthallium trifluoroacetates (e.g., $ArTl(OCOCF_3)_2$, Scheme 2.26a) have been shown to be a viable alternative for the photochemical intermolecular formation of aryl–carbon bonds. Irradiation in benzene caused photohomolysis of the Ar–Tl bond and allowed the preparation of the corresponding biphenyls in a good yield (>78%) [90, 91].

Moderate yields of biphenyls were likewise obtained when the aryl radical was generated in benzene by the photoinduced N–O bond cleavage in 1-aroyloxy-2(1*H*)-pyridones or 1-aroyloxy-3, 5-dinitro-2(1*H*)-pyridones. In analogy with Barton esters, the carboxylate radical formed splitted off carbon dioxide and formed an aryl radical [91].

Recently, the generation of phenyl radicals by cleavage of a Ar–N bond was demonstrated for the case of bisphenyl carbodiazone **68**. Thus, irradiation of **68** in dichloromethane generated a (phenyldiazenyl)carbonyl radical **70**˙ and a phenyldiazenyl radical (**69**˙, Scheme 2.26b). The latter intermediate liberated a phenyl radical upon nitrogen elimination, and subsequent trapping by an electron-poor olefin and by TEMPO led to the final product **71**, isolated in 62% yield [92]. The net process was the carbo-oxygenation of acrylonitrile.

Formation of an aryl–aryl bond is more versatile in the intramolecular version—as demonstrated, for example, by the photocyclization of 2-bromo-*N*-benzylpyridinium salts to give pyrido[2,1-*a*]isoindoles in 50–70% yields upon ultraviolet light photohomolysis of the Ar–Br bond [93]. This approach was found useful for the intramolecular construction of large rings (up to eight-membered rings), as in the preparation of both 6,7-dihydro-5*H*-dibenzo[*c*,*e*]azepine and 5, 6, 7, 8-tetrahydrodibenzo[*c*,*e*]azocine derivatives starting from substituted 2-iododibenzylamines and *N*-(2-halogenobenzyl)-β-phenethylamines, respectively [94].

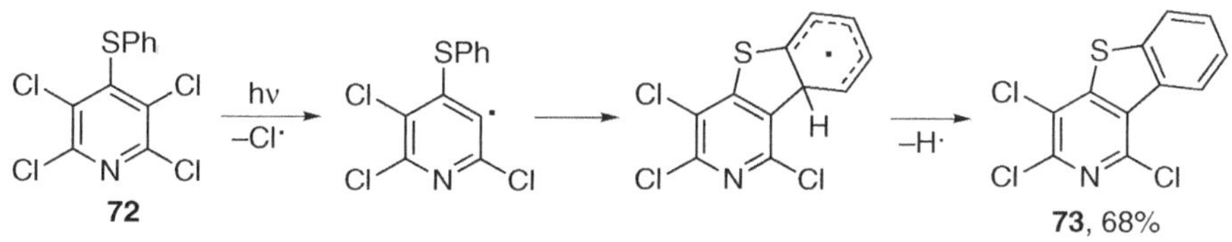

Scheme 2.27.

2-Halostilbenes have been largely used in photocyclization reactions for the formation of the phenanthrene ring (the photo-Pschorr reaction). The mechanism of the reaction is not always clear since an aryl radical formation followed by intramolecular attack may compete with the "usual" electrocyclic reaction followed by HX loss [85]. However, in some cases the role of an aryl radical has been clearly indicated, in particular in the photocyclization of 2-iodostilbenes [95].

The photochemical method allowed for the chemoselective activation of an Ar–Cl bond in polyhalogenoheteroaromatic compounds as shown in Scheme 2.27. Thus, the chemoselective cleavage of the 3-aryl-chlorine bond in phenylthiotetra-chloropyridine (**72**) has been reported. The ensuing intramolecular cyclization allowed for the synthesis of the pyridobenzothiophene **73** in 68% yield [96]. It is noteworthy that no further dechlorination of the primary product **73** took place.

Recently, a cyclization reaction via hydrogen halide elimination was applied to the synthesis of fluorenone **75** starting from 2-halobenzophenone **74**. Yields were excellent, and substituted 2-aroylaryl radicals were proposed as the intermediates (Scheme 2.28) [97].

An aryl radical was likewise generated starting from a 2-iodo-*N*-(2-arylethyl) imidazole (e.g., **76**) either photochemically or via a tin hydride-mediated reaction, with higher yields of cyclized heterocycle **77** in the former case. In fact, apart from avoiding the use of toxic and hazardous reagents, photochemistry prevented imid-azole reductive dehalogenation, in many cases the main (or the exclusive) thermal reaction (Scheme 2.29) [98].

Aryl iodides were likewise employed in the preparation of potentially bioactive phenanthridine derivatives (Scheme 2.30) [99]. Thus, irradiation of *N*-(2′-iodoben-zyl)anilines (**78a–c**) in acetonitrile at 254 nm smoothly led to dihydrophenanthri-dines that upon oxidation gave the end heteroaromatic derivatives (**79a–c**) in a very

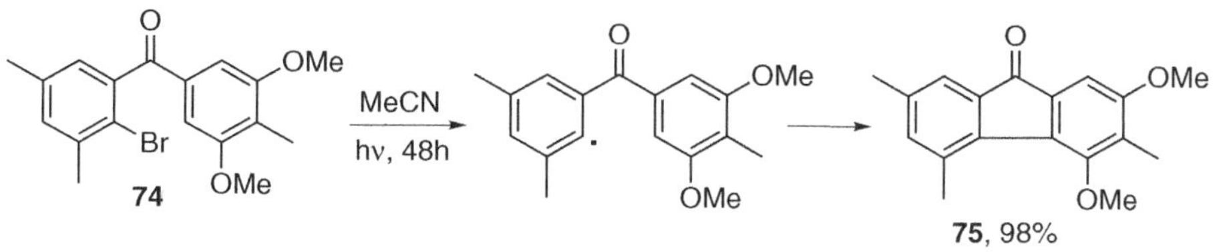

Scheme 2.28.

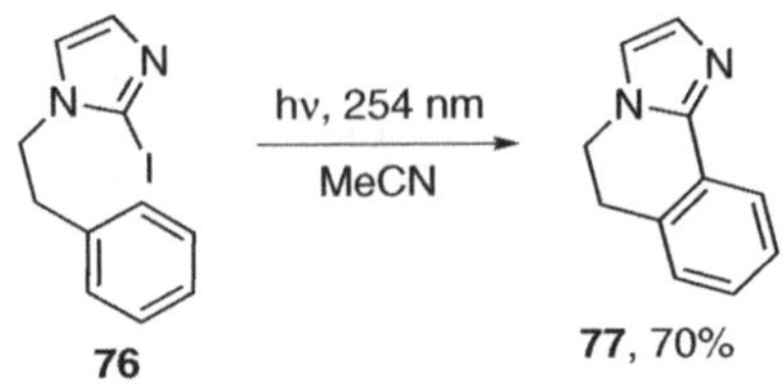

Scheme 2.29.

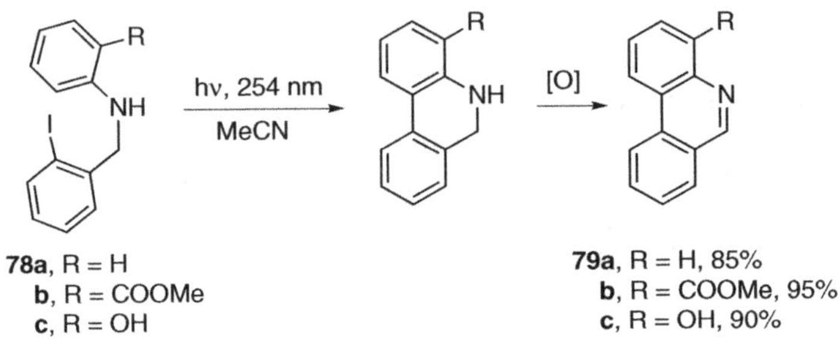

Scheme 2.30.

good yield. It was observed that when R was an electron-withdrawing group the yields were generally higher than with electron-donating groups [99].

A further example is shown in Scheme 2.31, where photohomolysis of hexamethyldistannane initiated a multistep reaction that was a part of the synthetic plan yielding racemic camptothecin [100]. Thus, the trimethylstannyl radical abstracted a iodine atom from *N*-propargyl-6-iodo-2-pyridone (**80**). The carbon radical thus formed readily added to phenyl isonitrile (**81**), and a radical cascade sequence followed to finally give the tetracyclic derivative **82** in almost quantitative yield [100].

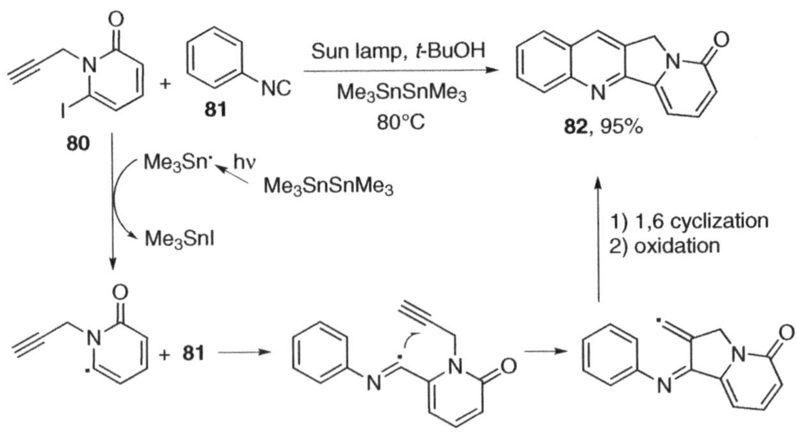

Scheme 2.31.

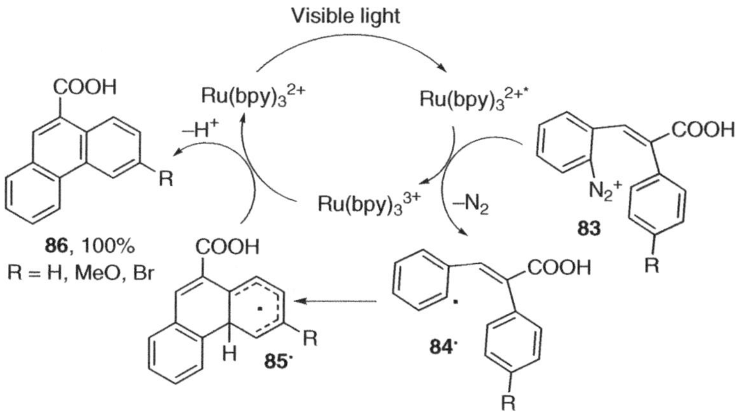

Scheme 2.32.

The cyclization of aryl radicals onto an alkene moiety could be coupled with reduction. Thus, in the photolysis of *ortho*-allyloxy and *ortho*-but-3-enyloxy-iodo- and bromo-benzenes, the 5-endo cyclization was followed by reduction of the radical adduct by $NaBH_4$ or $NaBH_3CN$ to give the corresponding methyldihydrobenzofurans. The reaction was rather efficient and was proposed to involve a photoinitiated chain reaction rather than a stoichiometric process after the initial Ar–X cleavage [101].

Arenediazonium salts have been used to some extent for the photoinduced generation of aryl radicals [102, 103]. The direct irradiation of these salts yielded either an aryl cation or an aryl radical, depending on conditions [104]. The monoelectronic reduction of these cations by a photogenerated reducing agent, however, is a clear-cut way to avoid the undesired formation of byproducts from the aryl cation path. Aryl radicals are formed by nitrogen loss and are efficiently trapped at least in the case of a phenyl ring in the intramolecular fashion. Such a photo Pschorr reaction is exemplified by the conversion of arene diazonium salts (**83**) into phenanthrene derivatives (**86**). This occurred in an almost quantitative yield upon irradiation in the presence of a visible light-absorbing photocatalyst such as $Ru(bpy)_3{}^{2+}$ (Scheme 2.32) [105]. Thus, the excited state of this ruthenium complex was able to reduce the salt **83** to the corresponding aryl radical **84·** (via nitrogen loss). This in turn cyclized, affording a radical tricyclic intermediate **85·** that was oxidized to the end product **86** by $Ru(bpy)_3{}^{3+}$ formed in the initial ET step. In this way, the ruthenium complex was restored in the original oxidation state and acted as a photocatalyst [105].

The reduction of aryl diazonium salts to form aryl radicals can be likewise accomplished by using an inexpensive visibile light-absorbing dye (eosin Y) in place of Ru-based complexes [106, 107]. The aryl radicals formed under these conditions have been used for the arylation of furans, *N*-Boc protected pyrroles, thiophenes, and enones in very good yields [106, 107]. Moreover, the aryl radicals could be trapped by an alkyne, and addition of the vinyl radical onto a vicinal SMe group led to the synthesis of benzothiophenes (**89**, Scheme 2.33) [108].

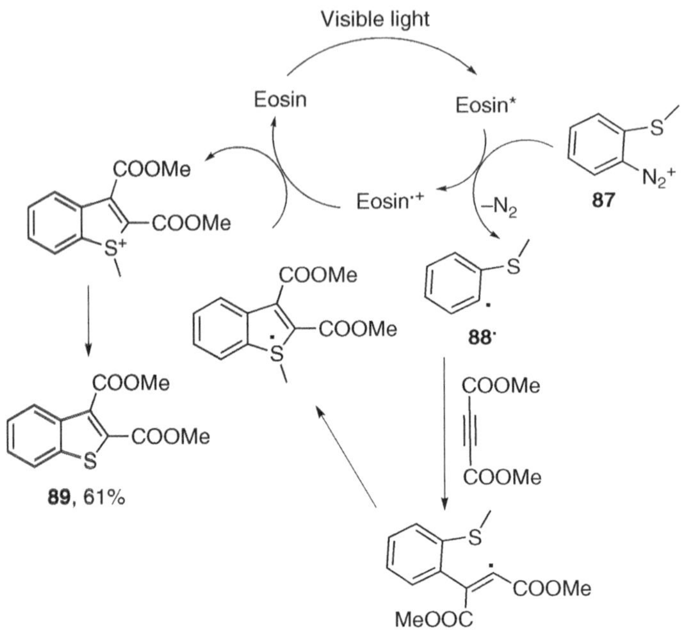

Scheme 2.33.

Photoreduction to the corresponding dehalogenated derivatives may be synthetically useful. This involves the reaction of the aryl radical with a suitable hydrogen donor (e.g., the solvent or better a thiol). As an example, some bromo-, iodoaceto-, and benzophenones were reduced by irradiation in cyclopentane to the corresponding dehalogenated phenyl ketones [109].

2.5 SUBSTITUTED CARBON-CENTERED RADICALS

2.5.1 α,α-Dioxy and α-Oxy Radicals

α-Oxy- and α,α-dioxyalkyl radicals can be thermally generated from ethers and, respectively, 1,3-dioxolanes by various methods [110–112]. As an example, such radicals were generated by treatment of ethers with pyrophoric Me_2Zn and $BF_3{\cdot}Et_2O$, with *N*-hydroxyphthalimide as the catalyst in the presence of benzoyl peroxide at 80°C or under initiation by the Et_3B/air system [110–112].

However, it will be apparent from the examples shown below that photocatalytic reactions are important alternatives for the production of radicals from unconventional donors by activation of C–H bonds [23–25, 53]. Classes of compounds that can be activated in this way are (substituted) dioxolanes (that form α,α-dioxyalkyl radicals), ethers, and alcohols (forming α-oxyalkyl radicals) [24]. The photocatalytic activation was conveniently carried out by using aromatic ketones for the hydrogen abstraction step, as has long been known [113, 114].

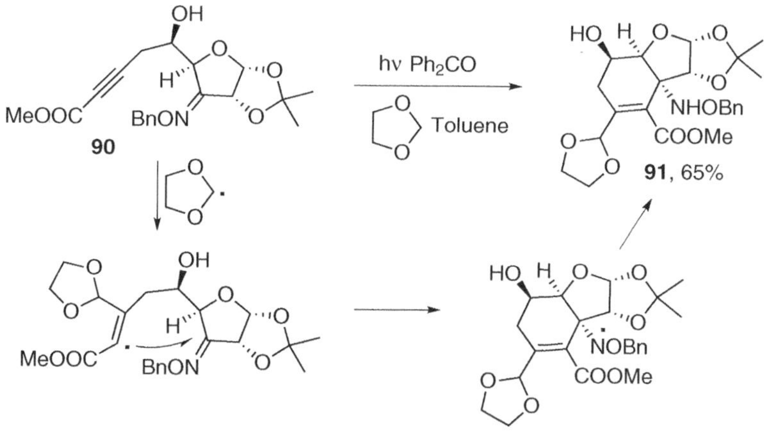

Scheme 2.34.

Thus, the methylene hydrogens in the 2-position of 1,3-dioxolane were activated by excited benzophenone (or its derivatives), and the resulting nucleophilic dioxolan-2-yl radical was employed in various addition reactions onto electron-poor olefins or alkynes [24, 115–117]. As a result, the introduction of a masked aldehyde function took place.

In a more elaborated example, a photogenerated 1,3-dioxolan-2-yl radical was used to initiate a tandem radical addition–carbocyclization sequence upon addition onto an electron-poor triple bond in ketoxime ethers bearing terminal δ-yne or internal γ-yne functions (**90**), as depicted in Scheme 2.34 [118]. The reaction required the use of equimolecular benzophenone and was completed in about 2 h.

In a further example, the 1,3-dioxolanyl group introduced was further elaborated for the synthesis of a *bis*-tetrahydrofuranyl ligand for the synthesis of HIV protease inhibitor UIC-94017. Thus, the benzyl protected 5(*S*)-hydroxymethyl-2(*5H*)-furanone (**92**) gave dioxolanyl derivative **93** in 82% yield with a high degree of stereoselectivity (Scheme 2.35a). Although the reaction was carried out in neat 1,3-dioxolane, benzophenone was used only in a catalytic amount (10% mol) [119]. Compound UIC-94017 was then prepared in a few steps.

Similarly to the nonsubstituted derivatives, 2-alkyl-1,3-dioxolanes have been used to introduce a protected carbonyl group by photochemical reaction. As an example, 2-ethyl-1,3-dioxolane was photoadded onto cyclopentenone to form the 1,4-monoprotected ketone **94** in 70% yield. (Scheme 2.35b) [120]. It is noteworthy that alkylation by alkyl dioxolanes of α,β-unsaturated aldehydes was likewise successful, despite the presence of a potentially labile formyl hydrogen (see Section 2.5.6) [121].

The steric hindrance exerted by long alkyl chains in 2-substituted-1,3-dioxolanes did not hamper the hydrogen abstraction process as demonstrated by the preparation of (−)-tetrahydrolipstatin (a potent pancreatic lipase inhibitor), where 2-undecyl-1,3-dioxolane was photochemically added to diketene in the key step [122]. Alkynes

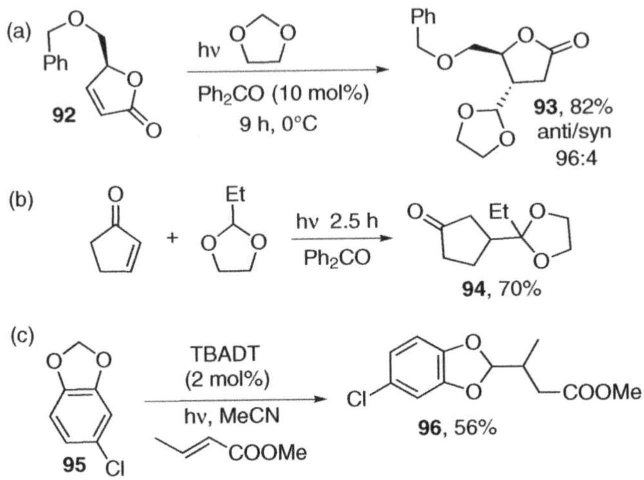

Scheme 2.35.

such as alkyl propiolates or dialkyl acetylenedicarboxylates were likewise useful as dioxolanyl radical traps [123].

The functionalization of benzodioxoles (e.g., **95**) is of great importance since the benzodioxole ring is present in many lead compounds having biological activity. The mild activation of the methylene hydrogens has recently been obtained by using TBADT (2 mol%) as the photocatalyst (Scheme 2.35c). The process allowed for the formation of potentially bioactive 2-substituted-1,3-benzodioxoles such as **96** in fair to good yields [124]. The success of the reaction was due to the selective absorption by TBADT of the radiation used (366 nm). Indeed, when this was shifted to 310 nm, competitive absorption by compound **95** caused its efficient photodechlorination [124].

In some instances, a trialkylorthoformate has also been used as radical precursor. The group $C(OAlk)_3$ introduced in the reaction was liable to alcoholysis to a COOR group upon purification on a silica gel column. This was reported to be the case in the functionalization of a hex-2-enopyranosid-4-ulose [125]. 1,2,4-Trioxane was likewise photoadded to diethyl maleate in 84% yield by using Ph_2CO as the photocatalyst [126].

The photochemical direct activation of C–H bonds in ethers has likewise been explored, and extension would be desirable. Cyclic ethers have been most often chosen and were used in a large excess, in some cases as the reaction medium [23–25]. 1,4-Dioxane is a good hydrogen donor as illustrated in Scheme 2.36a. Thus, photoactivation of a methylene hydrogen in this ether (used as the solvent) and addition of the resulting radical to 2-substituted-1,4-naphthoquinones (e.g., **97**) gave a stabilized semiquinone radical that upon hydrogen back-donation (from the ketyl radical) yielded hydroquinone **99**. The last compound was easily oxidized by oxygen to 2-anilino-3-(1,4-dioxan-1-yl)-1,4-naphthoquinone (**98**), thus restoring the quinone moiety. The net overall reaction thus amounted to the substitution of a vinylic hydrogen by an ether carbon [127].

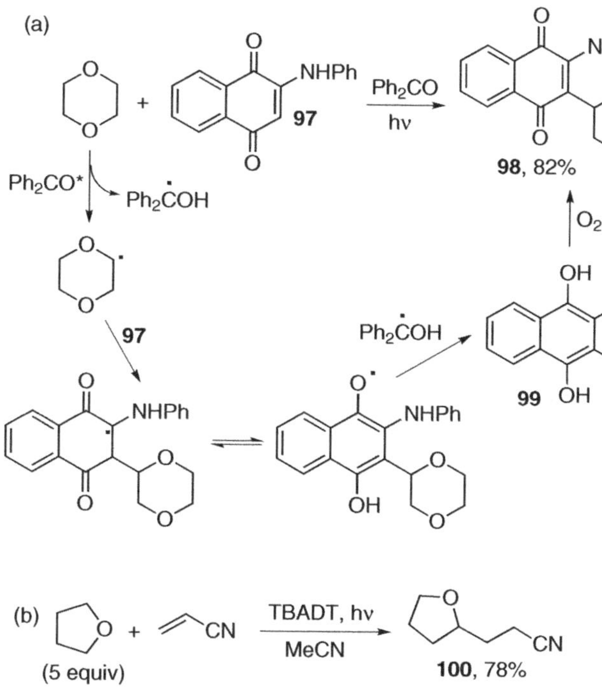

Scheme 2.36.

Recently, TBADT was successfully employed for the photoactivation of C–H bonds in ethers. Thus, when an acetonitrile solution of acrylonitrile and tetrahydrofuran (5 equiv) was irradiated in the presence of TBADT, the substituted nitrile **100** was synthesized in 78% yield (Scheme 2.36b) [63]. It is noteworthy that the catalyst was photoactivated also under solar light, and the reaction was accomplished in a few days. This reaction, exemplified by the addition of 1,4-dioxane onto dimethyl maleate, is a model green synthesis requiring no energy supply, an application of the "window ledge" chemistry [65].

Benzophenone was used as photomediator in the carbamoylation of ethers (e.g., **101**). Here the α-oxy radical was trapped by a phenyl isocyanate in good yield, provided that the aromatic bore electron-withdrawing groups, as it was the case for pentafluorophenyl isocyanate (Scheme 2.37) [128]. The carbamoylation was likewise effective on tetrahydrofuran and on larger ring homologues, such as tetrahydropyran and hexahydroxepane.

The benzophenone-photomediated generation of carbon-centered radicals from ethers was applied to the smooth synthesis of α-cyanoethers via a direct C–H cyanation process [129]. Some examples are reported in Scheme 2.38. Thus, the radical formed via hydrogen abstraction by Ph_2CO^* from R–H was trapped by tosyl cyanide to form the corresponding nitrile. Various cyclic ethers (e.g., crown ethers)

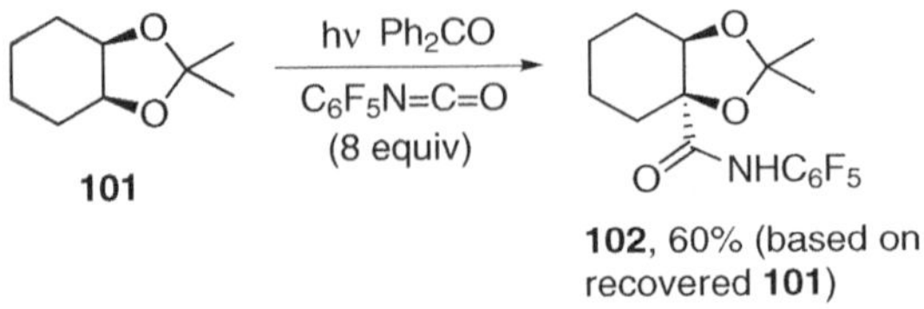

Scheme 2.37.

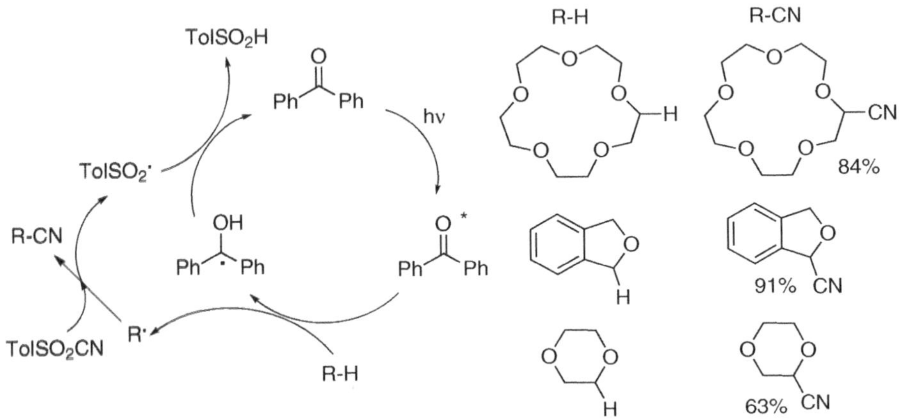

Scheme 2.38.

are particularly suitable for the reaction, although alcohols, substituted amides, and alkanes could be likewise functionalized (Scheme 2.38) [129].

Moreover, TBADT was used to functionalize [60]fullerene by activation of the methyl hydrogens in anisole [67] as well as of the methylene hydrogens in crown ethers to form [60]fullerene/crown ether conjugates [130]. The versatility of TBADT as photocatalyst has also been recently exploited for the straightforward PEGylation of SWCNTs by photomediated addition between various PEG derivatives (PEG200–600) and a nonionic surfactant onto the surface of carbon nanotubes, thus improving their dispersibility in water [131].

The generation of α-oxy radicals was likewise obtained via Barton esters. In this case, the radicals were mainly used for the preparation of derivatized nucleosides by forming C–C bonds [132–135]. The starting materials were uronic acids that were transformed into Barton esters and irradiated in the presence of an electron-poor olefin. A representative example is shown in Scheme 2.39.

Uronic acid (**103**) was converted into ester **104** by acylation with *N*-hydroxy-2-thiopyridone. The α-oxy radical generated upon photolysis readily added to vinyl phosphonate **105**, and the resulting adduct **106** was isolated in a high yield (70% yield starting from **103**) [132]. The radical addition reaction was highly stereoselective with retention of the configuration at the C-4′ (the position where the radical was generated), and this was ascribed to the bulk of the *t*-butyldiphenylsilyl

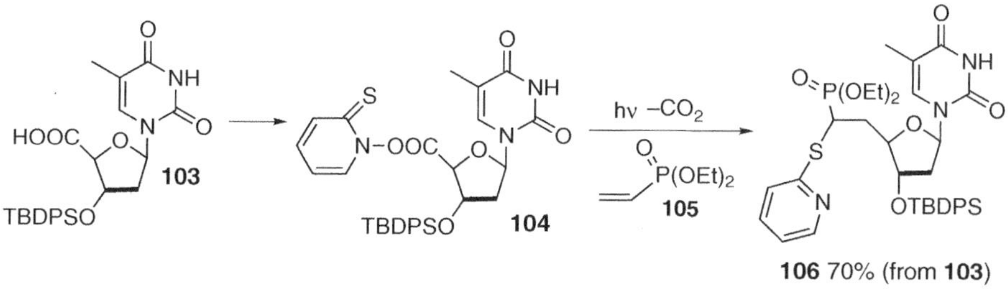

Scheme 2.39.

(TBDPS) protective group linked at the 3′ position. In such a way a phosphonate rather a phosphate group was attached, and further elaboration of compound **106** allowed for the preparation of a compound isosteric with 3′-azido-3′-deoxythymidine-5′-monophosphate that is promising for application in the treatment of AIDS [132]. Starting from the related deoxy sugar scaffold and applying the same protocol, various *C*-nucleosides have been prepared [135].

Barton esters were likewise applied for the generation of α-alkoxy bridgehead radicals as depicted in Scheme 2.40. Thus photolysis of the trioxadamantane deriving ester **108** in the presence of diphenyldiselenide gave *O,Se*-acetal **109** in 61% yield from the corresponding acid **107** [136]. Compound **109** was susceptible to further radical addition for the construction of highly substituted C–C bonds.

α-Oxy radicals could be indirectly generated by homolysis of a C–Te bond induced by photogenerated radicals (from a Barton ester). Thus, a hydroxyl group in a sugar was functionalized by transforming it into an anisyl sugar telluride. The last compound was irradiated in the presence of a Barton methyl ester that released a methyl radical upon photolysis. Addition of the methyl radical onto the telluride (induced by the strength of the Me–Te bond) and fragmentation gave a carbohydrate radical that then added onto a maleimide. The product obtained could be converted into the antibiotic Showdomycin in a few steps [137].

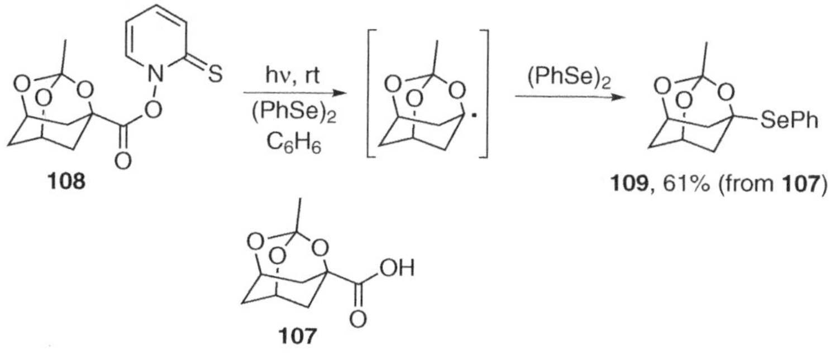

Scheme 2.40.

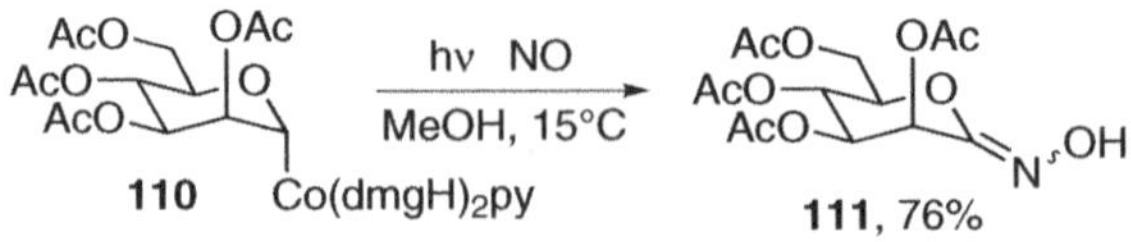

Scheme 2.41.

Carbohydrate cobaloxamines are alternative sources of α-oxyalkyl radicals. Thus, irradiation of a methanolic solution of mannosylcobaloxamine **110** in the presence of NO gave oxime **111** in a good yield (76%, Scheme 2.41) [138].

Similarly to the ethers case, C–H bonds adjacent to the OH group in alcohols are readily activated thermally. Among photochemical alternatives, the use of excited uranyl ions has been early reported. These species are smoothly reduced to $U_{aq}O_2^{+}$ ions with the concomitant formation of ketyl radicals by hydrogen atom abstraction from the alcohol (e.g., isopropanol, Eq. (2.1)). The heterogeneous version of the reaction was successfully tested by using mesoporous silica-supported uranyl [139]. The radicals could be further oxidized to the corresponding ketones whereas the $U_{aq}O_2^{+}$ ions were rapidly reoxidized by atmospheric oxygen to the starting uranyl ions.

$$^{*}U_{aq}O_2^{2+} + (CH_3)_2CHOH \rightarrow U_{aq}O_2^{+} + (CH_3)_2C{\cdot}OH \qquad (2.1)$$

A dedicated apparatus was devised for the mercury-photosensitized reactions of selected hydrogen donors (including alkanes). The substrate was placed in a quartz reactor with a drop of Hg and the mixture was heated to reflux. The upper vapor phase was irradiated at 254 nm in order to excite mercury atoms (Hg*) that abstracted hydrogen from the organic vapors. The carbon radical thus formed dimerized and the high-boiling products condensed so that they were protected from further photochemical conversions. This mercury-photosensitized dehydrodimerization in the vapor phase could have synthetic applications [140, 141] since various alcohols were converted into the corresponding glycols [141].

At any rate, the most widely employed method for the activation of C–H bonds in alcohols is such a time-honored reaction as that with an aromatic ketone. Selected examples are presented below [23–25]. Thus, with the aim of synthesizing building blocks useful for the preparation of analogues of glutamate and kainoids (e.g., kainic acid), methanol has been photoadded to chiral 5(*S*)-5-triisopropyisiloxymethyl-*N*-boc-dihydropyrrole-2(5*H*)-one upon benzophenone photocatalysis. The adduct was stereochemically pure, although it was formed in a modest yield (51%) [142].

Recently, benzophenone was employed for the photochemical generation of α-hydroxyalkyl radicals (from alcohols or diols) that were trapped by electron-poor alkynes forming hydroxy-substituted α,β-unsaturated esters (**112**), useful starting materials for the preparation of γ-butenolides (Scheme 2.42a) [143].

Changing the structure of the photocatalyst allows to carry out the reaction in a partial aqueous medium. Accordingly, α-oxy-substituted radicals could be produced from alcohols by means of benzophenone disodium disulfonate (**113**)-photosensitized hydrogen transfer (Scheme 2.42b). This salt was obtained (as a mixture of

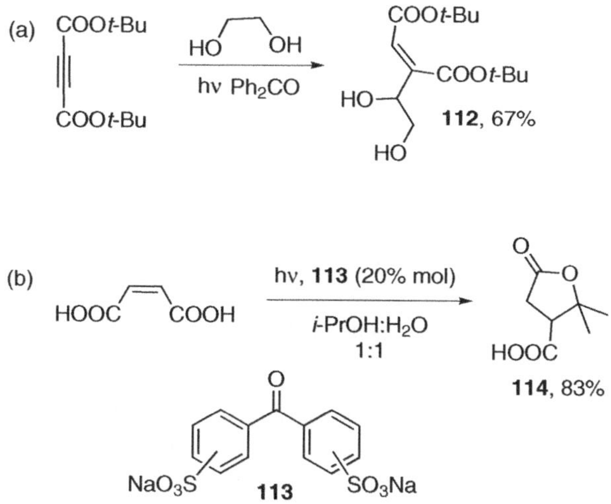

Scheme 2.42.

isomers) by sulfonation of benzophenone and showed a good solubility in water [121]. Applications of **113** as photocatalyst include the addition of alcohols onto unsaturated aldehydes, where the alkylated products cyclized to lactols (40–60% yield) [121]. The absorption of this compound in the UV-A is sufficient to make solar irradiation suitable for carrying out photoreactions; indeed the scaling up of the process was investigated at the Plataforma solar de Almeria (Spain) in a SOLFIN reactor purposely devised for green photochemical syntheses [144]. As an example, by using **113** as the catalyst, maleic acid was alkylated in a isopropanol–water 1:1 mixture as the solvent to give a 14-g amount of terebic acid (**114**, 83% yield) after 14 h (during three days) irradiation (Scheme 2.42b).

Recently, the dimethoxybenzophenone (**116**)-photocatalyzed isopropanol addition to furanone (**115**) in an LED-driven microchip reactor has been reported [145]. The reaction reached completion in only 5 min, a much better performance in comparison to the usual batch reaction. The improvement was ascribed both to the larger surface-to-volume ratio of this system and to the more uniform absorption of light within the reactor microchannels (Scheme 2.43). The same reaction was successfully carried out in a continuous-flow dual capillary microreactor by using a single 350-nm lamp as the light source (Scheme 2.43) [146] or in an LED-driven microchip apparatus [147]. Again, better results were obtained in comparison with a conventional batch reactor. Very recently, a multimicrocapillary flow reactor was applied for the same reaction with a lower energy consumption (~30%) than the batch reactor and avoiding the requirement of cooling water [148].

The synthetic significance of the photoaddition of alcohols onto an unsaturated compound via a α-alkoxyalkyl radical was demonstrated in the preparation of the immunosuppressant (+)-myriocin **120** (Scheme 2.44). In this case the C=N bond of ketoxime ethers (e.g., **118**) underwent radical addition giving access to a

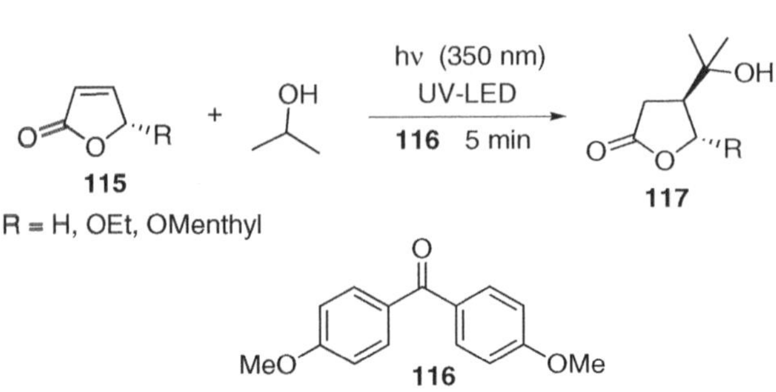

Scheme 2.43.

β-oxygenated quaternary α-amino acid derivative (**119a, b**) [149]. Further elaboration of lactone **119a** allowed the synthesis of the target natural compound.

α-Hydroxyalkyl radicals have been likewise used for the functionalization of carbon-made compounds, namely carbon nanotubes and fullerenes. In the first case, SWCNTs were suspended in a benzene solution containing benzophenone and benzhydrol (PhCH(OH)Ph) and irradiated by a 200-W mercury lamp for 4 h. That the alcohol had added onto the CNTs surface was confirmed by spectroscopic analysis, and the degree of functionalization was found to involve about one benzhydrol moiety for 52 sidewall carbons [150]. In the second case, isomerically pure hydroxyalkylated C_{60} monoadducts were formed by TBADT-photocatalyzed addition of various alcohols including methanol, propargyl, and benzyl alcohols onto C_{60} [151].

Rather persistent benzhydryl radicals dimerize, as reported for the photolysis of 4-oxo-4-phenylbutanamides **121** in diethylether (Scheme 2.45) [152]. Hydrogen abstraction from the solvent by the excited butanamides resulted in rather persistent benzhydryl radicals, and 4,5-dihydroxy-4,5-diphenyloctanediamides were formed by dimerization (**122a–c**, **123a–c**). The photopinacolization occurred with marked diastereoselectivity, with preference for racemic pinacols (**123a–c**) over the meso derivatives (**122a–c**) (Scheme 2.45). When amides **121** were photolyzed in *tert*-butyl alcohol and dichloromethane, however, intramolecular hydrogen abstraction took place (see also Chapter 4) and, as a result, photocyclisation to δ-lactams occurred [152].

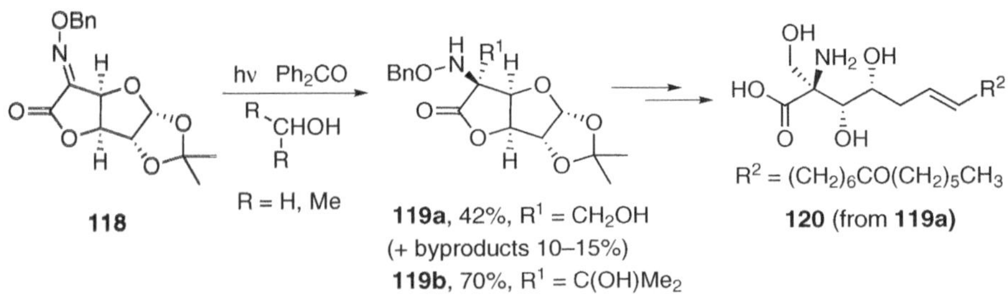

Scheme 2.44.

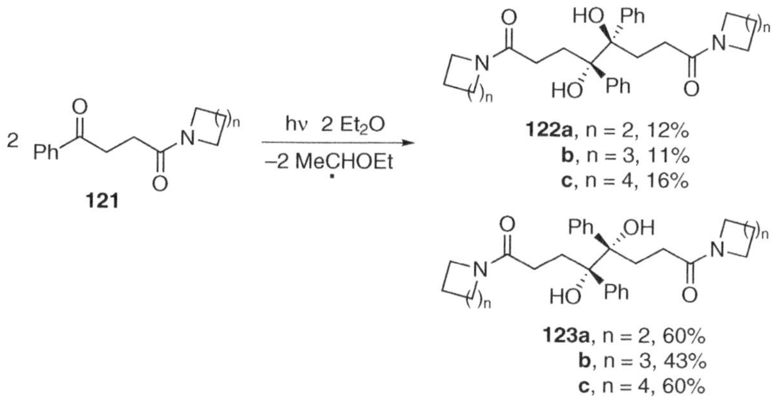

Scheme 2.45.

In rare instances, the radical adduct formed upon addition onto an olefin was able to abstract hydrogen from alcohols, as reported for the addition of secondary alcohols to maleimides to give 2-(substituted)succinimides under UV irradiation in 25–70% yield [153].

2.5.2 α-Thioalkyl Radicals

Another way to introduce a masked carbonyl functionality is the addition of a 1,3-dithian-2-yl radical onto an electron-poor olefin. Thus, irradiation of 2-phenylseleno-1,3-dithiane (**124**) in the presence of butyl acrylate caused cleavage of the C–Se bond, and radical addition led to compound **125** in 75% yield (Scheme 2.46) [154]. When the reaction was applied to highly substituted olefins, however, poor results were obtained.

2.5.3 α-Amino- and α-Amidoalkyl Radicals

α-Aminoalkyl radicals are most often generated by deprotonation of the corresponding radical cations obtained via a photoinduced electron transfer process (see Chapter 5). However, α-aminoalkyl radicals were likewise formed by irradiation of amino acids where the carboxylic group was protected as thiopyridone ester and were trapped by electron-poor olefins. Indeed, the side-chain decarboxylation of

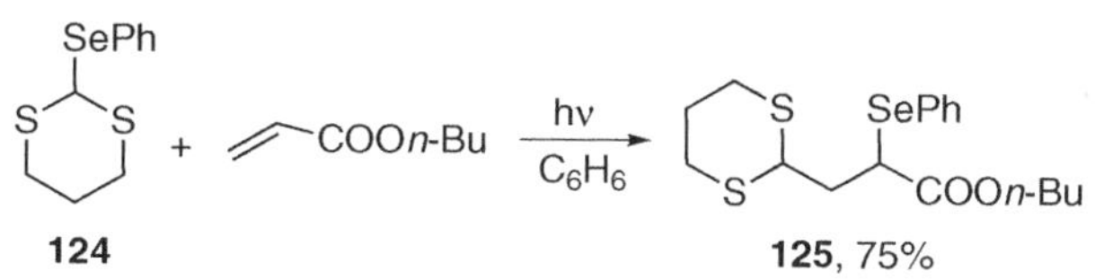

Scheme 2.46.

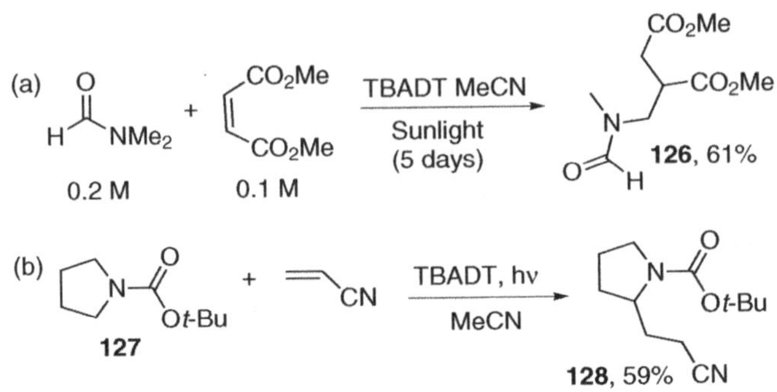

Scheme 2.47.

protected glutamic and aspartic acids allowed the preparation of α-aminoadipic and α-aminopimelic acid derivatives [155].

Alternatively, α-amidoalkyl radicals can be directly generated by TBADT photocatalyzed hydrogen abstraction from unusual H donors such as amides [65, 156]. As previously stated, TBADT can be excited under solar light irradiation, and indeed under such conditions *N,N*-dimethylformamide was photoadded onto dimethyl maleate to give the succinate derivative **126** in 61% yield. The reaction required 5 days' irradiation. It is noteworthy that no concurrent photoactivation of the formyl hydrogen was observed (Scheme 2.47a). The use of carbamates (actually protected amines) offered an interesting approach for functionalizing amines while avoiding the problems associated with the electron-donor properties of such functionality, which are incompatible with TBADT-photocatalyzed reactions. The hydrogens adjacent to the nitrogen atom in carbamates (e.g., **127**) were likewise cleaved under the same conditions, and the resulting radicals added to electron-poor olefins (e.g., acrylonitrile, Scheme 2.47b) [156].

2.5.4 Halomethyl Radicals

Halomethyl radicals are electrophilic radicals that can be exploited for the preparation of organochlorine or organofluorine derivatives. Compounds containing the difluoromethylene moiety have attracted much attention in pharmaceutical and agrochemical industry due to the peculiar properties of the fluorine atom (high electronegativity, small van der Waals radius, etc.) that affected favourably the biochemical profile. Photochemistry has contributed to the synthesis of such compounds. For example, difluoromethyl-substituted radicals were generated by photolysis of the PhS–C bond in compound **129** and added to a vinyl ether to form tetrahydrofuran **130** in 61% yield (Scheme 2.48) [157].

Perfluorinated alkyl radicals have found application for the sidewall functionalization of carbon nanotubes. In this case the weak C–I bond of heptadecafluorooctyl iodide (present in a 200-fold excess) was cleaved upon photolysis of a suspension of SWCNTs in 1,1,2,2-tetrachloroethane [158].

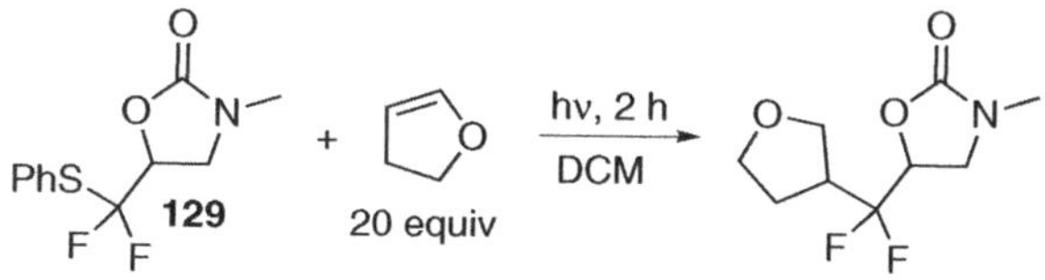

Scheme 2.48.

A trifluoromethyl radical was likewise obtained by the Ir^{III}-photocatalyzed monoelectronic reduction of Umemoto's reagent (**131**) and trapped by styrene derivatives. The resulting benzyl radical was reoxidized (thus regenerating the photocatalyst) and the cation was trapped by the alcoholic solvent, resulting in a three-component oxytrifluoromethylation of alkenes (compound **132**, Scheme 2.49) [159].

In rare instances, photogenerated dihalomethyl radicals have been used in addition reactions onto aromatics. Thus, substituted benzenes have been obtained via addition of difluoromethyl radicals bearing electron-withdrawing groups, which were in turn photogenerated from an alkyl aryl selenide as illustrated in Scheme 2.50. Dichloromethane resulted the most convenient solvent for the reaction that made use of a fivefold excess of benzene [160].

The lability of the C–Br bond in Cl_3C–Br was exploited in sparse instances for the photoaddition of this tetrahalide across the double bond of substituted olefins, resulting in the final formation of both C–C and C–Br bond after the initial addition of the Cl_3C radical. In some cases, the process has been photocatalyzed by $Mn_2(CO)_{10}$ [161]. Phenyl trichloromethyl selenide ($PhSeCCl_3$) is another suitable source of the Cl_3C radical. Irradiation of a benzene solution of this compound in the

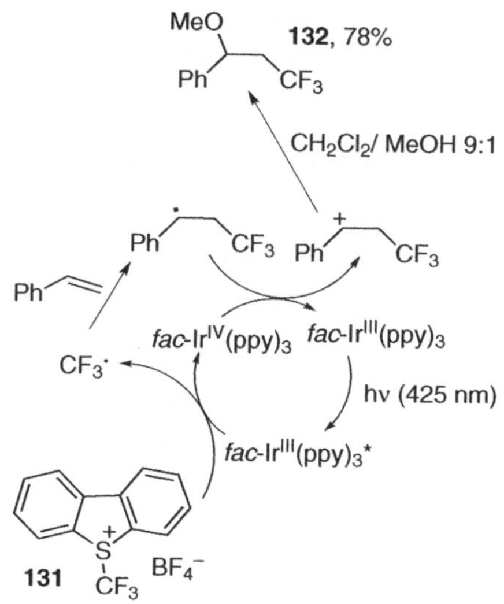

Scheme 2.49.

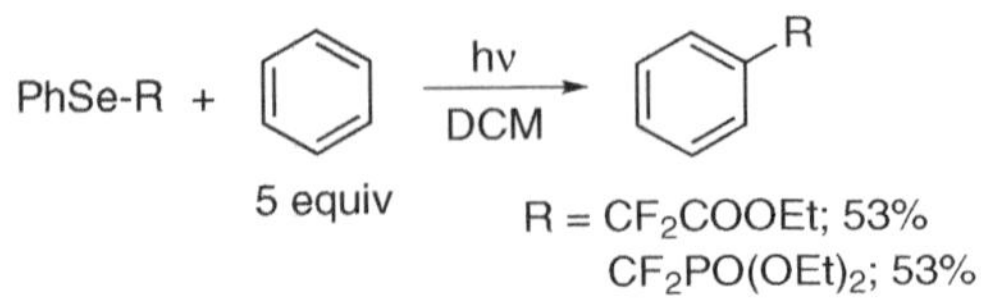

Scheme 2.50.

presence of a fivefold excess of an alkene allowed for the addition of the CCl_3 and PhSe groups across the double bond. It is noteworthy that the resulting 2-phenylseleno-1-trichloromethylalkanes can be transformed in two steps into α,β-unsaturated carboxylic acids or amides restoring the double C–C bond in the same position as in the starting alkene [162].

2.5.5 α-Keto, α-Carboxyl, and α-Carbamoylalkyl Radical

Another class of useful electrophilic carbon radicals is that of α-ketoalkyl radicals. Recently, a water-promoted metal-free protocol has been developed for the acetonyl radical addition to olefins as depicted in Scheme 2.51 [163]. The authors postulated that acetone (**133**, used as cosolvent) was photochemically excited and was able to abstract a hydrogen atom from ground state acetone to form a radical pair. The water present in solution (up to 40% v/v) then hydrated the acetone-derived radicals, thus driving the reaction toward the formation of a methyl ketone derivative (e.g., **135**) while suppressing any possible formation of acetone dimers or olefin-derived byproducts (Scheme 2.51). This green methyl ketone synthesis was extended by using various olefins as radical acceptors, and the end compounds were isolated in 71–96% yield [163]. The presence of Mg(II)-exchanged zeolite Y improved the efficiency of the hydrogen abstraction step by increasing the charge polarization of the carbonyl group in excited-state acetone [164].

Electrophilic radicals were conveniently photogenerated starting from 2-phenylselenomalonate (**137**) and trapped by a nucleophilic olefin (e.g., octene) to form diester **138** in 95% yield, with incorporation of the PhSe group (Scheme 2.52a) [165]. The

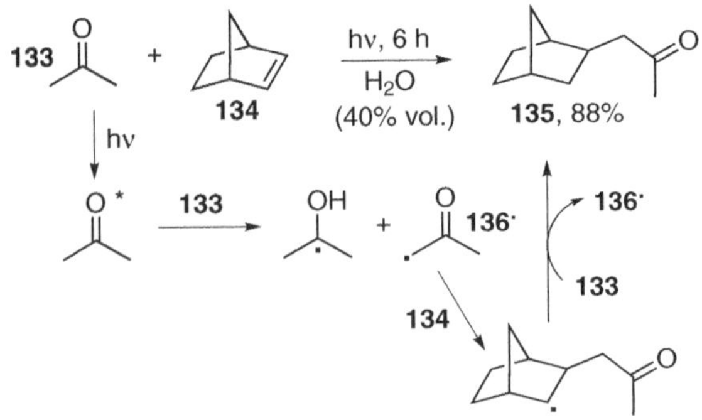

Scheme 2.51.

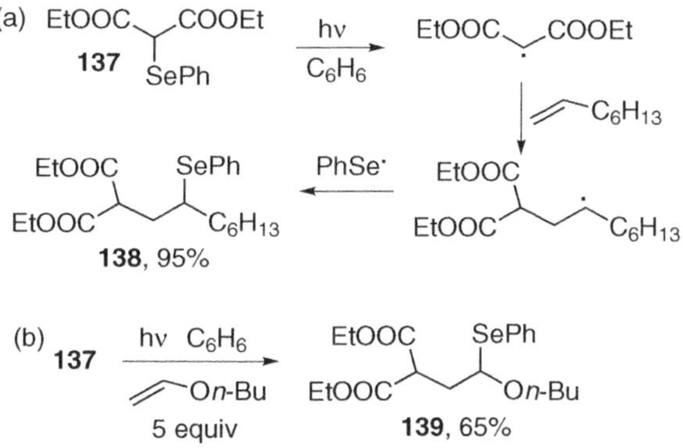

Scheme 2.52.

reaction, however, required an excess either of the starting diester or of the olefin. Accordingly, the functionalization of butyl vinyl ether was achieved in a satisfactory yield only when using a fivefold excess of the ether (Scheme 2.52b) [166].

Various compounds containing the PhSe group were employed as carbon radical precursors. Thus, various phenylselenyl ketones, esters, nitriles, and sulfones were used in the photochemical addition onto octene in moderate to good yields [167]. Moreover, both ester-substituted *O,Se*-acetals [168] and tetraalkyl phenylselenomethylenediphosphonate [169] were found to be convenient radical precursors in related reactions. The malonyl radical (obtained from the corresponding 2-phenylselenomalonate) was used in more elaborate strategies. As an example, the regioselective addition of this radical onto 7-oxabicyclo[2.2.1] hept-5-en-2-one followed by a 1,2-acyl shift was exploited for the synthesis of all-*cis*-Corey lactone [170]. Recently, the malonyl radical was likewise added to alkyl pent-4-enyl selenides in a tandem homolytic addition/substitution reaction to yield tetrahydroselenophenes along with the corresponding phenylselenides [171].

As for α-carbamoylalkyl radicals, photoinitiation by using dimanganese decacarbonyl has sparsely been applied for their generation. Thus, the C–I bond in iodoethanamide could be cleaved and the resulting carbamoylmethyl radical cyclized, forming a five-membered ring in pyrrolidinone derivatives [172].

2.5.6 Acyl Radicals

Acyl radicals are useful intermediates for the synthesis of unsymmetric ketones via formation of an acyl–C bond [173]. The thermal generation of such radicals is only possible by using precursors containing labile acyl–X bonds. Acyl selenides are commonly recognized as the elective substrates for this process under the usual AIBN/Bu_3SnH conditions [173]. The photogeneration of these radicals, however, has a larger scope and has been gaining an increasing role in organic synthesis [23–25]. As an example, extension to sulfonated derivatives is possible. Xanthate derivatives (e.g., *S*-acyl xanthates) were particularly suitable as starting materials due to the lability of the acyl-S

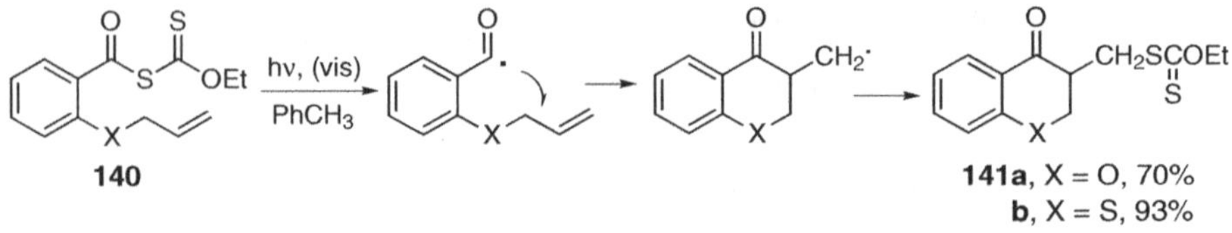

Scheme 2.53.

bond and to the absorption by these compounds in the visible region of the spectrum [174]. Thus, xanthate **140** generated a benzoyl radical upon visible light irradiation and this readily cyclized to give chromanone (**141a**) or thiochromanone (**141b**) as the end products in more than 70% yield (Scheme 2.53) [174]. Moreover, the xanthate functionality was easily removed to give the corresponding α,β-unsaturated ketones.

An analogous cyclization reaction mediated by acyl radicals has been reported starting from acyl tellurides [175].

An elegant way to prepare acyl radicals is via carbonylation of alkyl radicals by reaction with carbon monoxide [173, 176, 177]. In this case, alkyl radicals were generated by photolysis of a suitable alkyl iodide as illustrated in Scheme 2.54. The reaction of the acyl radical with the starting alkyl iodide (R-I, **142**) gave an acyl iodide that in turn reacted under basic conditions with an alcohol (e.g., benzyl alcohol) to give ester **143** as the end product in 87% yield [178].

A breakthrough in the acyl radical generation is starting directly from aldehydes via photoactivation of the C(=O)–H bond. This target has been indeed reached by having recourse to photocatalysis [23–25]. Aromatic ketones were first used in the role of photocatalysts [179], and the acyl radicals formed were exploited in the photoacylation of quinones to form the corresponding acylated hydroquinones in good yields [127,180–182].

Recently, TBADT has been applied as photocatalyst for the photogeneration of acyl radicals from both aliphatic [183] and aromatic [184] aldehydes and shown to act efficiently. Trapping of the thus formed radicals with electron-poor olefins opened the way to an alternative preparation of unsymmetrical ketones by using artificial light, but also sunlight (Scheme 2.55, path *a*) [65]. One of the drawbacks of the reaction is the competitive decarbonylation of acyl radicals deriving from

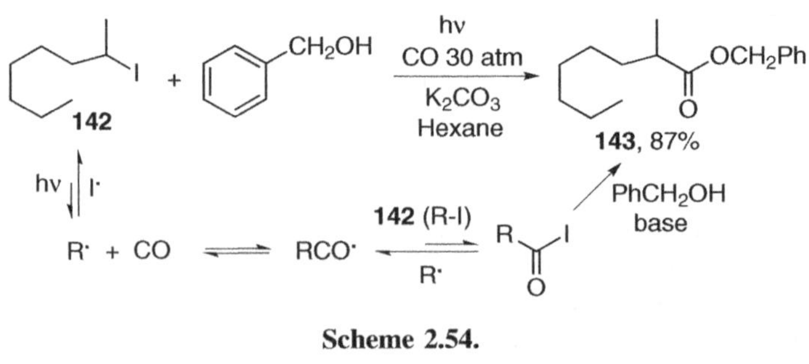

Scheme 2.54.

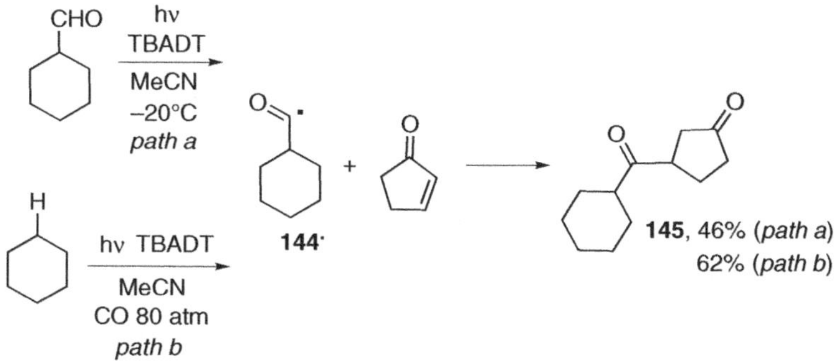

Scheme 2.55.

secondary or tertiary aldehydes. In order to avoid such a limitation, the reaction was carried out at a low temperature (−20°C), so that addition of the **144·** radical onto cyclopentenone gave diketone **145** in 46% yield [183]. Worth noting, when **144·** was obtained by the reverse reaction, namely the carbonylation of the cyclohexyl radical (80 atm CO required), the same diketone was isolated in a higher yield (62%, Scheme 2.55 path *b*) [185]. The TBADT-photocatalyzed acyl radical addition onto [60]fullerene has likewise been reported [186].

2.5.7 Carbamoyl Radicals

The photogeneration of carbamoyl radicals from amides was reported in the 1960s by Elad and co-workers [187]. Formamide (**147**) was the most investigated radical precursor and was often used both as the reagent and as the reaction solvent (or cosolvent). Heterocyclic bases were functionalized in a chemoselective fashion when a TiO_2 suspension of the base (H_2O_2 and H_2SO_4 in neat amide) was exposed to sunlight irradiation. Both the acid and the peroxide assisted the process, and under such conditions a carbamoyl radical was photocatalytically formed by reaction of an OH radical with the amide. Radical addition onto the heteroaromatic base (quinoline, **146**) to give **148** occurred in a high yield (Scheme 2.56) [188, 189].

The formyl hydrogen in **147** was likewise photochemically cleaved by using benzophenone as the catalyst as depicted in Scheme 2.57a. The photoamidation of dimethyl maleate took place in an excellent yield and was easily carried out under sunlight irradiation, although a large excess of both the amide (actually the reaction medium was **147**: *t*-BuOH 6:1) and the aromatic ketone (18 equiv) were used.

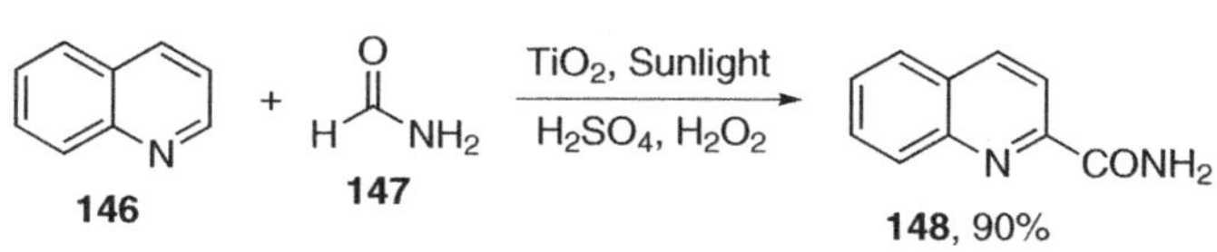

Scheme 2.56.

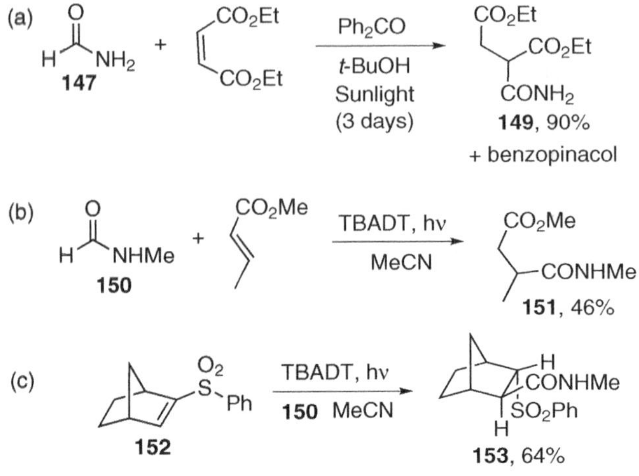

Scheme 2.57.

Moreover, benzopinacol was formed as a byproduct [187]. A greener process that allowed to use a lesser amount of both the amide (a fourfold excess) and the photocatalyst (TBADT, 2% equiv) has been recently developed. In this case, photoamidation of methyl crotonate by using *N*-methylformamide (**150**) was achieved (Scheme 2.57b). Although amide **151** was isolated in a modest yield, the process was completely regioselective since no competitive activation of the methyl hydrogen occurred in *N*-methylformamide (**150**) [156]. The same amide (**150**) was photoadded under similar conditions onto 2-(phenylsulfonyl)bicyclo [2.2.1]hept-2-ene, (**152**) and afforded the norbornane derivative **153** in 64% yield in a chemo- and diastereoselective way (Scheme 2.57c) [190].

An alternative route to carbamoyl radicals involved the photosensitized decomposition of oxime oxalate amides (e.g., **154**). The use of 4-methoxyacetophenone (**155**) as a photosensitizer was found to improve the efficiency of the reaction. The photohomolysis of **154** gave an iminyl radical as a byproduct and, after carbon dioxide loss, the carbamoyl radical **156˙** that upon cyclization gave the γ-lactam **157** (Scheme 2.58) [191].

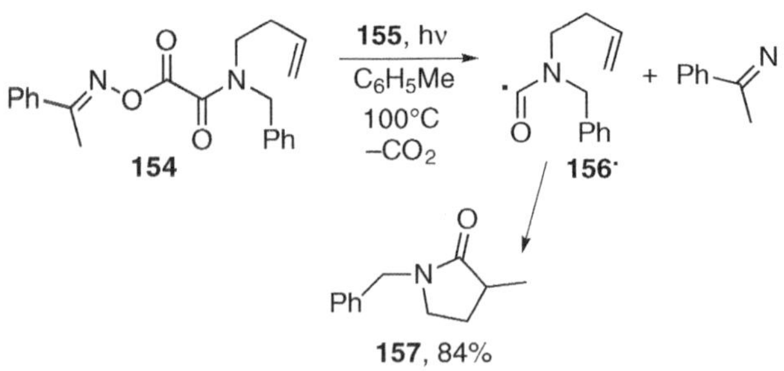

155 = 4-methoxyacetophenone

Scheme 2.58.

REFERENCES

1. Renaud, P., Sibi, M. P., Eds. *Radicals in Organic Synthesis*, Wiley-VCH, Weinheim, 2001.
2. Togo, H. *Advanced Free Radical Reactions for Organic Synthesis*, Elsevier Ltd., Oxford, 2004.
3. Moss, R. A., Platz, M. S., Jones, M., Jr, Eds. *Reactive Intermediates Chemistry*, John Wiley & Sons, Hoboken, NJ, 2004.
4. Gansäuer, A., Ed. (2006). Radicals in synthesis II. *Topics in Current Chemistry*, **264**, 1–236.
5. Gansäuer, A., Ed. (2006). Radicals in synthesis I. *Topics in Current Chemistry*, **263**, 1–189.
6. Forbes, M. D. E., Ed. *Carbon-Centered Free Radicals and Radical Cations. Structure, Reactivity, and Dynamics*, John Wiley & Sons, Hoboken, NJ, 2010.
7. Chatgilialoglu, C., Studer, A., Eds. *Encyclopedia of Radicals in Chemistry, Biology and Materials*, John Wiley & Sons, Hoboken, NJ, 2012.
8. Baguley, P. A., Walton, J. C. (1998). Flight from the tyranny of tin: The quest for practical radical sources free from metal encumbrances. *Angewandte Chemie International Edition*, **37**, 3072–3082.
9. Srikanth, G. S. C., Castle, S. L. (2005). Advances in radical conjugate additions. *Tetrahedron*, **61**, 10377–10441.
10. Fischer, H., Radom, L. (2001). Factors controlling the addition of carbon-centered radicals to alkenes. An experimental and theoretical perspective. *Angewandte Chemie International Edition*, **40**, 2001, **40**, 1340–1371.
11. Chatgilialoglu, C., Ferreri, C., Gimisis, T. Tris(trimethylsilyl)silane in organic synthesis. In Rappoport, Z., Apeloig, Y., Eds. *The Chemistry of Organic Silicon Compound*, Vol. 2, John Wiley & Sons, Hoboken, NJ, 1998, pp. 1539–1579.
12. Ollivier, C., Renaud, P. (2001). Organoboranes as a source of radicals. *Chemical Reviews*, **101**, 3415–3434.
13. Saraiva, M. F., Couri, M. R. C., Le Hyaric, M., de Almeida, M. V. (2009). The Barton ester free-radical reaction: A brief review of applications. *Tetrahedron*, **65**, 3563–3572.
14. Horspool, W. M., Song, P. S., Eds. *CRC Handbook of Organic Photochemistry and Photobiology*, CRC Press, Boca Raton, FL, 1995.
15. Lenci, F., Horspool, W., Eds. *CRC Handbook of Organic Photochemistry and Photobiology*, second edition, CRC Press, Boca Raton, FL, 2004.
16. *Molecular and Supramolecular Photochemistry* (Griesbeck, A. G., Mattay, J., Eds. *Synthetic Organic Photochemistry*, (Vol. 12), Dekker, New York, 2005.
17. *Molecular and Supramolecular Photochemistry* (Ramamurthy, V., Schanze, K., Eds. *Organic Photochemistry and Photophysics*, (Vol. 14), Dekker, New York, 2006, pp. 131–177.
18. Dichiarante, V., Fagnoni, M., Albini, A. Photochemical arylation reactions. In Ackermann, L., Ed. *Modern Arylation Methods*, Wiley-VCH Verlag GmbH & Co, 2009, pp. 513–535.
19. Albini, A., Fagnoni, M., Eds. *Handbook of Synthetic Photochemistry*, Wiley-VCH, Weinheim, 2010.

20. Griesbeck, A., Oelgemoeller, M., Ghetti, F., Eds. *CRC Handbook of Organic Photochemistry and Photobiology*, third edition, CRC Press, Boca Raton, FL, 2012.
21. Dalko, P. I. The photochemistry of Barton esters. In ref. 15, pp. 67-1–67-23.
22. Togo, H., Katohgi, M. (2001). Synthetic uses of organohypervalent iodine compounds through radical pathways. *Synlett*, 565–580.
23. Ravelli, D., Dondi, D., Fagnoni, M., Albini, A. (2009). Photocatalysis. A multi-faceted concept for green chemistry. *Chemical Society Reviews*, **38**, 1999–2011.
24. Fagnoni, M., Dondi, D., Ravelli, D., Albini, A. (2007). Photocatalysis for the formation of the C–C bond. *Chemical Reviews*, **107**, 2725–2756.
25. Palmisano, G., Augugliaro, V., Pagliaro, M., Palmisano, L. (2007). Photocatalysis: A promising route for 21st century organic chemistry. *Chemical Communications*, 3425–3437.
26. Ravelli, D., Fagnoni, M., Albini, A. (2013). Photoorganocatalysis. What for? *Chemical Society Reviews*, **42**, 97–113.
27. Kropp, P. J. Photobehavior of alkyl halides. In ref. 15, pp. 1-1–1-32.
28. Schutt, L., Bunce, N. J. Photodehalogenation of aryl halides. In ref. 15, pp. 38-1–38-18.
29. Hunther, N., Parsons, A. S. Photochemically induced alkylation reactions. In ref. 15, pp. 6-1–6-14.
30. Harendza, M., Junggebauer, J., Lessmann, K., Neumann, W.P., Tews, H. (1993). Mild photochemical generation of stannyl radicals $Bu_3Sn^\cdot$ from hexabutyl distannane for organic synthesis. *Synlett*, 286–288.
31. Roth, R. C., Binkley, R. W. (1985). A mechanistic study of the photoreactions of deoxy iodo sugars. *The Journal of Organic Chemistry*, **50**, 690–693.
32. Shimizu, N., Nishida S. (1977). Photochemical Reduction of gem-dibromocyclopropanes. *Chemistry Letters*, 839–842.
33. Thoma, G., Giese, B. (1989). Generation and synthetic use of alkyl radicals with $[CpFe(CO)_2]_2$ as mediator. *Tetrahedron Letters*, **30**, 2907–2910.
34. Giese, B., Thoma G. (1991). Dimeric metal complexes as mediators for radical C–C bond-forming reactions. *Helvetica Chimica Acta*, **74**, 1135–1142.
35. Giese, B., Thoma, G. (1991). C–C bond formation via carbon-centered radicals generated from dicarbonyl(η5-cyclopentadienyl) organoiron complexes. *Helvetica Chimica Acta*, **74**, 1143–1155.
36. Friestad, G. K., Qin, J. (2001). Intermolecular alkyl radical addition to chiral *N*-acylhydrazones mediated by manganese carbonyl. *Journal of the American Chemical Society*, **123**, 9922–9923.
37. Barton, D. H. R., Garcia, B., Togo, H., Zard, S. Z. (1986). Radical decarboxylative addition onto protonated heteroaromatic (and related) compounds. *Tetrahedron Letters*, **27**, 1327–1330.
38. Barton, D. H. R., Jaszberenyi, J., Theodorakis, E. A. (1991). Radical nitrile transfer with methanesulfonyl cyanide or *p*-toluenesulfonyl cyanide to carbon radicals generated from the acyl derivatives of *N*-hydroxy-2-thiopyridone. *Tetrahedron Letters*, **32**, 3321–3324.
39. Barton, D. H. R., Jaszberenyi, J., Theodorakis E. A. (1992). The invention of radical reactions. Part XXIII. New reactions: Nitrile and thiocyanate transfer to carbon radicals from sulfonyl cyanides and sulfonyl isothiocyanates. *Tetrahedron*, **48**, 2613–2626.

40. Barton. D. H. R., De Almeida, M. V., Lu, W., Shinada, T., Jaszberenyi, J., Dos Santos, H. F., Le Hyaric, M. (2001). Radical homologation of D-gluconic acid: Highly diastereoselective synthesis of D-gluco-KDO derivatives. *Tetrahedron*, **57**, 8767–8771.

41. Anaya, J., Barton, D. H. R., Caballero, M. C., Géro, S. D., Grande, M., Laso, N. M., Hernando J. I. M. (1994). The use of radical decarboxylation in the preparation of 1-methylcarbapenem antibiotic precursors from D-glucosamine. *Tetrahedron: Asymmetry*, **5**, 2137–2140.

42. Magnus, P., Payne, A. H., Hobson, L. (2000). Synthesis of the *Kopsia* alkaloids (±)-11,12-demethoxylahadinine B, (±)-kopsidasine and (±)-kopsidasine-*N*-oxide. *Tetrahedron Letters*, **41**, 2077–2081.

43. Jung Ko, E., Savage, G. P., Williams, C. M., Tsanaktsidis, J. (2011). Reducing the cost, smell, and toxicity of the Barton reductive decarboxylation: Chloroform as the hydrogen atom source. *Organic Letters*, **13**, 1944–1947.

44. Pimpalpalle, T. M., Yin, J., Linker, T. (2012). Barton radical reactions of 2-C-branched carbohydrates. *Organic and Biomolecular Chemistry*, **10**, 103–109.

45. Masterson, D. S., Porter, N. A. (2002). Diastereoselective free radical halogenation, azidation, and rearrangement of *β*-silyl Barton esters. *Organic Letters*, **4**, 4253–4256.

46. Barton, D. H. R., Géro, S. D., Holliday, P., Quiclet-Sire, B., Zard, S. Z. (1998). A practical decarboxylative hydroxylation of carboxylic acids. *Tetrahedron*, **54**, 6751–6756.

47. Kim, S., Jo Lim, C., Song, S.-E., Kang H.-Y. (2001). Decarboxylative acylation approach of thiohydroxamate esters. *Chemical Communications*, 1410–1411.

48. Keck, G. E., Kachensky, D. F., Enholm, E. J. (1985). Pseudomonic acid C from L-lyxose. *The Journal of Organic Chemistry*, **50**, 4317–4325.

49. Vismara, E., Torri, G., Pastori, N., Marchiandi, M. (1992). A new approach to the stereoselective synthesis of C-nucleosides via homolytic heteroaromatic substitution. *Tetrahedron Letters*, **33**, 7575–7578.

50. Togo, H., Aoki, M., Kuramochi, T., Yokoyama, M. (1993). Radical decarboxylative alkylation onto heteroaromatic bases with trivalent iodine compounds. *Journal of the Chemical Society, Perkin Transactions 1*, 2417–2427.

51. Togo, H., Aoki, M., Yokoyama, M. (1993). Reductive addition to electron-deficient olefins with trivalent iodine compounds. *Tetrahedron*, **49**, 8241–8256.

52. Fokin, A. A., Schreiner, P. R. (2002). Selective alkane transformations via radicals and radical cations: Insights into the activation step from experiment and theory. *Chemical Reviews*, **102**, 1551–1593.

53. Geraghty, N. W. A., Lohan, M. T. Intermolecular photochemical hydrogen abstraction: synthetic applications in relation to carbon–carbon bond formation. In ref. 20, pp. 445–488.

54. Cardarelli, A. M., Fagnoni, M., Mella, M., Albini, A. (2001). Hydrocarbon activation. Synthesis of β-cycloalkyl (di)nitriles through photosensitized conjugate radical addition. *The Journal of Organic Chemistry*, **66**, 7320–7327.

55. Gonzalez-Cameno, A. M., Mella, M., Fagnoni, M., Albini, A. (2000). Photochemical alkylation of ketene dithioacetal *S,S*-dioxides. An example of captodative olefin functionalization. *The Journal of Organic Chemistry*, **65**, 297–303.

56. Geraghty, N. W. A., Hannan, J. J. (2001). Functionalisation of cycloalkanes: The photomediated reaction of cycloalkanes with alkynes. *Tetrahedron Letters*, **42**, 3211–3213.
57. Doohan, R. A., Hannan, J. J., Geraghty, N. W. A. (2006). The photomediated reaction of alkynes with cycloalkanes. *Organic and Biomolecular Chemistry*, **4**, 942–952.
58. Doohan, R. A., Geraghty, N. W. A. (2005). A comparative analysis of the functionalisation of unactivated cycloalkanes using alkynes and either sunlight or a photochemical reactor. *Green Chemistry*, **7**, 91–96.
59. Hill, C. L. (1995). Introduction of functionality into unactivated carbon–hydrogen bonds. Catalytic generation and nonconventional utilization of organic radicals. *Synlett*, 127–132.
60. Tzirakis, M. D., Lykakis, I. N., Orfanopoulos, M. (2009). Decatungstate as an efficient photocatalyst in organic chemistry. *Chemical Society Reviews*, **28**, 2609–2621.
61. Lykakis, I. N., Evgenidou, E., Orfanopoulos, M. (2012). Photo-catalysis and polyoxoanion decatungstate in organic chemistry: A manifold concept for green chemistry. *Current Organic Chemistry*, **16**, 2400–2414.
62. Dondi, D., Fagnoni, M., Molinari, A., Maldotti, A., Albini, A. (2004). Polyoxotungstate photoinduced alkylation of electrophilic alkenes by cycloalkanes. *Chemistry—A European Journal*, **10**, 142–148.
63. Dondi, D., Fagnoni, M, Albini, A. (2006). Tetrabutylammonium decatungstate-photosensitized alkylation of electrophilic alkenes: Convenient functionalization of aliphatic C–H bonds. *Chemistry—A European Journal*, **12**, 4153–4163.
64. Dondi, D., Cardarelli, A. M., Fagnoni, M, Albini, A. (2006). Photomediated synthesis of β-alkylketones from cycloalkanes. *Tetrahedron*, **62**, 5527–5535.
65. Protti, S., Ravelli, D., Fagnoni, M., Albini, A. (2009). Solar light-driven photocatalyzed alkylations. Chemistry on the window ledge. *Chemical Communications*, 7351–7353.
66. Dondi, D., Ravelli, D., Fagnoni, M., Mella, M., Molinari, A., Maldotti, A., Albini, A. (2009). Regio- and stereoselectivity in the decatungstate photocatalyzed alkylation of alkenes by alkylcyclohexanes. *Chemistry—A European Journal*, **15**, 7949–7957.
67. Tzirakis, M. D., Orfanopoulos, M. (2008). Decatungstate-mediated radical reactions of C_{60} with substituted toluenes and anisoles: A new photochemical functionalization strategy for fullerenes. *Organic Letters*, **10**, 873–876.
68. Nakamura, T., Koga, Y. (1998). Perfluorooctylation of aromatic compounds with perfluoroazooctane. *Journal of the Chemical Society, Perkin Transactions* 2, 659–662.
69. Nakamura, T., Ishihara, M., Ohana, T., Tanaka, A. Koga, Y. (2004). Sidewall modification of single-walled carbon nanotubes using photolysis of perfluoroazooctane. *Chemical Communications*, 1336–1337.
70. Cadot, C., Cossy, J. Dalko, P. I. (2000). Radical carbon–carbon coupling reactions *via* organoboranes. *Chemical Communications*, 1017–1018.
71. Patel, V. F., Pattenden, G. (1990). Free radical reactions in synthesis. Homolysis of alkylcobalt complexes in the presence of radical-trapping agents. *Journal of the Chemical Society, Perkin Transaction 1*, 2703–2708.
72. Pattenden, G. (1988). Cobalt-mediated radical reactions in organic synthesis. *Chemical Society Reviews*, **17**, 361–382.

73. Tada, M. (1999). Reactions of alkyl–cobalt complexes. *Reviews on Heteroatom Chemistry*, **20**, 97–144.

74. Russell, G. A. (1989). Free radical chain reactions involving alkyl- and alkenylmercurials. *Account of Chemical Research*, **22**, 1–8.

75. Tada, M., Miura, K., Okabe, M., Seki, S., Mizukami, H. (1981). A mimic study on coenzyme-B_{12} using organocobaloximes, the rearrangement of 1-substituted-2-oxocyclopentylmethyl radical. *Chemistry Letters*, 33–36.

76. Baldwin, J. E., Li, C.-S. (1987). Cobalt-mediated cyclization of amino acid derivatives. Application to the kainoids. *Journal of Chemical Society, Chemical Communications*, 166–168.

77. Weiss, M. E., Kreis, L. M., Lauber, A., Carreira, E. M. (2011). Cobalt-catalyzed coupling of alkyl iodides with alkenes: Deprotonation of hydridocobalt enables turnover. *Angewandte Chemie International Edition*, **50**, 11125–11128.

78. Weiss, M. E., Carreira, E. M. (2011). Total synthesis of (+)-daphmanidin E. *Angewandte Chemie International Edition*, **50**, 11501–11505.

79. Goumans, T. P. M., van Alem, K., Lodder, G. (2008). Photochemical generation and structure of vinyl radicals. *European Journal of Organic Chemistry*, 435–443.

80. Kitamura, T. C–X bond fission in alkene systems. In ref. 14, pp. 1171–1180.

81. Wilson, R. M., Commons T. J. (1975). Photochemistry of β-iodoacrylamides. *The Journal of Organic Chemistry*, **40**, 2891–2897.

82. Kitamura, T., Kobayashi, S., Taniguchi, H. (1988). Photolysis of β-(o-methylthiophenyl) vinyl bromides. A versatile synthesis of benzo[b]thiophenes. *Chemistry Letters*, 1637–1638.

83. Campos-Gómez, E., Campos, P. J., González, H. F., Rodríguez. M. A. (2012). Light-induced carbocyclization of iodoalkenes. *Tetrahedron*, **68**, 4292–4295.

84. Pretali, L., Albini, A. Photochemistry of aryl halides. In ref. 20, pp. 369–391.

85. Grimshaw, J., de Silva, A. P. (1981). Photochemistry and photocyclization of aryl halides. *Chemical Society Reviews*, **10**, 181–203.

86. As for the case of photohomolysis of halogenoheterocyclic derivatives, see D'Auria, M. (1997). Photochemical behaviour of halogenoheterocyclic derivatives. *Heterocycles*, **45**, 1775–1780.

87. Wolf, W., Kharasch, N. (1961). Photolysis of aromatic iodo compounds as a synthetic tool. *The Journal of Organic Chemistry*, **26**, 283–284.

88. Matsuura, T., Omura, K. (1966). Photo-induced reactions. III. The synthesis of biphenyls from aromatic bromo-compounds. *Bulletin Chemial Society of Japan*, **39**, 944–947.

89. Robinson, G. E., Vernon, J. M. (1971). Photolysis of chlorobenzene derivatives in aromatic solvents. *Journal of the Chemical Society C*, 3363–3367.

90. Taylor, E. C., Kienzle, F., McKillop, A. (1970). Thallium in organic synthesis. XIX. Photochemical phenylation. An effective synthesis of unsymmetrical biphenyls from arylthallium ditrifluoroacetates. *Journal of The American Chemical Society*, **92**, 6088–6089.

91. Taylor, E. C., Altland, H. W., Kienzle, F. (1976). Thallium in organic synthesis. XLI. Synthesis of 1-substituted 2 (1H)-pyridones. A New synthesis of unsymmetrical biphenyls via photochemical N—O bond cleavage of 1-aroyloxy-2(1H)-pyridones. *The Journal of Organic Chemistry*, **41**, 24–27.

92. Heinrich, M. R. Kirschstein, M. D. (2006). Photochemical generation and trapping of radicals from bisphenyl carbodiazone. *Tetrahedron Letters*, **47**, 2115–2118.

93. Fozard, A. Bradsher C. K. (1967). The synthesis of the pyrido[2,1-*a*]isoindole system by an intramolecular photochemical cyclization. *The Journal of Organic Chemistry*, **32**, 2966–2969.

94. Jeffs, P. W., Hansen, J. F., Brine, G. A. (1975). Photochemical synthesis of 6,7-dihydro-5*H*-dibenz[*c*,*e*]azepine and 5,6,7,8-tetrahydrodibenz[*c*,*e*]azocine derivatives. *The Journal of Organic Chemistry*, **40**, 2883–2890.

95. Letcher, R. M., Wong, K.-M. (1977). Photochemical cyclisation of 2-iodostilbenes. *Journal of the Chemical Society, Perkin Transactions 1*, 178–180.

96. Bratt, J., Iddon, B., Mack, A. G., Suschitzky, H., Taylor, J. A., Wakefield, B. J. (1980). Polyhalogenoaromatic compounds. Part 41. Photochemical dehalogenation and arylation reactions of polyhalogenoaromatic and polyhalogenoheteroaromatic compounds. *Journal of the Chemical Society, Perkin Transactions 1*, 648–656.

97. Moorthy, J. N., Samanta, S. (2007). Photoinduced C–Br homolysis of 2-bromobenzophenones and Pschorr ring closure of 2-aroylaryl radicals to fluorenones. *The Journal of Organic Chemistry*, **72**, 9786–9789.

98. Clyne, M. A., Aldabbagh, F. (2006). Photochemical intramolecular aromatic substitutions of the imidazol-2-yl radical are superior to those mediated by Bu_3SnH. *Organic and Biomolecular Chemistry*, **4**, 268–277.

99. Linsenmeier, A. M., Williams, C. M., Bräse, S. (2011). Synthesis of phenanthridine derivatives via photolysis. *The Journal of Organic Chemistry*, **76**, 9127–9132.

100. Curran, D. P., Liu, H., Josien, H., Ko, S.-B. (1996). Tandem radical reactions of isonitriles with 2-pyridonyl and other aryl radicals: Scope and limitations, and a first generation synthesis of (±)-camptothecin. *Tetrahedron*, **52**, 11385–11404.

101. Liu, Q., Han, B., Zhang, W., Yang, L., Liu, Z.-L., Yu, W. (2005). Photo-induced radical cyclization of aromatic halides with sodium borohydride. *Synlett*, 2248–2250.

102. Galli, C. (1988). Radical reactions of arenediazonium ions: An easy entry into the chemistry of the aryl radical. *Chemical Reviews*, **88**, 765–792.

103. Grimshaw, J. Photochemistry of aryl diazonium salts, thiazoles and tetrazoles. In ref. 15, pp. 43-1, 43-17.

104. Milanesi, S., Fagnoni, M., Albini, A. (2005). (Sensitized) Photolysis of diazonium salts as a mild general method for the generation of aryl cations. Chemoselectivity of the singlet and triplet 4-substituted phenyl cations. *The Journal of Organic Chemistry* **70**, 603–610 and references cited therein.

105. Cano-Yelo, H., Deronzier, A. (1984). Photocatalysis of the Pschorr reaction by tris-(2,2′-bipyridyl) ruthenium (II) in the phenanthrene series. *Journal of the Chemical Society, Perkin Transactions 2*, 1093–1098.

106. Hari, D. P., Schroll, P., König, B. (2012). Metal-Free, Visible-light-mediated direct C–H arylation of heteroarenes with aryl diazonium salts. *Journal of the American Chemical Society*, **134**, 2958–2961.

107. Schroll, P., Hari, D. P., König B. (2012). Photocatalytic arylation of alkenes, alkynes and enones with diazonium salts. *ChemistryOpen*, **1**, 130–133.

108. Hari, D. P., Hering, T., König B. (2012). Visible light photocatalytic synthesis of benzothiophenes. *Organic Letters*, **14**, 5334–5337.

109. Wagner, P. J., Sedon, J., Waite, C., Gudmundsdottir, A. (1994). Photoinduced radical cleavage of acylphenyl halides: Different mechanisms for C–I and C–Br bonds. *Journal of the American Chemical Society*, **116**, 10284–10285.

110. Yamada, K., Maekawa, M., Akindele, T., Nakano, M., Yamamoto, Y., Tomioka, K. (2008). Chemoselective conjugate addition of dimethylzinc-mediated ether and acetal radicals to alkylidenemalonates and asymmetric reactions. *The Journal of the Organic Chemistry*, **73**, 9535–9538.

111. Tsujimoto, S., Sakaguchi, S., Ishii, Y. (2003). Addition of aldehydes and their equivalents to electron-deficient alkenes using *N*-hydroxyphthalimide (NHPI) as a polarity-reversal catalyst. *Tetrahedron Letters*, **44**, 5601–5604.

112. Yoshimitsu, T., Arano, Y., Nagaoka, H. (2005). Radical α-C–H hydroxyalkylation of ethers and acetal. *The Journal of Organic Chemistry*, **70**, 2342–2345.

113. Rosenthal, I., Elad, D. (1968). Photoalkylation of cyclic acetals. *The Journal of Organic Chemistry*, **33**, 805–811.

114. Fraser-Reid, B., Anderson, R. C., Hicks, D. R., Walker, D. L. (1977). Synthetic applications of the photochemically induced addition of oxycarbinyl species to α-enones. Part II. The addition of ketals, aldehydes, and polyfunctional species. *Canadian Journal of Chemistry*, **55**, 3986–3995.

115. Manfrotto, C., Mella, M., Freccero, M., Fagnoni, M., Albini, A. (1999). Photochemical synthesis of 4-oxobutanal acetals and of 2-hydroxycyclobutanone ketals. *The Journal of Organic Chemistry*, **64**, 5024–5028.

116. Campari, G., Mella, M., Fagnoni, M., Albini, A. (2000). Diastereoselective photosensitised radical addition to fumaric acid derivatives bearing oxazolidine chiral auxiliaries. *Tetrahedron: Asymmetry*, **11**, 1891–1906.

117. Dichiarante, V., Fagnoni, M. Intermolecular addition reactions onto C–C multiple bonds. In ref. 19, pp. 67–94.

118. Fernandez-Gonzalez, M., Alonso, R. (2006). Formation of five- and six-membered carbocycles with nitrogenated tetrasubstituted carbons by radical addition-carbocyclization of alkynyl ketoxime ethers. *The Journal of Organic Chemistry*, **71**, 6767–6775.

119. Ghosh, A. K., Leshchenko, S., Noetzel, M. (2004). Stereoselective photochemical 1,3-dioxolane addition to 5-alkoxymethyl-2(5*H*)-furanone: Synthesis of bis-tetrahydrofuranyl ligand for HIV protease inhibitor UIC-94017 (TMC-114). *The Journal of Organic Chemistry*, **69**, 7822–7829.

120. Mosca, R., Fagnoni, M., Mella, M., Albini, A. (2001). Synthesis of monoprotected 1,4-diketones by photoinduced alkylations of enones with 2-substituted-1,3-dioxolanes. *Tetrahedron*, **57**, 10319–10328.

121. Dondi, D., Caprioli, I., Fagnoni, M., Mella, M., Albini, A. (2003). A convenient route to 1,4-monoprotected dialdehydes, 1,4-ketoaldehydes, γ-lactols and γ-lactones through radical alkylation of α,β-unsaturated aldehydes in organic and organic–aqueous media. *Tetrahedron*, **59**, 947–957.

122. Parsons, P. J., Cowell, J. K. (2000). A rapid synthesis of (−)-tetrahydrolipstatin. *Synlett*, 107–109.

123. Geraghty, N. W. A., Lally, A. (2006). The rapid photomediated alkenylation of 2-alkyl-1,3-dioxolanes with alkynes: A stereoelectronically assisted chain reaction. *Chemical Communications*, 4300–4302.

124. Ravelli, D., Albini, A., Fagnoni, M. (2011). Smooth photocatalytic preparation of 2-substituted-1,3-benzodioxoles. *Chemistry—A European Journal*, **17**, 572–579.

125. Fraser-Reid, B., Hicks, D. R., Walker, D. L., Iley, D. E., Yunker, M. B., Tam, S. Y. K., Anderson, R. C., Saunders, J. (1975). Photoaddition of polyfunctional oxycarbinyl species to conjugated enones: A ready route to versatile γ-functionalized ketones. *Tetrahedron Letters*, **16**, 297–300.

126. Rosenthal, I., Elad, D. (1967). The photoalkylation of cyclic ethers. *Tetrahedron*, **23**, 3193–3204.

127. Kobayashi, K., Suzuki, M., Takeuchi, H., Konishi, A., Sakurai, H., Suginome, H. (1994). Photoinduced molecular transformations. Part 146. Photoacylation and photoalkylation of 2-arylamino- and 2-alkylamino-1,4-naphthoquinones. *Journal of the Chemical Society, Perkin Transactions 1*, 1099–1104.

128. Kamijo, S., Hoshikawa, T., Inoue, M. (2011). Carbamoylation of ethereal C–H bonds using pentafluorophenyl isocyanate. *Tetrahedron Letters*, **52**, 2885–2888.

129. Kamijo, S., Hoshikawa, T., Inoue, M. (2011). Photochemically induced radical transformation of C(sp^3)-H bonds to C(sp^3)-CN bonds. *Organic Letters*, **13**, 5928–5931.

130. Tzirakis, M. D., Orfanopoulos, M. (2010). Photochemical addition of ethers to C_{60}: synthesis of the simplest [60]fullerene/crown ether conjugates. *Angewandte Chemie International Edition*, **49**, 5891–5893.

131. Ravelli, D., Montanaro, S., Tomasi, C., Galinetto, P., Quartarone, E., Merli, D., Mustarelli, P., Fagnoni, M. (2012). One-step decatungstate-photomediated PEGylation of single-walled carbon nanotubes. *ChemPlusChem*, **1**, 210–216.

132. Barton, D. H. R., Géro, S. D., Quiclet-Sire, B., Samadi, M. (1989). Stereoselectivity in radical reactions of 2′-deoxynucleosides. A synthesis of an isostere of 3′-azido-3′-deoxythymidine-5′-monophosphate (AZT-5′ monophosphate). *Tetrahedron Letters*, **30**, 4969–4972.

133. Barton, D. H. R., Géro, S. D., Quiclet-Sire, B., Samadi, M. (1992). New synthesis of sugar, nucleoside and α-amino acid phosphonates. *Tetrahedron*, **48**, 1627–1636.

134. Barton, D. H. R., Géro, S. D., Quiclet-Sire, B., Samadi M. (1994). Stereocontrolled radical reactions in carbohydrate and nucleoside chemistry. *Tetrahedron: Asymmetry*, **5**, 2123–2136.

135. H. Togo, H., Ishigami, S., Fujii, M., Ikuma, T., Yokoyama, M. (1994). Synthesis of C-nucleosides *via* radical coupling reaction. *Journal of the Chemical Society, Perkin Transactions 1*, 2931–2942.

136. Urabe, D., Yamaguchi, H., Inoue, M. (2011). Application of α-alkoxy bridgehead radical for coupling of oxygenated carbocycles. *Organic Letters*, **13**, 4778–4781.

137. Barton, D. H. R., Ramesh, M. (1990). Tandem nucleophilic and radical chemistry in the replacement of the hydroxyl group by a carbon–carbon bond. A concise synthesis of showdomycin. *Journal of the American Chemical Society*, **112**, 891–892.

138. Veit, A., Giese, B. (1990). Synthesis of mannosamine via cobaloximes. *Synlett*, 166.

139. Nieweg, J. A., Lemma, K., Trewyn, B. G., Lin, V. S. Y., Bakac, A. (2005). Mesoporous silica-supported uranyl: Synthesis and photoreactivity. *Inorganic Chemistry*, **44**, 5641–5648.

140. Brown, S. H., Crabtree, R. H. (1989). Making mercury-photosensitized dehydrodimerization into an organic synthetic method: Vapor pressure selectivity and the behavior of functionalized substrates. *Journal of the American Chemical Society*, **111**, 2935–2946.

141. Lee, J. C., Jr., Boojamra, C. G., Crabtree, R. H. (1993). The synthesis of glycols by mercury-photosensitized alcohol dehydrodimerization. *The Journal of Organic Chemistry*, **58**, 3895–3900.

142. Drew, M. G. B., Harrison, R. J., Mann, J., Tench, A. J., Young, R. J. (1999). Photoinduced Addition of methanol to 5(*S*)-5-triisopropyisiloxymethyl-*N*-bocdihydropyrrole- 2(5*H*)-one: A new route to 4(*S*), 5(*S*)-disubstituted pyrrolidin-2-ones. *Tetrahedron*, **55**, 1163–1172.

143. Geraghty, N. W. A., Hernon, E. M. (2009). The reaction of photochemically generated α-hydroxyalkyl radicals with alkynes: A synthetic route to γ-butenolides. *Tetrahedron Letters*, **50**, 570–573.

144. Dondi, D., Protti, S., Albini, A., Mañas Carpio, S., Fagnoni, M. (2009). Synthesis of γ-lactols, γ-lactones and 1,4-monoprotected succinaldehydes under moderately concentrated sunlight. *Green Chemistry*, 1653–1659.

145. Shvydkiv, O., Yavorskyy, A., Nolan, K., Youssef, A., Riguet, E., Hoffmann N., Oelgemöller, M. (2010). Photosensitized addition of isopropanol to furanones in a 365 nm UV-LED microchip. *Photochemical Photobiological Sciences*, **9**, 1601–1603.

146. Yavorskyy, A., Shvydkiv, O., Nolan, K., Hoffmann, N., Oelgemöller, M. (2011). Photosensitized addition of isopropanol to furanones in a continuous-flow dual capillary microreactor. *Tetrahedron Letters*, **52**, 278–280.

147. Shvydkiv, O., Yavorskyy, A., Bee Tan, S., Nolan, K., Hoffmann, N., Youssef, A., Oelgemöller, M. (2011). Microphotochemistry: A reactor comparison study using the photosensitized addition of isopropanol to furanones as a model reaction. *Photochemical Photobiological Sciences*, **10**, 1399–1404.

148. Yavorskyy, A., Shvydkiv, O., Hoffmann, N., Nolan, K., Oelgemoeller, M. (2012). Parallel microflow photochemistry: Process optimization, scale-up, and library synthesis. *Organic Letters*, **14**, 4342–4345.

149. Torrente, S., Alonso, R. (2001). Photoinduced formation of β-oxy-α,α-disubstituted-α-amino acid derivatives from ketoximes. *Organic Letters*, **3**, 1985–1987.

150. Wei, L., Zhang, Y. (2007). Covalent sidewall functionalization of single-walled carbon nanotubes: A photoreduction approach. *Nanotechnology*, **18**, 495703.

151. Tzirakis, M. D., Alberti, M. N., Orfanopoulos, M. (2010). Hydroxyalkylation of [60] fullerene: Free radical addition of alcohols to C_{60}. *Chemical Communications*, **46**, 8228–8230.

152. Lindemann, U., Neuburger, M., Neuburger-Zehnder, M., Wulff-Molder, D., Wessig P. (1999). Photoinduced diastereoselective pinacolisation of 4-oxo-4-phenylbutanamides to 4,5-dihydroxy-4,5-diphenyloctanediamides. *Journal of the Chemical Society, Perkin Transactions 2*, 2029–2036.

153. Al-Amoudi, M. A. S., Vernon, J. M. (1999). Photochemical addition of secondary alcohols to maleimides. *Journal of the Chemical Society, Perkin Transactions 2*, 2667–2670.

154. Byers, J. H., Whitehead, C. C., Duff, M. E. (1996). Addition of a sulfur-stabilized radical to electron-deficient alkenes. *Tetrahedron Letters*, **37**, 2743–2744.

155. Barton, D. H. R., Hervè, Y., Potier, P., Thierry J. (1987). Synthesis of novel α-amino-acids and derivatives using radical chemistry: Synthesis of L- and D-α-amino-adipic acids, L-α aminopimelic acid and appropriate unsaturated derivatives. *Tetrahedron*, **43**, 4297–4308.

156. Angioni, S., Ravelli, D., Emma, D., Dondi, D., Fagnoni, M., Albini, A. (2008). Tetrabutylammonium decatungstate (chemo)selective photocatalyzed, radical C–H functionalization in amides. *Advanced Synthesis and Catalysis*, **350**, 2209–2214.

157. Nagura, H., Murakami, S., Fuchigami, T. (2007). Photochemical generation of difluoromethyl radicals having various functional groups and their highly regioselective addition to olefins and aromatic substitution. *Tetrahedron*, **63**, 10237–10245.

158. Holzinger, M., Vostrowsky, O., Hirsch, A., Hennrich, F., Kappes, M., Weiss, R., Jellen, F. (2001). Sidewall functionalization of carbon nanotubes. *Angewandte Chemie International Edition*, **40**, 4002–4005.

159. Yasu, Y., Koike, T., Akita, M. (2012). Three-component oxytrifluoromethylation of alkenes: Highly efficient and regioselective difunctionalization of C=C bonds mediated by photoredox catalysts. *Angewandte Chemie International Edition*, **51**, 9567–9571.

160. Murakami, S., Kim, S., Ishii, H., Fuchigami, T. (2004). Aromatic substitution with photochemically generated difluoromethyl radicals bearing electron-withdrawing group. *Synlett*, 815–818.

161. Huther, N., McGrail, P. T., Parsons, A. F. (2002). Biphasic manganese carbonyl reactions: A new approach to making carbon–carbon bonds. *Tetrahedron Letters*, **43**, 2535–2538.

162. Back, T. G., Minksztym K. (1997). The free-radical addition of phenyl trichloromethyl selenide to alkenes: A new method for the regioselective carboxylation of alkenes. *Chemical Communications*, 1759–1760.

163. Shiraishi, Y., Tsukamoto, D., Hirai, T. (2008). Highly efficient methyl ketone synthesis by water-assisted C—C coupling between olefins and photoactivated acetone. *Organic Letters*, **10**, 3117–3120.

164. Tsukamoto, D., Shiraishi, Y., Hirai, T. (2010). Highly efficient methyl ketone synthesis with photoactivated acetone and olefins assisted by Mg(II)-exchanged zeolite Y. *The Journal of Organic Chemistry*, **75**, 1450–1457.

165. Byers, J. H., Lane, G. C. (1993). Radical addition reactions of 2-(phenylseleno)propanedioates to alkenes and alkynes. *The Journal of Organic Chemistry*, **58**, 3355–3360.

166. Byers, J. H., Lane, G. C. (1990). Radical addition of diethyl (2-phenylseleno)propanedioate to olefins. *Tetrahedron Letters*, **31**, 5697–5700.

167. Byers, J. H., Harper, B. C. (1992). Radical addition of varied functionality via phenylselenenyl transfer. *Tetrahedron Letters*, **33**, 6953–6954.

168. Renaud, P., Abazi, S. (1996). Use of O,Se-acetals for radical-mediated phenylseleno group transfer reactions. *Synthesis*, 253–258.

169. Byers, J. H., Thissell, J. G., Thomas, M. A. (1995). The synthesis of geminal diphosphonates via phenyl selenide transfer radical addition. *Tetrahedron Letters*, **36**, 6403–6406.

170. Renaud, P., Vionnet J.-P. (1993). Radical additions to 7-oxabicyclo[2.2.1]hept-5-en-2-one. facile preparation of all-*cis*-corey lactone. *The Journal of Organic Chemistry*, **58**, 5895–5896.

171. Lobachevsky, S., Schiesser, C. H., Gupta V. (2007). Tandem homolytic addition/substitution sequences and their application to tin-free radical chemistry. *Tetrahedron Letters*, **48**, 9077–9079.

172. Gilbert, B. C., Kalz, W., Lindsay, C. I., McGrail, P. T., Parsons, A. F., Whittaker D. T. E. (2000). Initiation of radical cyclisation reactions using dimanganese decacarbonyl.

A flexible approach to preparing 5-membered rings. *Journal of the Chemical Society, Perkin Transactions 1*, 1187–1194.

173. Chatgilialoglu, C., Crich, D., Komatsu, M., Ryu, I. (1999). Chemistry of acyl radicals. *Chemical Reviews*, **99**, 1991–2070.
174. Zard S. Z. (1997). On the trail of xanthates: Some new chemistry from an old functional group. *Angewandte Chemie International Edition*, **36**, 672–685 and reference therein.
175. Chen, C., Crich, D., Papadatos, A. (1992). The chemistry of acyl tellurides: Generation and trapping of acyl radicals, including aryltellurium group transfer. *Journal of the American Chemical Society*, **114**, 8313–8314.
176. Ryu, I., Sonoda, N., Curran, D. P. (1996). Tandem radical reactions of carbon monoxide, isonitriles, and other reagent equivalents of the geminal radical acceptor/radical precursor synthon. *Chemical Reviews*, **96**, 177–194.
177. Ryu, I. (2001). Radical carboxylations of iodoalkanes and saturated alcohols using carbon monoxide. *Chemical Society Reviews*, **30**, 16–25.
178. Nagahara, K., Ryu, I., Komatsu, M., Sonoda N. (1997). Radical carboxylation: Ester synthesis from alkyl iodides, carbon monoxide, and alcohols under irradiation conditions. *Journal of the American Chemical Society*, **119**, 5465–5466.
179. Fraser-Reid, B., Anderson, R. C., Hicks, D. R., Walker, D. L. (1977). Synthetic applications of the photochemically induced addition of oxycarbinyl species to α-enones. Part II. The addition of ketals, aldehydes, and polyfunctional species. *Canadian Journal of Chemistry*, **55**, 3986–3995.
180. Oelgemöller, M., Jung, C., Ortner, J., Mattay, J., Schiel, C., Zimmermann, E. (2005). 'Green photochemistry' with moderately concentrated sunlight. *The Spectrum*, **18**, 10–15.
181. Kraus, G. A., Liu, P. (1994). Benzophenone-mediated conjugate additions of aromatic aldehydes to quinones. *Tetrahedron Letters*, **35**, 7723–7726.
182. Oelgemoeller, M., Schiel, C., Mattay, J., Froehlich, R. (2002). The "photo-Friedel–Crafts acylation" of 1,4-naphthoquinones. *European Journal of Organic Chemistry*, 2465–2474.
183. Esposti, S., Dondi, D., Fagnoni, M., Albini, A. (2007). Acylation of electrophilic olefins through decatungstate-photocatalyzed activation of aldehydes. *Angewandte Chemie International Edition*, **46**, 2531–2534.
184. Ravelli, D., Zema, M., Mella, M., Fagnoni, M., Albini, A. (2010). Benzoyl radicals from (hetero)aromatic aldehydes. Decatungstate photocatalyzed synthesis of substituted aromatic ketones. *Organic and Biomolecular Chemistry*, **8**, 4158–4164.
185. Ryu, I., Tani, A., Fukuyama, T., Ravelli, D., Fagnoni, M., Albini, A. (2011). Atom-economical synthesis of unsymmetrical ketones via photocatalyzed C–H activation of alkanes and coupling with CO and electrophilic alkenes. *Angewandte Chemie International Edition*, **50**, 1869–1872.
186. Tzirakis, M. D., Orfanopoulos, M. (2009). Acyl radical reactions in fullerene chemistry: Direct acylation of [60]fullerene through an efficient decatungstate-photomediated approach. *Journal of the American Chemical Society*, **131**, 4063–4069.
187. Rokach, J., Elad, D. (1966). Photoamidation. IV. The light-induced amidation of α,β-unsaturated esters. *The Journal of Organic Chemistry*, **31**, 4210–4215.

188. Caronna, T., Gambarotti, C., Palmisano, L., Punta, C., Recupero, F. (2003). Sunlight induced functionalisation of some heterocyclic bases in the presence of polycrystalline TiO_2. *Chemical Communications*, 2350–2351.

189. Gambarotti, C., Punta, C., Recupero, F., Caronna, T., Palmisano, L. (2010). TiO_2 in organic photosynthesis: Sunlight induced functionalization of heterocyclic bases. *Current Organic Chemistry*, **14**, 1153–1169.

190. Ravelli, D., Montanaro, S., Zema, M., Fagnoni, M., Albini, A. (2011). A tin-free radical photocatalyzed addition to vinyl sulfones. *Advanced Synthesis and Catalysis*, **353**, 3295–3300.

191. Scanlan, E. M., Walton, J. C. (2002). Preparation of oxime oxalate amides and their use in free-radical mediated syntheses of lactams. *Chemical Communications*, 2086–2087.

3

PHOTOGENERATION OF HETEROATOM-CENTERED RADICALS

3.1 INTRODUCTION

Heteroatom-centered radicals ($Y^\bullet$) are useful intermediates in organic synthesis and, in analogy to carbon-centered radicals, can be generated by a photohomolytic cleavage [1–7]. All of the radicals $Y^\bullet$ deriving from elements belonging to the IVA–VIIA are accessible by homolytic scission of a bond between two heteroatoms (Y–W) (Scheme 3.1, path *a*), as it happens in Barton esters (N–O) [8, 9], in *N*-haloamines or *N*-haloamides (N–X), in disulfides (S–S), or in dihalogens (X–X). Cleavage of a Y–H bond (path *b*) through the action of a suitable photoinitiator (PI, path *b*') is all important for P–H (in disubstituted phosphines) and S–H (in thiols) bonds.

The synthetic applications of heteroatom centered radicals are described in the following, excluding polymerization reactions [1]. Further note that in some cases radical $Y^\bullet$ promotes a process but it is not incorporated in the end product, as happens in the radical oxidation initiated by a halogen radical. This type of reaction will be mentioned briefly with some examples.

3.2 RADICALS FROM IVA GROUP ELEMENTS

3.2.1 Silicon-Based Radicals

Organosilyl radicals are easily obtained directly or indirectly by photochemical means and play various roles in organic chemistry, although their exploitment is

Photochemically-Generated Intermediates in Synthesis, First Edition.
Angelo Albini and Maurizio Fagnoni.

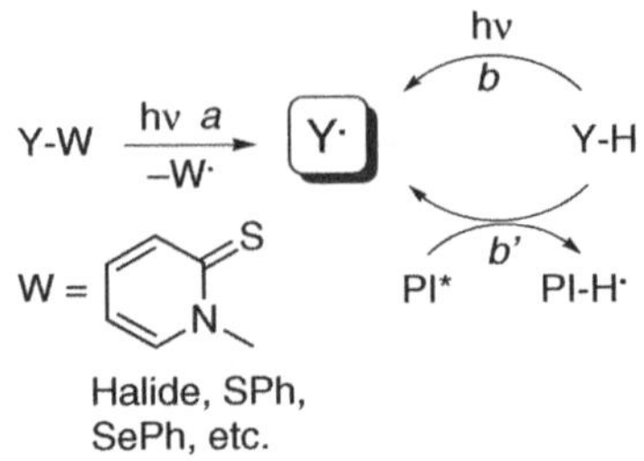

Scheme 3.1.

limited at present. In radical chain reactions they are used as chain carriers in the place of the usual stannyl radicals, with peculiar characteristics as reported in Chapter 2 [10]. In a few cases, $(TMS)_3Si$ radical-mediated alkylations have been photoinitiated by UV light [10]. In addition, silyl radicals have recently gained attention for photoinitiated polymerization reactions [11–13], because these make the process less liable to oxygen quenching.

Silicon-centered radicals have been generated photochemically for kinetic and spectroscopic measurements [14, 15]. Convenient methods involve hydrogen abstraction from a trialkylsilane R_3SiH either by another photogenerated radical (e.g., a *tert*-butoxyl radical) [11–15] or by an excited state, such as a ketone triplet [16, 17] or excited mercury atoms [14, 15, 18]. The direct photoejection of an electron from a silyl anion [14, 15] and the photolysis of a Si−Si bond (e.g., in $Me_3Si{-}SiMe_3$) [19] were likewise reported as alternative generation processes.

In a recent application, the hydrosilylation of both electron-poor and electron-rich olefins was reported. The silyl radicals were generated by photoinduced homolytic cleavage of the Si−B bond in bis(diisopropylamino)organosilylboranes (e.g., **1**) [20]. Thus, a $PhMe_2Si$ radical obtained from compound **1** added both to 1-octene and to methyl methacrylate and gave silylated products **2** and **3** in 56% and 46% yield, respectively (Scheme 3.2). It is noteworthy that no compound arising from the addition of the organoboryl radical to olefins has been detected, probably due to the steric hindrance of this species [20].

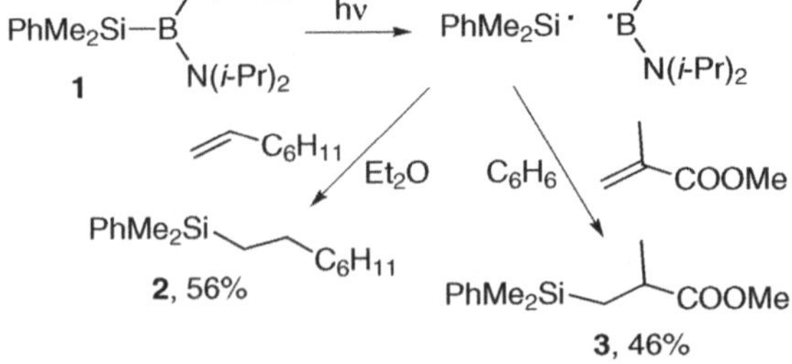

Scheme 3.2.

In addition, the silyl radicals formed by Si–Si bond cleavage in 1,1,2,2-tetraphenyl-1,2-di-*tert*-butyl-1,2-disilane were used for the functionalization of single-walled carbon nanotubes [21].

3.2.2 Germanium- and Tin-Based Radicals

In analogy to the case of silanes the photolability of the C–Ge bond can be exploited for initiating free radical polymerizations. It is noteworthy that bis(germyl)ketones (e.g., $Ph_3GeCOGePh_3$) were reported to undergo α-cleavage upon visible light irradiation [22].

The radical stannylation of alkenes is well known, but to our knowledge there are no examples where the photochemical version is advantageous. The importance of tin derivatives as chain carriers is difficult to underrate, and the effect of irradiation is briefly mentioned in Chapter 2.

3.3 RADICALS FROM VA GROUP ELEMENTS

3.3.1 Nitrogen-Based Radicals

Nitrogen-centered radicals—in particular, aminyl, amidyl, and iminyl radicals (see Scheme 3.3)—are important intermediates in organic synthesis and have been used to some extent as key steps in complex syntheses, in most cases involving an intramolecular radical addition [23–27].

Noteworthy properties are the high reactivity of aminyl radicals toward hydrogen abstraction—which indeed competes with any addition process—and the exclusive reaction at the nitrogen atom of electrophilic amidyl radicals, despite their nature of delocalized π-allyl radicals. Photochemical generation of these radicals is a mild process and mainly involves the photocleavage of a N–X bond.

3.3.1.1 Aminyl Radicals Amination via aminyl radicals has found limited application in organic synthesis despite its potential. These radicals are photochemically generated either by cleavage of the N–Cl bond in *N*-chloramines [27] or by deprotonation of an aminium radical cation [28]. Note, however, that in some instances the aminyl radicals were protonated under acidic conditions and that the resulting (more electrophilic) radical cations added faster to electron-rich double

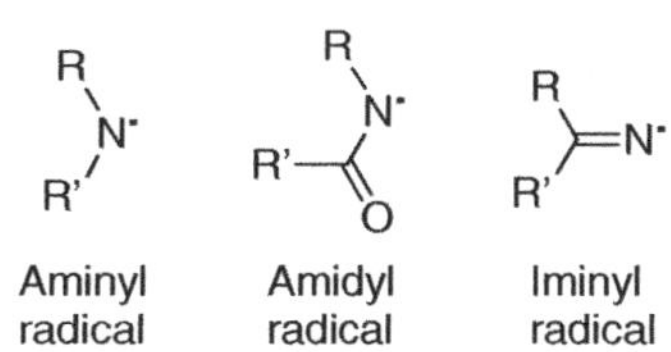

Scheme 3.3.

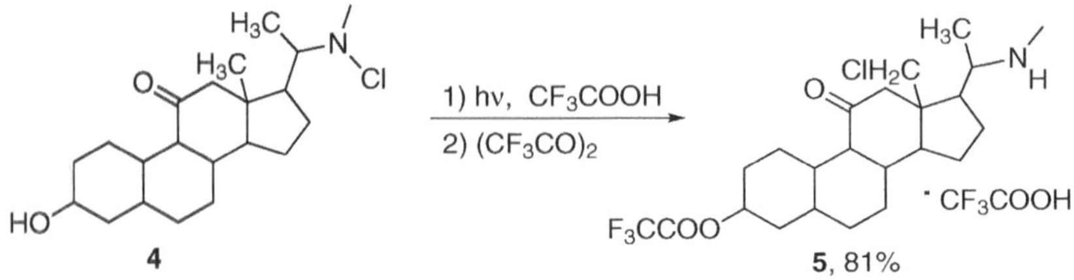

Scheme 3.4.

bonds. Recently, aminyl radicals were also invoked as intermediates in single-crystal to single-crystal photoinduced Orton rearrangement of *N*-chloro acetanilides [29].

Extensive kinetic studies on aminyl radicals explained the results and demonstrated that hydrogen abstraction is faster than addition to electrophilic olefins [28, 30–33]. A typical example of a synthetic application is illustrated in Scheme 3.4 and involves intramolecular hydrogen abstraction and recombination with the chlorine atom. Various steroid hormones precursors have been synthesized by irradiation in trifluoracetic acid of *N*-chloro derivatives (e.g., compound **4**). Subsequent treatment by trifluoroacetic anhydride gave product **5** that was isolated in a high yield and further elaborated [34].

3.3.1.2 Amidyl Radical Amidyl radicals are useful intermediates for building *N*-heterocycles by addition through the nitrogen atom to a double bond. The photogeneration of these radicals can be accomplished by the photohomolytic cleavage of a N–X bond (X = halogen, NO [35], oxygen in Barton esters or related compounds). *N*-Haloamides were the most often used radical precursors [36]. The reaction has been known since the seventies and, as an example, photolysis of *N*-chloro-*N*-[(3-cyclohexen-l-yl)methyl]-acetamide (**6**) gave bicyclic product **7** in 55% yield (Scheme 3.5). The cyclization was a good example of the selective course in the 5-*exo* mode of radical additions, with no competitive formation of a six-membered ring [37].

A related system was tested for the generation of the *N*-methyl-3-cyclohexene-1-carboxamido radical. A mixture of different lactams resulted from the homolytic photocleavage of either a N–NO or a N–Cl bond [38]. The cyclization of *N*-chloro-*N*-methyl-3-cyclohexenecarboxamide could be induced either photochemically or thermally upon initiation by decomposition of a peroxide (e.g., benzoyl peroxide) by

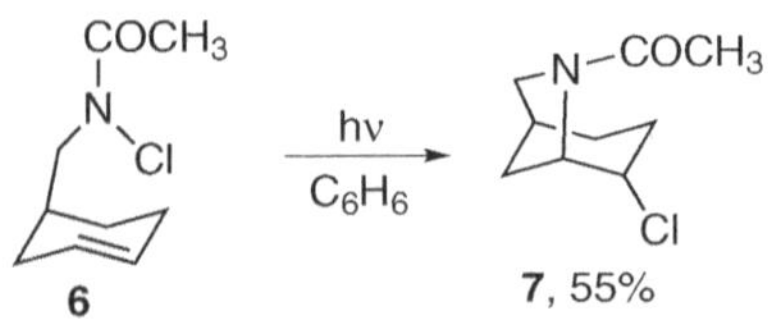

Scheme 3.5.

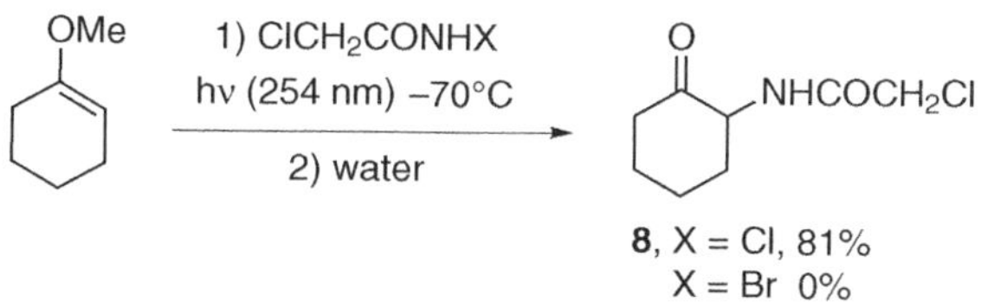

Scheme 3.6.

heating at 80°C. The photochemical approach gave comparable, and in some cases better, yields than the corresponding thermal reactions. *N*-haloamides were photochemically added onto electron-rich olefins such as vinyl ethers as depicted in Scheme 3.6. The reaction had to be carried out at a low temperature (−70°C) since thermal electrophilic halogenation of the alkene would otherwise compete. Under photochemical conditions, however, the addition was successful and α-functionalized ketone (**8**) was isolated in a very good yield upon hydrolysis. It is noteworthy that no concurrent photocleavage of the C−Cl bond was observed. On the contrary, when using a *N*-brominated amide, the thermal electrophilic bromination of the vinyl ether was much faster than the photocleavage of the *N*-Br bond even at such a low temperature and ketone **8** was not formed [39].

The photochemical amination is quite general in scope and has been extended to several vinyl ethers including oxygen-containing heterocycles such as 3,4-dihydro-2*H*-pyran [40]. The photohomolysis of *N*-bromoamides has been investigated to some degree. The nitrogen-centered radicals thus formed effectively and abstracted a hydrogen either intermolecularly from the reaction medium or intramolecularly from geometrical accessible C−H bonds. As an example, the fate of the radical obtained by irradiation of amide **9** depended on the solvent used. In cyclohexane, intermolecular hydrogen abstraction took place and the parent amide **10** was isolated in a high yield. On the contrary, when a poor H-atom-donating solvent such as benzene was used as the reaction medium, a geometrically accessible methyl hydrogen was abstracted and bromoamide **11** was exclusively formed after recombination with the bromine atom (Scheme 3.7) [41].

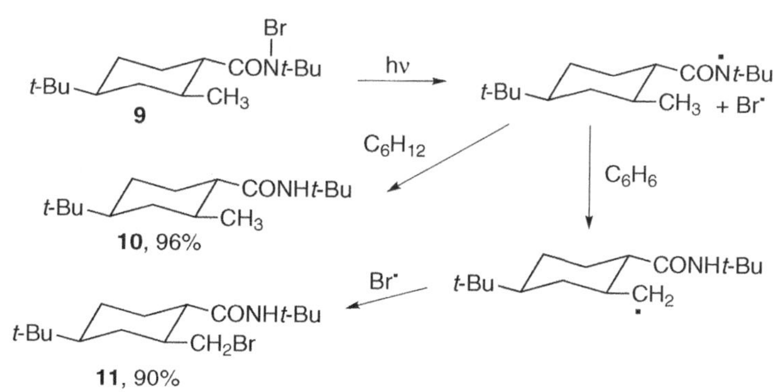

Scheme 3.7.

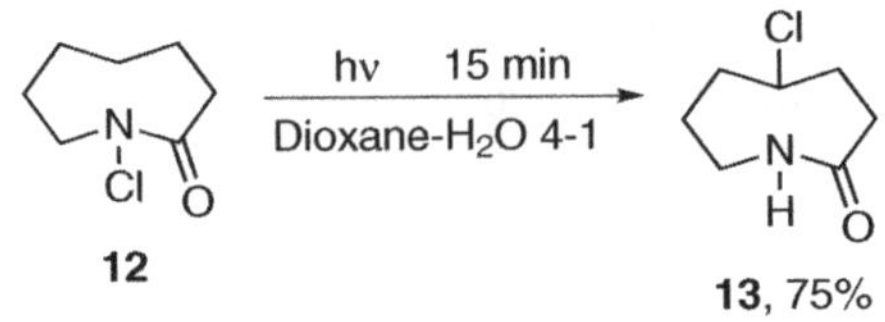

Scheme 3.8.

Configuration-determined intramolecular hydrogen abstraction has likewise been observed in the irradiation of *N*-chlorolactams (Scheme 3.8). Thus, compound **12** was efficiently converted to isomer **13** upon a mere 15-min irradiation in aqueous dioxane solution. Also in this case, the process is based on the following sequence: *N*-Cl photocleavage, intramolecular hydrogen abstraction, and recombination with the chlorine atom [42].

In recent reports, amidyl radicals were preferably formed starting from *N*-iodo amides prepared in situ from the corresponding halide-free amides in the presence of an oxidant (e.g., lead tetracetate) and iodine. In such a way, an amidyl radical mediated cyclization onto a vinyl halide moiety was accomplished [43, 44]. Interestingly, the cyclization step was controlled by the position of the heteroatom on the double bond as indicated in Scheme 3.9. Thus, photolysis of compound **14** where the chlorine atom was bonded at the terminal position of the alkene led to lactam **16** through a radical 6-*exo* cyclization (Scheme 3.9). The analogous compound having the chlorine atom at the adjacent position (**15**) underwent a different, but likewise regioselective, radical cyclization, in this case forming a seven-membered ring (compound **17**). The difference was mainly ascribed to the lone-pair–lone-pair repulsion between the intermediate N radical and the vinylic halogen atom [43, 44].

A phosphoramidate radical was involved in the preparation of *N*-dioxyphosphinyl 1,4-epimine derivatives via photolysis of dialkyl phosphoramidates in the presence of iodine and lead tetracetate [45]. Carboxyamidyl radicals were also used in the synthesis of bicyclic spirolactams or spirolactones starting from carbohydrate amides. In this case the nitrogen-centered radical initiated a tandem 1,5-hydrogen atom transfer/radical oxidation/nucleophilic cyclization where the *C*-radical intermediate was oxidized to an oxocarbenium ion [46].

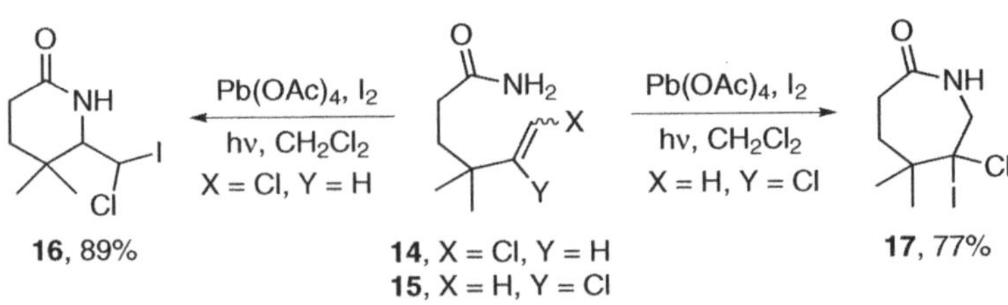

Scheme 3.9.

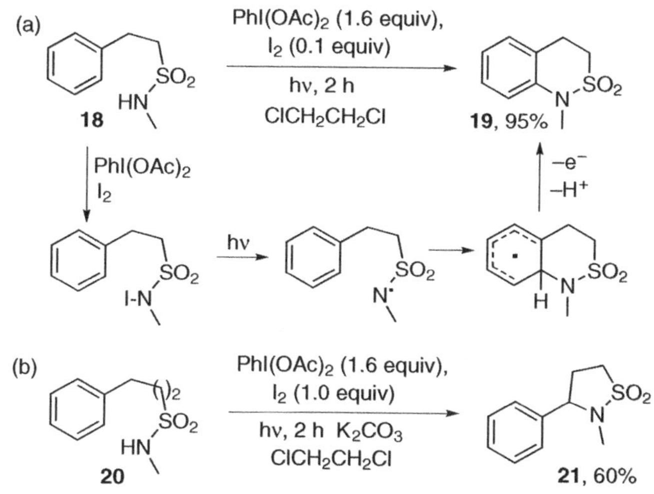

Scheme 3.10.

Recently, a route for the radical amidation of aromatic compounds has been developed [47–49]. Substituted sulfonamides were most often investigated as starting materials in view of the remarkable biological activity of the products formed. Thus, the preparation of 3,4-dihydro-2,1-benzothiazine 2,2-dioxides (e.g., **19**) was effected by treatment of sulfonamide **18** with (diacetoxyiodo)benzene and iodine followed by UV light irradiation. An *N*-iodosulfonamide was postulated as the intermediate and the cyclization allowed for the formation of a six-membered ring as shown in Scheme 3.10a [48]. The amount of iodine used (0.1 equiv) was determining for the course of the reaction, since a larger amount (1 equiv) caused a further iodination of the primarily formed benzothiazine. On the contrary, in the photolysis of the homologue **20** under similar conditions (1 equiv iodine and potassium carbonate added), *N*-methyl-3-phenyl-1,3-propanesultam **21** was prepared in a discrete yield by a Hofmann–Löffler–Freytag-type reaction (Scheme 3.10b) [48].

In a related study, it was found that the photolysis of (*E*)- and (*Z*)-6-iodo-2-methylhex-5-enamide in the presence of iodine and lead tetracetate gave 6-(diiodomethyl)-3-methylpiperidin-2-one in 83% yield as an equimolar mixture of diastereoisomers via a 6-*exo* cyclization reaction [49]. An alternative generation of amidyl radicals involved the use of pyridine-2-thioneoxycarbonyl (PTOC) imidate ester (e.g., **22**, Scheme 3.11a) [50]. Thus, visible-light-initiated reaction of these precursors afforded an amidyl radical **23**$^\bullet$ via N−O cleavage, and this cyclized in a 5-*exo* fashion to give radical **24**$^\bullet$. The last intermediate was then trapped by PhSeSePh and finally gave phenylselenyl-substituted *N*-acylpyrrolidine **25** in 61% yield. The PhSe$^\bullet$ radical formed in the last step was able to propagate the chain reaction by addition onto the starting thiopyridone **22** (Scheme 3.11a) [51, 52]. A related system was devised for the preparation of substituted *N*-acylpyrrolidines by using *N*-acyl PTOC carbamates as starting materials. Thus, compound **26** initially gave an amidyl

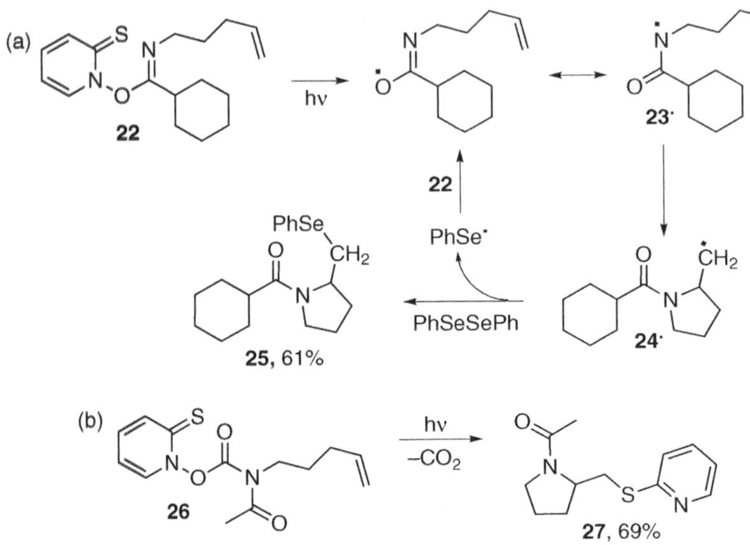

Scheme 3.11.

radical. Carbon dioxide loss followed by a *5-exo* radical cyclization and coupling with the pyridylthio radical gave amide **27** (Scheme 3.11b).

3.3.1.3 Iminyl Radical Iminyl radicals are further useful intermediates for the formation of N–C or other N–X bonds as well as for building *N*-heterocycles. Substituted oximes were the elective photochemical precursors of these radicals. An illustrative example is shown in Scheme 3.12, where a radical cascade sequence was applied to the synthesis of a substituted pyrroline starting from *O*-carboxymethyl oximes (e.g., **28**). An iminyl radical was formed by irradiation of compound **28** upon

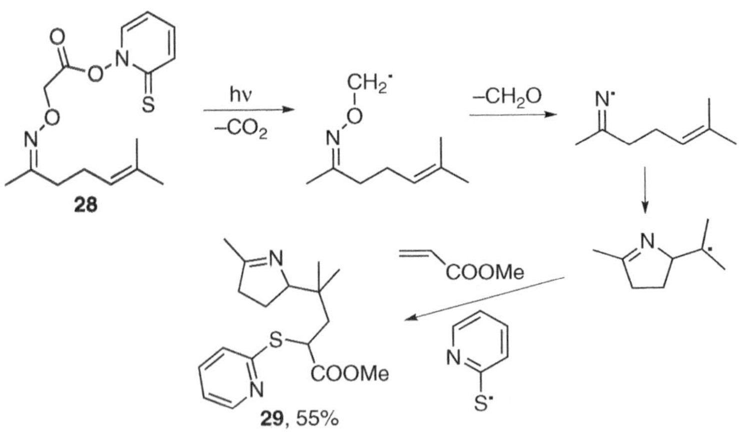

Scheme 3.12.

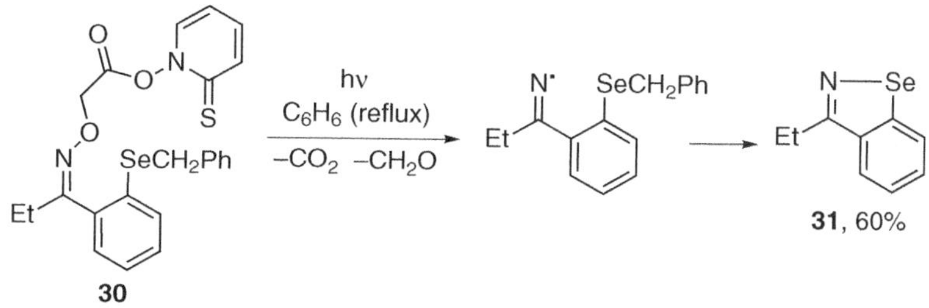

Scheme 3.13.

sequential carbon dioxide and formaldehyde elimination. This intermediate cyclized onto a double bond leading to a carbon-centered radical that readily added to methyl acrylate to give pyrroline **29** in a discrete yield (55%) [53, 54].

Starting from a related *O*-carboxymethyl oxime **30**, the intramolecular cyclization of the iminyl radical gave access to 1,2-benzoselenazole **31** in 60% yield as a result of the intramolecular free-radical homolytic substitution at the selenium atom that caused the loss of a benzyl radical (Scheme 3.13) [55].

The reactivity of iminyl radicals further made possible such an appealing process as the direct formation of an Ar–N bond. Thus, photolysis of *o*-phenylbenzophenone *O*-benzoyl oxime (**32**) in benzene resulted in the formation of phenanthridine **33** in 74% yield along with some byproducts. Noteworthy, the conceivable intermolecular addition of the iminyl radical to the solvent did not take place, and biphenyl **34** was not detected in the end mixture (Scheme 3.14) [56].

The photolysis of related carbaldehyde oxime carbonate derivatives afforded either again phenanthridine derivatives or alternatively substituted quinolines and isoquinolines [57]. A similar approach has recently been used for the construction of six-membered heterocyclic rings and involved the intermolecular addition of the initially formed iminyl radical onto a triple bond, followed by the intramolecular addition of the resulting vinyl radical onto the aromatic ring

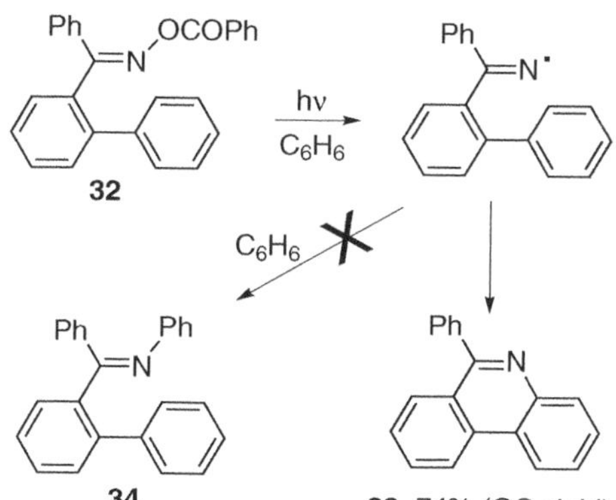

Scheme 3.14.

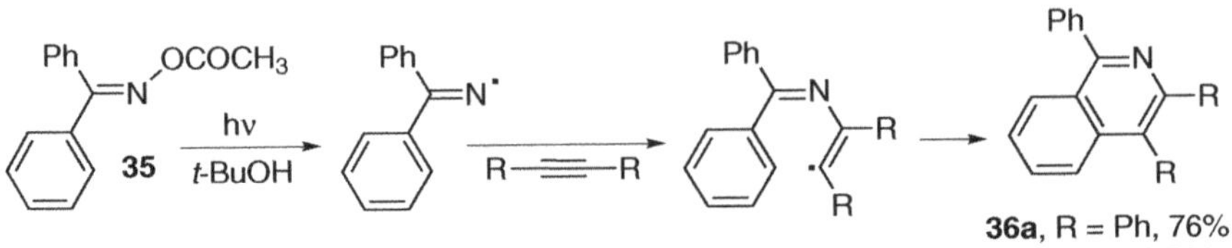

36a, R = Ph, 76%
b, R = COOMe, 74%

Scheme 3.15.

(Scheme 3.15) [58]. The reaction was successful with phenylalkynes and with methyl acetylenedicarboxylate. The intramolecular cyclization process of an iminyl radical was applied for the preparation of the alkaloid trisphaeridine as well as other natural compounds [59].

Recently, dioxime oxalates have been introduced as alternative iminyl radicals precursors. Contrary to the previous cases, a triplet sensitizer, namely 4-methoxyacetophenone, was added to the solution and improved the course of the process. This led to the preparation of 3,4-dihydro-2*H*-pyrroles, phenanthridines and isoquinolines [60, 61]. Alternatively, substituted *N*-bromo imidates were excellent precursors for the photoinduced generation of alkoxyiminyl radicals. These intermediates underwent either a 5-*exo* or a 6-*exo* cyclization onto the double bond in the *O*-alkyl side chains, and the formation of 4,5-dihydrooxazoles in high yields resulted [62].

A particular case is that of the photogeneration of a persistent radical such as NO [63]. This compound behaves as a radical trap and couples with alkyl radicals to form a diamagnetic product. The photonitrosation of cyclohexane to provide cyclohexanone oxime is the key step of the Toray process for the preparation of ε-caprolactame, based on the photocleavage of NOCl, hydrogen abstraction from cyclohexane by chlorine atoms, and final combination between NO and the cyclohexyl radical [64]. It has been shown that the process could be conveniently carried out under solar irradiation [65].

3.3.2 Phosporus-Based Radicals

P-centered radicals found limited application in the formation of C—P bonds for the synthesis of organophosphorus compounds, largely used as ligands, despite the seemingly straightforwardness of the process [66]. In the examples reported below, the P—H bond of a secondary phosphine was cleaved [67] and the phosphorus-centered free radical added onto an electron-rich olefin in the anti-Markovnikov sense. Both dialkyl and diaryl phosphines were used in the reaction. Scheme 3.16 illustrates the hydrophosphination of substituted vinyl sulfides and vinyl selenides, which gave the desired tertiary phosphines **37** (actually polydentate ligands designed for preparing metal complexes as catalysts) in nearly quantitative yields [68]. An additional advantage of this preparation is that it took place under solvent-free conditions by mere irradiation of an equimolar mixture of the two reagents.

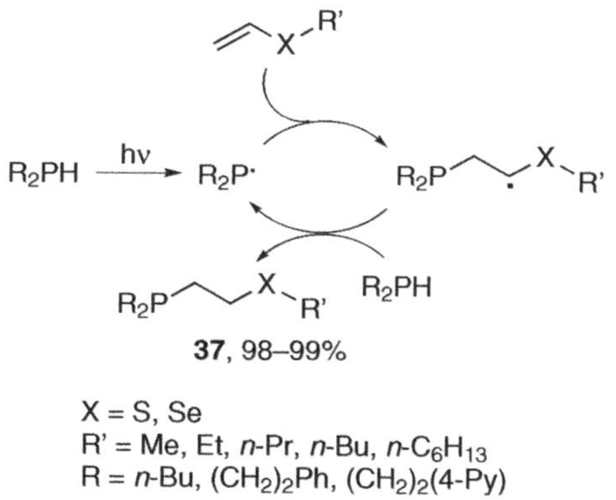

Scheme 3.16.

Another example involves the hydrophosphanation of 1-vinylimidazole by the *P*-centered radical from phosphine **38** to give the tertiary derivative **39** in 96% yield after 3 h of irradiation (Scheme 3.17, path *a*) [69]. This synthesis was also carried out thermally, having recourse to a radical initiator (AIBN, path *b*). The same yield was obtained, although a higher temperature and a longer reaction time were required [69].

The versatility of the method allowed for the polyaddition of phosphines onto substrates having more than one double bond available for functionalization. An exemplificative case is the synthesis of *octakis*-(dimethylphosphanoethyl)-octasilsesquioxane **41** where eight C–P bonds were formed in a single operation (Scheme 3.18) [70]. Compound **40** bearing eight vinyl functions was dissolved in pentane and mixed with a slight excess of $HPMe_2$. The reaction took 3 days to complete upon irradiation with a UV lamp, but compound **41** was easily isolated from the end mixture as a white solid (100% yield) after removal in vacuo of the unreacted phosphine [70].

Diaryl phosphines have been likewise used for photoaddition reactions onto double bonds as in the synthesis of both triazacyclononane derivatives having pendant phosphine arms [71] and of 2,7-bis(3-diphenylphosphanylpropoxy)naphthalene from bisallyloxynaphthalene [72]. Further examples are summarized in Scheme 3.19. First, new oligosilsesquioxane ligands were prepared starting from

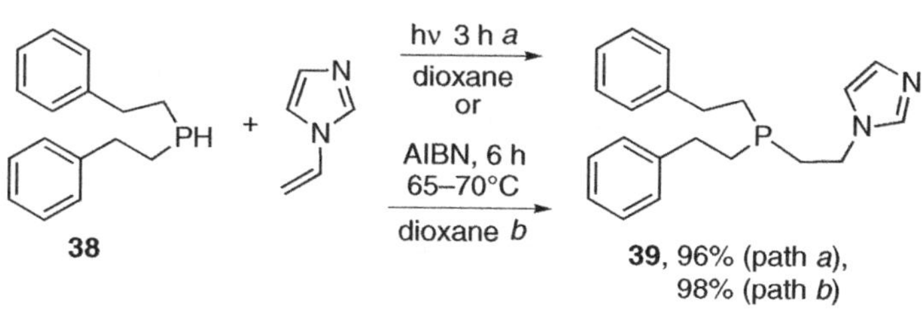

Scheme 3.17.

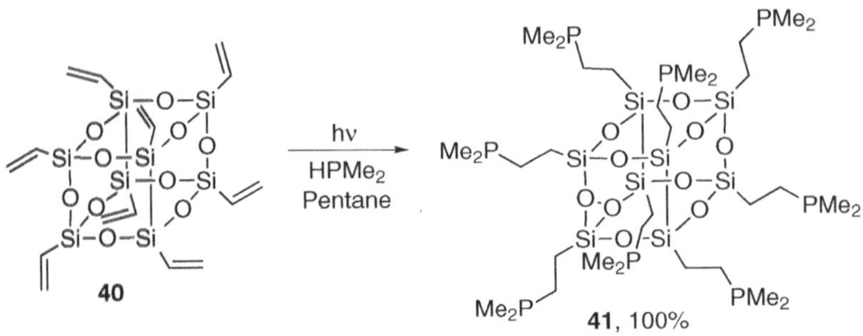

Scheme 3.18.

trichlorosilane **42** that in turn was prepared by photoaddition of diphenylphosphine onto vinyltrichlorosilane (Scheme 3.19a) [73]. A related approach (carried out under solvent-free conditions) was exploited for synthesizing a library of hemi-labile alkyldiphenyl phosphine ligands **44** for transition-metal-catalyzed applications (Scheme 3.19b) [74].

The introduction of a silyl ether moiety in phosphines yields ligands that are important for the preparation of Rh-based catalysts. A compound of such a structure was prepared by photolysis of a mixture of vinyltriethoxysilane and diphenylphosphine [75].

The hydroselenophosphorylation of olefins (Scheme 3.20) is based on the P–H bond homolytic cleavage in secondary phosphine selenides (e.g., compound **45**). The resulting radical (**46**$^{\cdot}$) added onto electron-rich alkenes in an *anti*-Markovnikov way and formed tertiary phosphine selenides **47** regiospecifically in up to 96% yield, depending on the stability of the radical adduct formed [76].

The lability of the P–H bond was sufficient also when selenium was replaced by sulfur in the reacting moiety. Thus, (2,2-diphenylethyl)diphenylphosphine sulfide was obtained in 55% yield from the reaction between the corresponding organophosphorus compound and 1,1-diphenylethylene [77].

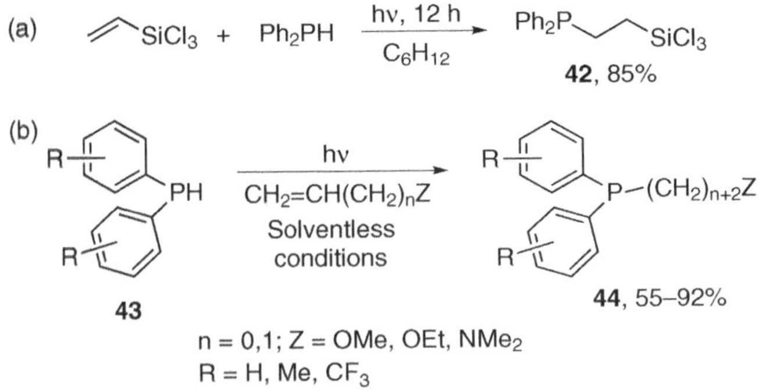

Scheme 3.19.

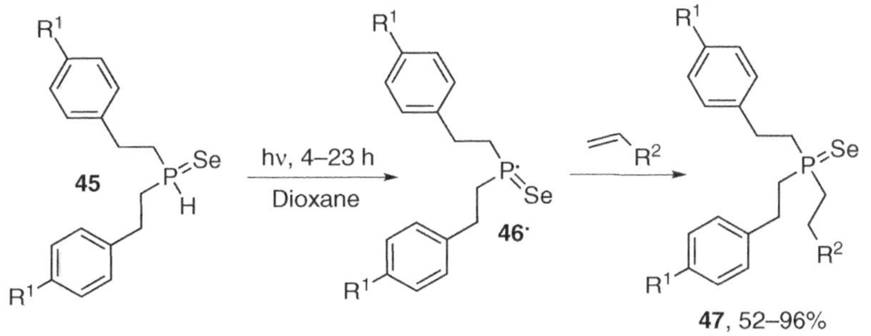

R_1 = H, *t*-Bu; R_2 = C_7H_{15}, Ph, Naph, OBu, SMe

Scheme 3.20.

3.4 RADICALS FROM VIA GROUP ELEMENTS

3.4.1 Oxygen-Based Radicals

The most important radical in this class is certainly the hydroxyl radical, one of to the so-called reactive oxygen species (ROS). These have a significant role in biology as well as in the chemistry of the atmosphere since the diradicalic nature of dioxygen molecule causes trapping of radicals whenever formed. The combination of carbon-centered radicals with oxygen gives peroxy radicals, thus initiating a variety of processes from fats peroxidation to combustion. The same mechanism is involved in the lipid oxidation reaction to form hydroperoxydes. Oxygen-based radicals are electrophilic species known to be effective in homolytic hydrogen abstraction reactions [4]. A radical pertaining to this class and often used in synthesis is that obtained from *N*-hydroxyphthalimide, usually designed by the acronym PINO (phtalimide-*N*-oxyl radical). Actually this radical is effective in the homolytic cleavage of labile C—H bonds (e.g., allylic, benzylic, α-oxy, etc.) and various organic transformations initiated by the resulting C-centered radical have been reported [78].

3.4.1.1 From the Barton Reaction A long-known method for the generation of an oxygen-centered radical is the Barton reaction, namely the photohomolysis of an O—N bond in an organic nitrite [5, 6]. The reaction has synthetic value, particularly when the reactive group is incorporated in rigid skeletons, such as the cyclopentaperhydrophenanthrene structure and related derivates. In such a case, the O radical abstracts a hydrogen atom intramolecularly and the C-centered radical thus formed adds the NO molecule formed in the initial cleavage to give an oxime function after tautomerization. This approach was employed about 50 years ago for the preparation of lanostane derivatives [79]. Recently, a related process was considered again for the synthesis of the steroidal substrate **49** (Scheme 3.21), a key intermediate in the synthesis of myriceric acid A (**50**), an endothelin receptor antagonist [80, 81]. The process was performed by irradiating with black light lamps

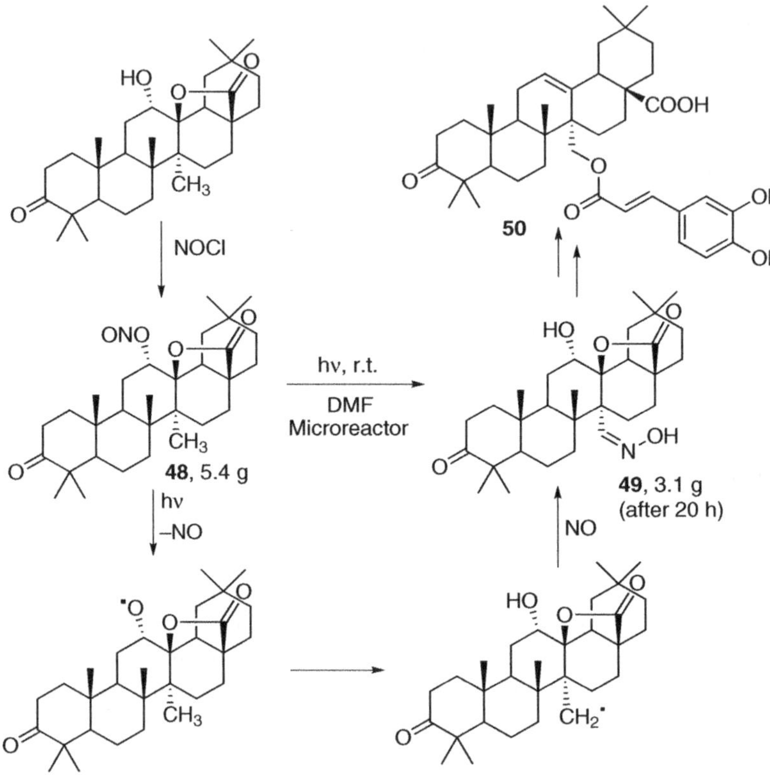

Scheme 3.21.

a DMF solution of **48** in two microreactors (flow rate 15 mL/h, residence time 32 min). The oxime **49** was isolated in 60% yield after purification by column chromatography (overall process time 20 h, Scheme 3.21) [81].

3.4.1.2 From O–I Bond Fragmentation A second approach involves the photohomolysis of an O–I bond in hypoiodides, in turn smoothly formed from alcohols by treatment with either HgO/I_2 or $PhI(OAc)_2/I_2$. Similarly to the case above (see Scheme 3.21), intramolecular hydrogen abstraction occurred when treating 5α-cholestane-3β,6β-diol 3-acetate in the presence of a hypervalent iodine derivative/iodine mixture under irradiation with a 100-W tungsten-filament lamp [82, 83].

This reaction was likewise applied to the synthesis of macrocyclic ketone-based natural products, such as (±)-muscone **54**, the main contributor to the odor of musk (Scheme 3.22) [84]. Thus, when alcohol **51** was iodinated and subjected to photolysis, alkoxy radical **52**$^{\bullet}$ was formed. The ensuing cleavage opened the cyclopentane moiety thus leading to the 15-membered iodoketone **53** that was isolated in 59% yield and was converted into the desired (±)-muscone by deiodination (Scheme 3.22) [84].

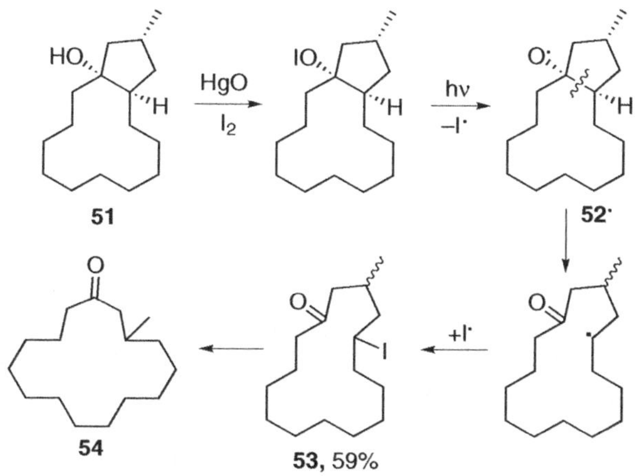

Scheme 3.22.

A similar strategy was followed in the preparation of (−)-lardolure (**57**, the aggregation pheromone of the acarid mite *Lardoglyphus konoi*), where the iodo-ketone intermediate **56** was photochemically obtained in 88% yield starting from the hemiacetal **55**, Scheme 3.23) [85].

The above reaction scheme via oxygen-centered radicals has been applied to various substituted alcohols—in particular, terpenic alcohols [86]—as well as to oxabicyclic hemiacetals [87]. In the latter case, the process gave access to medium-size spiro-lactones [87]. A further synthetic variation is based on the tandem β-fragmentation/cycloperoxyiodination reaction [88]. Thus, the visible light-induced photohomolysis of the O–I bond was carried out in the presence of molecular oxygen. Under these conditions, the β-fragmentation of the oxygen radical was followed by peroxidation of the carbon radical formed and the subsequent intramolecular addition onto a double bond gave cyclic ketones bearing a 1,2-dioxolane group [88]. In another example, a cascade radical fragmentation–transannulation–cyclization reaction occurred after the initial β-fragmentation process [89].

Allyl alcohols have been shown to be suitable starting substrates for the preparation of iodoepoxides (**59a–c**) as illustrated in Scheme 3.24 [90]. The process

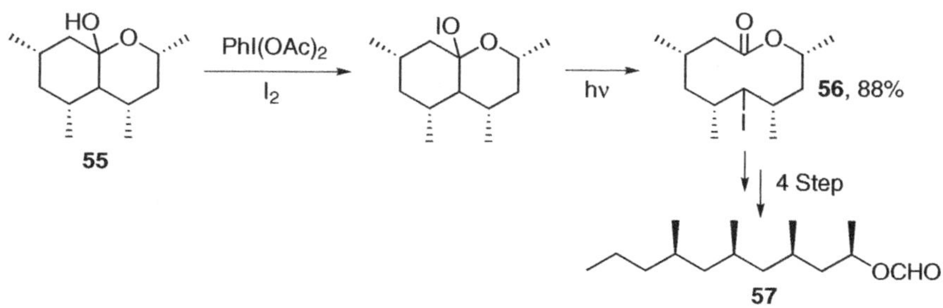

Scheme 3.23.

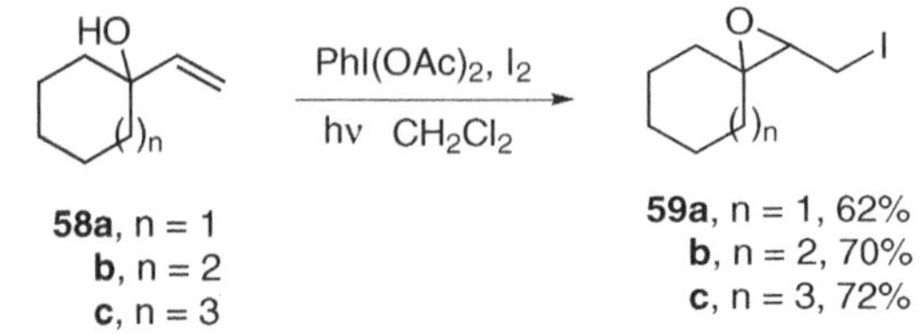

Scheme 3.24.

relies upon the radical addition by oxygen onto the allylic double bond followed by iodine atom addition. Interestingly, the overall reaction yield was not affected by the size of the ring in compounds **58a–c**.

3.4.1.3 From Barton Ester As in many other applications, Barton esters were suitable precursors for the photogeneration of O-centered radicals [8, 9]. A representative example is shown in Scheme 3.25, where the photocleavage of the N–O bond in ester **60** generated an electrophilic alkoxy radical prone to add intramolecularly to the C=C double bond. In the presence of perfluorobutyl iodide, iodine transfer to the radical intermediate occurred concomitantly with the generation of the radical chain carrier $C_4F_9^{\cdot}$. Thus, the stereoselective alkoxy radical ring closure led to the preparation of halogenated tetrahydrofurans (e.g., **61**), a class of natural compounds recently found as secondary metabolites in the marine environment [91].

N-Hydroxy-2-thiopyridone benzoate was employed as precursor of the benzoyloxy radical [92]. The latter electrophilic intermediate was trapped by several electron-rich olefins (mainly vinyl ethers) to give the corresponding benzoate esters in modest to good yield after 15 min of irradiation [92]. Alkoxycarbonyloxy radicals generated from Barton esters were likewise employed satisfactorily in synthesis [8, 9]. Thus, photolysis of ester **62** (prepared in situ) caused the liberation of an allyloxycarbonyloxy radical

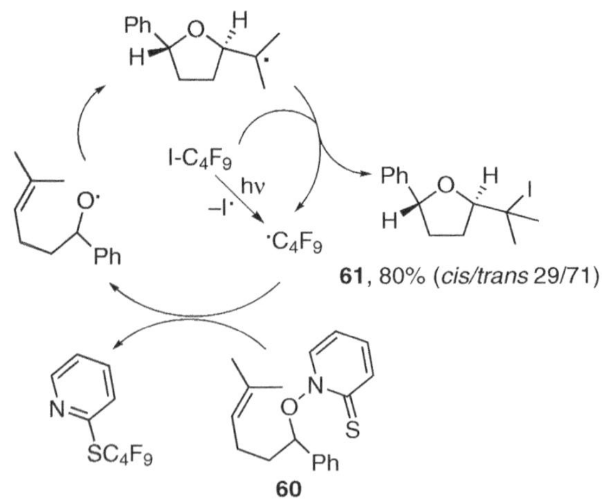

Scheme 3.25.

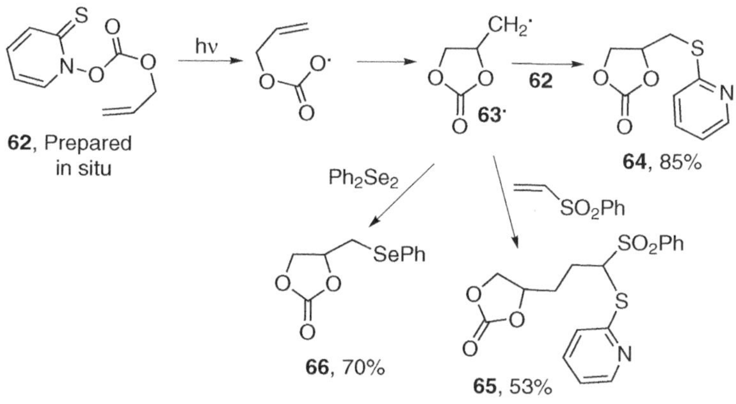

Scheme 3.26.

that readily cyclized in a 5-*exo* fashion to give radical **63˙** (Scheme 3.26). This intermediate was trapped by various substrates to form C–C or C–heteroatom bonds. In the absence of any purposely added radical trap, radical **63˙** added to the starting carbonate and thiopyridine **64** was the sole product (85% yield). On the other hand, in the presence of either phenyl vinyl sulfone or Ph_2Se_2, the sulfone **65** or the phenylseleno derivative **66** were respectively obtained (Scheme 3.26) [93].

Barton esters were likewise used for the generation of acyloxy radicals. These intermediates added to a triple bond and initiated a self-terminating, oxidative radical cyclization sequence, as illustrated in Scheme 3.27 for the case of compound **67** [94]. A modest yield of tetrasubstituted tetrahydrofuran **68** (a single diastereoisomer) was isolated.

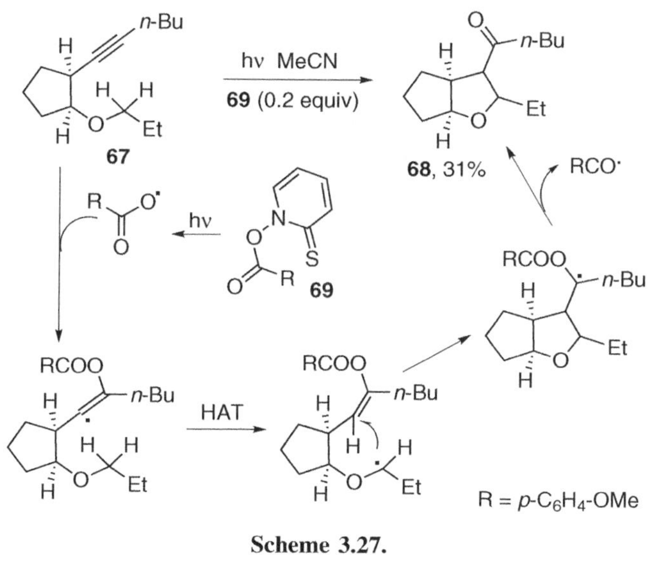

Scheme 3.27.

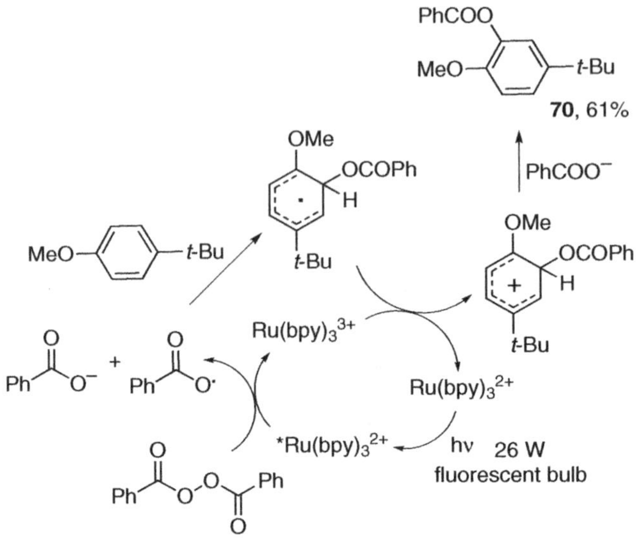

Scheme 3.28.

Recently, a mono-benzoyloxylation of electron-rich aromatics has been accomplished by the Ru(II)-based photocatalyzed generation of benzoyloxy radicals from benzoyl peroxide as depicted in Scheme 3.28 for the case of *p*-*t*-butylanisole. The activation of the Ar–H bond was thus achieved under visible light irradiation without the assistance of any directing group on the aromatic ring (often required in related transition-metal catalyzed reactions) [95].

3.4.2 Sulfur-Based Radicals

Sulfur-based radicals have long been known as convenient intermediates for introducing a sulfur-based functionality, as in the case of the thiol-ene coupling recently included as one of the few "click chemistry" processes [67, 96, 97]. This reaction involves the photoaddition of a thiol (RS–H) across a double bond and relies on the photochemical lability of the S–H bond that is photocleaved either directly or with the help of an initiator (R′–X, Scheme 3.29). The electrophilic sulfur-centered radical then attacks a nucleophilic double bond starting a radical chain reaction.

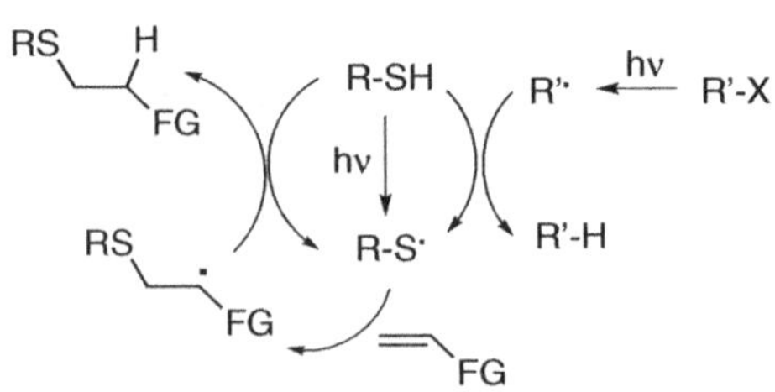

Scheme 3.29.

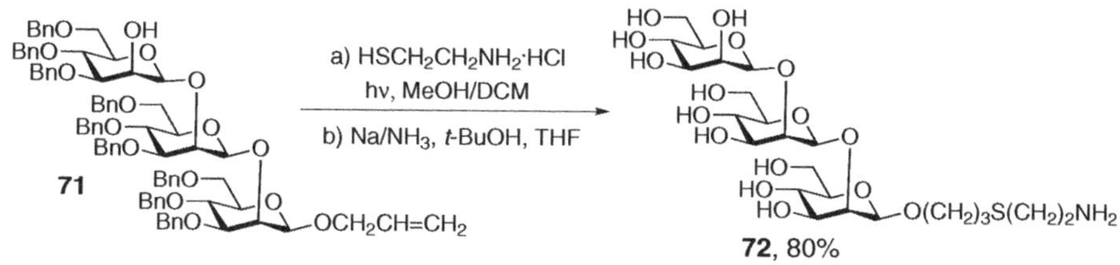

Scheme 3.30.

3.4.2.1 By Cleavage of the S–H Bond As mentioned above, thiols are convenient precursors of thiyl radicals and have been widely employed in this role for the synthesis of biological lead compounds, especially for glycoconjugations [7]. As an example, in the frame of a project aimed to the preparation of an effective vaccine against *C. albicans* infections, a multigram scale synthesis of *C. albicans* (1 → 2)-α-D-mannopyranan epitopes was recently reported. This involved the photoaddition of 2-aminoethanethiol hydrochloride to an *O*-allyl mannose polysaccharide (e.g., **71**) to give, upon debenzylation, the mannopyranoside **72** in 80% yield (Scheme 3.30) [98].

The presence of a photoinitiator allowed to carry out the reaction at longer wavelengths (up to the UV-A region). Actually, 2,2-dimethoxy-2-phenylacetophenone (**74**, Irgacure 651) is the most used initiator due to the efficient photoinduced α-cleavage. In such a way, the site-selective glycosylation of *N*- and *C*-protected cysteine derivatives as well as of cysteine-containing peptides (such as glutathione) was carried out in a high yield as shown in Scheme 3.31 [99].

The same approach was adopted in the preparation of thiodisaccharides by photoaddition of anomeric sugar thiols to sugar alkenes [100] and in the preparation of *S*-linked α-glucoconjugates [101].

If the molecule to be functionalized bore more than a single olefin moiety, as is the case for the *tris*-alkene core of 2,4,6-triallyloxy-1,3,5-triazine, the same number of thiol molecules could be added and gave access to poly(thioether) dendrimers [102]. This polyfunctionalization was exploited also for the preparation of multifunctional organosilicon thioethers starting from tetravinylsilane and various thiol derivatives [103].

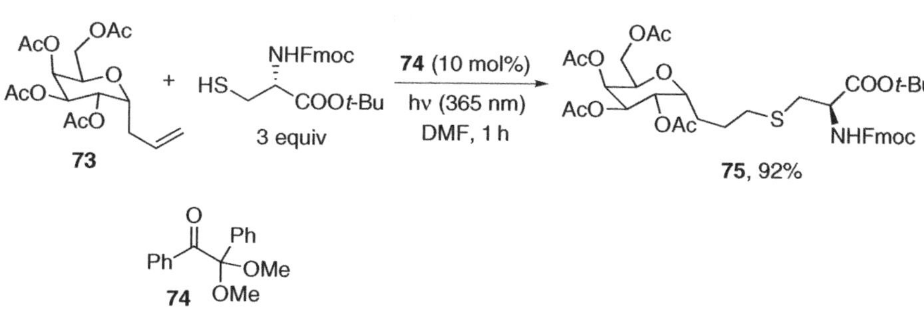

Scheme 3.31.

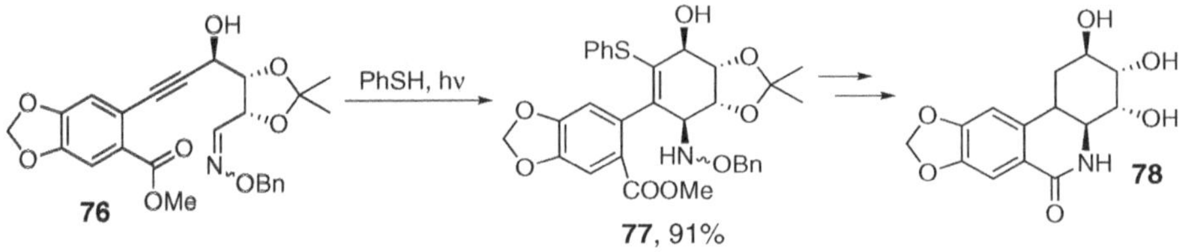

Scheme 3.32.

The addition of a photogenerated sulfur-centered radical onto a C–C triple bond (the thiol-yne reaction) [104] is likewise used in synthetic planning. The difference here is that the addition leads to a highly reactive vinyl radical that is prone to radical addition onto another double (whether C=C or C=N) or triple bond.

Accordingly, the synthesis of *ent*-lycoricidine (**78**) required the preparation of compound **77** and involved a diastereoselective thiyl radical addition/cyclization sequence as illustrated in Scheme 3.32 [105].

The versatility of the thiol-yne click reaction was evidenced in the synthesis of supramolecular structures, such as macrocycles and rotaxanes. This strategy was applied to the preparation of disulfide macrocycles [106].

Even in the thiol-yne reaction, the presence of the photoinitiator **74** was found to improve the overall photoclick process. Thus arylalkyne-tagged sugars were conveniently glicosylated by glutathione **80** to form adduct **81** in 82% yield (Scheme 3.33) [107].

In analogy with the thiol-ene case, the thiol-yne reaction could be applied to aromatic compounds bearing more than a single triple bond in the side-chains. Polyfunctional materials including dendrimers were prepared in such a case

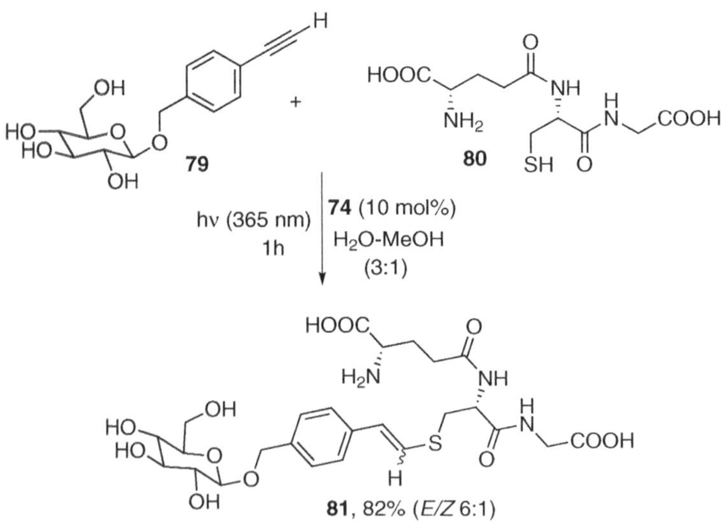

Scheme 3.33.

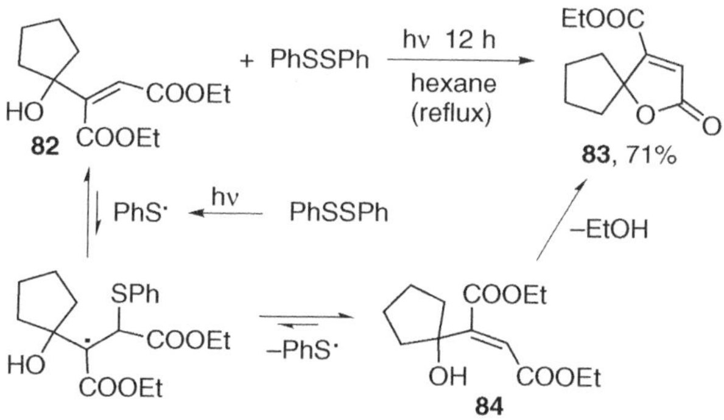

Scheme 3.34.

[108, 109]. In a related reaction, thiolacetic acid was photoadded in a high yield to butenyl aromatics [110].

3.4.2.2 By Cleavage of the S–S Bond The cleavage of S–S bonds offers an alternative entry to S-centered radicals. Since the addition of the PhS radical to olefins is reversible, in some cases the net result is a modification of the starting compound with no incorporation of the sulfur substituent, as shown in Scheme 3.34. Thus, addition of a thiyl radical onto substituted maleate **82** caused the isomerization to fumarate **84** that, under the reaction conditions, cyclized to butenolide **83** in 71% yield [111].

In other cases the sulfur radicals were incorporated. Thus, irradiation of a mixture containing C_{60} and a dendrimer disulfide in *o*-dichlorobenzene caused the formation of a fullerodendrimer, though in a low yield (16%) [112]. In order to have a successful reaction, however, a large excess of diphenyl diselenide had to be added to the reaction mixture. As an example, the four-component reaction based on the thiol-yne reaction initiated by S–S photocleavage is illustrated in Scheme 3.35 [113]. The radical cascade occurring was initiated by the chemoselective addition of the PhS$^{\cdot}$ radical onto the triple bond in phenyl acetylene (**85**, which is more reactive than the isonitrile group in compound **86**). The resulting vinyl radical was first trapped by the isonitrile to give an imidoyl radical that was then transformed in an amidyl radical by reaction with *m*-dinitrobenzene (**87**). A hydrogen exchange step closed the reaction sequence, and acrylamide **88** was isolated in 40% yield along with a nonnegligible amount of byproducts [113].

Recently, a sequential photoinduced thiol-ene/thiol-yne approach has been devised for the preparation of cyclic peptides containing the Arg–Gly–Asp motif [114].

A particular case of the use of sulfur radicals is depicted in Scheme 3.36. The radical intermediates were generated by monoelectronic oxidation of a thioimidate anion and used for a mild synthesis of substituted benzothiazoles. The process started by deprotonation of a thioanilide (e.g., **89**) to form anion **91**$^{-}$ which was photocatalytically oxidized by $Ru(bpy)_3^{3+}$ to the corresponding radical **92**$^{\cdot}$ under visible light irradiation. Cyclization onto the aromatic ring followed by aromatization (via hydrogen transfer) led to the end benzothiazoles **90a–c** in a good yield [115].

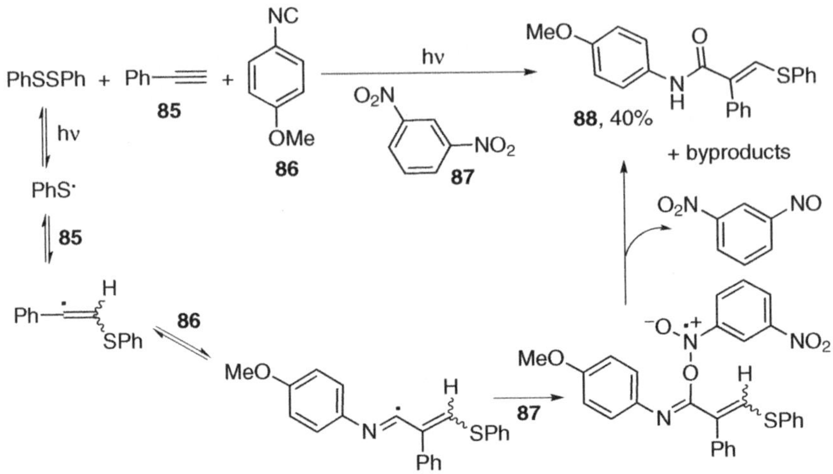

Scheme 3.35.

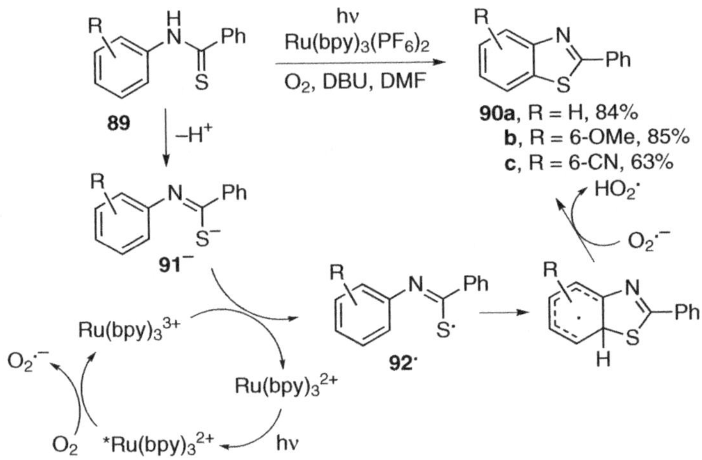

Scheme 3.36.

3.4.3 Selenium- and Tellurium-Based Radicals

The largest part of the reactions based on selenium and tellurium radicals relies on the photohomolytic dissociation of a Se—Se or a Te—Te bond that generates the corresponding chalcogen-centered radicals. Triple bonds have been most often used for trapping these intermediates. As an example, Scheme 3.37a shows the photoaddition of PhSeSePh (**93**) onto phenylacetylene (**85**) [116]. The reaction was efficient both by exposure to sunlight and by using a tungsten lamp (500 W) or UV light ($\lambda = 365$ nm). The process was initiated by addition of the PhSe$^{\cdot}$ radical onto the triple bond followed by reaction of the adduct radical with the starting diselenide forming the end product **94**

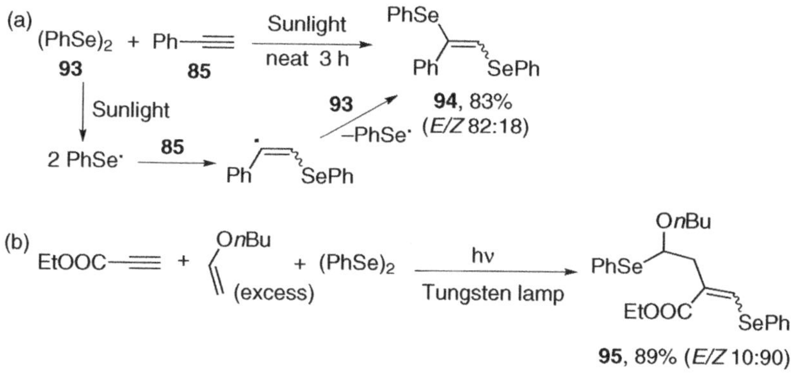

Scheme 3.37.

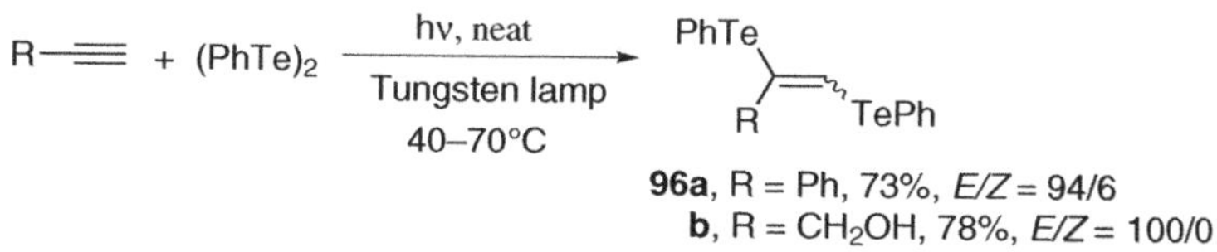

Scheme 3.38.

along with a new PhSe radical that continued the process [116]. The photoinitiated addition of diphenyl diselenide to methyl propiolate and other acetylenes took place smoothly under solar light irradiation [117]. Analogously, diphenyl ditelluride (PhTe-TePh) added to several acetylenes upon irradiation by visible light under solvent-free conditions and afforded the corresponding *vic-bis*(phenyltelluro)alkenes (**96**) in good yields (Scheme 3.38) [118, 119].

The reaction could be varied to a three-component addition in the presence of an electron-rich partner (e.g., a vinyl ether). A diselenide (**95**) incorporating both the alkyne backbone and the alkene resulted and involved the formation of a C–C bond along with two C–Se bonds (Scheme 3.37b) [120]. Somewhat similar was the case of isonitriles. Adding one equivalent had no effect on the photoaddition of diphenyldiselenide onto ethyl propiolate, but adding increasing amounts of an isonitrile (e.g., **97**) led to the three-component reaction, as shown in Scheme 3.39. A large

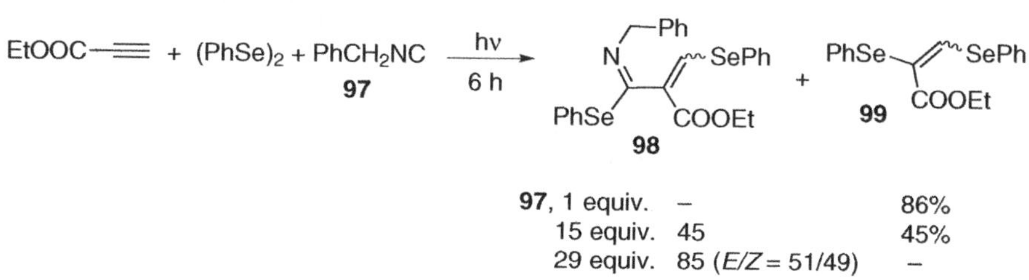

Scheme 3.39.

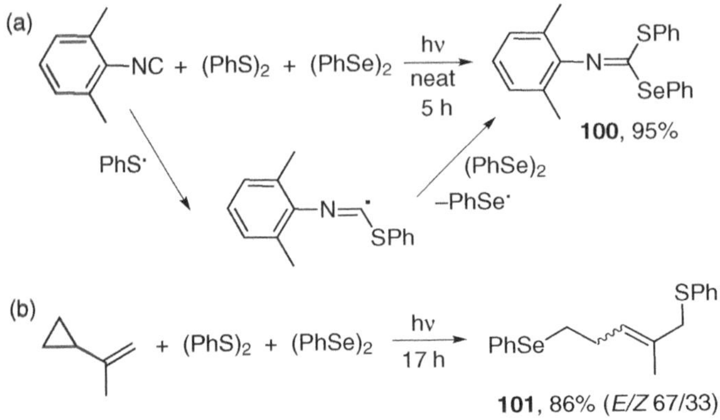

Scheme 3.40.

excess (29 equiv) of **97** was required for obtaining the incorporation of the isonitrile in the end compound **98** [120].

The thioselenation of a triple bond has been obtained by irradiation of a mixture containing the acetylene derivative along with $(PhSe)_2$ and $(PhS)_2$ [121–123]. In this case the PhS radical was first formed and added onto an isonitrile (Scheme 3.40a) [122] or a vinylciclopropane (Scheme 3.40b) [123]. The PhSe moiety was then incorporated. A regioselective and high yield synthesis of γ-(selenoethyl)allyl sulfides (**101**) was obtained with the latter substrates, and the process was likewise successful in the synthesis of the corresponding ring-opened thiotellurated analogues by using $(PhTe)_2$ in the place of $(PhSe)_2$ [123].

The use of the $(PhS)_2$–$(PhTe)_2$ mixed system in the radical cyclization of *o*-vinyl-substituted phenylisonitriles surprisingly led to the formation of a tellurium-free bisthiolated indole in moderate yield (Scheme 3.41). In fact, the addition of a PhS radical onto isonitrile **102** caused the formation of indole **104** that was further converted to compound **103** through a second photohomolytic step [124].

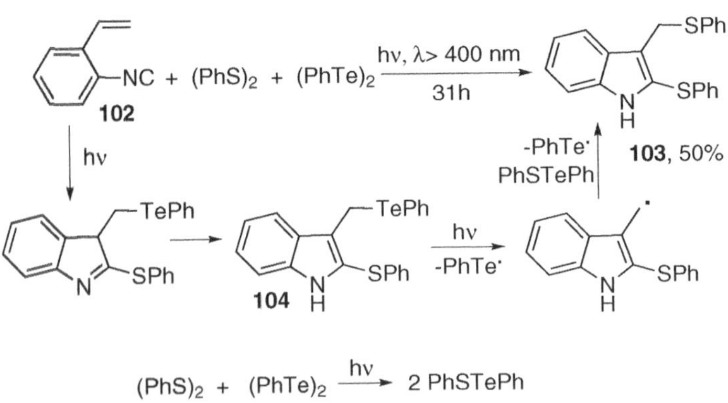

Scheme 3.41.

3.5 VIIA GROUP ATOMS

The photochemical generation of halogen atoms, most often from the corresponding biatomic molecules, is a well-known process; the photochemical halogenation of hydrocarbons is one of the few photochemical reactions universally dealt with in fundamental organic chemistry courses, as well as one that found a widespread application in the industry [64]. This section is thus only meant to give an idea of the diverse applications of photogenerated halogen atoms in organic synthesis by reporting a few selected examples.

3.5.1 Fluorine in Radical Reactions

Molecular fluorine has been rarely used in photofluorination reactions for the preparation of industrially useful perfluorinated compounds. An exception is the synthesis of perfluorotertiary amines (**106**) by treatment of a fluorinated enamine (**105**) with F_2 under photolytic conditions (Scheme 3.42a) [125].

In a related process, 2-phenyl-2-methoxyhexafluoropropane was photofluorinated to give a mixture of perfluoroalkyl ethers along with perfluorocumene (Scheme 3.42b) [126].

3.5.2 Chlorine Atoms

Most of the substrates that undergo photobromination reactions (see below) can likewise be photochlorinated by irradiation in the presence of a chlorine atom source such as Cl_2, *N*-chlorosuccinimide (NCS) or other sulfur-based chlorides [64]. Addition may be of interest, as in the case of the photochlorination of benzene to yield hexachlorocyclohexane, a reaction that had a great industrial importance since the γ stereoisomer of this polychlorinated hydrocarbon (known as lindane), now abandoned due to its toxicity, was largely used as a powerful insecticide [64]. Addition processes have been also used for the preparation of chlorofluoroalkanes by photochemical addition of chlorine onto perfluorinated olefins. Examples are (a) the

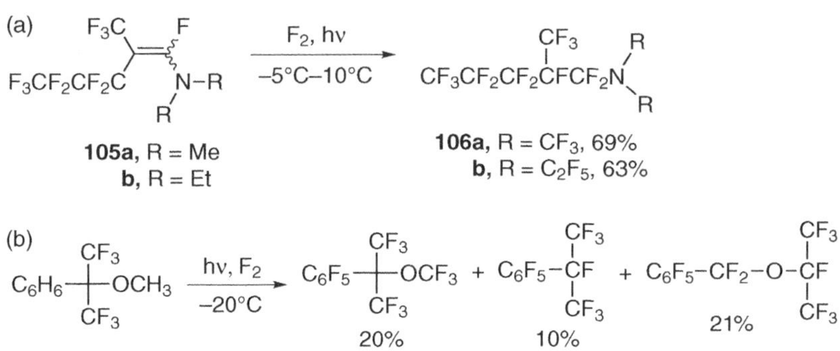

Scheme 3.42.

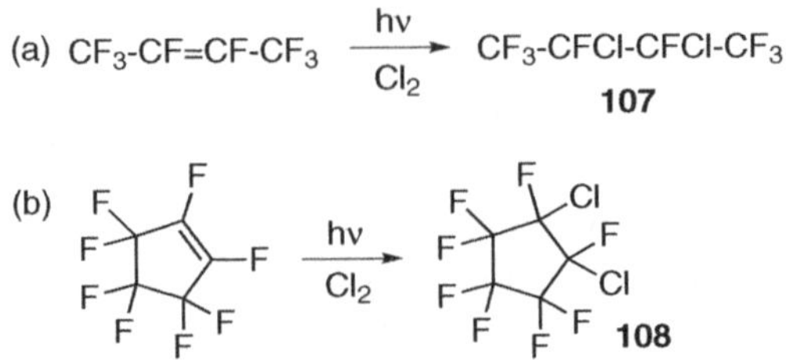

Scheme 3.43.

photochlorination in the gaseous phase of 2-perfluorobutene to give 2,3-dichlor-ooctafluorobutane (**107**) as the sole product (Scheme 3.43a) [127] and (b) the preparation of 1,2-dichlorooctafluorocyclopentane (**108**) from pefluorocyclopentene in a good yield (Scheme 3.43b) [128].

However, most of the processes that have been developed for large-scale industrial processes are radical substitutions. These reactions involve hydrogen abstraction (from a C–H bond) by the chlorine atom photogenerated from Cl_2 and chlorination of the carbon-centered radical. The selectivity of chlorination thus depends on the strength of the C–H bonds present in the reagent, although it is generally difficult to stop the process to monochlorination; mixtures of mono- and polychlorinated compounds are most often formed [64]. Some success on the way to monofunctionalization was met when using cycloalkanes as the substrates by maintaining a very low chlorine/cycloalkane molar ratio. On the other hand, irradiation at 43°C of a carbon disulfide solution of adamantane in the presence of chlorine afforded 1-chloroadamantane (**109**) in more than 85% yield, since the higher reactivity of the tertiary C–H bonds controlled the process (Scheme 3.44) [64].

The photochlorination of alkyl benzenes is likewise efficient. As an example, the conversion of toluene to the three chlorinated derivatives, namely benzyl chloride, benzylidene chloride, and α,α,α-trichlorotoluene, has been studied in depth and each of the products could be isolated in a good yield by a careful control of the reaction conditions [64]. Chlorosulfonation is likewise of great industrial importance and is obtained by carrying out the photochlorination reaction in the presence of SO_2 that in this case traps the alkyl radicals, as illustrated in Scheme 3.45 [64] for the chlorosulfonation of ethyl chloride (**110**) to give **111** in 60% yield [129].

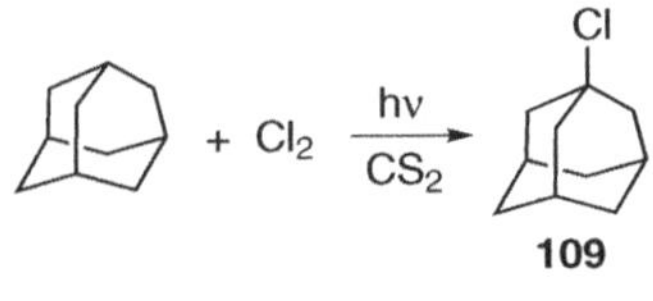

Scheme 3.44.

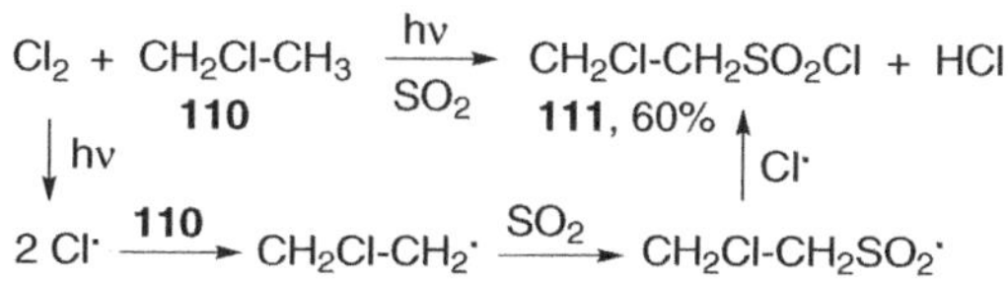

Scheme 3.45.

3.5.3 Bromine Atoms

The photochemical generation of bromine atoms is conveniently obtained by homolysis of a weak Br–X bond in precursors such as *N*-bromosuccinimide (NBS), HBr, CBr_4, and 1-bromo-3,5,5-trimethyl hydantoin, besides molecular Br_2. The course of the reaction depends on the reactivity or inertness of the co-generated radical X and of the solvent [64]. Both addition and benzylic or allylic substitution reactions have found large application. An illustrative example of the former class of reactions is the photobromination of β-styrenesulfonamides (e.g., **112**) to give the dibromo derivatives **113** in a very good yield, independently from the substituents present on the aromatic ring [130]. The reaction was simply carried out by exposing to sunlight an acetic acid solution of **112**, and the addition occurred with a significant selectivity toward the *threo* adduct (Scheme 3.46) [130].

Hydrobromination was likewise conveniently obtained photochemically—for example, in the so-called Phase-Vanishing (PV) reaction, where a multiphase system "lost" one of the phases during the process [131]. A representative example is shown in Scheme 3.47a. Thus, at the beginning a three-phase system was present, consisting of bromine at the bottom, a perfluorohexane phase (FC-72) in the middle, and a light hydrocarbon phase (isooctane) that contained the second reagent (e.g., an alkene) at the top. Part of the bromine and of the isooctane migrated in the freon phase, and upon irradiation of the system the bromine atoms formed abstracted a hydrogen from the isooctane (a reactive solvent here) and the resulting HBr diffused to the upper phase and underwent a photoinduced addition onto the double bond. As a result, a very efficient anti-Markovnikov hydrohalogenation of the alkene took place (Scheme 3.47a) [132]. During the reaction, the bromine phase slowly "vanished." The same approach was applied to the bromination of toluene (Scheme 3.47b) [133].

Substitution at a benzylic position was conveniently obtained also by photohomolysis of the N–Br bond in *N*-bromosuccinimide (NBS), as shown for the case of methyl naphthalene (**114**) that gave quantitatively bromide **115** (Scheme 3.48) [134].

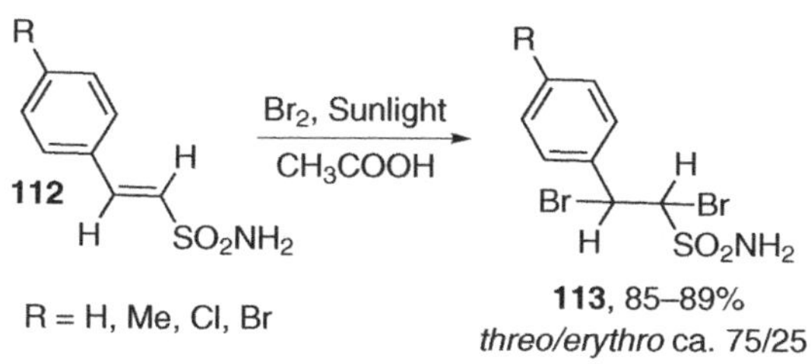

Scheme 3.46.

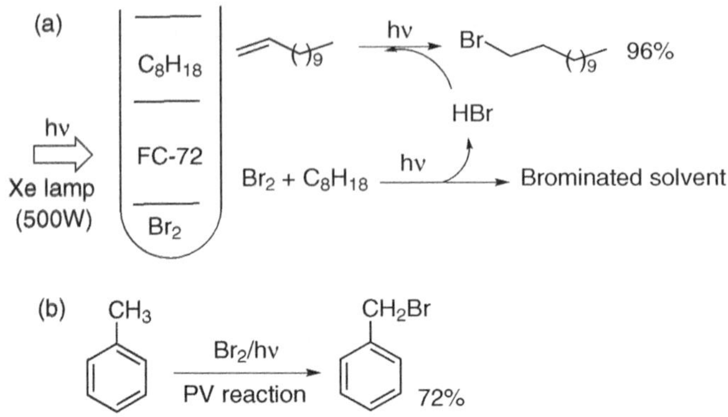

Scheme 3.47.

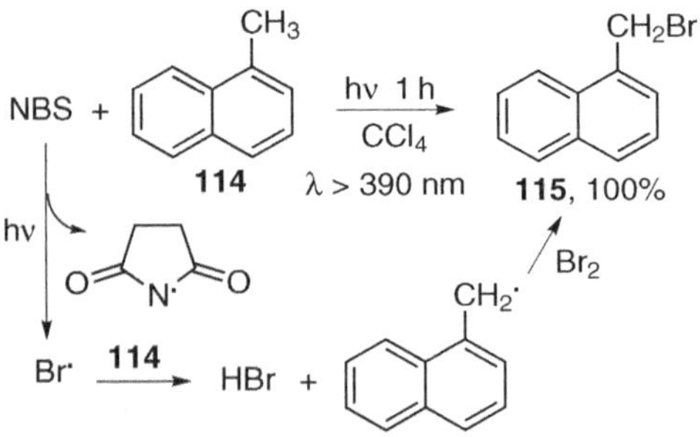

Scheme 3.48.

When using molecular bromine, CCl_4 was usually employed as the reaction medium, as exemplified by the photobromination of 2-difluoromethyl-substituted benzo-1,3-diazoles [135], 5-methoxyindane, and 5-methoxyindanone [136]. In the latter case, polybromination could be obtained by increasing the number of NBS equivalents employed [136]. The use of bromine in the place of NBS has been shown to be advantageous in some cases, particularly for benzylic functionalization and polyfunctionalization. A typical such reaction is the conversion of indane **116** into 1,1,3,3-tetrabromoindane **117** that was obtained in 95% yield upon 50 min of irradiation in the presence of Br_2 as reported in Scheme 3.49 [137]. Interestingly,

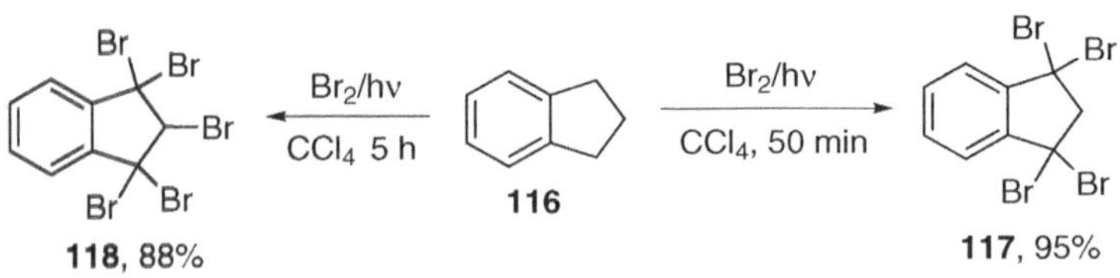

Scheme 3.49.

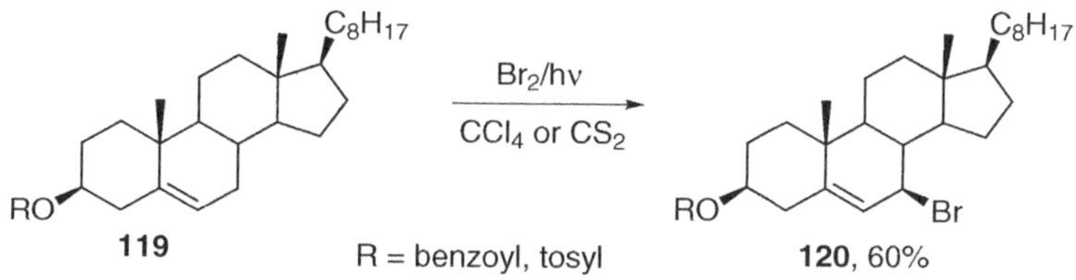

Scheme 3.50.

when the irradiation time was prolonged up to 5 h the pentabromoindane **118** was isolated in a slightly lower yield.

Under similar conditions, tetralin was photobrominated to 1,1,4,4-tetrabromo-1,2,3,4-tetrahydronaphthalene [138] and tetrabromonorbornane was isolated starting from norbornadiene [139]. In suitable cases, a chemo- and a stereoselective bromination was obtained, as shown with compound **119** that was efficiently photobrominated to give allyl bromide **120** in 60% yield (Scheme 3.50) [140].

Recently, photogenerated bromine atoms have been used for producing radicals that then gave nonbrominated final products. An example is the mild oxidation of primary alcohols to acids that involved trapping by molecular oxygen of the primarily formed α-hydroxyalkyl radicals. Various bromine sources could be used, and sunlight was effective in some cases. Scheme 3.51 shows an example [141] where NaBr was converted in situ to HBr by an Amberlyst 15 resin. The process was applied to various methylene groups oxidation [141].

A number of related procedures has been developed and involved the use of Br_2 [142], a CBr_4-PPh_3 mixture [143], LiBr [144], $MgBr_2$ [145, 146], and NBS [147] as the bromine source. This aerobic photooxidation reaction mediated by bromine radical was extended to the preparation of benzoic acids starting from the corresponding methylarenes [148–150]. In another application, ethylbenzenes were transformed into α-bromoketones or acetophenones in a proportion widely varying with the choice of the solvent [151]. The same procedure was adopted in the photooxidation of benzylamide to imides under visible light irradiation [152].

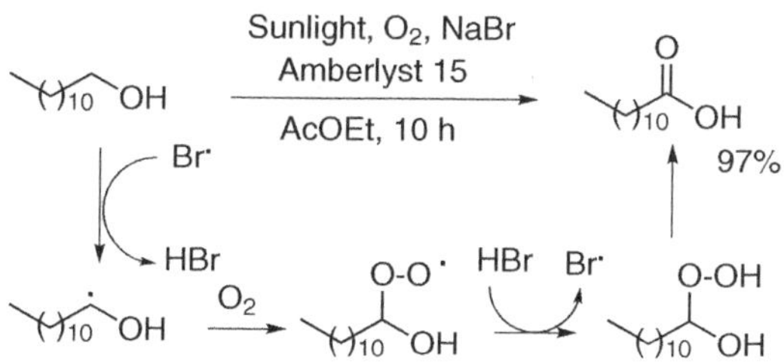

Scheme 3.51.

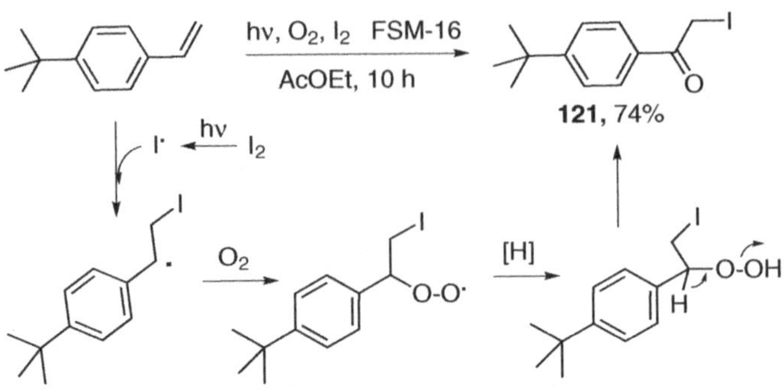

Scheme 3.52.

3.5.4 Iodine Atoms

Photoinduced iodination reactions are exclusively carried out by using elemental iodine as the iodinating agent. Actually, the lability of the I–I bond and the fact that I_2 absorbs in the visible region are substantial advantages of the method. Addition onto double bonds (e.g., in styrenes) was obtained under these conditions. The course of the reaction changed, however, when this was performed in the presence of oxygen and of FSM-16 (a mesoporous silica). In the latter case, phenacyl iodides (e.g., **121**) were formed independently from the ring substituents present (Scheme 3.52) [153]. Prolonging the reaction time could cause a partial oxidation of the phenacyl iodide to a benzoic acid. The reaction was unsuccessful when applied to a nonconjugated alkene, such as 1-dodecene [153].

The iodine radical was also useful to initiate the intramolecular cyclization reaction of *o*-alkynylaryl isonitriles, in analogy with that reported for tellurium derivatives in Scheme 3.41. Thus, irradiation of a chloroform solution of I_2 in the presence of these isonitriles afforded the corresponding 2,4-diiodoquinolines in a satisfactory yield [154]. Furthermore, iodine has been used in place of bromine in reactions such as the photooxidation of benzylamides [155] and of benzyl- and allyl alcohols (see Section 3.5.3) [156].

REFERENCES

1. Wille, U. Radical photochemistry. In Griesbeck, A., Oelgemoeller, M., Ghetti, F., Eds. *CRC Handbook of Organic Photochemistry and Photobiology*, third edition, CRC Press, Boca Raton, FL, 2012, pp. 329–345.
2. Albini, A., Fagnoni, M., Eds. *Handbook of Synthetic Photochemistry*, Wiley-VCH, Weinheim, 2010.
3. Chatgilialoglu, C., Studer, A., Eds. *Encyclopedia of Radicals in Chemistry, Biology and Materials*, John Wiley & Sons, Hoboken, NJ, 2012.
4. Suginome, H. Reactions and synthetic applications of oxygen-centered radicals photochemically generated from alkyl hypohalites. In Lenci, F., Horspool, W., Eds. *CRC*

Handbook of Organic Photochemistry and Photobiology, second edition, CRC Press, Boca Raton, FL, 2004, pp. 109-1–109-44.

5. Suginome, H. Remote functionalization by alkoxyl radicals generated by the photolysis of nitrite esters. The Barton reaction and related reaction of nitrite esters. In Lenci, F., Horspool, W., Eds. *CRC Handbook of Organic Photochemistry and Photobiology*, second edition, CRC Press, Boca Raton, FL, 2004, pp. 102-1–102-16.
6. Suginome, H. The Barton reaction. In Diker, G., Ed. *Handbook of C–H Transformations*, Vol. 2, John Wiley & Sons, Weinheim, 2005, pp. 579–589.
7. Dondoni, A., Marra, A. (2012). Recent applications of thiol-ene coupling as a click process for glycoconjugation. *Chemical Society Reviews*, **41**, 573–586.
8. Saraiva, M. F., Couri, M. R. C., Le Hyaric, M., de Almeida, M. V. (2009). The Barton ester free-radical reaction: A brief review of applications. *Tetrahedron*, **65**, 3563–3572.
9. Dalko, P. I. The photochemistry of Barton esters. In Lenci, F., Horspool, W., Eds. *CRC Handbook of Organic Photochemistry and Photobiology*, second edition, CRC Press, Boca Raton, FL, 2004, pp. 67-1–67-23.
10. Chatgilialoglu, C., Ferreri, C., Gimisis, T. Tris(trimethylsilyl)silane in organic synthesis. In Rappoport, Z., Apeloig, Y., Eds. *The Chemistry of Organic Silicon Compound*, Vol. 2, John Wiley & Sons, New York, 1998, pp. 1539–1579.
11. Lalevée, J., El-Roz, M., Allonas, X., Fouassier, J. P. Enhancement of radical and cationic photopolymerization reactions under air. In Fouassier, J. P., Allonas, X., Eds. *Basics and Applications of Photopolymerization Reactions*, Vol. 2, Research Signpost, Trivandrum, 2010, pp. 29–38.
12. Lalevée, J., El-Roz, M., Allonas, X., Fouassier, J. P. On the silyl radical chemistry. In Wyman, E. B., Skief, M. C., Eds. *Photopolymerization Reactions in Organosilanes*, Nova Science Publisher, Hauppauge, NY, 2010, pp. 177–189.
13. Brook, A. G. The photochemistry of organosilicon compounds. In Patai, S., Rappoport, Z., Eds. *The Chemistry of Organic Silicon Compounds*, John Wiley & Sons, Chichester, 1989, pp. 965–1005.
14. Chatgilialoglu, C. (1995). Structural and chemical properties of silyl radicals. *Chemical Reviews*, **95**, 1229–1251.
15. Steinmetz, M. G. (1995). Organosilane photochemistry. *Chemical Reviews*, **95**, 1527–1588.
16. Chatgilialoglu, C., Scaiano, J. C., Ingold, K. U. (1982). Absolute rate constants for the reactions of *tert*-butoxyl radicals and some ketone triplets with silanes. *Organometallics*, **1**, 466–469.
17. Lalevée, J., El-Roz, M., Allonas, X., Fouassier, J.-P. (2009). Surface modification of a UV curable acrylate coating: In situ introduction of hydrophobic properties. *Progress in Organic Coatings*, **65**, 457–461.
18. Nay, M. A., Woodall, G. N. C., Strausz, O. P., Gunning, H. E. (1965). The mercury $6(^3P_1)$ photosensitization of the methylsilanes and silane. *Journal of the American Chemical Society*, **87**, 179–187.
19. Brix, T., Bastian, E., Potzinger, P. J. (1989). Stationary and pulsed photolysis of hexamethyldisilane. *Journal of Photochemistry and Photobiology A: Chemistry*, **49**, 287–297.
20. Matsumoto, A., Ito, Y. (2000). New generation of organosilyl radicals by photochemically induced homolytic cleavage of silicon–boron bonds. *The Journal of Organic Chemistry*, **65**, 5707–5711.

21. Maeda, Y., Sato, Y., Kako, M., Wakahara, T., Akasaka, T., Lu, J., Nagase, S., Kobori, Y., Hasegawa, T., Motomiya, K., Tohji, K., Kasuya, A., Wang, D., Yu, D., Gao, Z., Han, R., Ye, H. (2006). Preparation of single-walled carbon nanotube-organosilicon hybrids and their enhanced field emission properties. *Chemistry of Materials*, **18**, 4205–4208.

22. Tehfe, M.-A., Blanchard, N., Fries, C., Lalevée, J., Allonas, X., Fouassier, J. P. (2010). Bis(germyl)ketones: Toward a new class of type I photoinitiating systems sensitive above 500 nm. *Macromolecular Rapid Communications*, **31**, 473–478.

23. Stella, L. (1983). Homolytic cyclization of *N*-chloroalkenyl amines. *Angewandte Chemie International Edition in English*, **22**, 337–350.

24. Neale, R. S. (1971). Nitrogen radicals as synthesis intermediates. *N*-Halamide rearrangements and additions to unsaturated hydrocarbons. *Synthesis*, 1–15.

25. Mackiewicz, P., Furstoss, R. (1978). Radicaux amidyl: Structure et réactivité. *Tetrahedron*, **34**, 3241–3260.

26. Esker, J. L., Newcomb, M. The generation of nitrogen radicals and their cyclizations for the construction of the pyrrolidine nucleus. In Katritzky, A., Ed. *Advances in Heterocyclic Chemistry*, Vol. 58, Academic Press, New York, 1993, pp. 1–45.

27. Wolff, M. E. (1963). Cyclization of *N*-halogenated amines (the Hofmann–Löffler reaction). *Chemical Reviews*, **63**, 55–64.

28. Morozova, O. B., Yurkovskaya, A. V. (2008). Aminium cation radical of glycylglycine and its deprotonation to aminyl radical in aqueous solution. *The Journal of Physical Chemistry B*, **112**, 12859–12862.

29. Pance, N., Kenji, S., Masahiko, T., Hideyuki, H. (2007). Direct observation of aminyl radical intermediate during single-crystal to single-crystal photoinduced Orton rearrangement. *The Journal of Physical Chemistry B*, **111**, 10373–10378.

30. Newcomb, M., Musa, O. M., Martinez, F. N., Horner, J. H. (1997). Kinetics of 5-exo cyclizations of *N*-alkyl-4-pentenaminyl radicals and β-fragmentations of β-(dialkylamino)alkyl radicals. *Journal of the American Chemical Society*, **119**, 4569–4577.

31. Newcomb, M., Horner, J. H., Shahin, H. (1993). Rate constants for aminyl radical reactions. *Tetrahedron Letters*, **34**, 5523–5526.

32. Lalevée, J., Gigmes, D., Bertin, D., Graff, B., Allonas, X., Fouassier, J.-P. (2007). Comparative reactivity of aminyl and aminoalkyl radicals. *Chemical Physics Letters*, **438**, 346–350.

33. Kuttappan-Nair, V., Samson-Thibault, F., Wagner, J. R. (2010). Generation of 2'-deoxyadenosine N6-aminyl radicals from the photolysis of phenylhydrazone derivatives. *Chemical Research in Toxicology*, **23**, 48–54.

34. Wolff, M. E., Kerwin, J. F., Owings, F. F., Lewis, B. B., Blank, B. (1962). The synthesis of C-18 functionalized steroid hormone analogs. III. Preparation and some reactions of 18-chloro steroids. *The Journal of Organic Chemistry*, **27**, 3628–3639.

35. Flesia, E., Croatto, A., Tordo, P., Surzur, J. M. (1972). Mise en evidence par R. P. E. de l'addition du radical amido sur les olefines. *Tetrahedron Letters*, **13**, 535–538.

36. Mackiewicz, P., Furstoss, R., Waegell, B., Cote, R., Lessard, J. (1978). Peroxide-initiated cyclizations of olefinic *N*-chloro amides. Electronic configuration of amido radicals. *The Journal of Organic Chemistry*, **43**, 3746–3750.

37. Kuehne, M. E., Horne, D. A. (1975). Photochemical cyclization of olefinic *N*-chloroamides. *The Journal of Organic Chemistry*, **40**, 1287–1292.

38. Chow, Y. L., Perry, R. A. (1972). Intramolecular reactions of amido radical. *Tetrahedron Letters*, **13**, 531–534.

39. Lessard, J., Mondon, M., Touchard, D. (1981). The photochemical addition of *N*-haloamides to olefins: the influence of various factors on the competition between 1,2-addition and hydrogen abstraction. *Canadian Journal of Chemistry*, **59**, 431–450.

40. Caron, G., Lessard, J. (1993). The regiochemistry of the radical addition of *N*-chloroamides to enol ethers. *Tetrahedron*, **49**, 8039–8058.

41. Chow, Y. L., Mojelsky, T. W., Magdzinski, L. J., Tichy, M. (1985). Chemistry of amido radicals: intramolecular hydrogen abstraction as related to amido radical configurations. *Canadian Journal of Chemistry*, **63**, 2197–2202.

42. Edwards, O. E., Paton, J. M., Benn, M. H., Mitchell, R. E., Watanatada, C., Vohra, K. N. (1971). Cyclization and fragmentation of *N*-chloro lactams. The transannular synthesis of bicyclic lactams with bridgehead nitrogen. *Canadian Journal of Chemistry*, **49**, 1648–1658.

43. Hu, T., Shen, M., Chen, Q., Li, C. (2006). Pushing radical cyclization from regioselective to regiospecific: Cyclization of amidyl radicals controlled by vinylic halogen substitution. *Organic Letters*, **8**, 2647–2650.

44. Yuan, X., Liu, K., Li, C. (2008). Development of highly regioselective amidyl radical cyclization based on lone pair–lone pair repulsion. *The Journal of Organic Chemistry*, **73**, 6166–6171.

45. Betancor, C., Concepción, J. I., Hernández, R., Salazar, J. A., Suárez, E. (1983). Intramolecular functionalization of phosphoramidate radicals. Synthesis of 1,4-epimine compounds. *The Journal of Organic Chemistry*, **48**, 4432–4433.

46. Martın, A., Perez-Martín, I., Suarez, E. (2005). Intramolecular hydrogen abstraction promoted by amidyl radicals. Evidence for electronic factors in the nucleophilic cyclization of ambident amides to oxocarbenium ions. *Organic Letters*, **7**, 2027–2030.

47. Togo, H., Hoshina, Y., Muraki, T., Nakayama, H., Yokoyama, M. (1998). Study on radical amidation onto aromatic rings with (diacyloxyiodo)arenes. *The Journal of Organic Chemistry*, **63**, 5193–5200.

48. Togo, H., Harada, Y., Yokoyama, M. (2000). Preparation of 3,4-dihydro-2,1-benzothiazine 2,2-dioxide skeleton from *N*-methyl 2-(aryl)ethanesulfonamides with (diacetoxyiodo)arenes. *The Journal of Organic Chemistry*, **65**, 926–929.

49. Zhuang, S., Liu, K., Li, C. (2011). Stereoselective 6-exo cyclization of amidyl radicals. An experimental and theoretical study. *The Journal of Organic Chemistry*, **76**, 8100–8106.

50. Esker, J. L., Newcomb, M. (1994). Multiple reaction channels of (*N*-acyl-*N*-alkylcarbamoyl)oxyl radicals from *N*-acyl PTOC carbamates. *The Journal of Organic Chemistry*, **59**, 2779–2786.

51. Esker, J. L., Newcomb, M. (1993). Chemistry of amidyl radicals produced from *N*-hydroxypyridine-2-thione imidate esters. *The Journal of Organic Chemistry*, **58**, 4933–4940.

52. Newcomb, M., Esker, J. L. (1991). Facile production and cyclizations of amidyl radicals. *Tetrahedron Letters*, **32**, 1035–1038.

53. Boivin, J., Fouquet, E., Schiano, A. M., Zard, S.Z. (1994). Iminyl radicals: Part III. Further synthetically useful sources of iminyl radicals. *Tetrahedron*, **50**, 1769–1776.

54. Boivin, J., Fouquet, E., Zard, S. Z. (1991). A new and synthetically useful source of iminyl radicals. *Tetrahedron Letters*, **32**, 4299–4302.

55. Fong, M. C., Schiesser, C. H. (1993). Homolytic substitution by iminyl radical at selenium: A free-radical route to 1,2-benzoselenazoles. *Tetrahedron Letters*, **34**, 4347–4348.
56. Sakuragi, H., Ishikawa, S.-i., Nishimura, T., Yoshida, M., Inamoto, N., Tokumaru, K. (1976). Homolytic aromatic substitution by iminyl radicals. Photolysis of aromatic ketone *O*-acyloximes in aromatic solvents. *Bulletin of the Chemical Society of Japan*, **49**, 1949–1954.
57. McBurney, R. T., Slawin, A. M. Z., Smart, L. A., Yu, Y., Walton, J. C. (2011). UV promoted phenanthridine syntheses from oxime carbonate derived iminyl radicals. *Chemical Communications*, **47**, 7974–7976.
58. Alonso, R., Campos, P. J., Garcia, B., Rodriguez, M. A. (2006). Cyclization reactions of acyloximes to isoquinolines. *Organic Letters*, **8**, 3521–3523.
59. Alonso, R., Caballero, A., Campos, P. J., Rodríguez, M. A. (2010). Photochemistry of acyloximes: Synthesis of heterocycles and natural products. *Tetrahedron*, **66**, 8828–8831.
60. Portela-Cubillo, F., Scanlan, E. M., Scott, J. S., Walton, J. C. (2008). From dioxime oxalates to dihydropyrroles and phenanthridines via iminyl radicals. *Chemical Communications*, 4189–4191.
61. Portela-Cubillo, F., Lymer, J., Scanlan, E. M., Scott, J. S., Walton, J. C. (2008). Dioxime oxalates; new iminyl radical precursors for syntheses of *N*-heterocycles. *Tetrahedron*, **64**, 11908–11916.
62. Glover, S. A., Hammond, G. P., Harman, D. G., Mills, J. G., Rowbottom, C.A. (1993). Cyclization of alkoxyiminyl radicals onto olefins: Formation of 2-alkoxy-Δ^1-pyrrolines, 4,5-dihydrooxazoles and 5,6-dihydro-4*H*-1,3-oxazines. *Australian Journal of Chemistry*, **46**, 1213–1228.
63. Hartung, J. (2009). Organic radical reactions associated with nitrogen monoxide. *Chemical Reviews*, **109**, 4500–4517.
64. Braun, A. M., Mauretta, M.-T., Oliveros, E. *Photochemical Technology*, John Wiley & Sons, Chichester, 1991.
65. Funken, K.-H., Müller, F.-J., Ortner, J., Riffelmann, K.-J., Sattler, C. (1999). Solar collectors versus lamps—A comparison of the energy demand of industrial photochemical processes as exemplified by the production of ε-caprolactame. *Energy*, **24**, 681–687.
66. Marque, S., Tordo, P. (2005). Reactivity of phosphorus centered radicals. *Topics in Current Chemistry*, **250**, 43–76.
67. Dichiarante, V., Fagnoni, M. Intermolecular addition reactions onto C–C multiple bonds. In ref. 2, pp. 67–94.
68. Trofimov, B. A., Gusarova, N. K., Malysheva, S. F., Ivanova, N. I., Sukhov, B. G., Belorgolova, N. A., Kuimov, V. A. (2002). Hydrophosphination of vinyl sulfides and vinyl selenides: First examples. *Synthesis*, 2207–2710.
69. Trofimov, B. A., Malysheva, S. F., Parshina, L. N., Gusarova, N. K., Belogorlova, N. A. (2011). Metal-free hydrophosphanation of 1-vinylimidazoles with secondary phosphanes: A straightforward atom-economic route to tertiary phosphanes with imidazolyl substituents. *Synlett*, 94–98.
70. Lücke, S., Stoppek-Langner, K., Kuchinke, J., Krebs, B. (1999). Octakis-(dimethylphosphanoethyl)-octasilsesquioxane—Synthesis, characterization and reactivity. *Journal of Organometallic Chemistry*, **584**, 11–15.

71. Ellis, D., Farrugia, L. J., Hickman, D. T., Lovatt, P. A., Peacock, R. D. (1996). A general route for the synthesis of triazacyclononane functionalised with one, two or three pendant phosphine arms: Crystal structure of $[Zn_2L_2Cl_3][ClO_4]$, L = *N*-(diphenylphosphinopropyl)-1,4,7-triazacyclononane. *Chemical Communications*, 1817–1818.

72. Armstrong, S. K., Cross, R. J., Farrugia, L. J., Nichols, D. A., Perry, A. (2002). Semi-rigid bis-phosphane ligands for metallamacrocycle formation. *European Journal of Inorganic Chemistry*, 141–151.

73. Ruffieux, V., Schmid, G., Braunstein, P., Rosk, J. (1997). T_8-OSS-ethyldiphenylphosphine: A new functional oligosilsesquioxane ligand. *Chemistry—A European Journal*, **3**, 900–903.

74. Jiménez, M. V., Pérez-Torrente, J. J., Bartolomé, M. I., Oro, L. A. (2009). Convenient methods for the synthesis of a library of hemilabile phosphines. *Synthesis*, 1916–1922.

75. Kröcher, O., Köppel, R. A., Fröba, M., Baiker, A. (1998). Silica hybrid gel catalysts containing group(VIII) transition metal complexes: Preparation, structural, and catalytic properties in the synthesis of *N,N*-dimethylformamide and methyl formate from supercritical carbon dioxide. *Journal of Catalysis*, **178**, 284–298.

76. Gusarova, N. K., Malysheva, S. F., Belogorlova, N. A., Sukhov, B. G., Trofimov, B. A. (2007). Radical addition of secondary phosphine selenides to alkenes. *Synthesis*, 2849–2852.

77. Semenzin, D., Etemad-Moghadam, G., Albouy, D., Diallo, O., Koenig, M. (1997). Dual radical/polar Pudovik reaction: Application field of new activation methods. *The Journal of Organic Chemistry*, **62**, 2414–2422.

78. Coseri, S. (2009). Phthalimide-*N*-oxyl (PINO) radical, a powerful catalytic agent: Its generation and versatility towards various organic substrates. *Catalysis Reviews*, **51**, 218–292.

79. Bentley, T. J., McGhie, J. F., Barton, D. H. R. (1965). The synthesis of 32-oxygenated lanostane derivatives. *Tetrahedron Letters*, **6**, 2497–2498.

80. Sugimoto, A., Sumino, Y., Takagi, M., Fukuyama, T., Ryu, I. (2006). The Barton reaction using a microreactor and black light. Continuous-flow synthesis of a key steroid intermediate for an endothelin receptor antagonist. *Tetrahedron Letters*, **47**, 6197–6200.

81. Sugimoto, A., Fukuyama, T., Sumino, Y., Takagi, M., Ryu, I. (2009). Microflow photo-radical reaction using a compact light source: Application to the Barton reaction leading to a key intermediate for myriceric acid A. *Tetrahedron*, **65**, 1593–1598.

82. De Armas, P., Concepción, J. I., Francisco, C. G., Hernandez, R., Salazar, J. A., Suárez, E. (1989). Intramolecular hydrogen abstraction. Hypervalent organoiodine compounds, convenient reagents for alkoxyl radical generation. *Journal of the Chemical Society, Perkin Transactions 1*, 405–411.

83. Concepción, J. I., Francisco, C. G., Hernández, R., Salazar, J. A., Suárez, E. (1984). Intramolecular hydrogen abstraction. Iodosobenzene diacetate, an efficient and convenient reagent for alkoxy radical generation. *Tetrahedron Letters*, **25**, 1953–1956.

84. Suginome, H., Yamada, S. (1987). A simple new general synthesis of macrocyclic ketones: A new entry to the synthesis of exaltone and (±)-muscone. *Tetrahedron Letters*, **28**, 3963–3965.

85. Kaino, M., Naruse, Y., Ishihara, K., Yamamoto, H. (1990). Stereospecific cyclization of vinyl ether alcohols. Facile synthesis of (−)-lardolure. *The Journal of Organic Chemistry*, **55**, 5814–5815.

86. Ceccherelli, P., Curini, M., Marcotullio, M. C., Mylari, B. L., Wenkert, E. (1986). Iodohydrins and tetrahydrofurans from lead tetraacetate–iodine oxidation of terpenic alcohols. *The Journal of Organic Chemistry*, **51**, 1505–1509.

87. Arencibia, M. T., Freire, R., Perales, A., Rodríguez, M. S., Suárez, E. (1991). Hypervalent organoiodine compounds: Radical fragmentation of oxabicyclic hemiacetals. Convenient synthesis of medium-sized and spiro lactones. *Journal of the Chemical Society, Perkin Transactions 1*, 3349–3360.

88. Boto, A., Betancor, C., Prangé, T., Suárez, E. (1994). Fragmentation of alkoxy radicals: Tandem β-fragmentation-cycloperoxyiodination reaction. *The Journal of Organic Chemistry*, **59**, 4393–4401.

89. Mowbray, C. E., Pattenden, G. (1993). Polycycle construction via cascade radical fragmentation transannulation-cyclisation processes. *Tetrahedron Letters*, **34**, 127–130.

90. Galatsis, P., Millan, S. D. (1991). Use of iodobenzene diacetate for the synthesis of α-iodoepoxides. *Tetrahedron Letters*, **32**, 7493–7496.

91. Hartung, J., Kneuer, R., Laug, S., Schmidt, P., Spehar, K., Svoboda, I., Fuess, H. (2003). A radical version of the bromo- and the iodocyclization of bis(homoallylic) alcohols—The synthesis of halogenated tetrahydrofurans by stereoselective alkoxyl radical ring closures. *European Journal of Organic Chemistry*, 4033–4052.

92. Barton, D. H. R., Jaszberenyi, J. C., Morrell, A. I. (1991). The generation and reactivity of oxygen centered radicals from the photolysis of derivatives of *N*-hydroxy-2-thiopyridone. *Tetrahedron Letters*, **32**, 311–314.

93. Newcomb, M., Kumar, M. U., Boivin, J., Crépon, E., Zard, S. Z. (1991). Cyclizations and intermolecular additions of alkoxycarbonyloxy radicals from *N*-hydroxypyridine-2-thione carbonates. *Tetrahedron Letters*, **32**, 45–48.

94. Wille, U. (2002). Self-terminating, oxidative radical cyclizations: A novel reaction of acyloxyl radicals. *Journal of the American Chemical Society*, **124**, 14–15.

95. Rao, H., Wang, P., Li, C.-J. (2012). Visible-light-triggered direct benzoyloxylation of electron-rich arenes at room temperature without chelation assistance. *European Journal of Organic Chemistry*, 6503–6507.

96. Hoyle, C. E., Lowe, A. B., Bowman, C. N. (2010). Thiol-click chemistry: A multifaceted toolbox for small molecule and polymer synthesis. *Chemical Society Reviews*, **39**, 1355–1387.

97. Hoyle, C. E., Bowman, C. N. (2010). Thiol-ene click chemistry. *Angewandte Chemie International Edition*, **49**, 1540–1573.

98. Wu, X., Bundle, D. R. (2005). Synthesis of glycoconjugate vaccines for *Candida albicans* using novel linker methodology. *The Journal of Organic Chemistry*, **70**, 7381–7388.

99. Dondoni, A., Massi, A., Nanni, P., Roda, A. (2009). A new ligation strategy for peptide and protein glycosylation: Photoinduced thiol-ene coupling. *Chemistry—A European Journal*, **15**, 11444–11449.

100. Fiore, M., Marra, A., Dondoni, A. (2009). Photoinduced thiol–ene coupling as a click ligation tool for thiodisaccharide synthesis. *The Journal of Organic Chemistry*, **74**, 4422–4425.

101. Lázár, L., Csávás, M., Herczeg, M., Herczegh, P., Borbás, A. (2012). Synthesis of S-linked glycoconjugates and S-disaccharides by thiol–ene coupling reaction of enoses. *Organic Letters*, **14**, 4650–4653.

102. Killops, K. L., Campos, L. M., Hawker, C. J. (2008). Robust, efficient, and orthogonal synthesis of dendrimers via thiol-ene "click" chemistry. *Journal of the American Chemical Society*, **130**, 5062–5064.

103. Rissing, C., Son, D. Y. (2008). Thiol–ene reaction for the synthesis of multifunctional branched organosilanes. *Organometallics*, **27**, 5394–5397.

104. Massi, A., Nanni, D. (2012). Thiol–yne coupling: Revisiting old concepts as a breakthrough for up-to-date applications. *Organic and Biomolecular Chemistry*, **10**, 3791–3807.

105. Keck, G. E., Wager, T. T. (1996). Total synthesis of *ent*-lycoricidine via a thiyl radical addition–cyclization sequence. *The Journal of Organic Chemistry*, **61**, 8366–8367.

106. Zhou, W., Zheng, H., Li, Y., Liu, H., Li, Y. (2010). Synthesis of sulfuric macrocycles and a rotaxane through thiol-yne click and dithiol coupling reactions. *Organic Letters*, **12**, 4078–4081.

107. Minozzi, M., Monesi, A., Nanni, D., Spagnolo, P., Marchetti, N., Massi, A. (2011). An insight into the radical thiol/yne coupling: The emergence of arylalkyne-tagged sugars for the direct photoinduced glycosylation of cysteine-containing peptides. *The Journal of Organic Chemistry*, **76**, 450–459.

108. Chen, G., Kumar, J., Gregory, A., Stenzel, M. H. (2009). Efficient synthesis of dendrimers via a thiol–yne and esterification process and their potential application in the delivery of platinum anti-cancer drugs. *Chemical Communications*, 6291–6293.

109. Chan, J. W., Hoyle, C. E., Lowe, A. B. (2009). Sequential phosphine-catalyzed, nucleophilic thiol–Ene/radical-mediated thiol–yne reactions and the facile orthogonal synthesis of polyfunctional materials. *Journal of the American Chemical Society*, **131**, 5751–5753.

110. Kanehara, M., Oumi, Y., Sano, T., Teranishi, T. (2004). Syntheses of the novel acidic and basic ligands and superlattice formation from gold nanoparticles through interparticle acid–base interaction. *Bulletin of the Chemical Society of Japan*, **77**, 1589–1597.

111. Harrowven, D. C., Hannam, J. C. (1999). Thiyl radical induced isomerisations of maleate esters provide a convenient route to fumarates and furanones. *Tetrahedron*, **55**, 9341–9346.

112. Takaguchi, Y., Katayose, Y., Yanagimoto, Y., Motoyoshiya, J., Aoyama, H., Wakahara, T., Maeda, Y., Akasaka, T. (2003). Photoinduced dithiolation of fullerene[60] with dendrimer disulfide. *Chemistry Letters*, **32**, 1124–1125.

113. Leardini, R., Nanni, D., Zanardi, G. (2000). Radical addition to isonitriles: A route to polyfunctionalized alkenes through a novel three-component radical cascade reaction. *The Journal of Organic Chemistry*, **65**, 2763–2772.

114. Aimetti, A. A., Feaver, K. R., Anseth, K. S. (2010). Synthesis of cyclic, multivalent Arg-Gly-Asp using sequential thiol-ene/thiol-yne photoreactions. *Chemical Communications*, **46**, 5781–5783.

115. Cheng, Y., Yang, J., Qu, Y., Li, P. (2012). Aerobic visible-light photoredox radical C–H functionalization: Catalytic synthesis of 2-substituted benzothiazoles. *Organic Letters*, **14**, 98–101.

116. Ogawa, A., Yokoyama, H., Yokoyama, K., Masawaki, T., Kambe, N., Sonoda, N. (1991). Photoinitiated addition of diphenyl diselenide to acetylenes. *The Journal of Organic Chemistry*, **56**, 5721–5723.

117. Back, T. G., Krishna, M. V. (1988). Free-radical additions of diselenides to dimethyl acetylenedicarboxylate, methyl propiolate, and dimethyl maleate. *The Journal of Organic Chemistry*, **53**, 2533–2536.
118. Ogawa, A., Yokoyama, K., Obayashi, R., Han, L.-B., Kambe, N., Sonoda, N. (1993). Photo-induced ditelluration of acetylenes with diphenyl ditelluride. *Tetrahedron*, **49**, 1177–1188.
119. Ogawa, A., Yokoyama, K., Yokoyama, H., Obayashi, R., Kambe, N., Sonoda, N. (1991). Photo-initiated addition of diphenyl ditelluride to acetylenes. *Journal of the Chemical Society, Chemical Communications*, 1748–1750.
120. Tsuchii, K., Doi, M., Ogawa, I., Einaga, Y., Ogawa, A. (2005). Diphenyl diselenide as a useful reagent for intermolecular domino reactions of various unsaturated compounds under photoirradiation conditions. *Bulletin of the Chemical Society of Japan*, **78**, 1534–1548.
121. Ogawa, A., Obayashi, R., Ine, H., Tsuboi, Y., Sonoda, N., Hirao, T. (1998). Highly regioselective thioselenation of acetylenes by using a $(PhS)_2$–$(PhSe)_2$ binary system. *The Journal of Organic Chemistry*, **63**, 881–884.
122. Tsuchii, K., Tsuboi, Y., Kawaguchi, S.-i., Takahashi, J., Sonoda, N., Nomoto, A., Ogawa, A. (2007). Highly selective double chalcogenation of isocyanides with disulfide–diselenide mixed systems. *The Journal of Organic Chemistry*, **72**, 415–423.
123. Ogawa, A., Ogawa, I., Obayashi, R., Umezu, K., Doi, M., Hirao, T. (1999). Highly selective thioselenation of vinylcyclopropanes with a $(PhS)_2$–$(PhSe)_2$ binary system and its application to thiotelluration. *The Journal of Organic Chemistry*, **64**, 86–92.
124. Mitamura, T., Tsuboi, Y., Iwata, K., Tsuchii, K., Nomoto, A., Sonoda, M., Ogawa, A. (2007). Photoinduced thiotelluration of isocyanides by using a $(PhS)_2$–$(PhTe)_2$ mixed system, and its application to bisthiolation via radical cyclization. *Tetrahedron Letters*, **48**, 5953–5957.
125. Ono, T., Yamanouchi, K., Scherer, K.V., Jr. (1995). Liquid-phase photofluorination with elemental fluorine. Part II. Synthesis of perfluorotertiary amines. *Journal of Fluorine Chemistry*, **73**, 267–272.
126. Ono, T., Yamanouchi, K., Fernandez, R. E., Scherer, K. V., Jr. (1995). Liquid-phase photofluorination with elemental fluorine. Part III. Synthesis of perfluorocycloalkyl ethers with/without a chlorine substituent. *Journal of Fluorine Chemistry*, **75**, 197–204.
127. Gonzales, M. C., Castellano, E. (1982). The kinetics of the photochlorination of perfluoro-2-butene under intermittent light. *Journal of Photochemistry*, **19**, 229–234.
128. Alberto, Z. R., Cosa, J. J., Staricco, E. H. (1973). The photochlorination of perfluorocyclopentene. *International Journal of Chemical Kinetics*, **5**, 321–328.
129. Walling, C., Pease, W. F. (1958). Chlorosulfonation of ethyl chloride. *The Journal of Organic Chemistry*, **23**, 478–479.
130. Hasegawa, K., Hirooka, S., Kawahara, H., Tanaka, A., Nomura, M., Hori, Y. (1977). The photobromination of β-styrenesulfonamides and syntheses of 2-arylacetylene-1-sulfonamides. *Bulletin of the Chemical Society of Japan*, **50**, 2346–2350.
131. Van Zee, N. J., Dragojlovic, V. (2010). Design of phase-vanishing reactions. *Chemistry—A European Journal*, **16**, 7950–7958.
132. Mitsubara, H., Tsukida, M., Ishihara, D., Kuniyoshi, K., Ryu, I. (2010). A Photoirradiative phase-vanishing method: Efficient generation of HBr from alkanes and

molecular bromine and its use for subsequent radical addition to terminal alkenes. *Synlett*, 2014–2018.

133. Podgoršek, A., Stavber, S., Zupan, M., Iskra, J. (2006). Association of fluorous "phase-vanishing" method with visible-light activation in benzylic bromination by bromine. *European Journal of Organic Chemistry*, 483–488.

134. Futamura, S., Zong, Z.-M. (1992). Photobromination of side-chain methyl groups on arenes with *N*-bromosuccinimide. Convenient and selective syntheses of bis(bromomethyl)- and (bromomethyl)methylarenes. *Bulletin of the Chemical Society of Japan*, **65**, 345–348.

135. Ge, F., Wang, Z., Wan, W., Lu, W., Hao, J. (2007). One-pot synthesis of 2-trifluoromethyl and 2-difluoromethyl substituted benzo-1,3-diazoles. *Tetrahedron Letters*, **48**, 3251–3254.

136. Tutar, A., Berkil, K., Hark, R.R., Balci, M. (2008). Bromination of 5-methoxyindane: Synthesis of new benzoindenone derivatives and ready access to 7*H*-benzo[*c*]fluoren-7-one skeleton. *Synthetic Communications*, **38**, 1333–1345.

137. Tutar, A., Cakmak, O., Balci, M. (2001). Photobromination of indane: Preparation of bromoindenones and ready access to benzo[*c*]fluorenone skeleton. *Tetrahedron*, **57**, 9759–9763.

138. Çakmak, O., Kahveci, I., Demirtaş, Í., Hökelek, T., Smith, K. (2000). Bromination of tetralin. Short and efficient synthesis of 1,4-dibromonaphthalene. *Collection of Czechoslovak Chemical Communications*, **65**, 1791–1804.

139. Adam, W., Cakmak, O., Saha-Moeller, C., Tutar, A. (2002). Highly brominated norbornanes by photobromination as precursors for the convenient synthesis of 2,3,5,6-tetrabromo- and 2,3,5,6-tetramethoxy-substituted norbornadienes. *Synlett*, 49–52.

140. Schalttegger, H. (1950). Die direkte photobromierung der cholesterylester in der allylstellung (C7-position). Sterine als ionoide systeme III. *Helvetica Chimica Acta*, **33**, 2101–2110.

141. Itoh, A., Hashimoto, S., Kuwabara, K., Kodama, T., Masaki, Y. (2005). Facile solar oxidation of alcohols with molecular oxygen. *Green Chemistry*, **7**, 830–832.

142. Hirashima, S.-i., Hashimoto, S., Masaki, Y., Itoh, A. (2006). Aerobic photo-oxidation of alcohols in the presence of a catalytic inorganic bromo source. *Tetrahedron*, **62**, 7887–7891.

143. Sugai, T., Itoh, A. (2007). Aerobic oxidation under visible light irradiation of a fluorescent lamp with a combination of carbon tetrabromide and triphenyl phosphine. *Tetrahedron Letters*, **48**, 9096–9099.

144. Itoh, A., Hashimoto, S., Masaki, Y. (2005). Facile aerobic photooxidation of alcohols in the presence of catalytic lithium bromide. *Synlett*, 2639–2640.

145. Hirashima, S., Itoh, A. (2007). Aerobic visible light-oxidation of aromatic methyl groups to carboxylic acids. *Photochemical & Photobiological Sciences*, **6**, 521–524.

146. Hirashima, S., Itoh, A. (2007). Aerobic oxidation of alcohols under visible light irradiation of fluorescent lamp. *Green Chemistry*, **9**, 318–320.

147. Kuwabara, K., Itoh, A. (2006). Facile aerobic photooxidation of alcohols in the presence of catalytic *N*-bromosuccinimide. *Synthesis*, 1949–1952.

148. Hirashima, S.-i., Itoh, A. (2006). Synthesis of benzoic acids by aerobic photooxidation with hydrobromic acid. *Synthesis*, 1757–1759.

149. Itoh, A., Hashimoto, S., Kodama, T., Masaki, Y. (2005). Aerobic photooxidation of methyl group at aromatic nucleus with LiBr. *Synlett*, 2107–2109.

150. Itoh, A., Kodama, T., Hashimoto, S., Masaki, Y. (2003). Oxidation of the methyl group at the aromatic nucleus with molecular oxygen in the presence of *N*-bromosuccinimide under photoirradiation. *Synthesis*, 2289–2291.

151. Tada, N., Ban, K., Hirashima, S.-i., Miura, T., Itoh, A. (2010). Direct synthesis of α-bromoketones from alkylarenes by aerobic visible light photooxidation. *Organic and Biomolecular Chemistry*, **8**, 4701–4704.

152. Tada, N., Ban, K., Yoshida, M., Hirashima, S.-i., Miura, T., Itoh, A. (2010). Aerobic photooxidation of benzylamide under visible light irradiation with a combination of 48% aq HBr and $Ca(OH)_2$. *Tetrahedron Letters*, **51**, 6098–6100.

153. Nakayama, H., Itoh, A. (2007). Facile synthesis of phenacyl iodides from styrenes under visible light irradiation with fluorescent lamps. *Tetrahedron Letters*, **48**, 1131–1133.

154. Mitamura, T., Ogawa, A. (2011). Synthesis of 2,4-diiodoquinolines via the photochemical cyclization of *o*-alkynylaryl isocyanides with iodine. *The Journal of Organic Chemistry*, **76**, 1163–1166.

155. Nakayama, H., Itoh, A. (2008). Aerobic photooxidation of benzylamide in the presence of catalytic iodine. *Synlett*, 675–678.

156. Nakayama, H., Itoh, A. (2006). Aerobic oxidation of benzyl- and allylic alcohols under visible light irradiation of a fluorescent lamp in the presence of catalytic iodine. *Chemical and Pharmaceutical Bulletin*, **54**, 1620–1621.

4

PHOTOGENERATION OF BIRADICALS AND RADICAL PAIRS

4.1 INTRODUCTION

Biradical intermediates have a limited role in thermal reactions, while they quite often intervene in photochemical reactions. Except that under pyrolytic conditions, formation of a radical pair is *thermally* viable only for weak bonds, as in the case of peroxides, or at any rate in favorable situations, as with azo derivatives containing stabilized radicals, such as the 2-cyano-2-propyl radical. In most of the applications, such intermediates initiate chain radical processes, but are not incorporated in the final products [1]. An important group of reactions that intrinsically produce aromatic biradicals involves the cycloaromatization of highly unsaturated hydrocarbons, as in the Bergman and the Myers–Saito reactions (see Section 4.4).

On the contrary, the high energy of excited states makes fragmentation of high-energy bonds viable under mild conditions [2]. Fragmentation of a single bond yields a radical pair or, when the fragments are connected by a tether, a biradical. A chemical transformation at one of the radical centers, or recombination in a different way, then leads to the products. A well-known example is the photo-Fries reaction [3] that, contrary to the thermal version, follows a homolytic path (Scheme 4.1).

A further possibility is that elimination of a neutral fragment precedes recombination, resulting in the extrusion of a molecule, such as CO, N_2, and SO_2 from ketones, azo compounds, and sulfones, respectively (Scheme 4.2) [4, 5].

Indeed, many excited states exhibit a radical chemistry. As an example, atom abstraction is quite common and, when occurring intramolecularly, leads to a

Photochemically-Generated Intermediates in Synthesis, First Edition.
Angelo Albini and Maurizio Fagnoni.

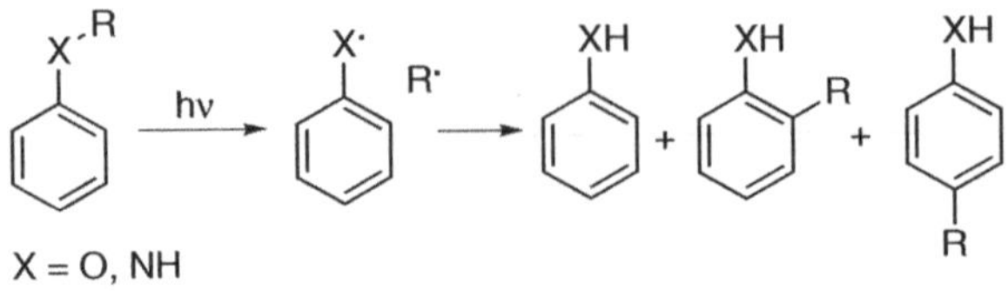

Scheme 4.1.

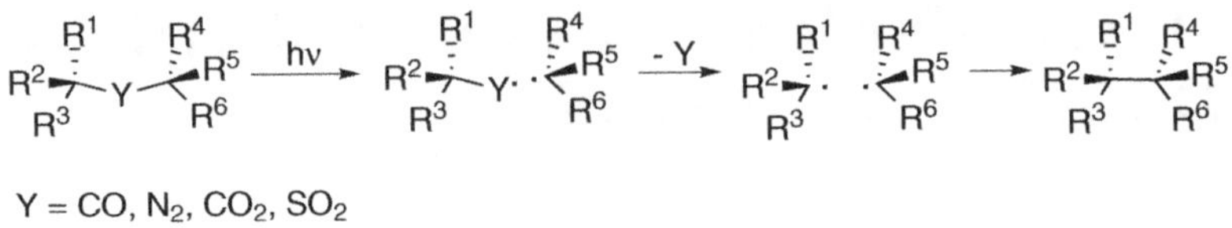

Scheme 4.2.

biradical and often to cyclization. As for cycloaddition reactions, a biradical path has a small role in thermal reactions, while it largely predominates among photochemical processes. The last class of reactions is one of the most important—and most typical—in photochemistry, including the 2 + 2 carbonyl + alkene (Paternò-Büchi) [6] and conjugated carbonyl + alkene cycloaddition reactions [7]. There is no relation in this case with thermal processes, and it has been thought nonappropriate to include these reactions here.

The reactions actually presented in this chapter are, for a large part, based on the rich photochemistry of the carbonyl moiety—in particular, on the generation of a biradical by a Norrish type I [4, 5] or type II reaction [8–11]. These are depicted in the Scheme 4.3 for the case of a cyclohexanone derivative. Intramolecular hydrogen abstraction from the geometrically favored γ position (path *a*) is usually efficient and

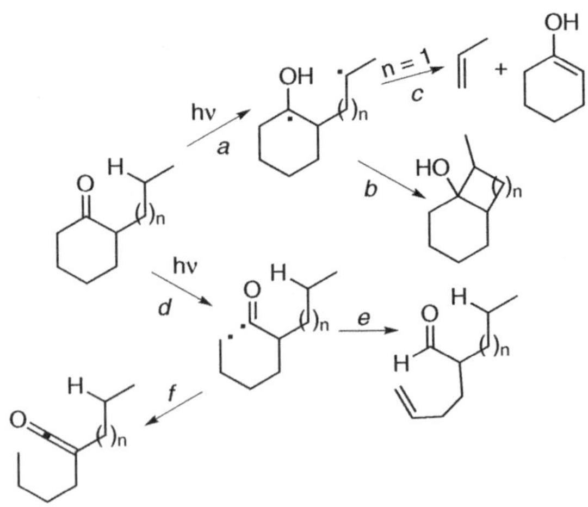

Scheme 4.3.

leads to a 1,4-biradical. C–C bond cleavage to form an alkene and an enol (path *c*, a Norrish II process) often ensues. Such a reaction has a limited synthetic utility and is outside the scope of this chapter. Alternatively, a cyclobutane is formed upon radical recombination (Norrish–Yang reaction, $n = 1$, path *b*) [8–11]. When the γ-position bears no hydrogen, or is not easily accessible, abstraction from the δ or ε position to give five- or six-membered rings is possible, as is also β-abstraction to form a cyclopropane ring when this position is activated. On the other hand, long-range abstraction to give macrocycles may be effective from appropriate structures. The Norrish I process consists in the homolytic scission of the C–C bond adjacent to the carbonyl group. Starting from a cyclic compound, a biradical is formed and undergoes intramolecular hydrogen abstraction to form an unsaturated aldehyde (path *e*) or disproportionation to give a ketene (Scheme 4.3, path *f*).

4.2 SYNTHESIS OF CYCLIC COMPOUNDS

In all of the processes considered here a ring is formed by radical–radical coupling from photogenerated biradicals. Various rings can be accessible in this way as discussed in the following.

4.2.1 Formation of a Three-Membered Ring

Three-membered rings can be formed by both Norrish I and Norrish–Yang processes. In the first case a photoextrusion of SO_2, nitrogen, and CO takes place, whereas in the latter case the usual γ-hydrogen abstraction is followed by a HX elimination to give a 1,3-biradical and by the desired cyclization. Reported examples are limited to the formation of cyclopropanes. As mentioned above, β-hydrogen abstraction is feasible only when no competitive pathways are presents. A typical example was reported for the case of β-substituted arylalkyl thiones [12, 13] such as 3-naphthylthiopivalophenone (**1**) and 2-benzyl-2-methyl-1-thiotetralone (**3**, Scheme 4.4). The 1,3-biradical appeared to be generated from both the $^1\pi\pi^*$ and $^3n\pi^*$ state. Although

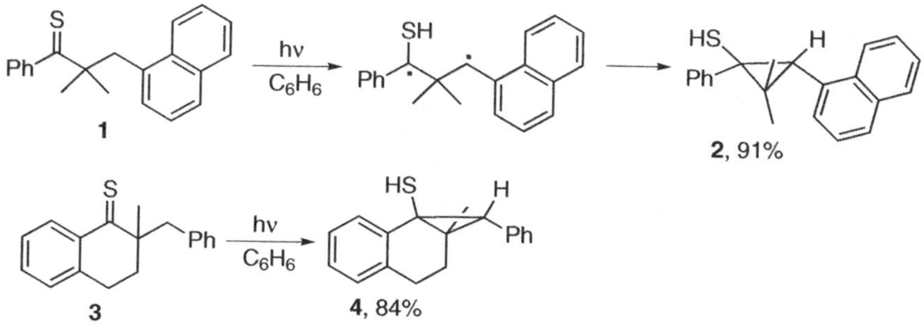

Scheme 4.4.

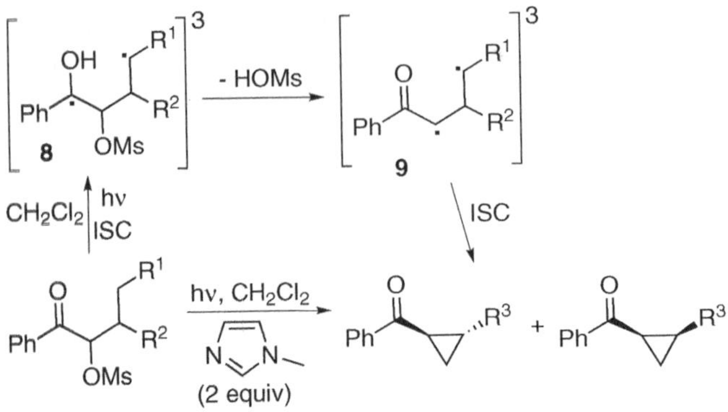

5a, R¹= H, R²= Me
b, R¹= Ph, R²= H
c, R¹= H, R²= COOMe

6a, R³= Me, 90%
b, R³= Ph, 47%
c, R³= COOMe, 59%

7a, 0%
b, 31%
c, 0%

Scheme 4.5.

disexcitation to the ground state thione was efficient, the chemical reaction was clean and cyclopropyl thiols (**2**,**4**) were formed in an excellent yield (Scheme 4.4) [12].

Most of the other reactions leading to cyclopropanes resulted from ketones bearing an electron-donating group in the β-position. In this case, the initial step was an electron transfer followed by a proton shift, rather than homolytic H-transfer [10].

Alternatively, a three-membered ring was easily accessed by an initial γ-hydrogen abstraction in derivatives bearing a leaving group (e.g., a mesylate) adjacent to the carbonyl (Scheme 4.5) [14–16]. Thus, photolysis of ketones **5a–c** initially formed 1,4-biradicals **8** that upon methansulfonic acid loss generated a 1,3-biradical **9**. Coupling then gave the corresponding cyclopropylketones **6**,**7a–c** (Scheme 4.5). The reaction gave the best result in methylene chloride, since in different media a partial fragmentation of **9** took place. Since the acidity liberated in the reaction was incompatible with the stability of the cyclopropane ring, an acid scavenger (*N*-methylimidazole, 2 equiv) was required. The reaction gave exclusively the *trans*-benzoylcyclopropanes (**6a–c**), except for the case of the β-phenyl-substituted butyrophenone (**5b**) where a 40:60 *cis–trans* mixture was obtained instead, probably due to a secondary photochemical *cis–trans* isomerization (Scheme 4.5) [14].

This process was applied to the synthesis of bicyclic derivatives with a modest ee (up to 52%) by exploiting the 1,2-chirality transfer based on the preferred collinear arrangement of the OMs group and the π system of the photochemically excited carbonyl group (a stereoelectronic effect) [16].

As mentioned before, the photoextrusion of small molecules is another widely used method to form a three-membered ring via biradicals. The smooth preparation of substituted cyclopropanes was reported by irradiation (30 min) of 2-phenylthietan-1,1-dioxides (**10**) in methylene chloride or methanol (Scheme 4.6) [17]. The presence of the phenyl group adjacent to the sulfonyl moiety assured both the absorption of light by the sulfone moiety at a reasonable

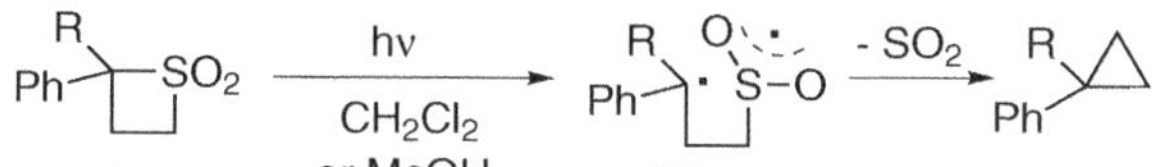

12a, R = Me; 95%
b, R = $C_6H_5CH_2$; 92%
c, R = H; 95%

Scheme 4.6.

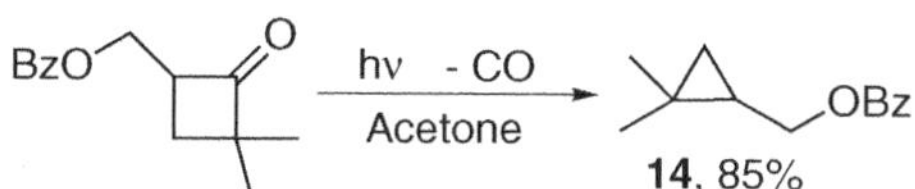

14, 85%

Scheme 4.7.

wavelength (254 nm) and the stabilization of the biradical intermediate **11**. The end phenyl cyclopropanes (**12a–c**) were isolated by bulb-to-bulb distillation in a high yield [17].

The photoinduced ring opening of a cyclobutanone offers another entry to cyclopropanes through CO elimination (Scheme 4.7). Upon direct irradiation, ring opening to the corresponding decarbonylated alkenes took place; but when the photolysis was carried out in acetone, and thus under triplet sensitization, the cyclopropane was formed as the main or exclusive product, accompanied in some cases by small amounts of the olefin [18]. The success of the photodecarbonylation depended on the α-substituents and, when these exerted a stabilizing effect, as with methyl groups in compound **13**, benzyl ether **14** was obtained in 85% yield as the only isolated product (Scheme 4.7).

The photochemically induced loss of nitrogen from cyclic azoalkanes has likewise been used for the synthesis of cyclopropanes [19–22]. Scheme 4.8 reports a representative example regarding the photolysis of *anti*-hexahydro-1,4:5,8-dimethanophthalazine **15** [20], the course of which depended on the conditions. Thus, triplet sensitization by benzophenone led to a mixture of products (including a rearranged azirane), while addition of a triplet quencher such as *trans*-piperylene drove the reaction toward the cleavage of the C–N bond from ^{1}**15**. Housane **16** was then isolated as the exclusive product (Scheme 4.8).

The photochemistry of compound **15** (Scheme 4.8, R = H) was likewise investigated in confined media or in the crystalline state [21]. In the latter case, a clean formation of **16** was observed and was controlled by the close packing in the crystal

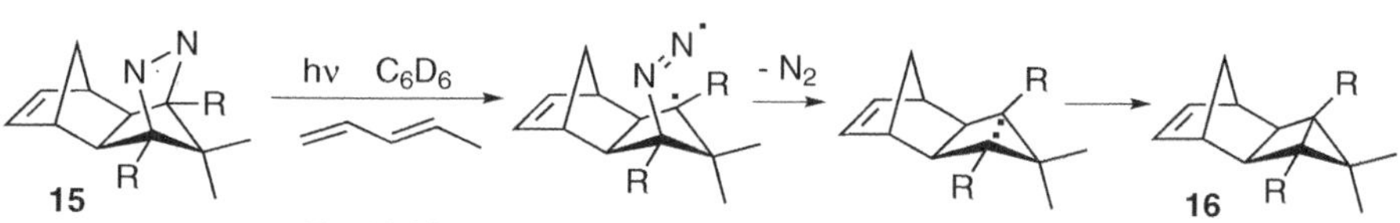

R = H, Me

Scheme 4.8.

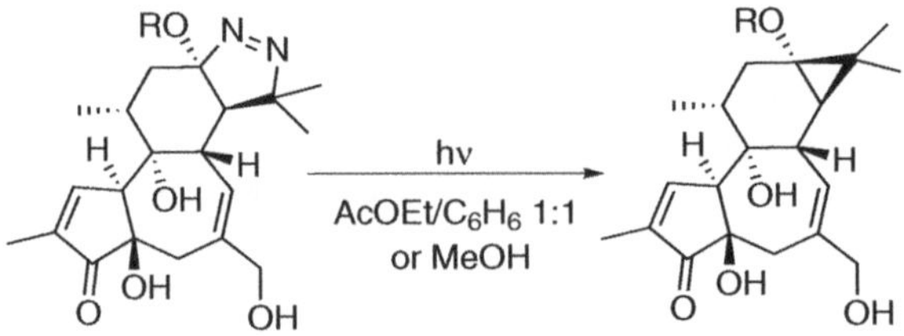

17, R = Ac
18, R = PhCH$_2$CO

19, R = Ac (67% in MeOH, 92% in AcOEt/C$_6$H$_6$ 1:1)
20, R = PhCH$_2$CO, 90%

Scheme 4.9.

lattice. On the other hand, the irradiation in zeolites promoted the formation of the azirane at the expense of the housane [21]. A related clean reaction was reported in the photochemical denitrogenation of 2,3-diazabicyclo[2.2.1]hept-2-ene to give the corresponding bicyclo[2.1.0]pentane [19].

Interestingly, the photoextrusion of nitrogen was likewise used in the key step of the preparation of prostratin (**19**, 12-deoxyphorbol-13-acetate) and DPP (**20**, 12-deoxyphorbol-13-phenylacetate), compounds exhibiting a potent in vitro activity in inducing HIV expression in latently infected cell lines and primary cells [22]. Diazene precursors **17** and **18** (easily prepared from crotophorbolone, a component of the *Jatropha curcas* seed oil) gave **19** and **20** respectively in a very high yield (>90%, Scheme 4.9).

4.2.2 Formation of a Four-Membered Ring

Building a four-membered ring, whether homo- or heterocyclic, is viable by both the Norrish I and the Norrish–Yang reactions [8–11]. In a representative example of the former case, sulfone **21** underwent photochemical extrusion of SO_2 when irradiated in an acetone–benzene 1:1 mixture, thus affording 9,10-diphenyltetracyclo [6.2.0.0^{4,10}.0^{5,9}]decane **22** in 95% yield (Scheme 4.10) [23]. Similarly, nitrogen loss upon photolysis of 2-(2,6-dimethylpheny1)-2*H*-1,2,3,4-benzothiatriazine 1,1-dioxide led to ring contraction and to the corresponding benzothiazete 1,1-dioxide [24].

The Norrish–Yang reaction is particularly suitable for the preparation of (nitrogen containing) four-membered rings and results from the intramolecular γ-hydrogen abstraction process. In such a way, enantiomerically pure highly functionalized

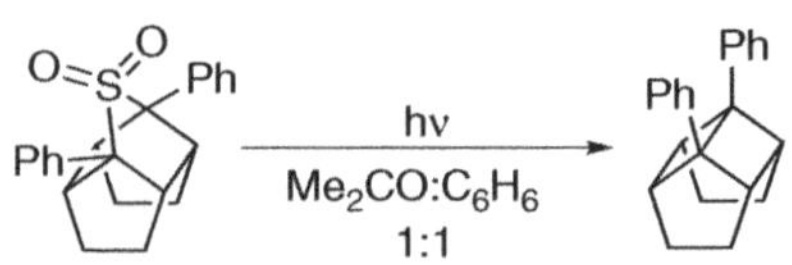

21 **22**, 95%

Scheme 4.10.

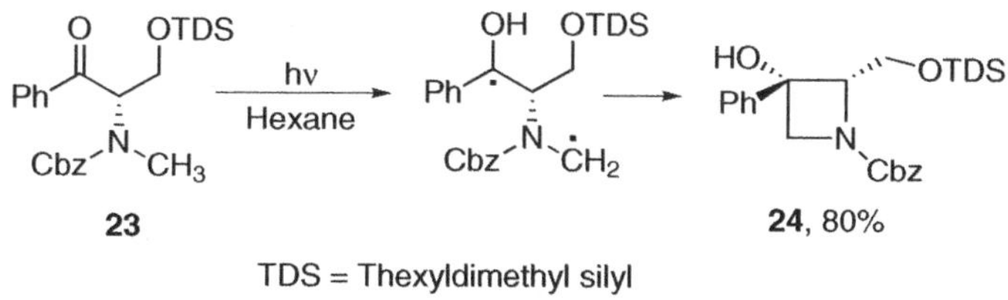

Scheme 4.11.

azetidinols **24** were obtained starting from amino ketones **23** (Scheme 4.11) [25]. It was shown that for the success of the photocyclization, the role of H-bonding and the preferred conformation of the triplet biradical intermediate were of key importance.

The valuable β-lactam moiety could likewise be obtained by a Norrish–Yang reaction in variable yields by irradiation of phenylglyoxylamides in acetonitrile. Interestingly, the presence of a small amount of acid in the reaction mixture was required to drive the reaction to completion and also affected the diastereoselectivity of the photocyclization [26]. In some cases, the reaction was carried out either in the solid state [27] or from inclusion molecular complexes, in order to improve the yield of formation of the β-lactam ring. For instance, the irradiation in solution of *N,N*-dialkylpyruvamides did not lead to the corresponding β-lactams. On the contrary, when desoxycholic acid (**25**) was crystallized by using pyruvamides **26** as the solvent, irradiation of the resulting inclusion complex formed (**25**:**26** 4:1 molar ratio) gave the desired four-membered ring **27** as a mixture of two stereoisomers (Scheme 4.12) [28].

Moreover, different chiral hosts were adopted to impart chirality during the photochemical cyclization. Thus, an inclusion complex was prepared by mixing achiral *N,N*-dimethylphenylglyoxylamide guests with chiral (*R,R*)-(−)-*trans*-4,5-bis(hydroxydiphenylmethyl)-2,2-dimethyl-1,3-dioxacyclopentane [29]. A suspension of this complex in water containing a surfactant (hexadecyltrimethylammonium bromide) was irradiated under stirring and afforded the corresponding β-lactam in 70% yield and with a 61% ee. It is noteworthy that the dioxacyclopentane chiral host could be largely recovered from the end mixture by filtration and recrystallization [29].

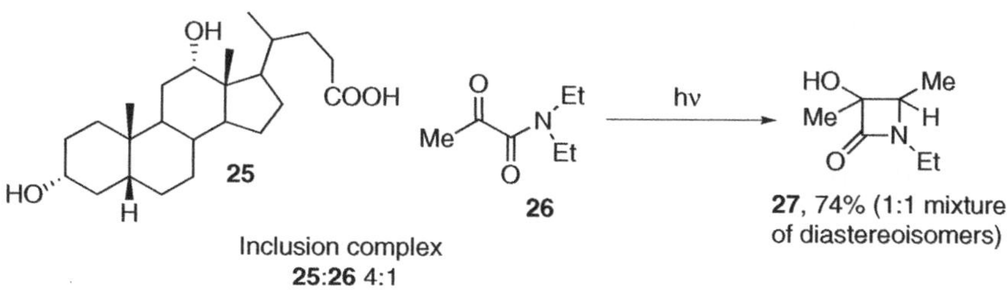

Scheme 4.12.

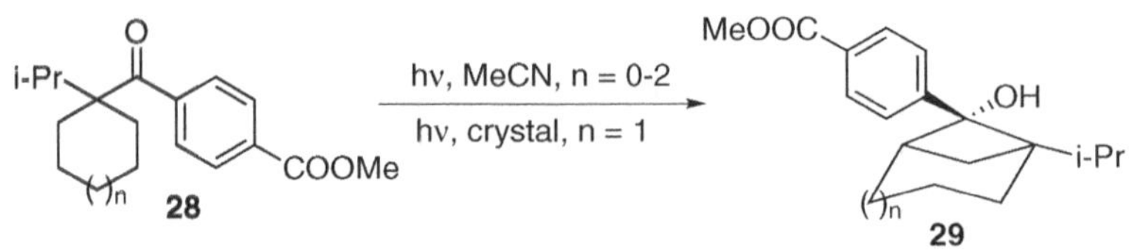

Scheme 4.13.

The photogeneration of 1,4-biradicals can likewise be carried out in the solid state in the absence of host compounds [27]. An interesting example is shown in Scheme 4.13 for the preparation of aryl substituted cyclobutanols **29** from the photolysis of 1-isopropylcycloalkyl aryl ketones **28** [30]. The course of the reaction depended both on the size of the cycloalkyl ring and on the reaction conditions. In each case, the photoreaction in acetonitrile gave the desired cyclobutanes in excellent chemical yields (>95% by GC, Scheme 4.13). A similar behavior was reported for the photolysis in the solid state only for the case of the cyclohexyl derivative ($n = 1$). In fact, under these conditions when $n = 0$ no reaction occurred, whereas for $n = 2$ no cyclobutanol-type photoproducts were formed [30]. Such a behavior was attributed to the unusually large distance to be confronted for 1,4-cyclization for $n = 0$, 2 that made back hydrogen transfer predominant in those cases.

A diastereospecific Norrish–Yang cyclization occurred upon photolysis of crystals of 2,4,6-triisopropylbenzophenone derivatives having a (*S*)-phenylethylamide chiral auxiliary group (e.g., **30**). This gave a (*R*,*S*)-cyclobutenol **31a** as the exclusive product in quantitative yield (Scheme 4.14) [31]. However, when this compound was irradiated in solution, the (*R*,*S*)-isomer was accompanied by an equimolar amount of the (*S*,*S*)-isomer **31b**.

A cyclobutanol ring was likewise formed through a Norrish–Yang type cyclization and used for the synthesis of highly enantiomerically enriched tricyclic tetralin derivatives [32]. The reaction was carried out in the solid state at −43°C on the chiral

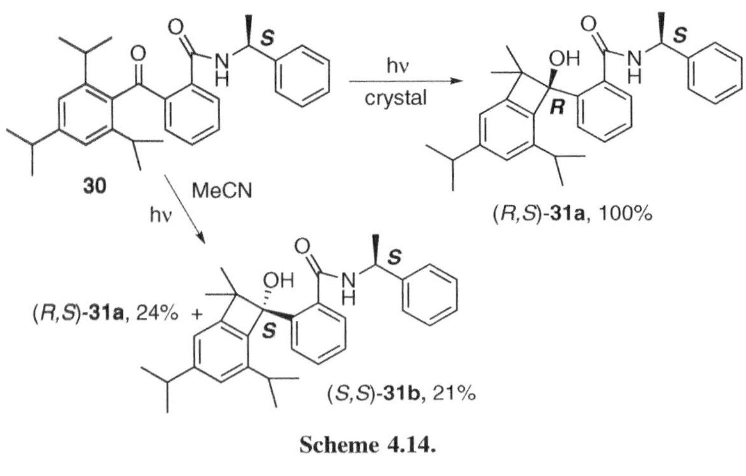

Scheme 4.14.

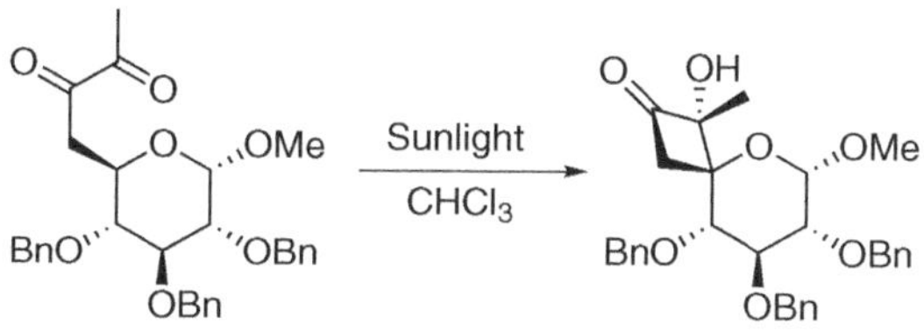

32 **33**, 100% (*2R, 5R*)

Scheme 4.15.

ammonium salts of 2,2-diethyl-1-tetralone-6-carboxylic acid. The presence of the ionic chiral auxiliary allowed for a single crystal-to-single crystal reaction and the product was formed with an enantiomeric excess of up to 99%.

Biradical intermediates were likewise involved in the building of the cyclobutanone moiety. The reaction was applied to carbohydrate models for the preparation of *C*-ketosides [33, 34]. Here the Norrish–Yang photocyclization of 1-glycosyl-2,3-butandione derivatives was well suited for the stereocontrolled synthesis of chiral cyclobutanols fused to the carbohydrate backbone as shown in Scheme 4.15. The 1,2-diketone moiety in compound **32** absorbed at a relatively long wavelength. This permitted the use of sunlight for the synthesis of the optically pure 1-hydroxy-1-methyl-5-oxaspiro[3.5]nonan-2-one carbohydrate **33** in a quantitative yield [33, 34].

A related reaction is shown in Scheme 4.16 and again involves α-diketones. The intramolecular hydrogen abstraction by one of the carbonyls in compound **34** led to the formation of α-hydroxy-cyclobutanones **35a–d** in a good yield that was little affected by the size of the initial cycloalkyl ring [35]. Photosensitization by benzophenone was adopted here to improve the yield.

Sparse examples in the literature concern a domino process involving two consecutive Norrish–Yang reactions, as in the photolysis of *N*-(1-adamantyl)phthalimide **36** to form the hexacyclic benzazepine derivative **38** (Scheme 4.17) [36]. Thus, compound **36** underwent initial γ-hydrogen abstraction and finally gave the tetracyclic compound **37** upon radical recombination. Opening of the four-membered ring gave an azepine derivative that in turn underwent a further intramolecular hydrogen abstraction and formed again a cyclobutane ring, thus giving the substituted adamantane **38** in 82% yield.

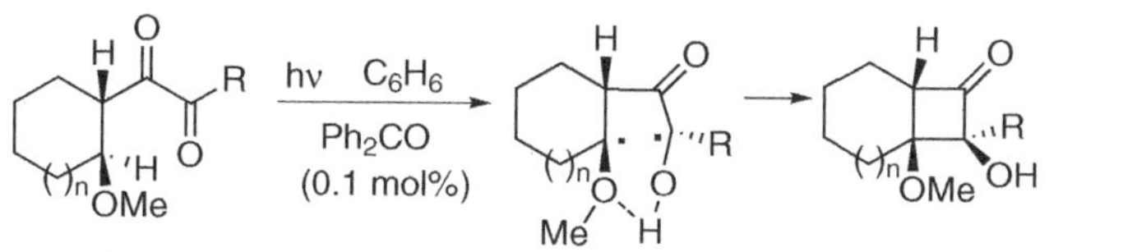

34

35a, n = 1, R = Me, 74%
b, n = 0, R = n-Pr, 51%
c, n = 1, R = n-Pr, 70%
d, n = 2, R = n-Pr, 74%

Scheme 4.16.

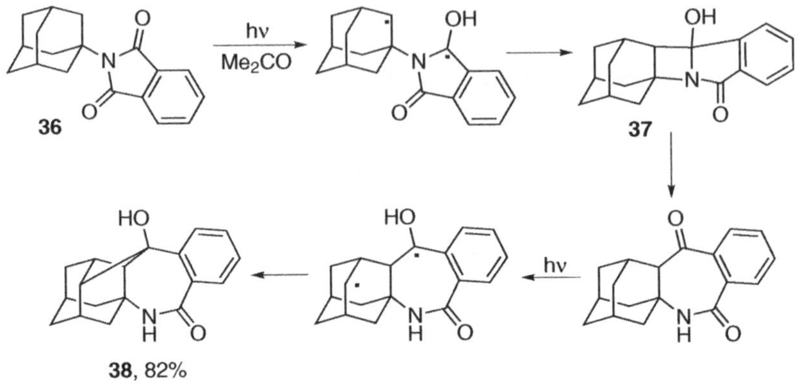

Scheme 4.17.

The photochemical formation of a heterocycle for the synthesis of fused alkylideneoxetanols has been likewise considered (Scheme 4.18). This is based on a [5 + 2] photocycloaddition/Norrish–Yang cascade sequence from a maleimide [37]. *N*-Pentenyl-substituted maleimide (**39**) was irradiated under standard continuous flow conditions (see Chapter 1) to give fused azepine **40** along with a small amount of isomeric **41**. Compound **40** underwent a second photochemical reaction and substituted oxetanol **42** was isolated in 76% yield [37].

As the last example, an initial ring opening could lead to ring contraction. As an example, the photolysis of aliphatic imide **43** at 254 nm gave the azetidine-2,4-dione **45** in 50% yield. This arose from a further photochemical reaction of the primarily formed aldehyde **44** (Scheme 4.19) [38, 39].

4.2.3 Formation of a Five-Membered Ring

The preparation of five-membered rings by decarbonylation of cyclohexanone derivatives via a Norrish I process is sparsely documented in the literature. Some

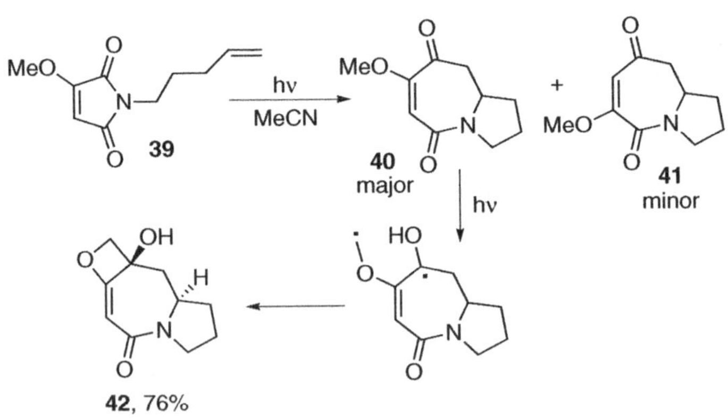

Scheme 4.18.

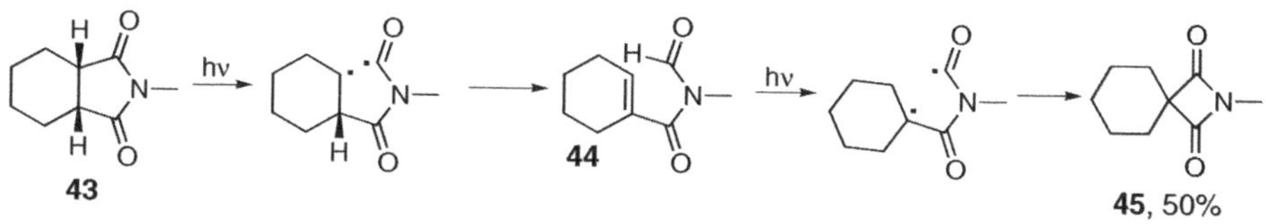

Scheme 4.19.

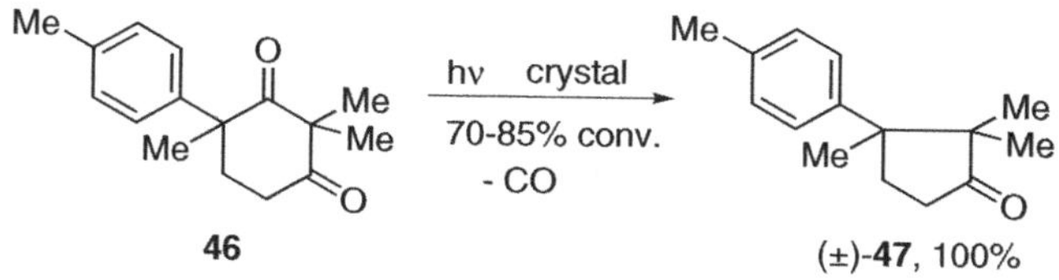

Scheme 4.20.

of these reactions were carried out in the solid state [4, 27] and some natural products were accessed in this way. Scheme 4.20 shows the synthesis of racemic (±) α-cuparenone (**47**) from cyclohexanedione **46** [40]. The chiral version of the reaction was likewise performed by having recourse to chiral difluorodioxaborinane complexes [40].

A similar approach was employed in the total synthesis of (±)-herbertenolide. The interest for this reaction relied on the efficient generation of the radical pair and on the good stereochemical control obtained in the solid state [41]. A photochemical decarbonylation was likewise the key step in the synthesis of carbohydrate derivatives starting from compounds containing the bicyclo[3.2.1]octane backbone [42].

Five-membered rings were conveniently formed by a Norrish–Yang reaction via an initial intramolecular δ-hydrogen abstraction [8–11]. Most of the products formed in this way contained a pyrrolidine moiety, as indeed present in natural occurring compounds. As an example, the photolysis of α-keto ester **48** in *tert*-butyl alcohol (where the keto tautomer was predominant) gave **49** in 70% yield (1:1 mixture of two diastereoisomers, Scheme 4.21) [43]. This lactam was further elaborated for the preparation of the pyrrolizidine alkaloid (±)-isoretronecanol (**50**).

The formation of γ-lactams was efficient even when the starting material was adsorbed on silica gel. This was the case for β-oxo amides such as *N,N*-dibenzyl-2-benzoylacetamide that, upon irradiation by a 100-W high-pressure mercury lamp,

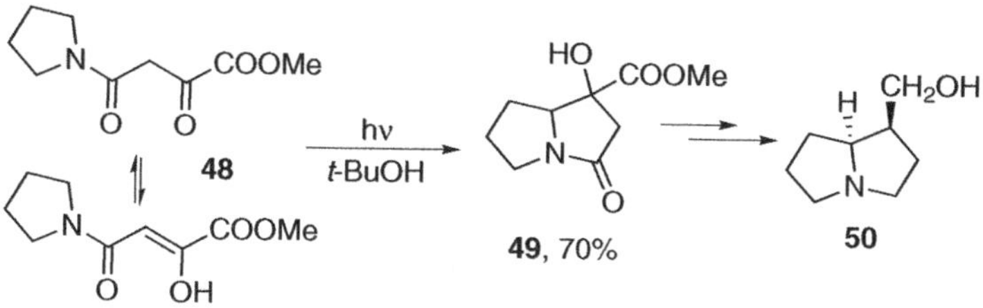

Scheme 4.21.

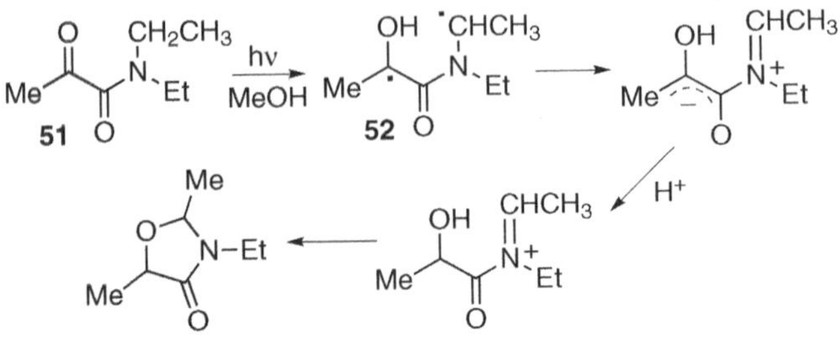

Scheme 4.22.

afforded the corresponding five-membered lactam in a yield comparable to, or even higher than, that obtained in benzene solution [44]. The success of the reaction was attributed to the large surface area of the silica gel adsorbent.

On the other hand, 1,4-biradicals could be initially formed and then converted to a zwitterion. This was the case for α-oxo amide **51**, where the zwitterion was protonated and underwent intramolecular nucleophilic cyclization to yield oxazolidin-4-one **53** in a quantitative yield (Scheme 4.22) [45]. Depending on the reaction conditions and on the substituents present in the oxoamide, other competing pathways could likewise operate, namely cyclization to β-lactam (from **52**) and photoelimination. The adoption of a protic solvent was found to enhance the formation of the oxazolidinone, whereas the presence of radical-stabilizing substituents directed the process toward the formation of β-lactams [45].

The Norrish–Yang cyclization was found likewise useful for the synthesis of *cis*-3-hydroxyproline esters in a high diastereomeric and enantiomeric excess starting from chiral *N*-(2-benzoylethyl)-*N*-tosylglycine esters [46]. Moreover, a photocyclization took place when irradiating *o*-carboxymethylaminophenylketones **54a,b** to give predominantly *cis*-dihydroindolinols **55a,b**. The *cis* orientation in compounds **55a,b** was rationalized as being due to the lower bulkiness of the OH group with respect to the phenyl group and to the possible role of hydrogen bonding interactions with the -COR substituents (Scheme 4.23) [47].

An alternative access to a nitrogen heterocycle via photocyclization is illustrated in Scheme 4.24 for the tetrafluoroborate salt of *ortho*-benzoyl *N*-trialkylanilinium cation **57** [48]. The reactions gave quantitatively the dihydroindolium derivative **58**

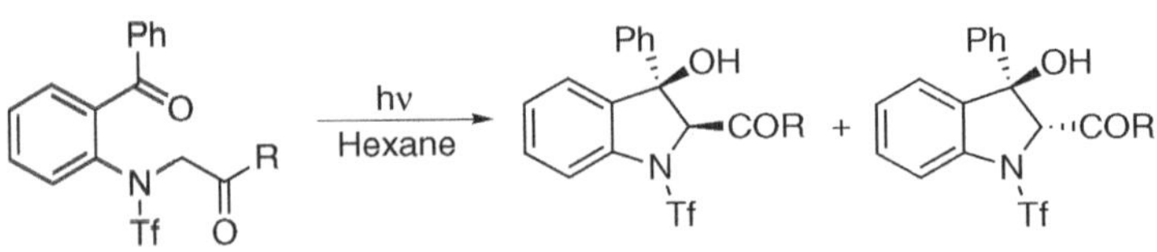

54a, R = OMe
b, R = N(CH_2)$_4$

55a, 92
b, 98

56a, 8 83%
b, 2 90%

Scheme 4.23.

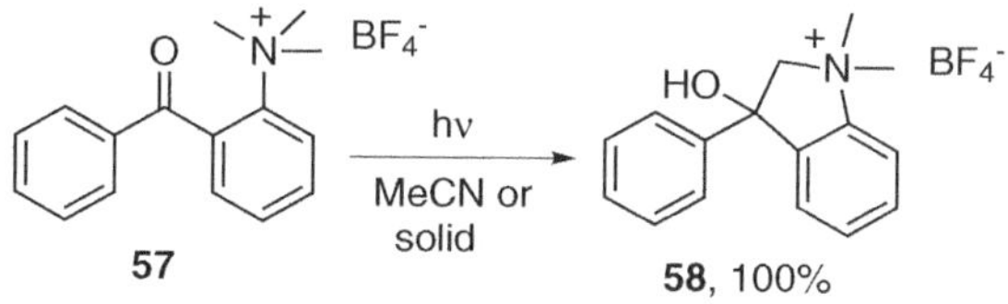

Scheme 4.24.

when ketone **57** was irradiated both in diluted acetonitrile solution and in the crystal state. The reaction was rationalized as resulting from δ-hydrogen abstraction by the triplet state of the ketone and cyclization of the 1,5-biradical (Scheme 4.24) [48].

The versatility of the Norrish–Yang reaction was conveniently applied for building oxygen heterocycles—in particular, those containing the tetrahydrofuran-3-ol moiety present in important biologically active compounds, such as lignans. This radical annulation route was applied to the synthesis of racemic (±)-paulownin (**61**) and (±)-isogmelinol (**62**, Scheme 4.25). Both compounds were accessible from the photolysis in benzene of the 5-aryl-tetrahydrofuran-3-ones **59** and **60**, respectively [49, 50]. In this way, compound **61** has been prepared in a seven-step route from piperonal [49].

This approach was also used as one of the key steps in the preparation of Aflatoxin M_2 via the photocyclization of 2,6-disubstituted acetophenones [51] and in the preparation of dihydrobenzofuranol by irradiation of ethyl 2-formylphenoxyacetates or ethyl acetylphenoxyacetates [52]. The effect of the reaction medium on the selectivity in the photocyclization of related 2-benzoylphenoxyacetates was investigated [53]. A decrease in stereoselectivity was found when substituting benzene by more polar solvents, such as acetonitrile and methanol. This change was attributed to the establishment of an intermolecular hydrogen bond between the hydroxyl group of the 1,5-biradicals and the solvent [53]. The reactions likewise worked when acyclic, nonaromatic derivatives were used. The photoinduced δ-hydrogen abstraction was efficient in the irradiation of β-allyloxy-carbonyl compounds (prepared from terpenoid ether adducts of geraniol and nerol), where a mixture of tetrahydrofuran-3-ols was obtained [54].

The synthetic significance of the Norrish–Yang reaction in the preparation of five-membered rings via 1,5-biradical cyclization was demonstrated also by the synthesis

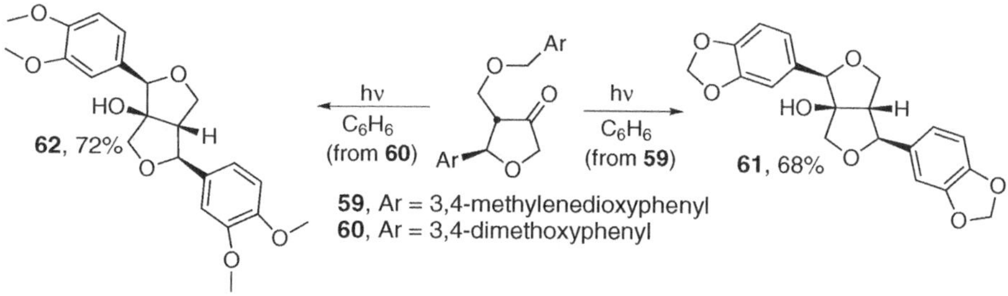

Scheme 4.25.

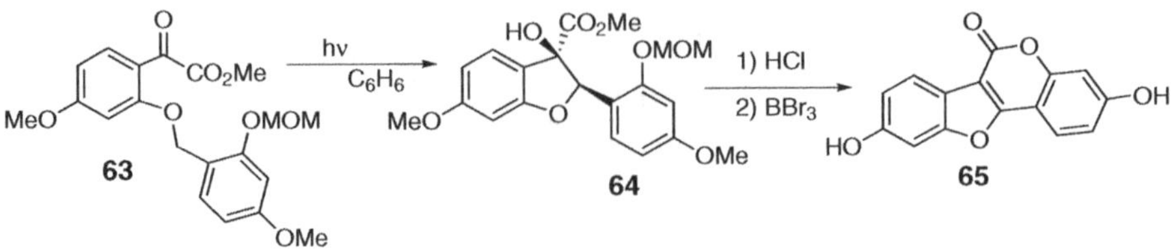

Scheme 4.26.

of coumestrol (**65**), a phytoestrogen found in plants (Scheme 4.26) [55]. The synthetic plan was based on the synthesis of glyoxylate ester **63** in a few steps from 1,3-dimethoxybenzene. Photolysis of **63** caused hydrogen abstraction from the benzylic position and led to hydroxyester **64** that was not isolated from the reaction mixture but treated with HCl to allow the deprotection of the MOM group. Cyclization to the lactone followed by treatment with BBr_3 converted the two methyl ethers into the corresponding hydroxy groups (Scheme 4.26) [55].

Hydrogen abstraction from a benzylic position was likewise exploited in the synthesis of tetrahydropyrrolo[2,1-*a*]isoindol-5-ones [56]. The stereoselectivity observed in the process depended on the acyl substituent R in the starting 3-(acylmethyl)-2-alkylisoindol-1-ones (**66a,b**, Scheme 4.27). When R = Me, the OH and the Ph groups in compound **67** were at the opposite side of the acylpyrrolidine ring, whereas when R = Ph the corresponding diastereoisomer **68** was obtained [56].

The initial formation of a 1,4-biradical from a Norrish–Yang reaction can likewise lead to the formation of a five-membered ring. A recent example is shown in Scheme 4.28. The course of the reaction depended on the nature of the substituent R in aromatic γ,δ-epoxy ketone **69**. In fact, the irradiation of substituted derivatives (e.g., R = Me, path *a*) led to epoxide ring opening from 1,4-biradical **71** leading to indanone **70** that was isolated in 62% yield. On the contrary, in unsubstituted epoxyketones (R = H, path *b*) a cyclization/semi-pinacol rearrangement cascade led to benzocyclobutanones (compound **72**, 73% yield, Scheme 4.28) [57].

Recently, an efficient intramolecular γ-H-atom abstraction was observed by irradiating γ-azido butyrophenone derivatives. The 1,4 biradical here lost a nitrogen molecule and gave a 1,5-ketyl iminyl biradical that upon cyclization/dehydration yielded a pyrrole [58].

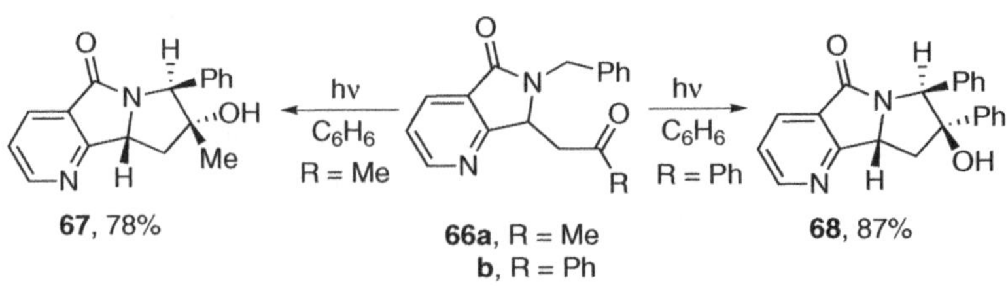

Scheme 4.27.

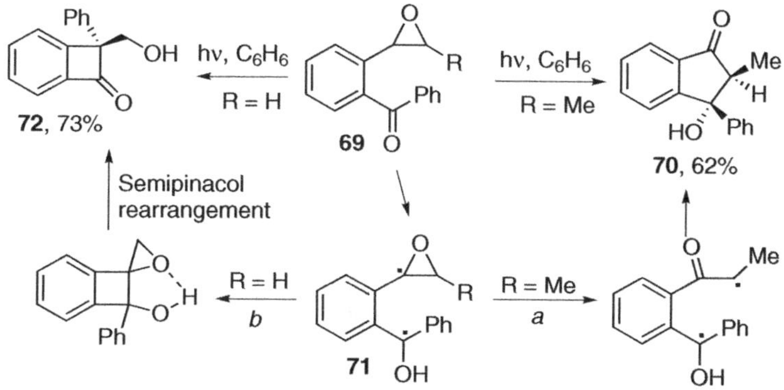

Scheme 4.28.

4.2.4 Formation of a Six-Membered Ring

The formation of six-membered rings via biradicals relies on the intramolecular ε-hydrogen abstraction by an excited carbonyl and Norrish–Yang cyclization through the intermediacy of a 1,6-biradical [8–11]. The reaction was efficient when no hydrogen was present in either γ- or δ-positions. Thus, β-(*o*-tolyl)- or β-mesityl-propiophenone **73** and -isobutyrophenone underwent photocyclization to 2-tetralols **74** in up to nearly quantitative yield (Scheme 4.29) [59].

o-Methylphenyl-substituted 1,3-dicarbonyl compounds likewise formed tetralones in yields not exceeding 50% [60].

The preparation of nitrogen-containing six-membered rings was obtained by a Norrish–Yang cyclization, as exemplified in Scheme 4.30. In particular δ-lactams **76a,b** were easily accessible in an almost diastereomeric pure form by following the above procedure and irradiating chiral α-acylamino γ-ketoamides **75a,b** [61]. The high selectivity observed was explained by the chair-like conformation adopted by the 1,6-biradical intermediate. As a result, in the synthesis of **76a** the orthogonal

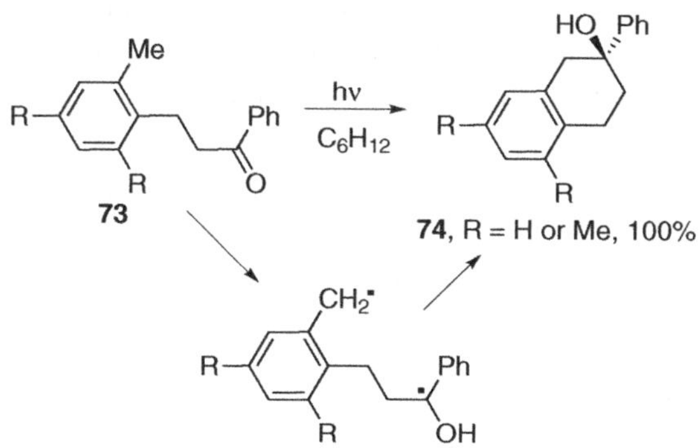

Scheme 4.29.

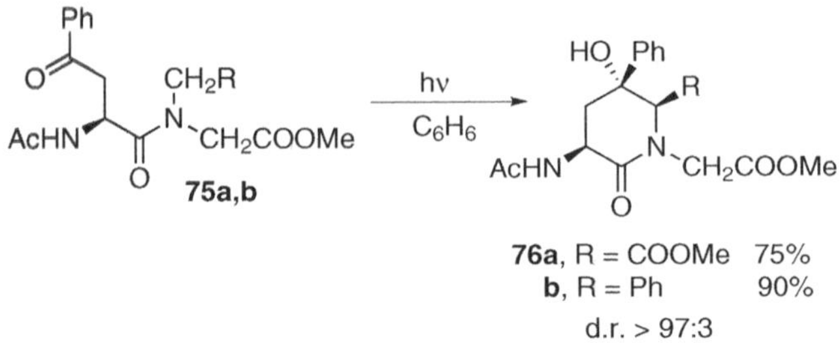

Scheme 4.30.

biradical conformation was sufficient to direct the radical coupling despite the strong steric interaction between the –OH and the –COOMe groups [61].

Under similar conditions (DCM was used as the solvent in place of benzene), the irradiation of 4-oxo-4-phenyl-butanoyl amines gave the corresponding δ-lactams with a remarkable diastereoselectivity (>99%) via hydrogen abstraction from the ε-position [62]. Moreover, bicyclic derivatives based on the indolizinone skeleton were synthesized in satisfying yields starting from proline containing dipeptides. This approach represented an interesting route toward bicyclic β-turn dipeptides [63].

In favorable cases, very complex structures were prepared via photogenerated 1,6-biradicals in a fully diastereoselective fashion, as was the case for the preparation of compounds **78a,b** having the methanotropane skeleton (Scheme 4.31) [64]. This was a key step in the design of unnatural amino acids and an example of novel chimerical amino acids. The stereoselectivity observed has been attributed to a different stability of the triplet biradicals conformers, as supported by conformational analysis studies on simplified models [64].

In rare instances, the Norrish–Yang cyclization led to the formation of a six-membered ring via a C–O bond formation as in the photoreaction of α-mesyloxy-β-keto amides (e.g., **79**, Scheme 4.32) [65]. Contrary to the previous cases, the reaction started with a δ-hydrogen transfer that gave an ambident enolate biradical upon methanesulfonic acid release. It is noteworthy that 3,4-dihydro-2*H*-1,3-oxazin-4-ones **80a–e** were finally obtained in 40–60% yields by radical coupling involving an oxygen atom [65]. Investigations by using DFT and *ab initio* methods did not give a fully satisfactory rationalization of the actual mechanism.

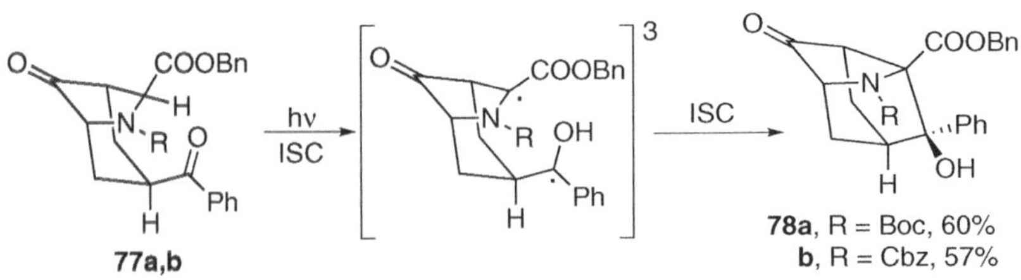

Scheme 4.31.

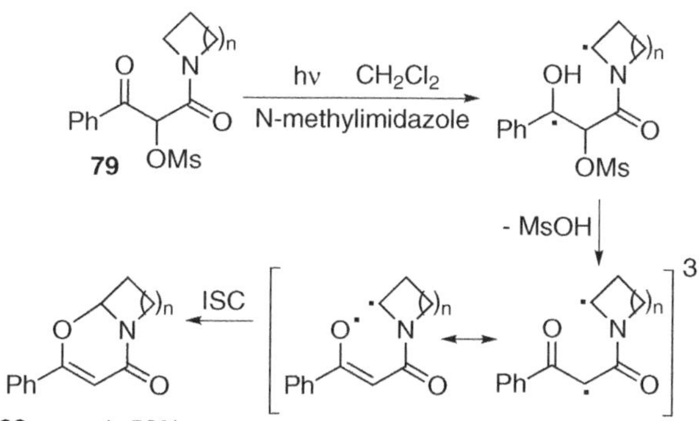

80a, n = 1, 58%
b, n = 2, 42%
c, n = 3, 57%
d, n = 4, 64%

Scheme 4.32.

4.2.5 Formation of a >6-Membered Ring

Building large rings is of obvious importance, albeit not an easy task. A few examples have been reported, and both Norrish type I and Norrish–Yang processes have been exploited for this purpose [10]. The photoextrusion of CO_2 was applied in one case for the synthesis of 15-membered cyclic ketones having interest for the production of perfumes, such as muscone. Photolysis of either of the cyclic tetraacyl peroxides **81** induced O–O bond cleavage that led to the elimination of four molecules of carbon dioxide. Radical–radical coupling then gave macrocarbocycles, namely cyclotetradecane (**82a**) or an acetal of cyclopentadecanone (**82b**), respectively (Scheme 4.33) [66]. The same compounds could be formed by thermal decomposition of the peroxides, albeit in a lower yield (~40 rather than ~70%).

A related approach was adopted in the synthesis of [2.2]cyclophanes by photoextrusion of sulfur dioxide from the corresponding bis sulfones [67].

Formation of 1,*n* biradicals was found to be feasible by 1,*n* hydrogen atom abstraction for $n > 5$ as in the photocyclization of ketoesters **83**. In fact, the

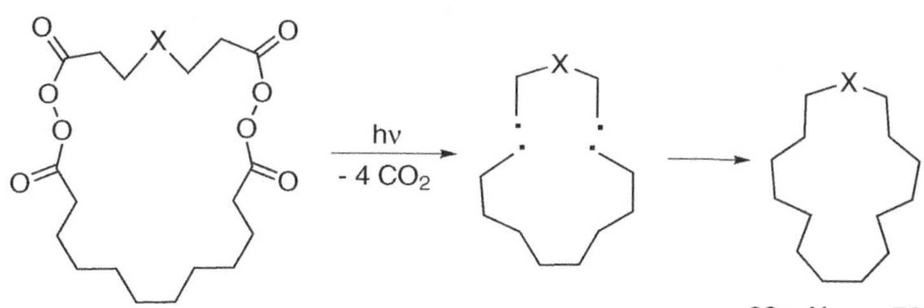

81

82a, X = --; 73%
b, X = C(OCH2CH2O) 65%

Scheme 4.33.

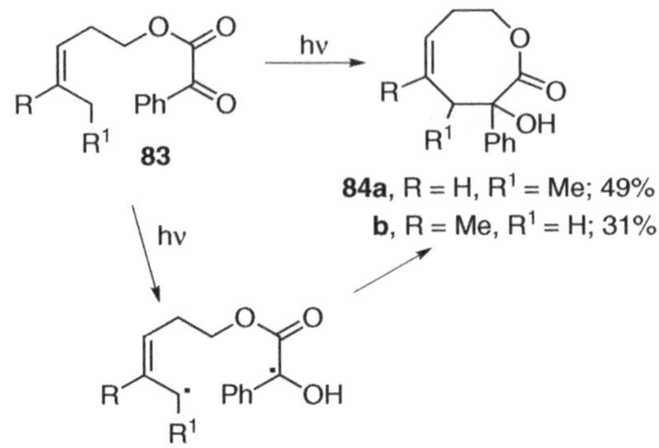

Scheme 4.34.

eight-membered ring lactones **84a,b** were formed in a modest yield in place of the expected six-membered analogues (Scheme 4.34) [68]. Such an unusual 1,9-hydrogen abstraction was rationalized by attributing the photoreaction to the *syn* rotamer of **83** and postulating a quick closure of the proximate biradical.

A useful case of macrocyclization that has been reported is the series of photocyclization of (*N*-acetylglycyl)oligopeptides bearing a quinone ring in the side chain (Scheme 4.35) [69, 70]. The occurrence of the Norrish–Yang reaction depended on the length and sequence of the oligopeptide chain. A precise distance was required for intramolecular attack to the acetylglycine chain by anthraquinone. Thus, no photoadducts were formed when $n = 0$ or 3 in compounds **85**, whereas intramolecular attack occurred and yielded derivative **86** in a discrete yield (44%) when $n = 2$ (Scheme 4.35). Compound **86** was accompanied by small amounts of an isomer resulting from hydrogen abstraction by the other C=O group of the quinone moiety [69, 70].

4.3 SYNTHESIS OF OPEN-CHAIN COMPOUNDS

In other instances, reactions via photogenerated biradicals did not lead to ring formation. Often, the end compounds were obtained by cleavage of a preexisting ring. A typical case is the Norrish I-mediated ring opening of cycloalkanones that is reviewed here for the case where the photoextrusion does not lead to the formation of

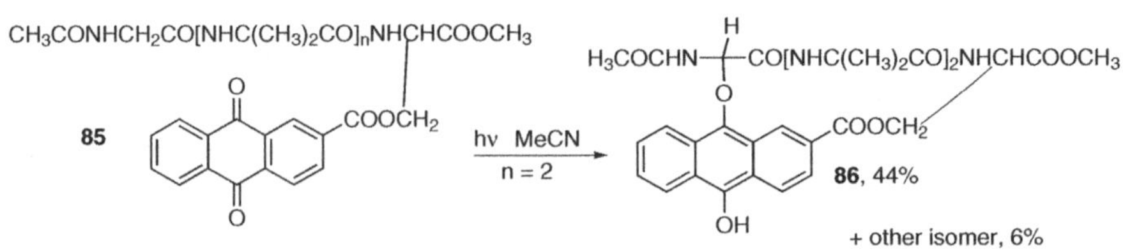

Scheme 4.35.

a new ring. The elimination of an alkene induced after a Norrish II process (Scheme 4.3, path *c*) will be likewise mentioned. Various rearrangements that involve a biradical intermediate such as the photo-Fries reaction [3] are likewise considered here.

4.3.1 By Ring Opening

Various paths are accessible to the biradical formed by Norrish I reactions, and this often causes the formation of mixtures. However, in some cases the process was sufficiently clean to be adopted in the synthetic planning of natural compounds. A photoinduced fragmentation was actually used in the diastereocontrolled preparation of (+)-juvabione (**89**), a natural sesquiterpene having insect juvenile hormone activity (Scheme 4.36) [71]. Here the biradical formed by fragmentation of the C–C bond in triethylsilyl protected ketoalcohol **87** gave unsaturated aldehyde **88** in 66% yield by selective hydrogen transfer. Elaboration of the latter compound afforded the desired sesquiterpene in a few steps.

The same scheme was applied in the total synthesis of the iridoid monoterpene (±)-dimethyl secologanoside **92**. This has been accomplished by including the photofragmentation of the bicyclo[2.2.1]heptanone **90** as a key step (Scheme 4.37) [72]. Accordingly, the irradiation of an acetone solution of **90** gave the cyclopentene derivative **91** as the exclusive product in 83% yield [72].

The bicyclo[2.2.1]heptanone skeleton and related structures were ideal models for the clear-cut photoinduced ring opening to give cyclopentene derivatives. In fact, 5-*endo*-7-*anti*-disubstituted bicyclo[2.2.1] heptan-2-ones were used as starting compounds for the preparation of Prostaglandin C_2 and their analogues [73], as well as for the synthesis of the monoterpene (±)-hop ether [74]. Similarly, a formal synthesis of the antimicrobial diterpene (±)-udoteatrial hydrate was accomplished via the ring opening of a 4-oxa-tricyclo[5.2.1.0^{2,6}]-decan-8-one derivative [75].

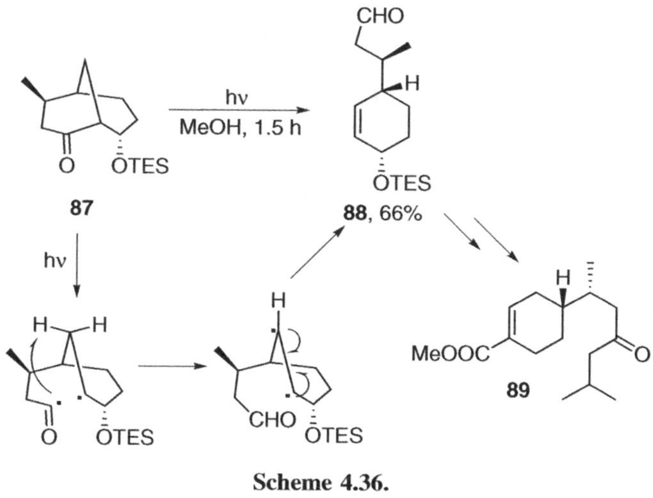

Scheme 4.36.

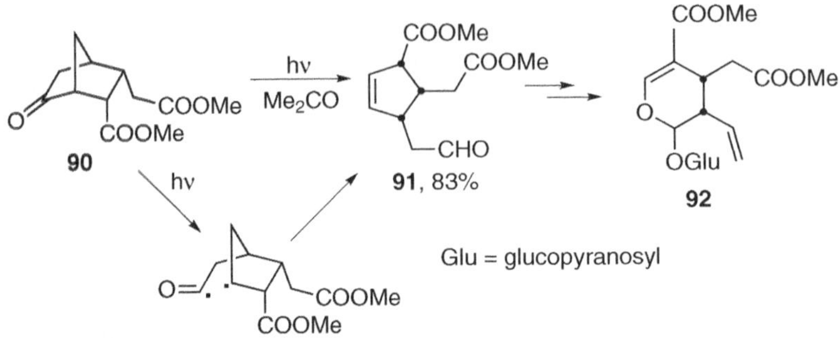

Scheme 4.37.

The alternative hydrogen transfer to give a ketene was exploited in other cases. This is exemplified by the irradiation of bicyclo[3.3.0]octanone **93** (Scheme 4.38), where the biradical resulting from the fragmentation underwent hydrogen transfer and gave ketene **94**. In turn, this gave the substituted cyclopentane **95** in 82% yield upon solvent addition, along with some unreacted **93** [76]. Compound **95** was employed for the synthesis of 1,22-dihydroxynitianes in a synthetic route toward the sesterterpenoid nitiol.

A substituted bicyclononan-3-one closely related to **93** was likewise irradiated in MeOH to give the corresponding cyclopentane in a good yield [77]. The yield of this photochemical step was found to improve when using diluted solutions of the bicyclic compounds (0.015 M), so that a prolonged irradiation could be avoided [77].

A synthetic application of the Norrish type II ring cleavage is depicted in Scheme 4.39 dealing with the photolysis of bicyclo[3.1.0]hexyl aryl ketones. The geometry of compound **96** allowed for the hydrogen abstraction from the bicyclic ring by the excited carbonyl. The cleavage of the cyclopropane ring ensued forming a cyclopentene ring in compound **97** (53% yield) [78]. Related *exo*-bicyclo[2.1.1]hexyl derivatives bearing chiral auxiliaries were likewise irradiated and the ring opened derivatives were formed in an ee up to 98% [78].

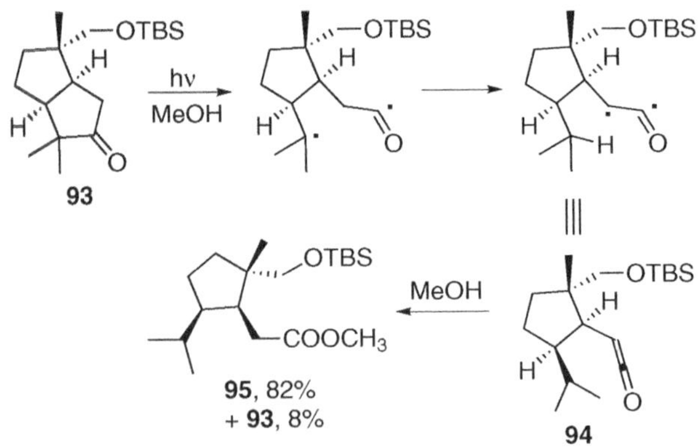

Scheme 4.38.

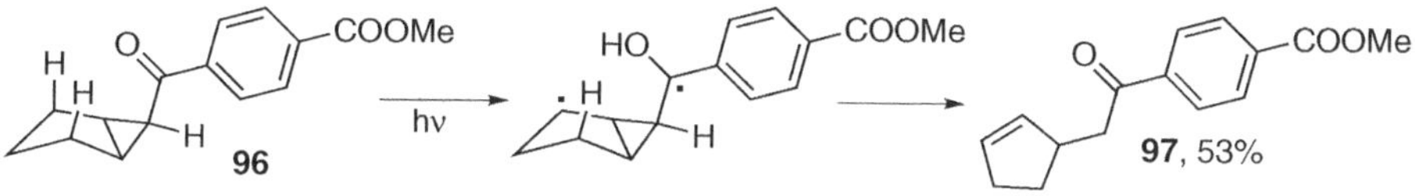

Scheme 4.39.

A benzoyl group was tethered at the 3-position in *N*-methyl-2-piperidones with the aim of inducing ring opening. Photolysis of these substituted lactams gave the corresponding *N*-allyl-3-oxo propionic amide derivatives [79]. The methyl-substituted cyclobutane ring in bicyclo[4.2.0]octan-2-one systems was likewise opened by a Norrish II process. This strategy was applied to the synthesis of spirosesquiterpenes of the acorane family [80].

Recently, the stereoselective synthesis of cyclopentitols was accomplished by a tandem Norrish type II photoelimination/intramolecular aldol cyclization reaction (Scheme 4.40) [81]. Also in this case, the stereochemistry of the starting nono-2,3-diulose **98** was crucial for the success of the reaction. Here the course of the process was different from that typical of compounds containing the 1,2-diketone moiety, that is, the exclusive formation of 2-hydroxycyclobutanones by a Norrish–Yang photocyclization. Cleavage of the pyrane ring via the 1,4-biradical intermediate instead took place. The triketone **100** was then prone to undergo an intramolecular aldol reaction to bicyclic **99**, obtained in a discrete yield (Scheme 4.40) [81].

4.3.2 By Photoelimination

As mentioned in the introduction, the 1,4-biradical intermediates resulting from the Norrish II reaction can undergo fragmentation, thus leading to an overall photoelimination. When no competing pathways were available, this elimination found synthetic applications. An intriguing case is shown in Scheme 4.41 for the photoreaction of 2-methyl-2-isobutylindan-1-one (**101**). Photoinduced intramolecular

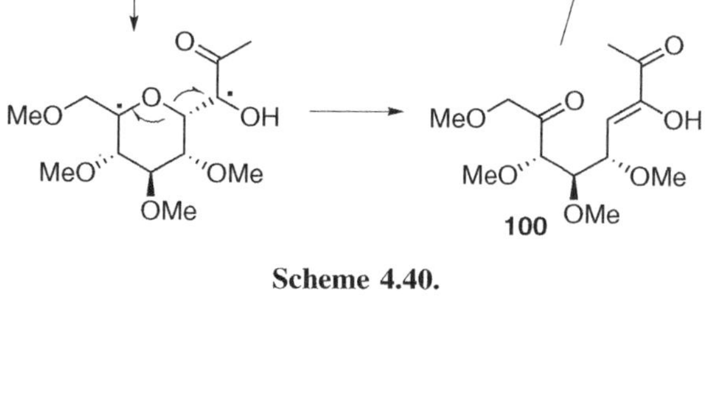

Scheme 4.40.

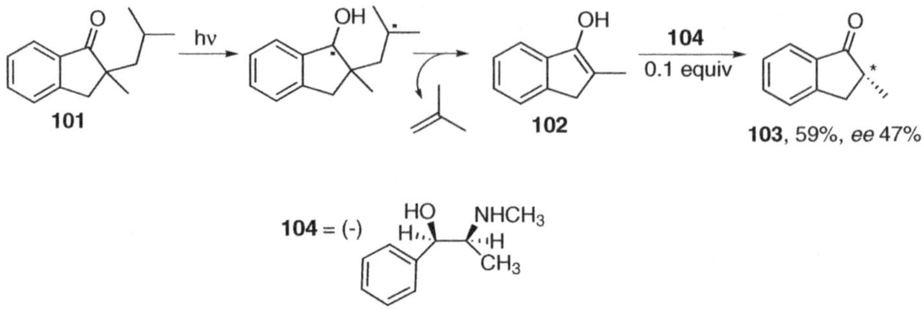

Scheme 4.41.

hydrogen abstraction led to the elimination of isobutene with the concomitant formation of the 2-methylinden-3-ol (**102**). The lifetime of enol **102** was long enough for allowing protonation in the presence of a chiral agent such as (−)-ephedrine (**104**, 0.1 equiv) and formation of indanone **103** with a significant ee (Scheme 4.41) [82].

A clear-cut photoelimination was also observed upon the irradiation of α-alkyl-β-oxoamides to give the corresponding benzoylacetamides, albeit in some cases γ- and δ-hydrogen abstraction competed [83].

The photoelimination sequence was exploited in the photoremoval of the 2,5-dimethylphenacyl group used as ester for protecting carboxylic acids. The irradiation at 254 or 366 nm of these aryl ketones caused enolization involving the *o*-alkyl group that finally led to the liberation of the desired acid in close to quantitative isolated yields [84].

Scheme 4.42 illustrates a representative case where an acyclic compound was formed by the photofragmentation of acyl peroxides [85]. Chiral amino acids containing the diacyl peroxide moiety (e.g., **105**) were prepared and stored at −20°C to avoid thermal decomposition. The irradiation was carried out at either −78°C or −196°C by using a 254-nm UV lamp for 2–5 days. The photochemical elimination of carbon dioxide followed by radical recombination smoothly gave various bis(amino acids) (**106**). This approach was applied to the preparation of (4*R*)-5-propyl-L-leucine, a component of HUN-7293 (a potent inhibitor of vascular cell adhesion molecule 1) [85].

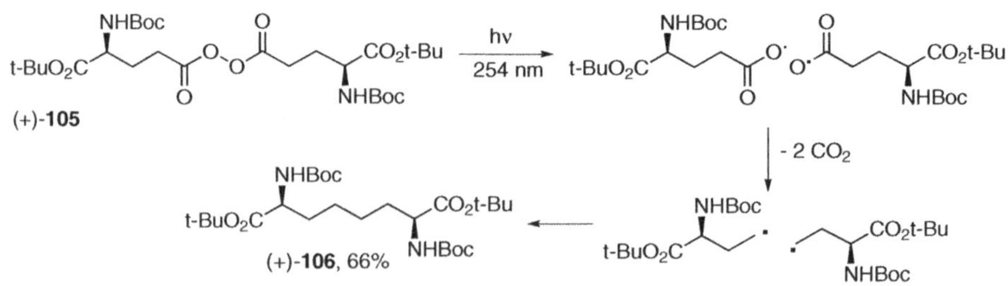

Scheme 4.42.

4.3.3 By Rearrangement

Most of the reactions pertaining to this category are photo-Fries [3] and photo-Claisen rearrangements, but only the former reactions will be presented here since the latter process mainly involve aryl or allyl ethers and gives rise to mixtures [86].

A recent case not pertaining to the above mentioned class is the metal-free transformations of an aryl–O bond into an Ar–C bond through a photochemical rearrangement/oxidation cascade sequence initiated by the generation of a 1,5-biradical (Scheme 4.43) [87]. In fact, excitation of *O*-acetyl aryloxy benzene derivative **107** initially gave epoxyalcohol **109** that benefited from some stabilization from the intramolecular hydrogen bond between the hydroxyl group and the epoxy ring. Then, the aromaticity of the ring was restored by rearrangement of the epoxide and benzyl alcohol **110** was formed. The end diketone **108** resulted via oxidation by TEMPO either directly from **110** or via the ketal intermediate **111**.

A particular case is the photochemical analogue of the Arbuzov rearrangement on organophosphorus molecules, involving the photocleavage of a C–O bond in substituted phosphites [88–90]. The reaction met some success when a sufficiently weak C–O bond was cleaved, as it is the case for benzyl phosphite **112**. Potentially useful phosphonates (e.g., **113**) were readily prepared in this way by the formation of a C–P bond upon rearrangement of the first formed oxygen centered radical as illustrated in Scheme 4.44 [89].

The photochemical version of the Fries reaction, contrary to its thermal counterpart, follows a radical path [3]. The photo-Fries reaction of phenyl esters or amides proceeds homolytically and forms a phenoxy (or an aminyl) radical along with an acyl radical (Scheme 4.1). Radical recombination gives a mixture of the *ortho*- and *para*-substituted ketones. The reaction has preparative significance with *para*-

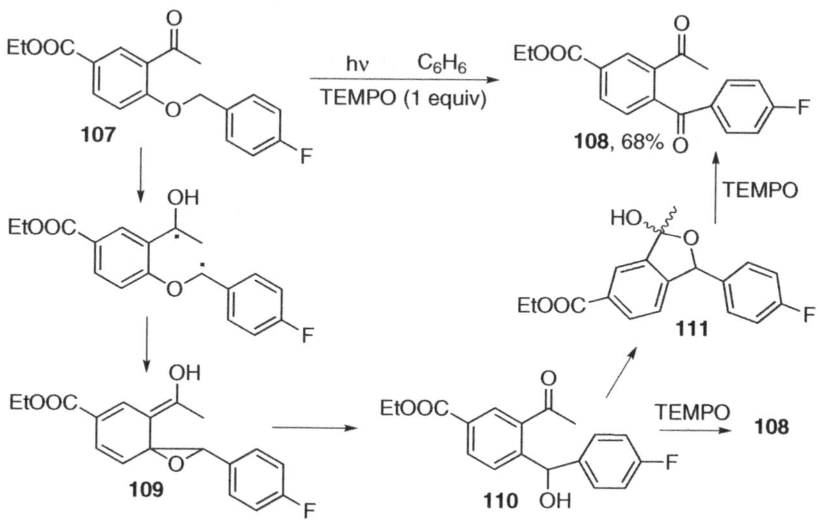

Scheme 4.43.

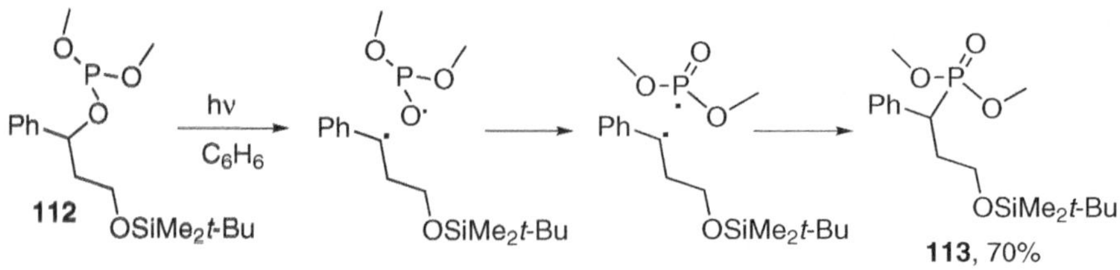

Scheme 4.44.

substituted aromatics, although the yields of rearranged products are in most cases modest. As an example, the irradiation of either the *p*-methoxyphenyl or the *p*-methylphenyl ester of propynoic acid **114** afforded the corresponding ketones **115a,b** in ~50% yield. These were in turn smoothly converted into chromones **116** under acidic or basic conditions (Scheme 4.45) [91]. Lower yields of rearranged products were reported when the same reaction was performed by using the esters of phenylpropynoic acid [92].

Similar yields were reported in the photo–Fries reaction of some hydroquinone monosuccinates that upon lactonization gave the corresponding enol lactones in what can be considered an indirect succinoylation of hydroquinone [93]. The photochemical (radical) and the thermal (ionic, $AlCl_3$-catalyzed) reactions have been compared for the case of aryl succinates. The two approaches were shown to be complementary—the photochemical activation was more general, while the elective method was for alkoxy- or hydroxy-substituted phenyl esters—whereas the thermal rearrangement was more effective for aryl esters bearing alkyl or halogen substituents [94].

An interesting route for the synthesis of flavonoids was found in the photo-Fries reaction of aryl dihydrocinnamates to give the corresponding 2′-hydroxydihydrochalcones in discrete yields [95]. The photorearrangement of aryl benzoylacetates gave poor yields that were enhanced, however, when the carbonyl group was masked

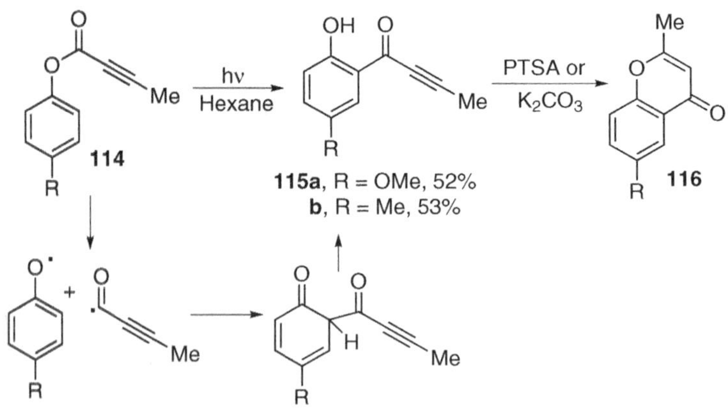

Scheme 4.45.

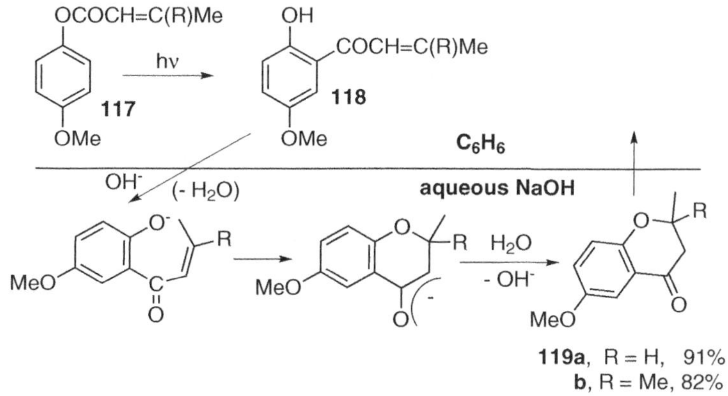

Scheme 4.46.

as acetal. In such a way, some flavones were prepared in acceptable yields from the corresponding *o*-hydroxydibenzoylmethane intermediates [96].

Improved yields were obtained by carrying out the photo–Fries reaction in a biphasic basic aqueous/benzene system, as shown in the preparation of 4-chromanones according to Scheme 4.46 [97]. Aryl esters of α,β-unsaturated carboxylic acids **117** underwent the usual photo-Fries rearrangement in the organic phase to give substituted phenols **118**. Migration of the latter compounds in the aqueous phase and basic cyclization afforded chromanones **119a,b** in excellent yields [97].

The same approach has been exploited for the one-pot synthesis of benzochroman-4-one and of heteroarylchroman-4-one derivatives, again by making use of an intramolecular oxa-Michael addition reaction as the key step and replacing the highly toxic benzene by cyclohexane as the organic phase [98]. The photo-Fries reaction was also used as the key step for the preparation of natural occurring derivatives, and in some cases the two radical sites formed are bonded and a biradical results. The yield of such a step was low in the synthesis of the antitumor compound bikaverin [99, 100], but satisfactory in that of the antibiotic kendomycin **122** as shown in Scheme 4.47. Irradiation of the macrocyclic lactone **120** in cyclohexane gave exclusively the acyl phenol **121** in 75% yield, since the alternative positions in compound **120** bore a substituent (Scheme 4.47) [101].

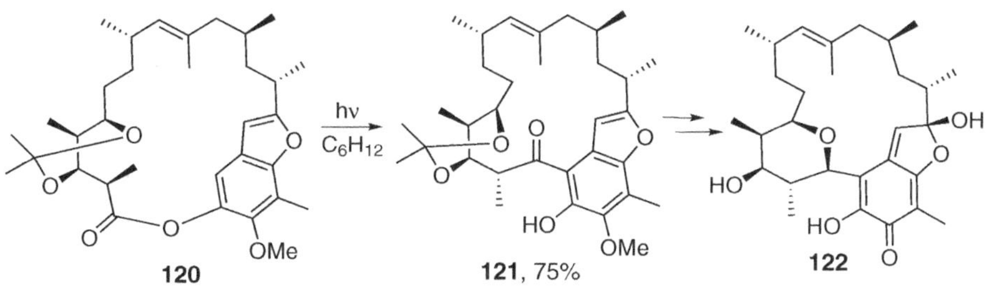

Scheme 4.47.

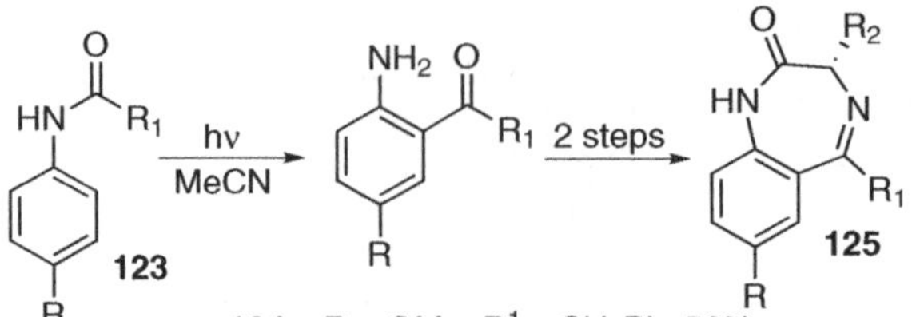

124a, R = OMe, $R^1 = CH_2Ph$, 56%
b, R = H, $R^1 = CH_2Ph$, 60%
c, R = H, $R^1 = C_{11}H_{25}$, 61%
d, R = H, R^1 = Me, 87%
e, R = NO_2, R^1 = Me, 87%

Scheme 4.48.

Substituted anilides were recently subjected to the photo-Fries reaction for the preparation of 3,5-disubstituted 1,4-benzodiazepines (Scheme 4.48). Thus, *para*-substituted anilides **123** underwent a photoinduced acyl migration and the resulting *ortho*-aminoacylbenzene derivatives **124a–e** were first derivatized with a protected naturally occurring amino acid and then cyclized to the desired benzodiazepines **125** [102].

Alternatively, compounds **124a–e** were acylated and treated with ammonium formate under microwave activation. A rapid cyclization took place and 2,4-disubstituted quinazolines (and benzoquinazolines) were isolated [103]. Furthermore, the *o*-amino ketones obtained from the anilides could be treated in situ with acetylenic Michael acceptors, namely dimethyl acetylenedicarboxylate, to afford 6,4-disubstituted quinoline 2,3-dicarboxylates that upon further elaboration gave various naphthyridine derivatives [104].

The result was more complex with unsubstituted phenyl acetate **126**. Photolysis in hexane gave a mixture of phenol **127**, *ortho*-acetyl phenol **128** and of the corresponding *para* isomer **129**. However, carrying out the irradiation in a confined media such as in a zeolite changed the product distribution significantly. Thus, in a NaX zeolite some phenol was still detected, but in a NaY zeolite the *ortho* isomer largely predominated (Scheme 4.49) [105].

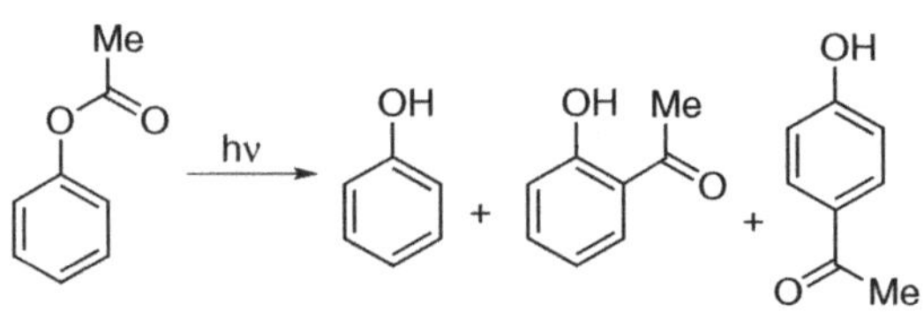

	126	**127**	**128**	**129**
Hexane		20	53	27
NaX-Hexane		21	79	
NaY-Hexane		5	91	4

Scheme 4.49.

A similar result was reported for the irradiation of three 1-naphthyl phenylacylates in cation-exchanged Y zeolites, where the *ortho* isomer was formed in a selectivity up to 99% [106]. Again, a single photoproduct resulted from the irradiation of the same 1-naphthyl phenylacylates included in γ-cyclodextrin [107].

4.4 *o*-, *p*-BENZYNES AND α, *n*-DIDEHYDROTOLUENES (BERGMAN, MYERS–SAITO CYCLIZATIONS AND RELATED REACTIONS)

In this section the role of aromatic biradical intermediates, such as *o*-benzynes, *p*-benzynes, and α,*n*-didehydrotoluenes, will be briefly mentioned, albeit their application in synthesis has been scarcely documented.

The photogeneration of *ortho*-benzyne has been obtained from a few precursors. As one may expect, the approach involved the consecutive elimination of two adjacent groups in the aromatic ring, as shown in Scheme 4.50. Benzenediazonium-2-carboxylate was largely used for the thermal generation of benzyne, but early reports evidenced that the process could be induced photochemically (path *a*) [108]. The photolysis of either 1,2-diiodobenzene (path *b*) [109] or phthaloyl peroxide (path *c*) [110] likewise gave benzyne to some extent. In the last case, when the photolysis was carried out in the presence of propyne, allene, or cyclopropane, low amounts of the ene adducts phenylallene, benzylacetylene, or allylbenzene, respectively, were formed [110].

An interesting way to form *ortho*-benzyne was found in the photolysis of *o*-nitrobenzaldehyde *N*-acetyl-*N*-alkylhydrazone (**130**) [111, 112]. Irradiation led initially to *o*-(*N*-acetyl-*N*-alkyltriazeno)-benzoic acid (**131**) that upon a second photochemical reaction gave the desired benzyne (path *d*, Scheme 4.50). Trapping experiments in methanol were carried out in the presence of tetracyclone and the benzyne adduct 1,2,3,4-tetraphenylnaphthalene was obtained in ~10% yield [111, 112].

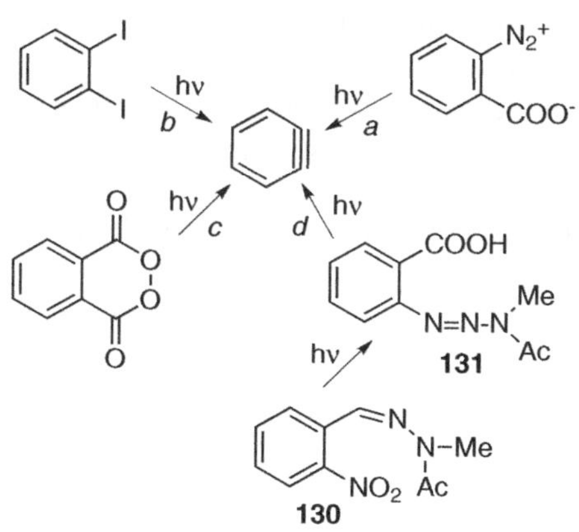

Scheme 4.50.

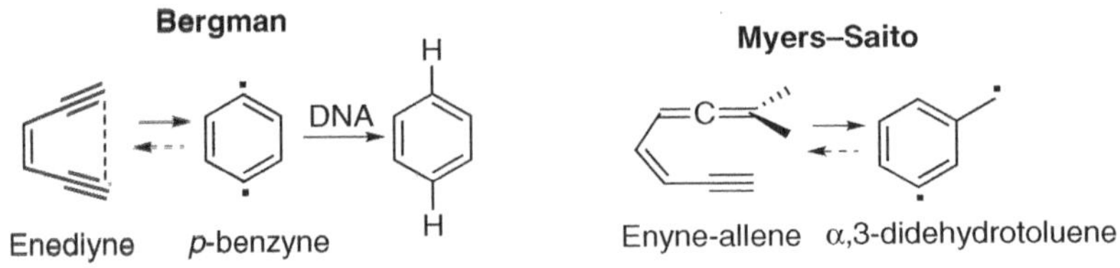

Scheme 4.51.

A particular case is represented by *p*-benzyne (a σ,σ-biradical) and α,3-didehydrotoluene (α,3-DHT, a σ,π-biradical). These intermediates have mainly been studied as simplified models for the mechanism of action of natural occurring anticancer agents, of which these mimic the active moiety [113]. These biradicals were obtained by the Bergman and Myers–Saito cyclizations by means of a cycloaromatization of, respectively, (*Z*)-hexa-1,5-diyn-3-enes (Bergman reaction) [114] or of enyne-allenes (Myers–Saito reaction, Scheme 4.51). The reactivity of the biradicals formed is sufficient to induce DNA damage, mainly by abstracting hydrogen atoms from the DNA sugars of opposite strands. The cycloaromatization can be induced photochemically and is important from the biological application, since it allows for the spatial and temporal control of the enyne reactivity [115, 116].

Here the stability imparted by the creation of the aromatic system drives the process. In the Bergman reaction the enediyne moiety is usually incorporated in a strained ring, and this favors both the cyclization and the retro-Bergman cleavage [114]. The reactions were in most cases unsatisfactory for a large-scope application in synthesis. A few selected examples where the reactions proceeded successfully will be mentioned here. Thus, the photolysis in hexane of the 1,2-diethynylcyclopentene derivatives **132** gave the aromatic derivatives **134a,b** via the 4,7-didehydroindane biradical intermediates **133** [117]. Interestingly, the reaction was almost completely suppressed (3% yield) when terminal alkynes were present in place of the methyl-substituted analogues (**134b**, 71% yield, Scheme 4.52) in the starting compounds.

A carbinol group adjacent to the triple bond favored the cyclization step, as shown in Scheme 4.53. In this case, the double bond of the enediyne had been incorporated into biologically relevant heterocycles such as 10-membered pyrimidine enediynes **135**. Irradiation of compound **135** under stirring at 40°C and at 310 nm (light filtered through a potassium chromate solution) afforded the tricyclic compound **136** in 83% yield (Scheme 4.53) [118]. The same reaction worked under thermal conditions as well, by heating at reflux the starting reaction mixture.

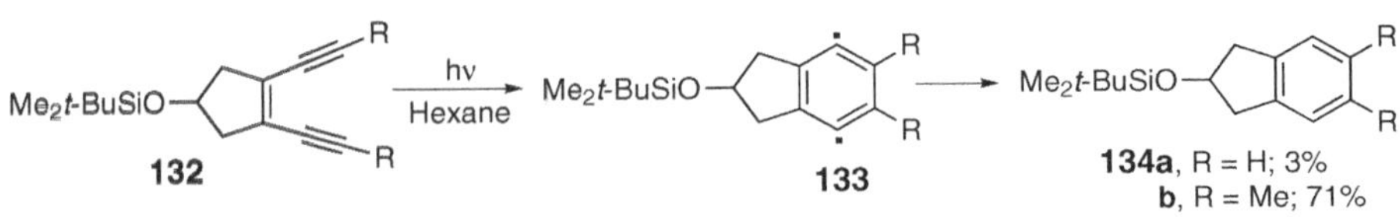

Scheme 4.52.

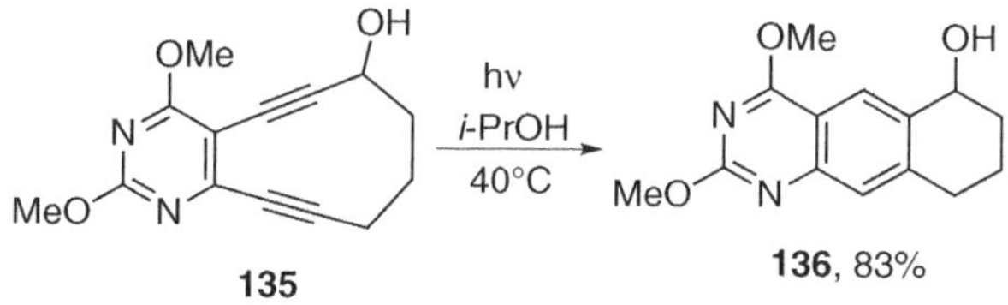

Scheme 4.53.

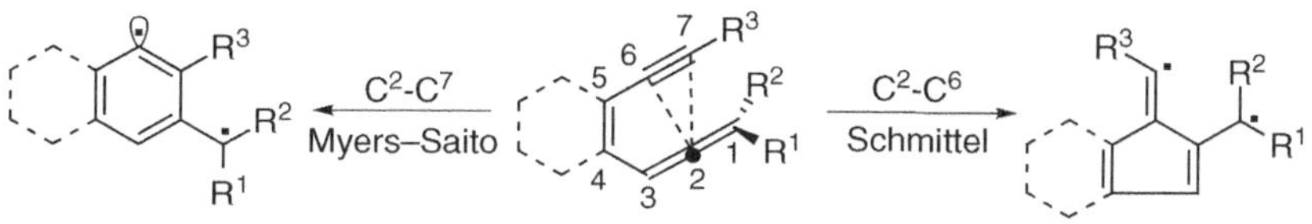

Scheme 4.54.

As for the generation of didehydrotoluene biradicals from an open-chain (*Z*)-hepta-1,2,4-trien-6-yne derivative, the Myers–Saito reaction (that followed the C^2–C^7 cyclization) competed with a second biradical rearrangement occurring from a C^2–C^6 cyclization (discovered by Schmittel, Scheme 4.54) [119]. In a few cases, the photochemical version of this reaction has been carried out [120].

The original contribution of photochemistry, however, instead consists in the introduction of alternative ways of arriving at the enyne–allene precursor, namely the generation of a triple bond by CO elimination from a cyclopropenone moiety, as shown in Scheme 4.55 [115, 116]. Thus, compound **137** was stable in the dark, but upon irradiation CO extrusion demasked the triple bond (compound **138**) allowing the Myers–Saito cyclization and forming **139**. Alternatively, compound **140** underwent a photo-Wolff rearrangement to give enediyne-β-keto ester **141** that easily tautomerized to the more reactive allene **138** (Scheme 4.55).

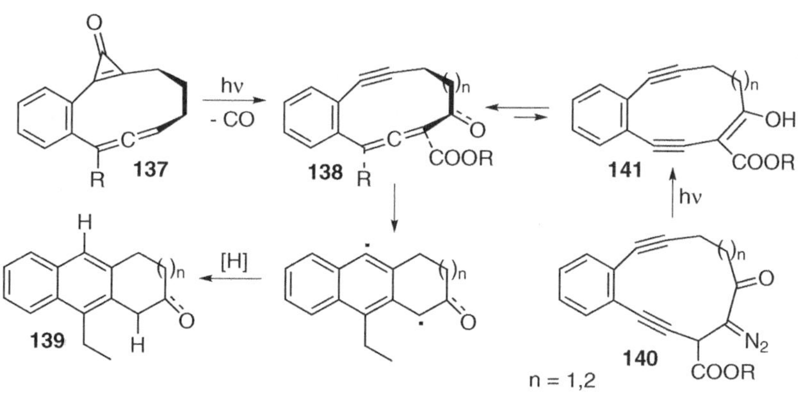

Scheme 4.55.

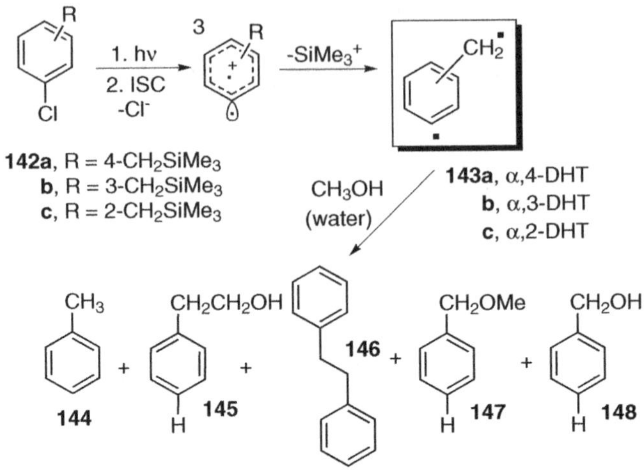

Scheme 4.56.

Very recently, a fully different photochemical approach to generate didehydrotoluenes has been reported that makes recourse to the phenyl cation chemistry (see Section 6.3.3) [121]. Thus, isomeric (chlorobenzyl)trimethylsilanes **142a–c** underwent heterolytic cleavage in the triplet state in protic solvents and afforded triplet phenyl cations that in turn suffered spontaneous desilylation to give DHTs (**143a–c**, Scheme 4.56). In this approach, the two radical centers are sequentially formed (contrary to the Myers–Saito cyclization, where they are formed simultaneously) and all of the α,*n*-DHT isomers (rather that only the α,3-) are generated. In the case of α,3-DHT, the identity of the intermediate was supported by the closely similar product distribution obtained from the irradiation of **142b** and from the thermally induced cyclization of (*Z*)-1,2,4-heptatrien-6-yne (Scheme 4.56) [121].

REFERENCES

1. Chatgilialoglu, C., Studer, A., Eds. *Encyclopedia of Radicals in Chemistry, Biology and Materials*, John Wiley & Sons, Hoboken, NJ, 2012.
2. Albini, A., Fagnoni, M., Eds. *Handbook of Synthetic Photochemistry*, Wiley-VCH, Weinheim, 2010.
3. Miranda, M. A., Galindo, F. Photo–Fries reaction and related processes. In Lenci, F., Horspool, W., Eds. *CRC Handbook of Organic Photochemistry and Photobiology*, second edition, CRC Press, Boca Raton, FL, 2004, pp. 42-1–42-11.
4. Shiraki, S., Garcia-Garibay, M. A. Carbon–carbon bond formation by the photoelimination of small molecules in solution and in crystals. In ref. 2, pp. 25–66.
5. Garcia-Garibay, M. A. Photochemical decarbonylation of ketones: Recent advances and reactions in crystalline solids. In Lenci, F., Horspool, W., Eds. *CRC Handbook of Organic Photochemistry and Photobiology*, second edition, CRC Press, Boca Raton, FL, 2004, pp. 48-1–48-41.

6. Abe, M. Formation of four-membered ring: Oxetanes. In ref. 2, pp. 217–286.
7. Hehn, J. P., Müller, C., Bach, T. Formation of four-membered ring: From a conjugate alkene. In ref. 2, pp. 171–216.
8. Wagner, P. J. Abstraction of γ-hydrogens by excited carbonyls. In Griesbeck, A. G., Mattay, J., Eds. *Molecular and Supramolecular Photochemistry. Synthetic Organic Photochemistry*, Vol. 12, Marcel Dekker, New York, 2005, pp. 11–40.
9. Wessig, P., Mühling, O. Abstraction of ($\gamma \pm n$)-hydrogens by excited carbonyls. In Griesbeck, A. G., Mattay, J., Eds. *Molecular and Supramolecular Photochemistry. Synthetic Organic Photochemistry*, Vol. 12, Marcel Dekker, New York, 2005, pp. 41–88.
10. Wessig, P. Regioselective photochemical synthesis of carbo- and heterocyclic compounds: The Norrish/Yang reaction. In Lenci, F., Horspool, W., Eds. *CRC Handbook of Organic Photochemistry and Photobiology*, second edition, CRC Press, Boca Raton, FL, 2004, pp. 57-1–57-20.
11. Wagner, P. J. Yang photocyclization: Coupling of biradicals formed by intramolecular hydrogen abstraction of ketones. In Lenci, F., Horspool, W., Eds. *CRC Handbook of Organic Photochemistry and Photobiology*, second edition, CRC Press, Boca Raton, FL, 2004, pp. 58-1–58-70.
12. Couture, A., Gomez, J., de Mayo, P. (1981). Thione photochemistry: Abstraction and cyclization at the β-position of aralkyl thiones from two excited states. *The Journal of Organic Chemistry*, **46**, 2010–2016.
13. Couture, A., Hoshino, M., de Mayo, P. (1976). Thione photochemistry: Cyclopropanethiol formation from arylalkyl thiones. *Journal of the Chemical Society, Chemical Communication*, 131–132.
14. Wessig, P., Mühling, O. (2001). A new photochemical route to cyclopropanes. *Angewandte Chemie International Edition*, **40**, 1064–1065.
15. Wessig, P., Mühling, O. (2003). Photochemical synthesis of highly functionalized cyclopropyl ketones. *Helvetica Chimica Acta*, **86**, 865–893.
16. Wessig, P., Mühling, O. (2005). 1,2-Chirality tansfer in the synthesis of cyclopropanes. *Angewandte Chemie International Edition*, **44**, 6778–6781.
17. Finlay, J. D., Smith, D. J. H., Durst, T. (1978). The photolysis of 2-phenylthietan-1,1-dioxides; preparation of substituted cyclopropanes. *Synthesis*, 579–580.
18. Ramnauth, J., Lee-Ruff, E. (2001). Photochemical preparation of cyclopropanes from cyclobutanones. *Canadian Journal of Chemistry*, **79**, 114–120.
19. Adam, W., Oppenlander, T., Zang, G. (1985). Photochemistry of the azoalkanes 2,3-diazabicyclo[2.2.l]hept-2-ene and spiro[cyclopropane-1,7′-[2,3]diazabicyclo[2.2.1] hept-2-ene]: On the questions of one-bond vs. two-bond cleavage during the denitrogenation, cyclization vs. rearrangement of the 1,3-diradicals, and double inversion. *The Journal of Organic Chemistry*, **50**, 3303–3312.
20. Adam, W., Nau, W. M., Sendelbach, J. (1994). Temperature dependence of the α versus β bond cleavage in the direct and triplet-sensitized photolysis of azoalkanes of the 2,3-diazabicyclo[2.2.1]hept-2-ene type. *Journal of the American Chemical Society*, **116**, 7049–7054.
21. Adam, W., García, H., Martí, V., Moorthy, J. N., Peters, K., Peters, E.-M. (2000). Photochemical denitrogenation of norbornene-annelated 2,3-diazabicyclo[2.1.1]hept-2-ene-type azoalkanes: Crystal-lattice versus zeolite-interior effects. *Journal of the American Chemical Society*, **122**, 3536–3537.

22. Wender, P. A., Kee, J.-M., Warrington, J. M. (2008). Practical synthesis of prostratin, DPP, and their analogs, adjuvant leads against latent HIV. *Science*, **320**, 649–652.
23. Sponholtz, W. R., III, Trujillo, H. A., Gribble, G. W. (2000). Tandem 1,3-dipolar cycloaddition of mesoionic 2,4-diphenyl-1,3-oxathiolium-5-olate with cycloocta-1,5-diene. A new synthesis of the tetracyclo[6.2.0.0^{4,10}.0^{5,9}]decane ring system. *Tetrahedron Letters*, **41**, 1687–1690.
24. Ao, M. S., Burgess, E. M. (1971). Benzothiazete 1,l-dioxides. *Journal of the American Chemical Society*, **93**, 5298–5299.
25. Wessig, P., Schwarz, J. (1998). Enantioselective preparation of (2*R*)- and (2*S*)-azetidin-2-carboxylic acids. *Helvetica Chimica Acta*, **81**, 1803–1814.
26. Griesbeck, A. G., Heckroth, H. (2002). Reversible diastereoselective photocyclization of phenylglyoxylamides of α-amino acid methyl esters to 3-hydroxy β-lactams. *Synlett*, 131–133.
27. Yang, C., Xia, W. (2009). Solid-state asymmetric photochemical studies using the ionic chiral auxiliary approach. *Chemistry—An Asian Journal*, **4**, 1774–1784.
28. Aoyama, H., Miyazaki, K., Sakamoto, M., Omote, Y. (1983). Photochemical reaction of inclusion molecular complexes of *N,N*-dialkylpyruvamides with desoxycholic acid: Host-controlled reaction of guest compounds in the solid state. *Journal of the Chemical Society, Chemical Communication*, 333–334.
29. Toda, F., Miyamoto, H., Kanemoto, K. (1995). Formation by mixing of chiral host–Achiral guest inclusion complexes in which the guest molecules are arranged in a chiral form. Production of optically active photocyclisation compounds by irradiation of the inclusion complexes. *Journal of the Chemical Society, Chemical Communication*, 1719–1720.
30. Xia, W., Scheffer, J. R., Botoshansky, M., Kaftory, M. (2005). Photochemistry of 1-isopropylcycloalkyl aryl ketones: Ring size effects, medium effects, and asymmetric induction. *Organic Letters*, **7**, 1315–1318.
31. Koshima, H., Fukano, M., Uekusa, H. (2007). Diastereospecific photocyclization of a isopropylbenzophenone derivative in crystals and the morphological changes. *The Journal of Organic Chemistry*, **72**, 6786–6791.
32. Yang, C., Xia, W. J., Scheffer, J. R. (2007). Asymmetric synthesis of tricyclic tetralin derivatives via an intramolecular photoreaction. *Tetrahedron*, **63**, 6791–6795.
33. Herrera, A. J., Rondón, M., Suárez, E. (2007). Stereocontrolled photocyclization of 1,2-diketones applied to carbohydrate models: A new entry to *C*-ketosides. *Synlett*, 1851–1856.
34. Herrera, A. J., Rondón, M., Suárez, E. (2008). Stereocontrolled photocyclization of 1,2-diketones: Application of a 1,3-acetyl group transfer methodology to carbohydrates. *The Journal of Organic Chemistry*, **73**, 3384–3391.
35. Kamijo, S., Hoshikawa, T., Inoue, M. (2010). Regio- and stereoselective acylation of saturated carbocycles via Norrish–Yang photocyclization. *Tetrahedron Letters*, **51**, 872–874.
36. Basaric, N., Horvat, M., Mlinaric-Majerski, K., Zimmermann, E., Neudörfl, J., Griesbeck, A. G. (2008). Novel 2,4-methanoadamantane-benzazepine by domino photochemistry of *N*-(1-adamantyl)phthalimide. *Organic Letters*, **10**, 3965–3968.
37. Hickford, P. J., Baker, J. R., Bruce, I., Booker-Milburn, K. I. (2007). Acid-catalyzed rearrangement of fused alkylideneoxetanols. *Organic Letters*, **9**, 4681–4684.

38. Maruyama, K., Ishitoku, T., Kubo, Y. (1981). Photochemistry of aliphatic imides. Synthesis of azetidine-2,4-diones via photochemical isomerization of succinimides and *N*-formyl-*N*-methyl α,β-unsaturated amides. *The Journal of Organic Chemistry*, **46**, 27–34.

39. Maruyama, K., Ishitoku, T., Kubo, Y. (1979). Azetidine-2,4-diones via photoinduced ring contraction of succinimides. *Journal of the American Chemical Society*, **101**, 3670–3671.

40. Natarajan, A., Ng, D., Yang, Z., Garcia-Garibay, M. A. (2007). Parallel syntheses of (+)- and (−)-α-cuparenone by radical combination in crystalline solids. *Angewandte Chemie International Edition*, **46**, 6485–6487.

41. Ng, D., Yang, Z., Garcia-Garibay, M. A. (2004). Total synthesis of (±)-herbertenolide by stereospecific formation of vicinal quaternary centers in a crystalline ketone. *Organic Letters*, **6**, 645–647.

42. Kadota, K., Ogasawara, K. (2001). A carbohydrate synthesis employing a photochemical decarbonylation. *Tetrahedron Letters*, **42**, 8661–8664.

43. Gramain, J. C., Remuson, R., Vallee, D. (1985). Intramolecular photoreduction of α-keto esters: Total synthesis of (±)-isoretronecanol. *The Journal of Organic Chemistry*, **50**, 710–712.

44. Hasegawa, T., Moribe, J., Yoshioka, M. (1988). The photocyclization of *N,N*-dialkyl-β-oxo amides adsorbed on silica gel. *Bulletin of the Chemical Society of Japan*, **61**, 1437–1439.

45. Aoyama, H., Sakamoto, M., Kuwabara, K., Yoshida, K., Omote, Y. (1983). Photochemical reactions of α-oxo amides. Norrish type II reactions via zwitterionic intermediates. *Journal of the American Chemical Society*, **105**, 1958–1964.

46. Steiner, A., Wessig, P., Polborn, K. (1996). Asymmetric synthesis of 3-hydroxyprolines by photocyclization of C(1′)-substituted *N*-(2-benzoylethyl)glycine esters. *Helvetica Chimica Acta*, **79**, 1843–1862.

47. Seiler, M., Schumacher, A., Lindemann, U., Barbosa, F., Giese, B. (1999). Diastereoselective photocyclization to dihydroindolinols. *Synlett*, 1588–1590.

48. Wagner, P. J., Cao, Q. (1991). Photocyclization of ortho-benzoyl *N*-alkylanilinium ions. *Tetrahedron Letters*, **32**, 3915–3918.

49. Kraus, G. A., Chen, L. (1990). A total synthesis of racemic paulownin using a type II photocyclization reaction. *Journal of the American Chemical Society*, **112**, 3464–3466.

50. Roy, S. C., Adhikari, S. (1993). Stereoselective total synthesis of (±)-paulownin and (±)-Isogmelinol through radical annulation route. *Tetrahedron*, **49**, 8415–8422.

51. Kraus, G. A., Thomas, P. J., Schwinden, M. D. (1990). An approach to aflatoxins using type II photocyclization reactions. *Tetrahedron Letters*, **31**, 1819–1822.

52. Suzuki, T., Tanemura, K. (1991). Photocyclization reactions. Part 2 synthesis of dihydrobenzofuranols using photocyclization of ethyl 2-formylphenoxyacetates and ethyl 2-acetylphenoxyacetates. *Journal of Heterocyclic Chemistry*, **28**, 1273–1280.

53. Sharshira, E. M., Okamura, M., Hasegawa, E., Horaguchi, T. (1997). Photocyclization reactions. Part 6. Solvent and substituent effects in the synthesis of dihydrobenzofuranols using photocyclization of 2-alkoxybenzophenones and ethyl 2-benzoylphenoxyacetates. *Journal of Heterocyclic Chemistry*, **34**, 861–869.

54. Carless, H. A. J., Swan, D. I., Haywood, D. J. (1993). Stereoselective synthesis of tetrahydrofuran-3-ols by photochemical δ-hydrogen abstraction of β-allyloxy-carbonyl compounds. *Tetrahedron*, **49**, 1665–1674.

55. Kraus, G. A., Zhang, N. (2000). Hydrogen-atom abstraction/cyclization in synthesis. Direct syntheses of coumestan and coumestrol. *The Journal of Organic Chemistry*, **65**, 5644–5646.

56. Wu, J., Zhang, W., Wang, C. (2009). Synthesis of benzopyrrolizidinones by Norrish–Yang cyclization of 3-(acylmethyl)-2-alkylisoindol-1-ones. *Synthesis*, 1821–1828.

57. Shao, Y., Yang, C., Gui, W., Liu, Y., Xia, W. (2012). Photochemical studies on aromatic γ,δ-epoxy ketones: Efficient synthesis of benzocyclobutanones and indanones. *Chemical Communications*, **48**, 3560–3562.

58. Muthukrishnan, S., Sankaranarayanan, J., Klima, R. F., Pace, T. C. S., Bohne, C., Gudmundsdottir, A. D. (2009). Intramolecular H-atom abstraction in γ-azidobutyrophenones: Formation of 1,5 ketyl iminyl radicals. *Organic Letters*, **11**, 2345–2348.

59. Zhou, B., Wagner, P. J. (1989). Long-range triplet hydrogen abstraction: Photochemical formation of 2-tetralols from β-arylpropiophenones. *Journal of the American Chemical Society*, **111**, 6796–6799.

60. Hornback, J. M., Poundstone, M. L., Vadlamani, B., Graham, S. M., Gabay, J., Patton, S. T. (1988). Photochemical cyclization of *o*-methylphenyl 1,3-diketones. *The Journal of Organic Chemistry*, **53**, 5597–5601.

61. Griesbeck, A. G., Heckroth, H., Schmickler, H. (1999). Regio- and stereoselective 1,6-photocyclization of aspartic acid-derived chiral γ-ketoamides. *Tetrahedron Letters*, **40**, 3137–3140.

62. Lindemann, U., Reck, G., Wulff-Molder, D., Wessig, P. (1998). Photocyclization of 4-oxo-4-phenylbutanoyl amines to δ-lactams. *Tetrahedron*, **54**, 2529–2544.

63. Wessig, P. (1999). An efficient synthesis of bicyclic β-turn dipeptides via a photochemical key step. *Tetrahedron Letters*, **40**, 5987–5988.

64. Wessig, P. (1999). Stereoselective synthesis of novel chimerical amino acids via a photochemical key step. *Synlett*, 1465–1467.

65. Wessig, P., Schwarz, J., Lindemann, U., Holthausen, M. C. (2001). Photochemical synthesis of 3,4-dihydro-2*H*-1,3-oxazin-4-ones. *Synthesis*, 1258–1262.

66. Feldhues, M., Schäfer, H. J. (1986). Selective mixed coupling of carboxylic acids—III: Synthesis of cyclopentadecanone from cyclic tetraacyl diperoxides. *Tetrahedron*, **42**, 1285–1290.

67. Givens, R. S., Olsen, R. J., Wylie, P. L. (1979). Photoextrusion of sulfur dioxide: General route to [2.2]cyclophanes. *The Journal of Organic Chemistry*, **44**, 1608–1613.

68. Kraus, G. A., Wu, Y. (1992). 1,5- and 1,9-Hydrogen atom abstractions. Photochemical strategies for radical cyclizations. *Journal of the American Chemical Society*, **114**, 8705–8707.

69. Tamiaki, H., Hashimoto, M., Maruyama, K. (1994). Intramolecular photocyclization in quinone-bearing oligopeptides. *Bulletin of the Chemical Society of Japan*, **67**, 1987–1989.

70. Maruyama, K., Hashimoto, M., Tamiaki, H. (1992). Intramolecular photoreaction of synthetic oligopeptide-linked anthraquinone molecules. *The Journal of Organic Chemistry*, **57**, 6143–6150.

71. Itagaki, N., Iwabuchi, Y. (2007). Enantio- and diastereocontrolled synthesis of (+)-juvabione employing organocatalytic desymmetrisation and photoinduced fragmentation. *Chemical Communications*, 1175–1176.

72. Tai, H.-M., Huang, M.-H., Yang, C.-C. (2003). Formal total synthesis of (±)-dimethyl secologanoside. *Journal of the Chinese Chemical Society*, **50**, 441–444.

73. Crossland, N. M., Roberts, S. M. (1979). Photosynthetic routes to prostaglandins. Synthesis of prostaglandin-C_2 and analogues. *Journal of the Chemical Society, Perkin Transactions 1*, 2397–2400.

74. Lin, C.-H., Su, Y.-L., Tai, H.-M. (2006). A total synthesis of (±)-hop ether. *Heterocycles*, **68**, 771–777.

75. Chang, M.-Y., Lin, C.-H., Chang, N.-C. (2004). Formal synthesis of (±)-udoteatrial hydrate. *Tetrahedron*, **60**, 1581–1585.

76. Wilson, M. S., Woo, J. C. S., Dake, G. R. (2006). A synthetic approach toward nitiol: Construction of two 1,22-dihydroxynitianes. *The Journal of Organic Chemistry*, **71**, 4237–4245.

77. Dake, G. R., Fenster, E. E., Patrick, B. O. (2008). A synthetic approach to the fusicoccane A–B ring fragment based on a Pauson–Khand cycloaddition/Norrish type 1 fragmentation. *The Journal of Organic Chemistry*, **73**, 6711–6715.

78. Zhao, G., Yang, C., Chen, Q., Jin, J., Zhang, X., Zhao, L., Xia, W. (2009). Photochemical studies on exo-bicyclo[2.1.1]hexyl and bicyclo[3.1.0]hexyl aryl ketones: Two approaches for synthesis of enantiomerically enriched cyclopentene derivatives. *Tetrahedron*, **65**, 9952–9955.

79. Azzouzi, A., Dufour, M., Gramain, J.-C., Remuson, R. (1988). Photocyclisation of keto-lactams. A new synthesis of functionalized 1-azabicyclo(x.y.o)alkanes. *Heterocycles*, **27**, 133–148.

80. Khac Manh, D. D., Ecoto, J., Fetizon, M., Colin, H., Diez-Masa, J.-C. (1981). A new approach to spirosesquiterpenes of the acorane family. *Journal of the Chemical Society, Chemical Communication*, 953–955.

81. Álvarez-Dorta, D., León, E. I., Kennedy, A. R., Riesco-Fagundo, C., Suárez, E. (2008). Sequential Norrish type II photoelimination and intramolecular aldol cyclization of 1,2-diketones in carbohydrate systems: Stereoselective synthesis of cyclopentitols. *Angewandte Chemie International Edition*, **47**, 8917–8919.

82. Henin, F., Muzart, J., Pete, J.-P., M'boungou-M'passi, A., Rau, H. (1991). Enantioselective protonation of a simple enol: Aminoalcohol-catalyzed ketonization of a photochemically produced 2-methylinden-3-ol. *Angewandte Chemie International Edition*, **30**, 416–417.

83. Hasegawa, T., Arata, Y., Mizuno, K., Masuda, K., Yoshihara, N. (1986). Photochemical reaction of α-alkyl-β-oxo amides: Competitive γ- and δ-hydrogen abstraction by excited ketone carbonyl groups. *Journal of the Chemical Society, Perkin Transactions 1*, 541–543.

84. Klán, P., Zabadal, M., Heger, D. (2000). 2,5-Dimethylphenacyl as a new photoreleasable protecting group for carboxylic acids. *Organic Letters*, **2**, 1569–1571.

85. Spantulescu, M. D., Jain, R. P., Derksen, D. J., Vederas, J. C. (2003). Photolysis of diacyl peroxides: A radical-based approach for the synthesis of functionalized amino acids. *Organic Letters*, **5**, 2963–2965.

86. Galindo, F. (2005). The photochemical rearrangement of aromatic ethers: A review of the photo-Claisen reaction. *Journal of Photochemistry and Photobiology C: Photochemistry Reviews*, **6**, 123–138.

87. Gou, B., Li, D., Yang, C., Xia, W., Li, Y., Chen, X. (2012). Conversion of aryl C–O to C–C bond through a UV light activation/TEMPO oxidation cascade reaction. *Journal of Photochemistry and Photobiology A: Chemistry*, **233**, 46–49.

88. Omelanzcuk, J., Sopchik, A. E., Lee, S.-G., Akutagawa, K., Cairns, S. M., Bentrude, W. G. (1988). Photo–Arbuzov rearrangements of benzyl phosphites. *Journal of the American Chemical Society*, **110**, 6908–6909.
89. Mullah, K. B., Bentrude, W. G. (1991). Photo–Arbuzov rearrangement route to acyclic nucleoside benzylphosphonates. *The Journal of Organic Chemistry*, **56**, 7218–7224.
90. Bhanthumnavin, W., Bentrude, W. G. (2001). Photo-Arbuzov rearrangements of 1-arylethyl phosphites: Stereochemical studies and the question of radical-pair intermediates. *The Journal of Organic Chemistry*, **66**, 980–990.
91. Alvaro, M., García, H., Iborra, S., Miranda, M. A., Primo, J. (1987). New photochemical approaches to the synthesis of chromones, cyclodextrin-mediated regioselective photo–Fries reaction of 1-naphthyl phenyl acylates. *Tetrahedron*, **43**, 143–148.
92. García, H., Iborra, S., Primo, J., Miranda, M. A. (1986). 6-Endo-Dig vs. 5-Exo-Dig ring closure in *o*-hydroxyaryl phenylethynyl ketones. A new approach to the synthesis of flavones and aurones. *The Journal of Organic Chemistry*, **51**, 4432–4436.
93. Martínez-Utrilla, R., Miranda, M. A. (1980). Indirect hydroquinone succinoylation via a photo–Fries rearrangement: Application to the synthesis of enol lactones. *Tetrahedron Letters*, **21**, 2281–2282.
94. Fillol, L., Martínez-Utrilla, R., Miranda, M. A., Morera, I. M. (1989). Photochemical versus aluminium chloride-catalyzed Fries rearrangement of aryl hydrogen succinates. synthesis of 2(3*H*)-furanones. *Monatshefte fuer chemie*, **120**, 863–870.
95. García, H., Iborra, S., Miranda, M. A., Primo, J. (1985). Application of the photo–Fries rearrangement of aryl dihydrocinnamates to the synthesis of flavonoids. *Heterocycles*, **23**, 1983–1989.
96. García, H., Iborra, S., Miranda, M. A., Primo, J. (1986). Photolysis of cyclic acetals of aryl benzoyl acetates as the key step in a new synthesis of flavones. *Heterocycles*, **24**, 2511–2517.
97. Miranda, M. A., Primo, J., Tormos, R. (1982). A new synthesis of 4-chromanones. *Heterocycles*, **19**, 1819–1822.
98. Samaniego López, C., Erra-Balsells, R., Bonesi, S. M. (2010). A mild and convenient one-pot photochemical synthesis of chroman-4-one derivatives. The photo-Fries rearrangement of (hetero)aryl 3-methyl-2-butenoate esters under basic catalysis. *Tetrahedron Letters*, **51**, 4387–4390.
99. Lewis, J. R., Paul, J. G. (1981). Oxidative coupling. 11. Approaches to the synthesis of bikaverin. *Journal of the Chemical Society, Perkin Transactions 1*, 770.
100. Katagiri, N., Nakano, J., Kato, T. (1981). Synthesis of bikaverin. *Journal of the Chemical Society, Perkin Transactions 1*, 2710.
101. Magauer, T., Martin, H. J., Mulzer, J. (2009). Total synthesis of the antibiotic kendomycin by macrocyclization using photo-Fries rearrangement and ring-closing metathesis. *Angewandte Chemie International Edition*, **48**, 6032–6036.
102. Ferrini, S., Ponticelli, F., Taddei, M. (2006). Rapid approach to 3,5-disubstituted 1,4-benzodiazepines via the photo–Fries rearrangement of anilides. *The Journal of Organic Chemistry*, **71**, 9217–9220.
103. Ferrini, S., Ponticelli, F., Taddei, M. (2007). Convenient synthetic approach to 2,4-disubstituted quinazolines. *Organic Letters*, **9**, 69–72.
104. Guerrini, G., Taddei, M., Ponticelli, F. (2011). Synthesis of functionalized quinolines and benzo[*c*][2,7]naphthyridines based on a photo–Fries rearrangement. *The Journal of Organic Chemistry*, **76**, 7597–7601.

105. Pitchumani, K., Warrier, M., Ramamurthy, V. (1996). Remarkable product selectivity during photo–Fries and photo–Claisen rearrangements within zeolites. *Journal of the American Chemical Society*, **118**, 9428–9429.

106. Gu, W., Warrier, M., Ramamurthy, V., Weiss, R. G. (1999). Photo-Fries reactions of 1-naphthyl esters in cation-exchanged zeolite Y and polyethylene media. *Journal of the American Chemical Society*, **121**, 9467–9468.

107. Koodanjeri, S., Pradhan, A. R., Kaanumalle, L. S., Ramamurthy, V. (2003). Cyclodextrin-mediated regioselective photo–Fries reaction of 1-naphthyl phenyl acylates. *Tetrahedron Letters*, **44**, 3207–3210.

108. Berry, R. S., Spokes, G. N., Stiles, R. M. (1960). Spectroscopic evidence for gaseous benzyne. *Journal of the American Chemical Society*, **82**, 5240–5241.

109. Kampmeier, J. A., Hoffmeister, E. (1962). The photolysis of 1,2-diiodobenzene: A photochemical source of benzyne. *Journal of the American Chemical Society*, **84**, 3787–3788.

110. Jones, M., Jr., DeCamp, M. R. (1971). Photochemically generated benzyne. *The Journal of Organic Chemistry*, **36**, 1536–1539.

111. Maki, Y., Furuta, T., Kuzuya, M., Suzuki, M. (1976). Photochemistry of *o*-nitrobenzaldehyde *N*-acetyl-*N*-alkylhydrazones resulting in the formation of benzyne. *Journal of the Chemical Society, Chemical Communication*, 616–617.

112. Maki, Y., Furuta, T., Suzuki, M. (1979). Photochemical reactions. Part 20. Photolysis of *o*-nitrobenzaldehyde *N*-acylhydrazones leading to benzyne and 5-nitrophthalazines. *Journal of the Chemical Society, Perkin Transactions 1*, 553–557.

113. Gredičak, M., Jerič, I. (2007). Enediyne compounds—New promises in anticancer therapy. *Acta Pharmaceutica* **57**, 133–150.

114. Alabugin, I. V., Yang, W.-Y., Pal, R. Photochemical Bergman cyclization and related photoreactions of enediynes. In Griesbeck, A., Oelgemoeller, M., Ghetti, F., Eds. *CRC Handbook of Organic Photochemistry and Photobiology*, third edition, CRC Press, Boca Raton, FL, 2012, pp. 549–592.

115. Polukhtine, A., Karpov, G., Popik, V. V. (2008). Towards photoswitchable enediyne antibiotics: Single and two-photon triggering of Bergman cyclization. *Current Topics in Medicinal Chemistry* **8**, 460–469.

116. Polukhtine, A., Karpov, G., Pandithavidana, D. R., Kuzmin, A., Popik, V. V. (2010). Photochemical triggering of the Bergman and Myers–Saito cyclizations. *Australian Journal of Chemistry*, **63**, 1099–1107.

117. Kaneko, T., Takahashi, M., Hirama, M. (1999). Photochemical cycloaromatization of non-benzenoid enediynes. *Angewandte Chemie International Edition*, **38**, 1267–1268.

118. Choy, N., Blanco, B., Wen, J., Krishan, A., Russell, K. C. (2000). Photochemical and thermal Bergman cyclization of a pyrimidine enediynol and enediynone. *Organic Letters*, **2**, 3761–3764.

119. Schmittel, M., Strittmatter, M., Kiau, S. (1995). Switching from the Myers reaction to a new thermal cyclization mode in enyne–allenes. *Tetrahedron Letters*, **36**, 4975–4978.

120. Schmittel, M., Mahajan, A. A., Bucher, G. (2005). Photochemical Myers–Saito and C^2–C^6 cyclizations of enyne–allenes: Direct detection of intermediates in solution. *Journal of the American Chemical Society*, **127**, 5324–5325.

121. Protti, S., Ravelli, D., Mannucci, B., Albini, A., Fagnoni, M. (2012). α,*n*-Didehydrotoluenes by photoactivation of (chlorobenzyl)trimethylsilanes. An alternative to enyne–allenes cyclization. *Angewandte Chemie International Edition*, **51**, 8577–8580.

5

PHOTOCHEMICAL GENERATION OF RADICAL IONS

5.1 PHOTOCHEMICAL ACCESS TO RADICAL IONS

The redox chemistry of organic molecules has a relatively limited scope, because of the not so easily accessible redox potentials (see also Chapter 8). Electrochemical methods certainly play a significant role in synthesis, but, possibly because of the specific instrumentation required, are not extensively used in organic chemistry. A redox reaction caused by a chemical reagent is also not that frequent (and popular among synthesis practitioners) and often requires rather drastic conditions, though many synthetically useful procedures pertain to this class [1]. As an example, the Birch reduction of aromatic molecules requires the use of a strong reducing agent, such as sodium. However, the very structure of electronically excited states make them both easier to oxidize (the electron promoted to a high-lying antibonding orbital can be easily donated) and to reduce (an electron can be easily transferred to a semi-occupied bonding orbital). As a result, redox reactions are quite common in photochemistry and occur under mild conditions, for example, the photoreduction of aromatics can be carried out by using amines not sodium, as donors. As a matter of fact, photoinduced electron transfer appears to be the best—or at least one of the best—ways for generating radical ions [1–9]. The advantage is that under photochemical conditions a precise control on sequential electron transfer, generally not available through other methods, can be achieved [1–9]. The limitation, on the other hand, is that electron transfer in the reverse direction is likewise facile. In electrochemistry, placing a compound into either the cathodic or the anodic compartment

Photochemically-Generated Intermediates in Synthesis, First Edition.
Angelo Albini and Maurizio Fagnoni.

establishes in which direction the redox process will occur and either oxidized or reduced species will accumulate while remaining in an oxidizing or, respectively, reducing environment. The same thing will occur when treating a compound by either an oxidizing or a reducing chemical reagent. Photoinduced electron transfer, however, generates a *pair* of oppositely charged radical ions that are in close contact. Therefore, back electron transfer is facile and leads back to the ground state. As a matter of fact, this has been considered a characteristic of photoinduced ET (PET) as opposed to photochemical reactions involving atom transfer. The latter processes are slower, but generally less easily reversed, while PET reactions, if thermodynamically allowed, are faster by one order of magnitude or more. The PET processes, however, are followed by an equally fast reverse process, so that even when ET is efficient, the quantum yield of the end products will be low or, at any rate, generally lower than that of atom transfer reactions. Fortunately, there are many exceptions to this generalization, and the chance to control the competition between direct and back transfer allows efficient applications of this approach.

In order that back electron transfer is avoided, some other process has to be competitive. Thus, the efficiency of the chemical reaction depends on the path accessible by each radical ion under the conditions involved, as summarily indicated in Schemes 5.1–5.3. A possibility is that the radical anion bears nucleofugal groups and fragments to give a radical. This is exemplified by a large scope process such as the $S_{RN}1$ reaction (Scheme 5.1, path *a*). This is typical of aromatic halides and particularly of weakly bonded iodides. In the key step, the radical anion $AY^{\bullet -}$ thus cleaves to give an aryl radical ($A^{\bullet}$). The last species is then trapped by a nucleophile Nu^- that most often serves also as the donor in the initial ET step. This leads to a new radical anion that in turn transfers an electron to the starting material initiating a chain process and giving the product of a formal aromatic substitution [10–14]. When the radical $A^{\bullet}$ is an aliphatic, rather than an aromatic species, this can be trapped by a nucleophilic olefin (W-J) and alkylation of the alkene results (product A-W, path *b*). In all of the other cases, the radical anion is a rather stable species (e.g.,

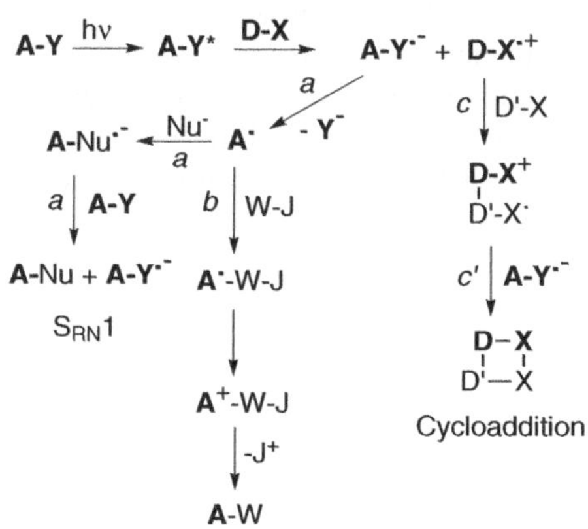

Scheme 5.1.

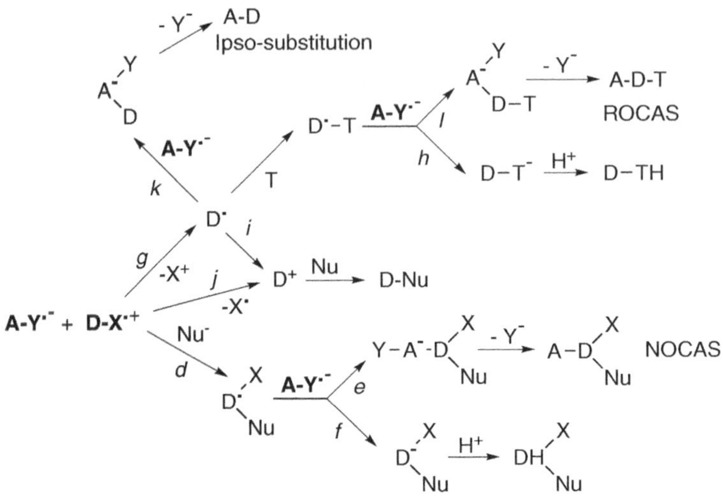

Scheme 5.2.

when it is generated from an aromatic nitrile) and is thus the co-formed radical cation $DX^{\bullet +}$ that reacts. The fact that a donor is transformed into a cation reverts the electronic character of the starting species (an Umpolung process) and makes viable addition and cycloaddition reactions of this intermediate with donors, whether the starting molecule or a purposely added one D′X (path *c*). In the following course of the reaction, the sequence is completed by back electron transfer from $AY^{\bullet -}$. The acceptor thus acts as a nonconsumed photocatalyst (path *c*′).

Further chemical reactions may take place. Thus, a nonfragmenting radical cation $DX^{\bullet +}$ could add a nucleophile Nu^- (Scheme 5.2, path *d*) and the resulting radical X-D•-Nu may incorporate the (aromatic) radical anion $AY^{\bullet -}$. Loss of anion Y^- finally gives a three-component adduct X-A-D-Nu (NOCAS process, short for nucleophile olefin combination aromatic substitution, path *e*) [15]. Alternatively, an $AY^{\bullet -}$-mediated reduction of the X-D•-Nu intermediate may occur to give the adduct X-DH-Nu (path *f*). When the $DX^{\bullet +}$ intermediate contains a good electrofugal group, however, unimolecular fragmentation is efficient and gives radical D• (path *g*). This applies generally to aliphatic derivatives and gives access to several paths. These include the general reactions of radicals, namely dimerization and, in the presence of a suitable trap T, addition (e.g., for the case of a nucleophilic–alkyl radical, to

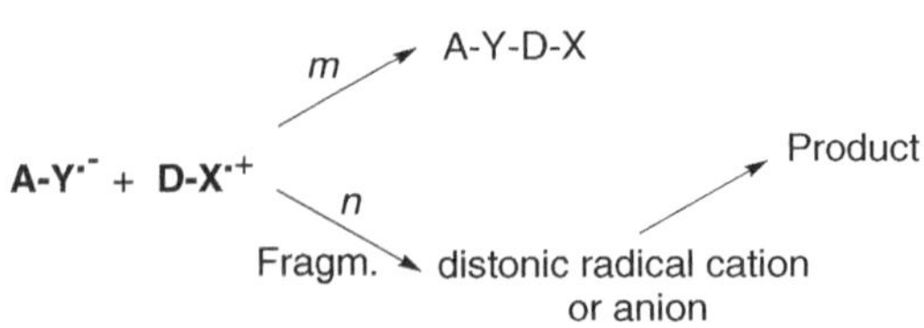

Scheme 5.3.

electrophilic alkenes, path *h*) where the acceptor acts again as a photocatalyst. Alternatively, radical $D^•$ can be oxidized to the corresponding cation (e.g., an α-aminoradical to an iminium ion, D^+, path *i*) that is in turn trapped by a (carbon) nucleophile. Cation D^+ may likewise be formed by fragmentation of $DX^{•+}$ (path *j*).

Other reactions from radical $D^•$ involve the radical anion. A typical case is coupling, when $AY^{•-}$ is a persistent aromatic species [16]. The resulting anion adduct D-A-Y^- loses Y^-, resulting in overall *ipso*-substitution by a (substituted) alkyl group (path *k*). In the presence of a trap, capture of the radical $D^•$ precedes coupling and then the radical adduct couples with the aromatic radical anion $AY^{•-}$. Loss of the Y^- group finally yields the three-components adduct A-D-T through a ROCAS process (short for radical olefin combination aromatic substitution, path *l*). The reverse ET process may be operative in the aromatic compound/olefin system, which is using an electron-rich arene and an electron-deficient (or at least less electron-rich) alkene. In such a case, the radical anion of the latter species is formed and protonated. Depending on the chemical nature, the neutral radical that is formed either dimerizes, is oxidized to the cation by back ET to the radical cation of the aromatic (that thus acts as a photocatalyst), or reacts with it through what is the third three-component reaction in this group, the EOCAS process (short for electrophile olefin combination aromatic substitution) [17].

Sparse examples are reported where the oppositely charged radical ions couple, a case usually observed in intramolecular PET (Scheme 5.3). The adduct thus formed A–Y–D–X contains all, or most, of the atoms of the starting reagents and is an example of atom economy (path *m*). Yet another possibility, occurring when one of the starting molecules contains a ring, is that this opens up at the radical ion stage to give a distonic radical ion (path *n*).

From the point of view of the end result, the reactions via radical ions can be classed as (overall) substitution, addition, or cycloaddition. Furthermore, (cyclo)addition reactions via an excited donor–acceptor complex (exciplex), rather than via a radical ion pair, have often been observed with the same or closely related reagents. Whether complexation or full electron transfer will happen in a specific reaction depends on the reagents structure, but also on the conditions (solvent polarity, presence of inorganic ions, secondary electron transfer, etc.). Distinguishing the exciplex reactions from those that involve a ion pair—and in the latter case in contact or as separated, solvated species—is important from the mechanistic point of view and also in synthetic applications, but a thorough examination of this issue is beyond the scope of this presentation. A few references are indicated [18–21].

5.2 SUBSTITUTION

Unimolecular nucleophilic substitution has an important role in photochemistry. Typically this involves a chain process initiated by electron transfer (ET) from a suitable donor and proceeding with the fragmentation of the thus formed radical anion ($S_{RN}1$ process). This has found many applications of synthetic interest [10–14]. The process may begin upon simply adding the donor (usually the nucleophile) to the solution of the other reagents; but in many cases irradiation, cathodic reduction, or chemical reduction (e.g., by Fe^{2+} salts, SmI_2, alkali metals dissolved in liquid ammonia) are required for getting the process started. The reaction has been repeatedly reviewed elsewhere [10–14], and only a few photostimulated reactions will be presented here.

5.2.1 Aliphatic

Photostimulated $S_{RN}1$ reactions have been reported for aliphatic substitution. The initial observation has been the substitution of 2-chloro-2-nitropropane by the anion of 2-nitropropane [22, 23]. A recent example is offered by the stannylation of 1,1-dichlorocyclopropanes **1a–c**, readily available from the addition of dichlorocarbene to olefins. This process afforded the desired products (**2a–c**) in good to excellent yield by irradiation of a solution in liquid ammonia in the presence of Me_3SnCl (Scheme 5.4) [24].

Substitution at the benzylic position has been observed with *p*-nitrobenzyl derivatives and has found several applications with (hetero)arylmethyl derivatives, including non-nitrated terms—for example, with 3-bromo-2-(chloromethyl)-4*H*-pyrido[*1,2-a*]pyrimidin-4-one (**3**) that undergoes substitution by various C- and S-centered nucleophiles [25]. As reported in Scheme 5.5, when a nitronate anion is used as the nucleophile, substitution is often accompanied by HNO_2 elimination leading to styrenes and hetero analogues.

The same synthetically useful procedure has been reported for the preparation of other heterocycles—for example, derivatives of quinazolines [26]. On the other hand, photochemical initiation is not necessarily the best way for carrying out a substitution. For example, with 4-[4-(chloromethyl)phenyl]-1,2-dimethyl-5-nitro-1*H*-imidazole, the reaction with nitronate at room temperature under the classical $S_{RN}1$ conditions led to electron transfer rather than bond formation and finally

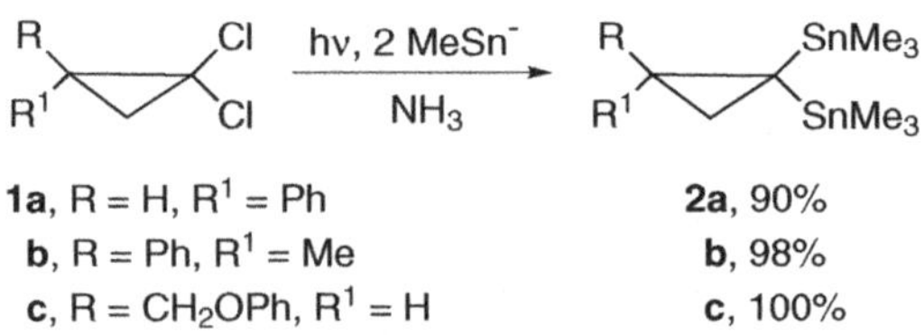

Scheme 5.4.

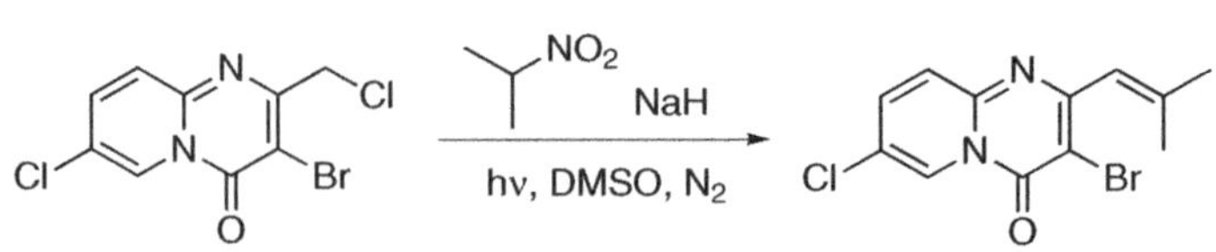

3 **4**, 49%

Scheme 5.5.

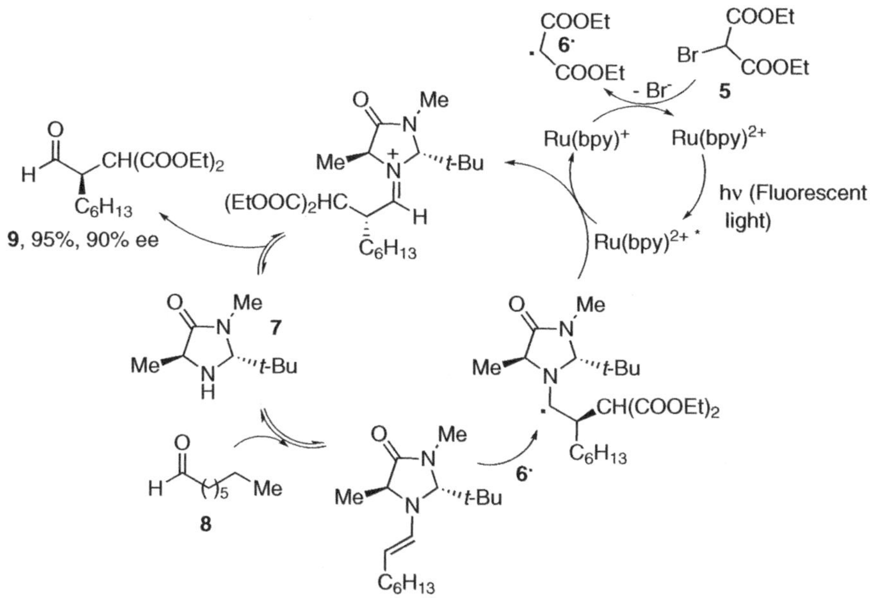

Scheme 5.6.

afforded the corresponding benzaldehydes [27]. The $S_{RN}1$ process was instead obtained upon heating.

Recently, a breakthrough for the functionalization of the α-CH bonds in aldehydes has been reported, which is based on visible light photocatalysis [28–34]. Thus, the enantioselective intermolecular α-alkylation of aldehydes has been smoothly accomplished by combining a photoredox catalyst (e.g., a Ru(II)-based complex) and a chiral imidazolidinone organocatalyst **7** (Scheme 5.6) [35]. Thus, the starting aldehyde (e.g., octanal **8**) was converted into an enamine by a thermal reaction with chiral imidazolidinone **7**. In turn, this trapped an electrophilic radical **6**• generated by the monoelectronic reduction of bromomalonate **5** (after bromide anion loss) by the excited photocatalyst. The end product, alkylated aldehyde **9**, was formed in excellent yield and ee [35].

Various related reactions have been developed following this approach [28–34] and by using different photocatalysts. Accordingly, the strong reductive properties of the excited state of the $Ir(ppy)_2(dtb\text{-}bpy)^+$ complex were exploited for the enantioselective organocatalytic α-trifluoromethylation and α-perfluoroalkylation of aldehydes [36]. Moreover, organic compounds, such as dyes (typical is the case of Eosin Y), could likewise be used in the role of photocatalyst in the place of toxic and expensive inorganic complexes [37–39].

5.2.2 Nonaliphatic

5.2.2.1 Vinyl and (Hetero)aryl via $S_{RN}1$ Vinyl substitution usually operates successfully, in particular for β-halostilbenes. This is the case for some

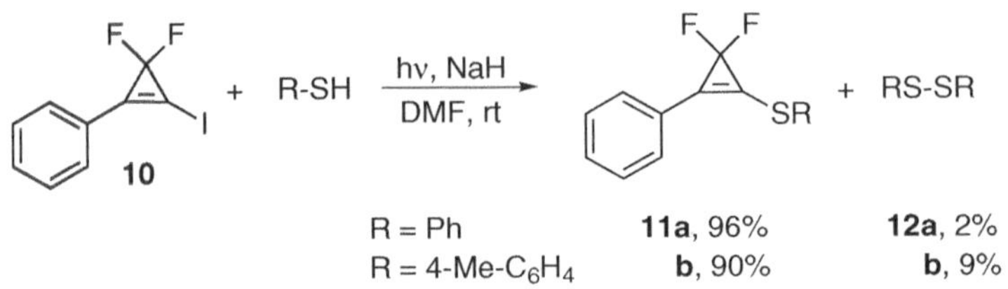

Scheme 5.7.

iodo-phenylcylopropenes **10** where substitution was effective with thiolate anions (Scheme 5.7) [40].

As one may expect, an electron-withdrawing substituent on the phenyl ring activated the substitution at the vinylogous position. Thus, 3-(4-nitrophenyl)-propargyl chloride underwent substitution by nitronate anions through this mechanism [41].

Aromatic substitution has largely been applied, and has become a main synthetic path, in particular for heterocycles [10–14, 42]. Several nucleophiles such as carbanions and heteroatom-based anions were used for $S_{RN}1$ reactions to form new C–C or C–heteroatom bonds in good yield. In the reaction of aromatic amide ions with aromatic substrates, both C–N and C–C bond formation was achieved. For instance, naphthylamide ions initiated the photo $S_{RN}1$ process of PhI, 4-$MeOC_6H_4I$, and 1-iodonaphthalene in liquid ammonia. 1-Aryl 2-naphthylamines were formed regioselectively from the last compound in 45–63% yields, with only 3–6% of *N*-arylation [43].

Various 6-substituted 2,4-dimethoxy pyrimidines have been prepared by irradiation of the corresponding 6-chloro derivatives and hydrolyzed to the uracils, obtained in 45–90% yield in this one-pot procedure [44]. A group of reactions that has been extensively developed is stannylation, a choice that is explained also because of the interest of such compounds as reagents for palladium-catalyzed coupling. Indeed, the two processes could be advantageously carried out in a one-pot procedure. A further feature of this process is that two groups can be substituted in a single operation. Thus, distannylated pyridines were obtained from the irradiation of the corresponding dichloro derivatives in liquid ammonia [45]. On the other hand, the presence of two different nucleofugal groups was exploited for selective processes. As an example, some chloropyridinyl phosphates were found to give distannylated products by irradiation in liquid NH_3 at 30°C, but yielded mono-stannylated piridinyl phosphates upon heating in the dark at 60°C. This result was explained by the fact that chlorine substitution occurred first and formed, as is generally the case, the corresponding radical anion. Under thermal conditions, this intermediate transferred an electron to the starting material and the chain process of monosubstitution proceeded further. Under photochemical conditions, on the other hand, the radical anion fragmented faster than ET took place, and the radical produced entered the second substitution cycle [46]. Having the same substituent in a different position may again lead to a different reactivity and thus to a regiospecific reaction. As an example, the 3,5-dichlorobenzoate **13** (as well as

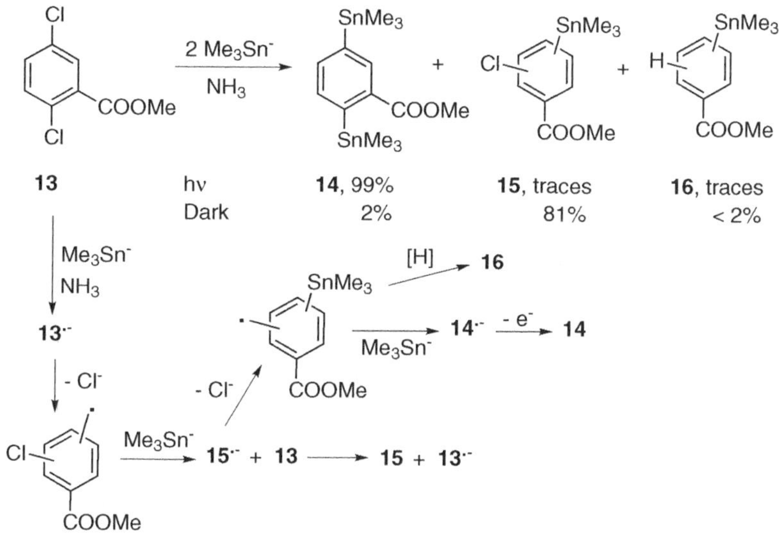

Scheme 5.8.

the related 3,5-dichloro-2-methoxybenzoate) gave the corresponding distannanes **14** upon irradiation in liquid ammonia, but monostannanes **15** in the dark (Scheme 5.8) [47].

Intermolecular substitution and condensation was applied to the preparation of heterocycles, as in the synthesis of benzothiazine *S,S*-dioxides **18a,b** from the irradiation of *o*-iodobenzenesulfonamide **17** in the presence of ketone enolates (Scheme 5.9). A competing reaction was reductive dehalogenation, which had a role proportional to the number of β-hydrogens in the enolate and predominated when the enolates of esters were used in the place of those of ketones [48].

The competition between aryl and benzyl substitution of the bromine atom in position 8 in 6,8-dibromo-2-chloromethyl-3-nitroimidazo[1,2-*a*]pyridine (**19**) has been found to depend on the nucleophile nature (C- and S-centered anions have been compared; see Scheme 5.10) [49].

The $S_{RN}1$ process has largely been applied to the synthesis of biaryls by coupling of an aryl halide and an electron-rich arene, such as phenolates or anilides [50, 51]. A similar mechanism is reasonably involved in the photochemical synthesis of

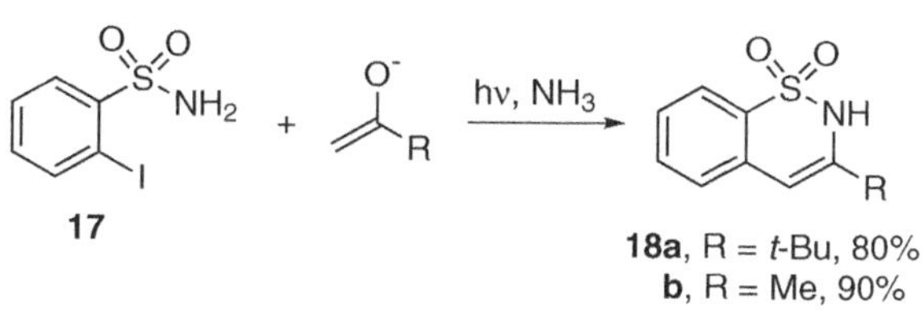

Scheme 5.9.

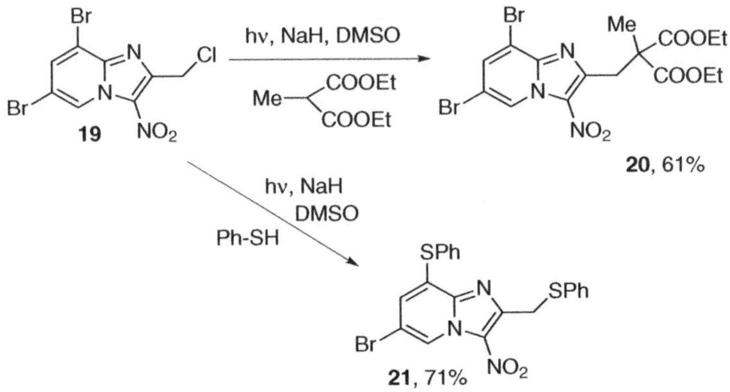

Scheme 5.10.

heterobiaryl derivatives by irradiation of five-membered heterocyclic halides bearing an electron-withdrawing group, such as thiophene, furan, or pyrrole bearing in position C5 a iodo- or bromo-substituent and in position C2 an aldehyde, a ketone, or a nitro group (**22**, **25**; see Scheme 5.11) [52, 53]. The cleavage of the halide apparently involved assistance by the aromatic ring and gave an aryl–iodine complex along with the key intermediate, the heteroaryl radical [54].

Intramolecular reactions can be conveniently carried out. As an example, 3-substituted 2,3-dihydrobenzofurans, 1,2-dihydronaphto(2,1-*b*)furans, and *N*-substituted 2,3-dihydro-1*H*-indoles **29** were synthesized in very good yields in liquid ammonia, starting from aryl allyl ethers or amines (**28**) and using Me_3Sn^-, Ph_2P^-, or $^-CH_2NO_2$ anions as nucleophiles (Scheme 5.12, path *a*, $n = 1$). The photogenerated phenyl radical underwent selective 5-*exo* cyclization and addition of the nucleophiles to give the five-membered heterocycles. The same approach was applied to the preparation of six-membered derivatives such as 4-substituted chromanes and 4-substituted benzo[*f*]chromanes in good yields by lengthening

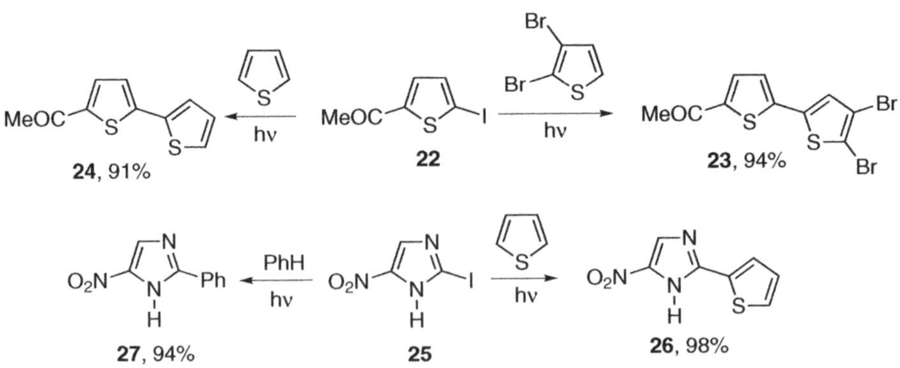

Scheme 5.11.

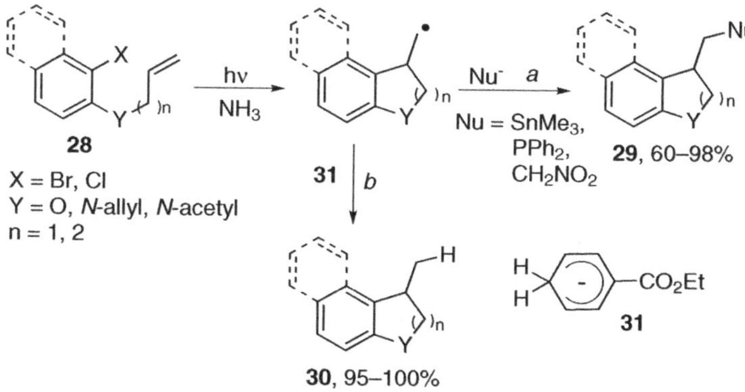

Scheme 5.12.

the tether (path *a*, $n=2$) [55]. Moreover, the radical intermediate could be reduced, rather than functionalized, as illustrated in path *b* of Scheme 5.12. This tandem cyclization/reduction sequence could be successfully achieved by using as the hydrogen donor a hydrobenzoate anion **31**, a reagent that is available from nontoxic, reusable, and inexpensive ethyl benzoate [56].

Intramolecular reactions offer a useful access for building a heteroaromatic ring [42]. As an example, all of the four carbolines **33a–d** have been obtained from the photostimulated cyclization of the corresponding anilinohalopyridines **32a–d** in good to excellent yields (Scheme 5.13) [57]. The oxazole ring has been built by *O*-cyclization of *N*-(2-haloaryl)benzamides [58]. The irradiation of substituted acetophenones **34a–c** in the presence of excess base yielded seven-memberd heterocycles **35a–c** in good yields (Scheme 5.14) [59].

Another example where a substrate bears both the leaving group and the nucleophilic center is illustrated in Scheme 5.15 for the synthesis of (benzo) phenanthridines [60]. Thus, benzo-fused azaheterocycles such as phenanthridine (90%), 4-phenylphenanthridine (87%), benzo[*a*]phenanthridine (**37**, 98%), and benzo[*c*]phenanthridine (84%) were synthesized by a photostimulated reaction of *N*-(2-halobenzyl)aryl amines **36** with *t*-BuOK in liquid ammonia by the $S_{RN}1$ mechanism [60].

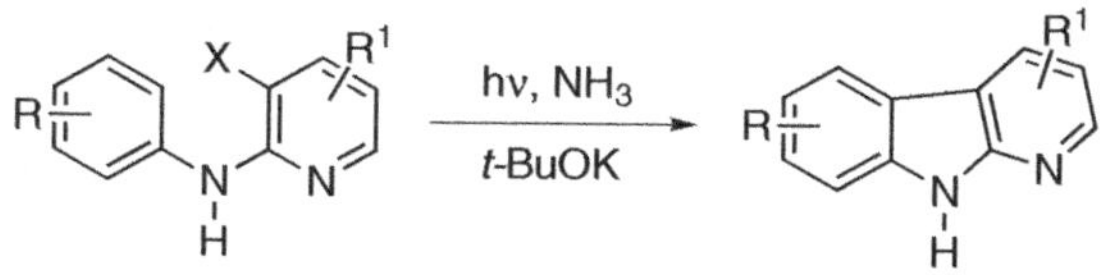

32a, X = Cl, R = R¹ = H — **33a**, 75%
b, X = Br, R = H, R¹ = 2-OEt — **b**, 63%
c, X = Br, R = H, R¹ = 3-Me — **c**, 90%
d, X = Br, R = 6-COO*t*-Bu, R¹ = H — **d**, 79%

Scheme 5.13.

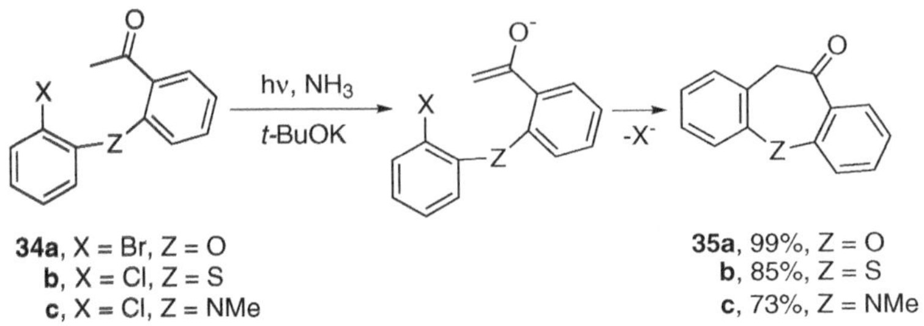

Scheme 5.14.

Aporfines **40a,b** have been obtained by irradiation of tetrahydroisoquinolines bearing a 2-bromobenzyl (or phenethyl) group in 1 (**39a,b**) through an *ortho*-cyclization process (Scheme 5.16). The reaction was applied only to nonbasic compounds (bearing an electron-withdrawing substituent on the nitrogen atom) [61].

Electron transfer to an aromatic acceptor and fragmentation of the radical anion do not necessarily lead to a chain process. If the radical arising from the acceptor couples with the donor radical cation an even electron cation results. The last intermediate deprotonates and recover aromaticity. This non-chain reaction has been reported for good acceptors—for example, for some polyfluorinated derivatives. Thus, pentafluorophenylsulfonates (e.g., **41**) and iodide photochemically reacted with alkylbenzenes, phenyl ethers, and anilines, giving the corresponding *ortho*- (**42**) and *para*- (**43**) pentafluorophenyl derivatives. The reaction has been applied also to pyrroles and indoles (Scheme 5.17) [62].

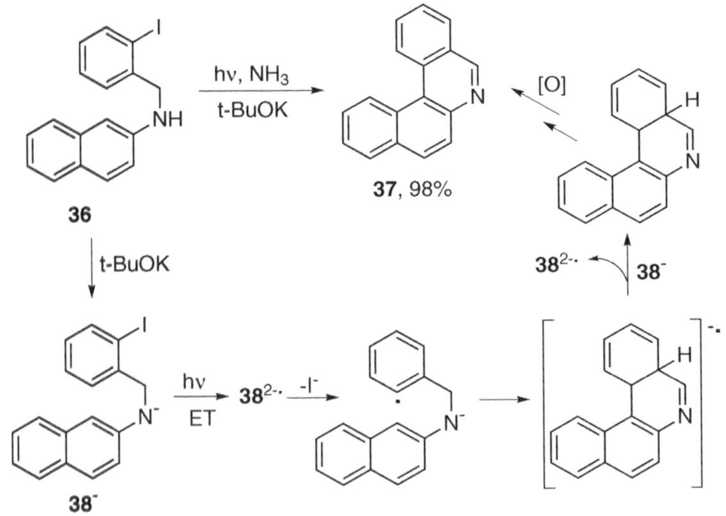

Scheme 5.15.

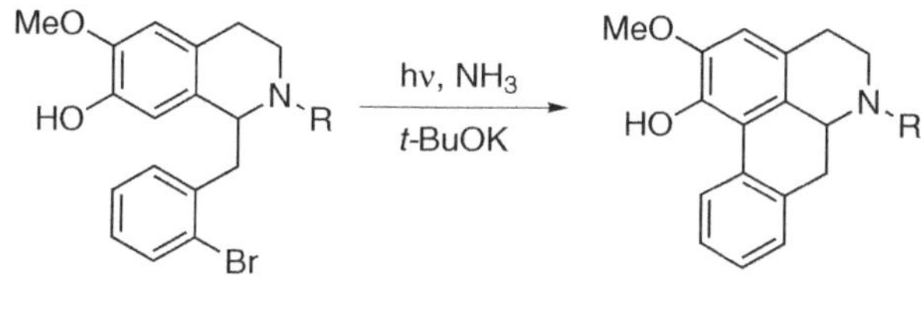

39a, R = COOMe
b, R = COMe

40a, 75%
b, 82%

Scheme 5.16.

5.2.2.2 Aromatic via Radicals. Ipso Substitution On the other hand, the radical anion of the electron-withdrawing substituted arenes not bearing a good leaving group are persistent species. Their odd-electron character makes coupling with radicals a significant outcome, despite the low concentration of both species. It may happen that the radical cation fragments and reacts with the acceptor radical anion. The resulting anion undergoes either (a) rearomatization by loss of a nucleofugal group (and then the overall result is substitution) or (b) protonation to give a dihydroaromatic molecule. Typical is the case of aromatic nitriles, whose radical anion is a rather persistent, nonbasic and non-nucleophilic species. In the presence of suitable donors [16], these undergo ipso substitution of the cyano group (see Scheme 5.2, path *k*), a process that would otherwise require the use of organometallic reagents. As seen above, this process is reciprocal to the $S_{RN}1$ reaction in some aspects, but it is no chain process.

The fragmentation of a sigma bond to give a neutral radical and a cation after monoelectronic oxidation is often a facile process [2, 4, 63]. The thermodynamic feasibility of fragmentation in a radical ion can be predicted by using a thermochemical cycle (Scheme 5.18) and introducing the relevant redox potentials and bond dissociation energies (see Chapter 8).

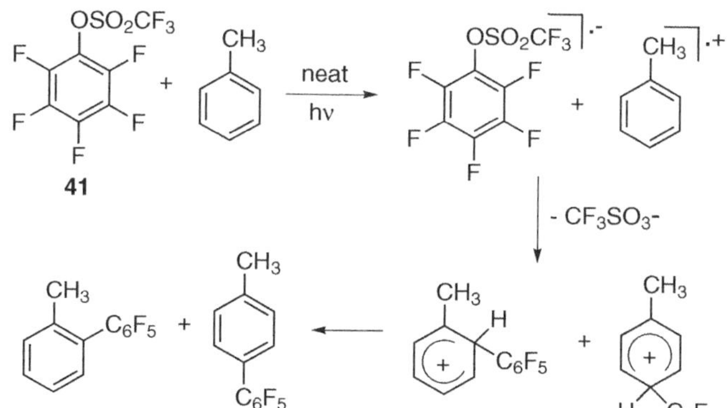

41

42, 54%

43, 32%

Scheme 5.17.

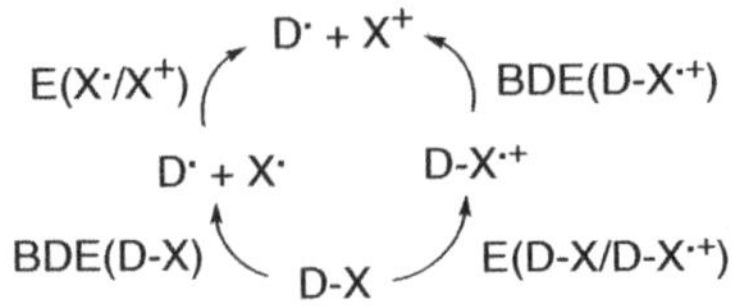

$$BDE(D\text{-}X^{\cdot+}) = BDE(D\text{-}X) - [E(D\text{-}X/D\text{-}X^{\cdot+}) - E(X^{\cdot}/X^{+})]$$

Scheme 5.18. The bond dissociation energy of a radical cation is equal to that of the neutral molecule minus the difference between the oxidation potential of the molecule and that of the electrofugal group. The BDE for a radical anion is calculated analogously.

The kinetics of the process have been studied in a number of significant cases. The role of nucleophiles in assisting the cleavage [64] has been evidenced, and 5-hexenyl derivatives have been used as radical clocks in order to obtain some information about the timing of the steps [65, 66]. The electrofugal group is often a proton, as established both with π donors, such as alkenes [67] and alkyl aromatics [68, 69], and with heteroatom-centered donors, such as carboxylic acids [70] and *tert*-butyl esters [71], ethers [72], ketals [73], and amines [74]. Further fragmentable radical cations are those formed from organometallic donors such as silanes, silyl ethers, and silyl amines [65, 75] (as well as cyclopropanone silyl ketals) [76] and borates [77]. Some examples of the radical generation from these donors are collected in Scheme 5.19.

As an example, *o*-dicyanobenzene (**44**) was benzylated to give *o*-benzylbenzonitrile (**45**) when irradiated in the presence of either benzyltrimethylsilane (75% yield) [78] or benzyltributylstannane (81%) [79] via the benzyl radical formed after fragmentation of the organometallic radical cations. The reaction led to a clean monobenzylation and the formation of disubstituted products was prevented, because the reduction potential of the mono-benzylated benzonitrile in the excited state was not sufficient for allowing a further oxidation of either the silane or the stannane (Scheme 5.20a).

Even alkanes have been used for the alkylation reaction of strong acceptors such as 1,2,4,5-benzenetetracarbonitrile (**46**, TCB) and yielded the corresponding alkylbenzenetricarbonitriles, where the alkyl radical incorporated was formed with remarkable

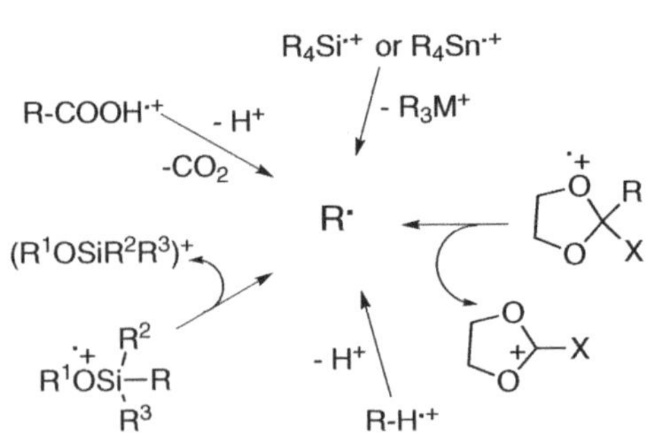

Scheme 5.19.

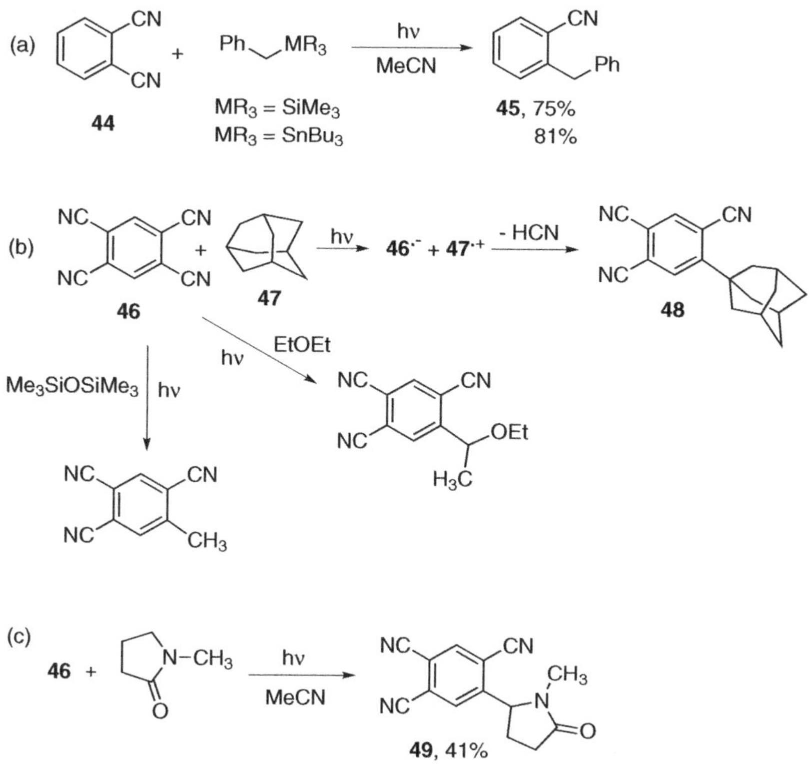

Scheme 5.20.

regioselectivity (e.g., the 1-adamantyl derivative **48** was exclusively obtained from adamantane (**47**) along with some diarylated derivative, see Scheme 5.20b) [80, 81]. Simple aliphatic derivatives, such as the above-mentioned ethers and siloxanes, were effective radical precursors under these conditions [75]. Substituted amides and lactams, such as DMF or 2-pyrrolidones, were also used in this reaction [82]. It has been found that deprotonation was chemoselective also in the case of *N*-methylpyrrolidone, as illustrated in Scheme 5.20c, and occurred exclusively at the methylene, not at the methyl group. Tricyano benzene **49** was thus isolated in 41% yield. Likewise, acetonitrile could function as a donor, and tricyanophenylacetonitrile was formed by irradiation of the tetracyano derivative **46** in this solvent [83].

While with benzene derivatives the overall process was consistently substitution, with naphthalene (poly)nitriles and larger aromatics, both substitution and addition leading to dihydroaromatics were observed, as well as the introduction of the new group into an unsubstituted position (see Scheme 5.21). The donors used included alkenes [84], allylsilanes [85], benzylsilanes (or germanes or stannanes [86]), and alkyltriphenyl borate salts [87].

N-Alkylacridinium ions and other cationic heteroarenes have likewise been photoalkylated and yielded acridanes by addition when using either tetraalkylstannanes [88]

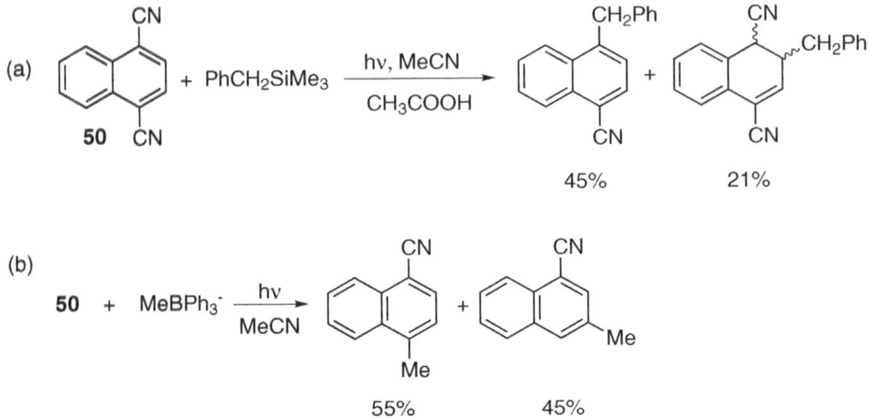

Scheme 5.21.

or the more environmentally benign 4-alkyl-1-benzyl-1,4-dihydronicotinamides as the source of alkyl radicals [88]. More generally, the role of photoinduced ET in heterocyclic chemistry has been reviewed [63].

5.2.2.3 ***Aromatic via Radicals. Three-Component Reactions*** More varied functionalizations could be obtained by elaborating the radical cation or the radical before attack onto the aromatic. As an example, a variation of the above alkylation reactions could be obtained by trapping the radical by an electron-poor olefin. As an example, photolysis of a mixture of **46**, 2-methyl-2-phenyl-1,3-dioxolane (**51**), and methyl vinyl ketone formed a 3-aryl substituted 2-pentanone, which cyclized during purification on silica gel to afford isocoumarine **52** in 75% yield [89]. This ROCAS reaction (radical olefin combination aromatic substitution reaction, Scheme 5.22,

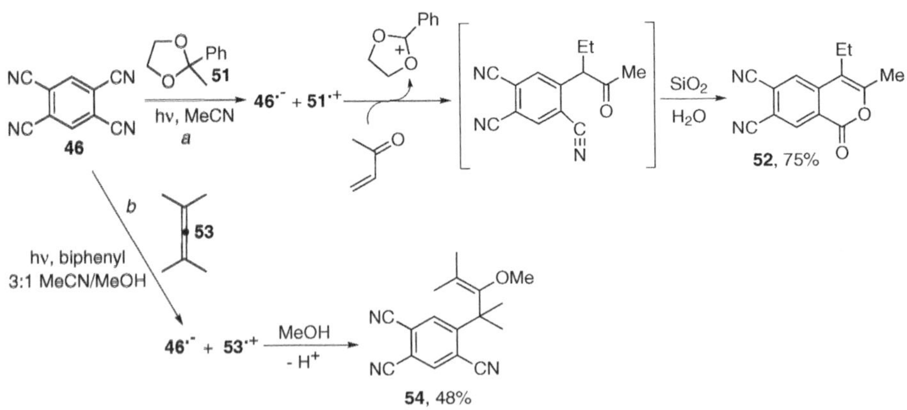

Scheme 5.22.

path *a*) [89, 90] is based on the trapping of the methyl radical resulting from the fragmentation of radical cation **51**$^{\bullet +}$ by the enone (see also Scheme 5.2, path *l*) and coupling with the aromatic radical anion. It is noteworthy that in this reaction two C–C bonds were cleaved and two C–C bonds were formed in a one-pot procedure.

With nonfragmentable radical cations a neutral radical can be formed by addition of a nucleophile. This is the key feature of another three-component sequence, the nucleophile olefin combination aromatic substitution (NOCAS) process [15, 16] (see Scheme 5.2, path *e*). The nucleophile is often the solvent, and addition to the acceptor radical anion is followed by rearomatization by loss of an anion leading again to an overall *ipso*-substitution. A nice example is found in the case of allenes [91]. Thus, irradiation of **46** in the presence of tetramethylallene (**53**) in a 3:1 MeCN/MeOH mixture afforded 1:1:1 arene-allene-methanol adduct **54** in 48% yield (Scheme 5.22, path *b*). It is noteworthy that addition of methanol took place exclusively at the central allene carbon and that aromatic substitution occurred through the terminal carbons. The two reacting moieties could also be combined in a single molecule by using ω-alkenols. In this case, initial ET was followed by a cyclization, yielding aryl-substituted tetrahydrofurans or tetrahydropyrans as the final products via a tandem Ar–C, C–O bond formation [92]. Water or diols could likewise be employed as the nucleophile [93]. The course of the reaction depended on the conditions. As for the regiochemistry, interesting results have been obtained for the case of 1,4-dicyanobenzene (**55**) with 2-methylpropene (**56**). When the reaction was carried out in methanol, the radical cation was stabilized by the solvent, and this allowed a protic equilibration of the regioisomeric radicals [94–96]. Thus, the main product (**58**) resulted from attack by the most stable radical (see Scheme 5.23).

On the other hand, when a charged nucleophile was used (e.g., fluoride [97] or cyanide [98]), no bridged intermediate was involved and the regiochemistry was controlled by the rate of addition of the anion, so that the main product arose from the least substituted radical. Finally, in neat acetonitrile the radical cation underwent

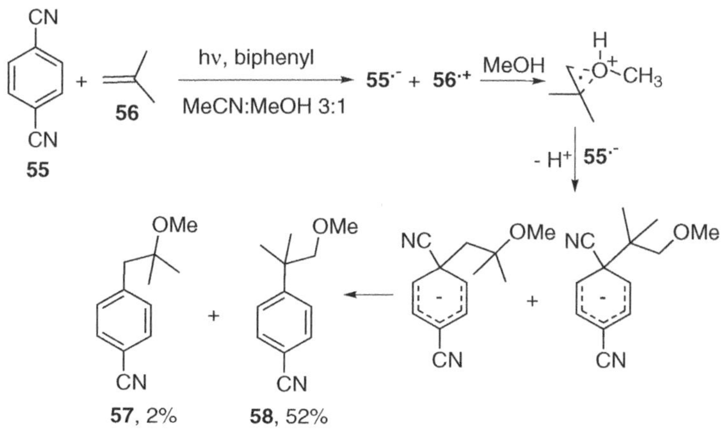

Scheme 5.23.

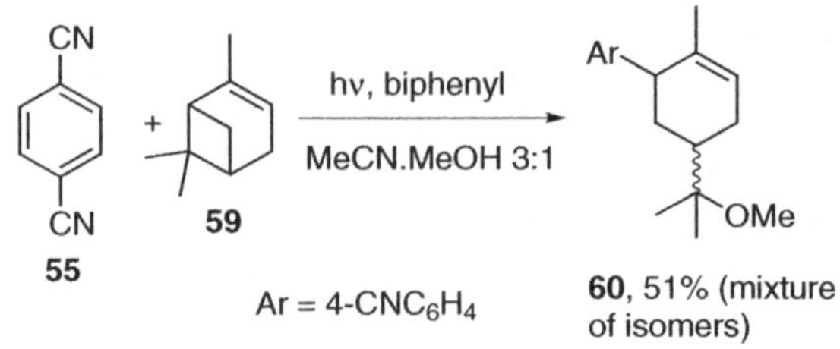

Scheme 5.24.

solvent addition rather than deprotonation. In the last case, coupling of the resulting distonic radical cation with the dicyanobenzene radical anion was followed by electrophilic attack by the nitrilium moiety on the ring leading to a dihydroisoquinoline. The same process took place in the reaction between tetracyanobenzene and 1-hexene in acetonitrile [67, 97].

Strained alkenes have been observed to undergo carbon–carbon bond cleavage at the radical cation stage, as shown in the case of α-pinene (**59**) where the four-membered ring was cleaved in the NOCAS product obtained (**60**, Scheme 5.24). Cleavage of a strained donor and a NOCAS-type photosubstitution process have also been obtained from 2-carene and from a saturated strained donor such as tricyclene [99, 100].

Ammonia or amines are active nucleophiles even when in solution rather than neat [101]. As an example, photoinduced electron transfer between an aromatic compound S (e.g., triphenylbenzene) and 1,4-dicyanobenzene (**55**) led to the arene radical cation $S^{\bullet +}$ and the radical anion **55**$^{\bullet -}$. In the presence of a further donor D, secondary ET to $S^{\bullet +}$ produced a new radical cation ($D^{\bullet +}$) that added ammonia and deprotonated to form a neutral radical. In turn, coupling with **55**$^{\bullet -}$ and cyanide loss led to the corresponding NOCAS product. Suitable donors were arylcyclopropanes, quadricyclane, and dienes that gave 4-(1-aryl-3-aminopropyl)benzonitriles, 7-amino-5-(*p*-cyanophenyl)bicyclo[2.2.1]hept-2-ene [102], and 4-(4-aminobut-2-enyl)-benzonitriles, respectively [103].

Further choices for the donor were cyanide anion and an enolate. In the former case, since cyanide was liberated in the substitution step, a catalytic amount of tetrabutylammonium cyanide was sufficient in the reaction of dicyanobenzene and dimethylhexadiene [104]. In the latter one, malonodinitrile and dimethylhexadiene reacted with both dicyano- and tetracyanobenzene to give the corresponding NOCAS products. With the latter acceptor, a monosubstituted product was obtained in the absence of phenanthrene, while a disubstituted derivative was formed when omitting the secondary donor. This demonstrated the potential of a secondary donor for accelerating the process (by preventing BET and thus prolonging the lifetime of the donor radical cation). With dicyanonaphthalene, addition rather than substitution was observed [105] and the quantum yield of the NOCAS process remarkably increased at a lower temperature, because of the disturbing role of an exciplex at a high temperature [106]. Monosubstitution occurred cleanly by using **46**, an aromatic alkene and the enolate of 1,3-dicarbonyls [107]. Note that loss of a -COMe group

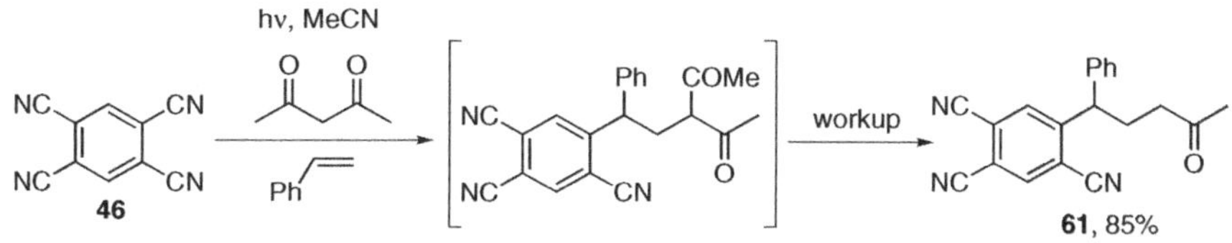

Scheme 5.25.

occurred during the workup, probably via an acid-catalyzed decomposition upon chromatographic separation on silica gel (Scheme 5.25).

A variation of this reaction involved the use of the alkene itself as nucleophile. In this case, a dimeric radical cation was formed by attack of the alkene radical cation onto the neutral alkene, forming a distonic structure, with radical and cation sites separated (Scheme 5.26). In the presence of α-methylstyrene (**62**), cyclization and deprotonation respectively occurred and the obtained neutral radical coupled with the radical anion of the acceptor, in this case **46**. This led to the NOCAS adduct **63** as a diastereoisomeric mixture in an overall 90% yield [108].

The irradiation of aromatic nitriles in the presence of alkenes may lead to different products, particularly when carried out in an apolar medium. As an example, 1,4-dicyanobenzene gave isoquinolines by a [4 + 2]-cycloaddition with a cyano group through irradiation in the presence of diphenylethylenes in benzene via a polar exciplex [109].

Sparse examples were reported involving the complementary photo-EOCAS reaction, where charges were reversed [17]. Typical reactions involved a methoxyarene [e.g., 1,4-dimethoxybenzene or 1-methoxynaphthalene (**64**)] as the donor and involved an electron-poor olefin [e.g., acrylonitrile (**65**)] as the acceptor [110]. As an example, the photochemically produced radical anion of acrylonitrile was protonated by MeOH used as the solvent and the thus formed radical coupled with the radical cation of the aromatic ether to form a 2-aryl-propionitrile **66**. With regard to the donor, the reaction bore some similarity to a Friedel–Crafts alkylation, while with regard to the alkene, it resulted in the arylation at the α-position of the α,β-unsaturated nitrile, rather than at the more commonly functionalized β-position. In some instances, the aromaticity was partially lost in the reaction as reported for the photoaddition of acrylonitrile onto dimethoxynaphthalene (Scheme 5.27) [110].

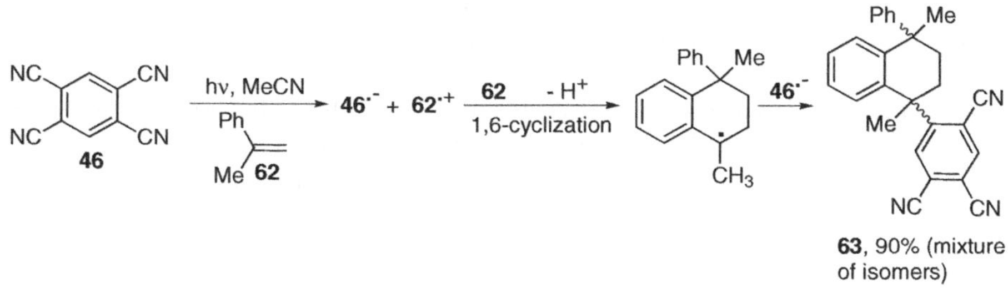

Scheme 5.26.

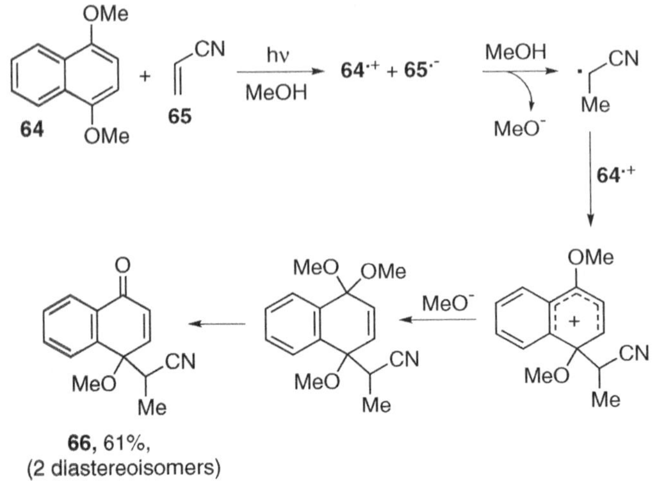

Scheme 5.27.

5.2.2.4 Aromatic, Direct Substitution Attack at an unsubstituted position is also viable. A typical example is the photocyanation of aromatics by addition of cyanide to the radical cation of the arene and subsequent rearomatization. A recent example of this strategy is given by the photocyanation of pyrene in the presence of **55** as the electron acceptor and NaCN in an oil-in-water emulsion system to give 1-cyanopyrene in 83% yield [111]. Phenyl ethers were likewise obtained, though in a low yield, by irradiation of benzene and alcohols via photoinduced electron transfer to 3-cyano-1-methylquinolinium salts and trapping of the benzene radical cation by an alcohol [112]. Photocatalytic reactions between quinoline (or quinaldine) and an amide, namely formamide, *N,N*-dimethylformamide and *N,N*-dimethylacetamide, have been carried out by using TiO_2 under solar light and allowed the introduction of α-*N*-acylalkyl groups (see also Scheme 2.56) [113].

Intramolecular versions of such reactions have been recently reported, as in the case of the preparation of chroman from phenylpropanol (hydrogen peroxide was the byproduct) and of 3,3-disubstituted oxindoles **68** from 2-bromo-anilides bearing an electron-withdrawing group in position 2 (**67**). The former reaction was photocatalyzed by a 3-cyanoquinolinium salt, the latter one by *fac*-Ir(ppy)$_3$ (in this case under visible light irradiation) (Scheme 5.28) [28–34, 114]. The last reaction involved the photoinduced monoelectronic reduction of the anilide moiety, followed by intramolecular radical addition onto the aromatic ring as illustrated in Scheme 5.28. The end compounds were formed in a high yield under very mild reaction conditions.

5.2.2.5 Aromatic, Substituent Modification The reduction of aromatic nitriles to arylmethylamines has been obtained by a photostimulated reaction with SmI_2 [115].

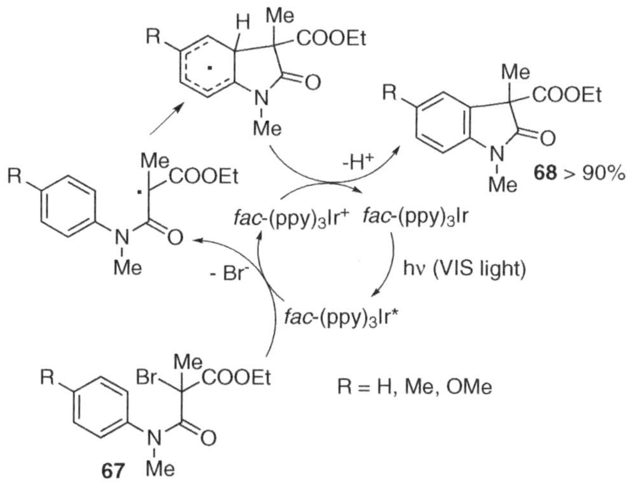

Scheme 5.28.

5.3 ADDITION TO ALKENES AND ALKYNES

Radical cations usually react as electrophiles and undergo addition reactions, in typical examples by cyanide (Schemes 5.1 and 5.2, paths *c,d*) and by alkenes. With alkenes and dienes a variety of addition and cycloaddition processes [116, 117] has been developed, as summarized in Scheme 5.1 (see also Section 5.5).

5.3.1 Intermolecular Addition onto Double or Triple C–C Bonds

5.3.1.1 Addition of an Alkyl or Phenyl Group In Section 5.2.2.1 the chemistry occurring upon electron transfer involving an aromatic compound has been briefly reviewed. Such molecules are particularly well suited for photoinduced ET, because of the relatively accessible redox potential and the convenient absorption in the near UV and in some cases in the visible, coupled, at least with many classes, with a remarkable unimolecular photostability. Irradiation in the presence of acceptor- or donor-generated radical ions and fragmentation of one of them led to a radical that was then trapped by the aromatic. The last compound thus played a twofold role, both of inducing ET and trapping the radicals. This does not extend to other π-molecules, such as alkenes and alkynes, that are excellent traps, but generally absorb at shorter wavelengths and at any rate would exhibit interfering chemical reactions when excited. Thus, a light-absorbing, nonconsumed photocatalyst is used and generates the radical ions of suitable precursors, often liable to fragmentation reactions. Both radical ions and radicals are then trapped by alkenes and alkynes. The mechanism may be more complex; for example, the compounds containing a multiple bond may be involved in secondary ET and form, in turn, radical ions or radicals from them. In fact, some reactions involving such compounds appear to proceed via coupling of the derived radical ions or radicals rather than via trapping by a neutral species. The end result is the same, but different mechanisms have been invoked, as will be mentioned when appropriate (see Section 5.4.2).

Most of the reactions involved a PET-produced (nucleophilic) alkyl radical and an electrophilic alkene. Direct alkylation by an alkane has been reported for the case of adamantane, which was found to add to an unsaturated nitrile under TiO_2 photocatalysis, although the conversion of this hydrocarbon was not extensive [118]. Deprotonation of a nonpolar species such as an alkane is inefficient and back electron transfer predominates, but the presence of a better electrofugal group makes the generation of an alkyl radical from the initially formed radical cation more competitive (see Scheme 5.19).

Thus, PET from carboxylic acids causes deprotonation and CO_2 loss, as shown for the alkylation of dimethyl acetylenedicarboxylate and dimethyl maleate by pivalic and isobutyric acid that was obtained in low to moderate yields (22–45%) under 1,2,4,5-tetracyanobenzene (**46**) photocatalysis. The reaction was suggested to involve the radical cation of the carboxylic acid [119]. Similarly, alkylation in position 4 was obtained with 1-alkylpyridinium perchlorate upon photocatalysis by Pt-coated commercial TiO_2 powder suspended in an aqueous solution of an aliphatic acid [120].

Tetraalkylstannanes showed to be effective precursors. With asymmetric derivatives, the most stable alkyl radical was selectively produced under PET conditions. Thus, dimethylmaleate (close to 80% isolated yield) and acrylonitrile (40–50%) underwent addition of a butyl or a *t*-butyl group under photocatalysis by aromatic nitriles in the presence of tetrabutylstannane and *t*-butyltrimethylstannane, respectively [121]. The result depended on the nitriles used as photocatalysts. Introducing more cyano groups in the benzene ring made these more potent catalysts, able to oxidize weaker donors, but also better traps able to scavenge both the alkyl radical and the radical adducts before their reduction to the desired final product, the alkylated alkene (compare Section 5.2.2.3). When using **46**, however, some alkyl tricyanobenzene was formed, but this did not stop the reaction. Indeed, the alkyltricyanobenzene was practically as good an oxidant as the starting material and thus the catalyzed alkylation continued. On the other hand, limitation due to competitive reactions could be lifted either by using biphenyl as a cocatalyst [90] or by substituting benzenecarboxylic esters for nitriles. Both choices hindered back electron transfer within the original ion pair and favored diffusion out of cage, thus giving a better chance for fragmentation to the radical ions, as discussed in some detail in Section 5.2.2.3. The analogous alkylation via radical cation/radical has been reported for a silane in the reaction of dimethyl maleate with *tert*-butyltributylsilane under **46** photocatalysis [121].

Electrophilic alkynes such as dimethyl acetylenedicarboxylate also underwent alkylation with *tert*-butyl- and *i*-propyl- trimethylstannane (yields obtained near 50%, tetramethylpiromellitate **70** used as the catalyst) [119]. Cyclic enones (cyclopentenone and cyclohexenone) and methyl vinyl ketone (**69**) were likewise alkylated under these conditions. In this case, however, a singlet excited-state catalyst such as **46** performed poorly, due to the competitive light absorption by ketones. When the

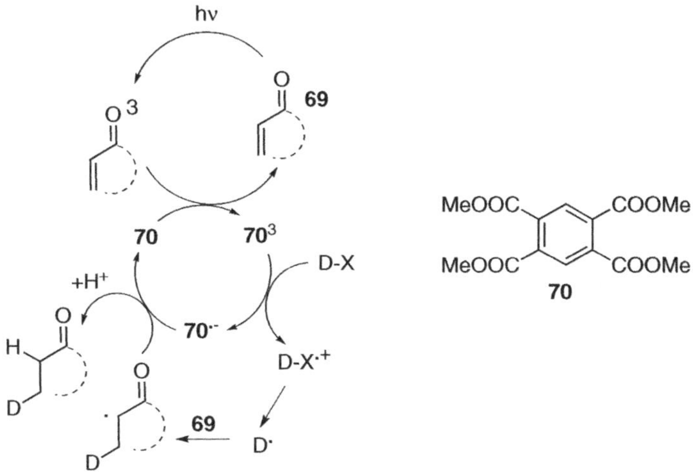

Scheme 5.29.

catalyst was an ester such as **70**, active from the triplet, this limitation was overcome since the unsaturated ketones acted as photosensitizer by absorbing light. Intersystem crossing and energy transfer to the aromatic ester, ensued thus leading to catalyzed alkylation (see Scheme 5.29) [122].

Alkyl radicals were conveniently obtained also by fragmentation of 2,2-dialkyl-1,3-dioxolanes or of 2-alkyl-2-phenyl-1,3-dioxolanes (Scheme 5.30). The latter donors were found to perform quite conveniently due to the easy oxidation and the selective fragmentation. Furthermore, when using the phenylalkyldioxolanes, *ipso* substitution of the catalyst (**46**) occurred to a minimal extent, while on the contrary this was a main competitive path when the dialkyl derivatives were used in the alkylation of alkynes [89]. This suggests that the above mechanism is oversimplified. Possibly, the radical cation fragmentation occurs when this species is complexed with the radical anion, rather than a free solvated intermediate. The different complexation of the donating dioxolane moiety in the presence/absence of the phenyl group would give a different chance to the alkyl radicals to react with the radical anion (see also Scheme 5.22).

Additional precursors for the generation of alkyl radicals via C–C bond fragmentation are some cyclopropane derivatives. Thus, cyclopropanone silyl acetal radical

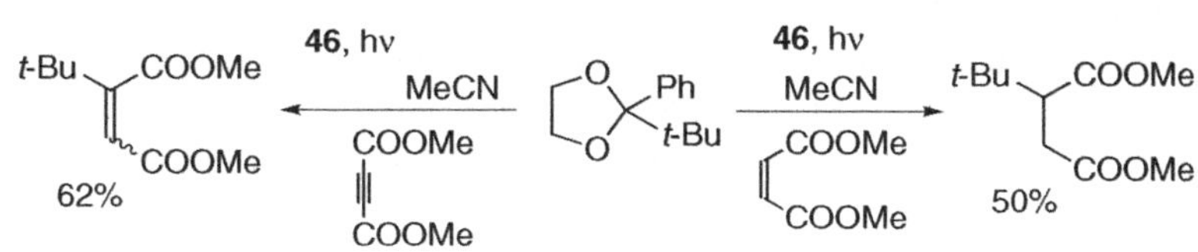

Scheme 5.30.

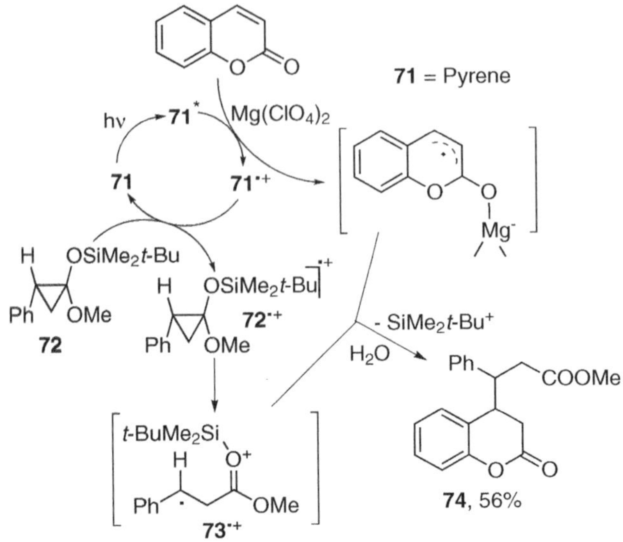

Scheme 5.31.

cations ($72^{•+}$ in Scheme 5.31) [123] likewise underwent C–C bond fragmentation (to $73^{•+}$) and, when stabilized by magnesium perchlorate, acted as efficient alkylating agents of conjugated esters and ketones under photocatalysis by phenanthrene or pyrene (**71**) (10% mol equiv), as depicted in Scheme 5.31. The functionalized coumarine **74** was formed in ~56% yield [124]. This reaction was also carried out in an intramolecular way by tethering the cyclopropanone unit to a cyclohexenone ring and using pyrene or 9,10-dicyanoanthracene as the photocatalysts [125]. The 1-silyloxy derivatives of bicyclo[4.1.0]heptane and bicyclo[5.1.0]octane and some related compounds also underwent cleavage of the three-membered ring under PET conditions, giving radicals that were subsequently trapped by acrylates [126].

A variety of precursors were available for the generation of stabilized allyl and benzyl radicals under PET conditions. However, migration of these relatively stable species and trapping by the alkene had to compete with coupling with the radical anion of the catalyst. Thus, the activation of toluene by aromatic nitriles in acetonitrile produced benzylation of the latter and not of an added alkene. The use of a nucleophilic solvent such as an alcohol favored both diffusion out of cage and deprotonation, however. Thus, dimethyl maleate was efficiently benzylated under **50** photocatalysis in the presence of methanol and of *tert*-butanol (although in the first case this alkene was in part hydroxymethylated) [127]. The same reaction was carried out by using a polyoxometalate (namely tetrabutylammonium decatungstate, $[W_{10}O_{32}]^{4-}$) as the photoredox catalyst, benzyl silanes (**75**) as donors, and electron-poor olefins in the double role of radical traps and electron acceptors as shown in Scheme 5.32 [128]. Good yields were obtained.

Both the use of more easily cleaved precursors (involving a C–Si rather than a C–H bond) and of heterogeneous rather than homogeneous conditions (in order to minimize

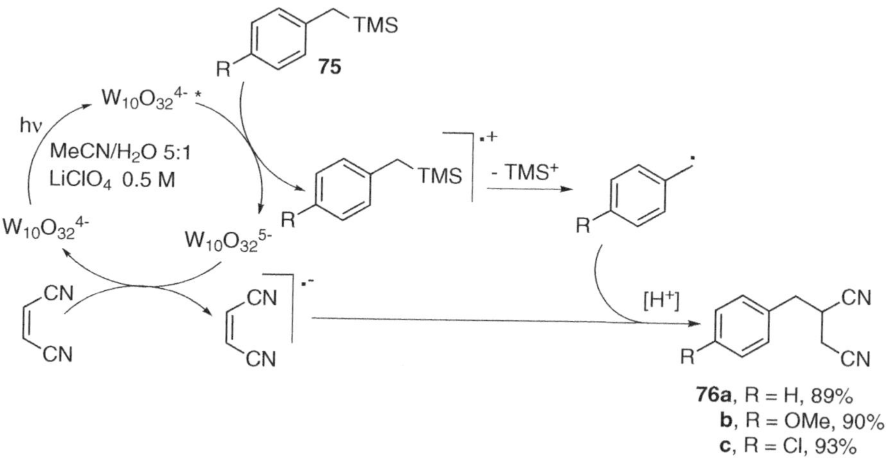

Scheme 5.32.

ion pairing and thus both BET and coupling) led to higher alkylation yields. Thus, a satisfactory benzylation of conjugated acids, esters, and carbonyls was obtained by using TiO_2 as heterogeneous photocatalyst and benzyltrimethylsilanes or, to a lesser extent, phenylacetic acids as precursors of benzyl radicals (Scheme 5.33) [129–131].

The result could be further improved when the heterogeneous catalyst was loaded with a small amount of Pt (which facilitates ET) [130] or when adding an inorganic oxidant (e.g., Ag_2SO_4) [131]. Under these conditions, benzylation was obtained also from toluene. However, in this case a large proportion of the dibenzylated, rather than monobenzylated, derivatives was obtained, reasonably because of the formation of a larger local concentration of benzyl radicals, which coupled with the adduct radical before diffusion. The TiO_2-catalyzed benzylation has been carried out on the multigram scale by exposure to solar light [132].

Allylation reactions have been similarly carried out, as in the alkylation of β,β-dicyanostyrene by allylsilanes. This was obtained selectively (at position α) in

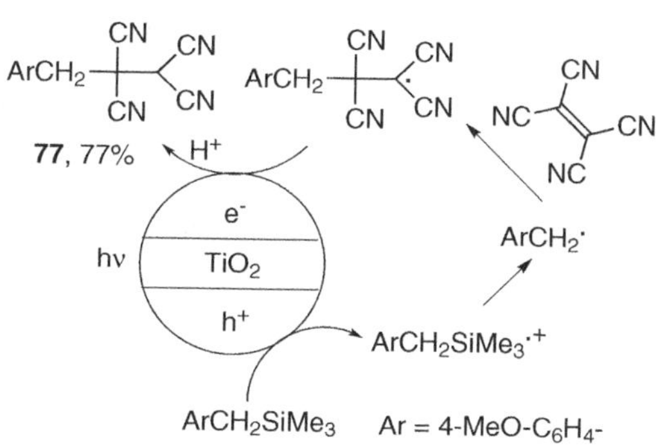

Scheme 5.33.

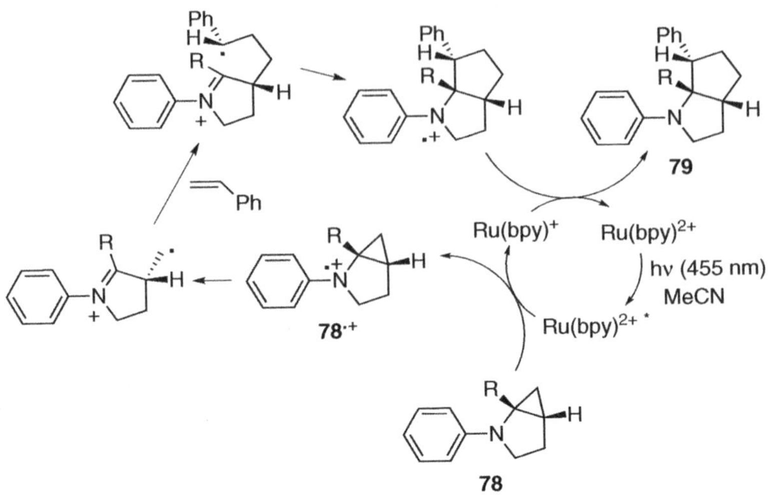

Scheme 5.34.

84% yield upon photocatalysis by phenanthrene (~20% mol equiv) [133]. Various substituted cyclohexylidenemalononitrile derivatives as well as the related tetrahydronaphthalene derivatives have been likewise employed for the synthesis of the corresponding α-allylnitriles. The diastereoselectivity of the reaction depended on the structure and on the medium characteristics, in particular on the presence of acids [134–136]. The result was rationalized with reference to the spin and charge distribution in the radical anion intermediate.

Alkyl-substituted radicals were recently obtained by a PET oxidation of cyclopropylamines as illustrated in Scheme 5.34. Electron donation from the amine moiety led to the radical cation **78**$^{\bullet+}$ that underwent cleavage of the cyclopropyl ring. Trapping of the resulting radicals by styrene derivatives or electron-poor olefins and cyclization onto the thus formed iminium ion led to 5,5-fused bicyclic heterocycles **79** in a good yield and with a good diastereomeric excess (Scheme 5.34) [137].

Fullerenes (in the role of electron acceptor) were conveniently derivatized by having recourse to a PET process with suitable donors such as ketene silyl acetals [138] or allylstannanes [139]. Thus, addition onto a double bond of C_{60} was smoothly attained in this way (Scheme 5.35).

5.3.1.2 Addition of a α-Oxy- or Amino-Substituted Alkyl Group Heteroatom-containing compounds, such as ethers and amines, are smoothly oxidized and the resulting radical cations deprotonate, or at any rate lose an electrofugal group, from the α-position, offering another major entry to radicals. Thus, α-trimethylsilyl alcohols [140] and ethers (e.g., **81**) [141] were used successfully in the addition to unsaturated esters and imides by irradiation in the presence of 9,10-dicyanoanthracene (**82**)/biphenyl (Scheme 5.36).

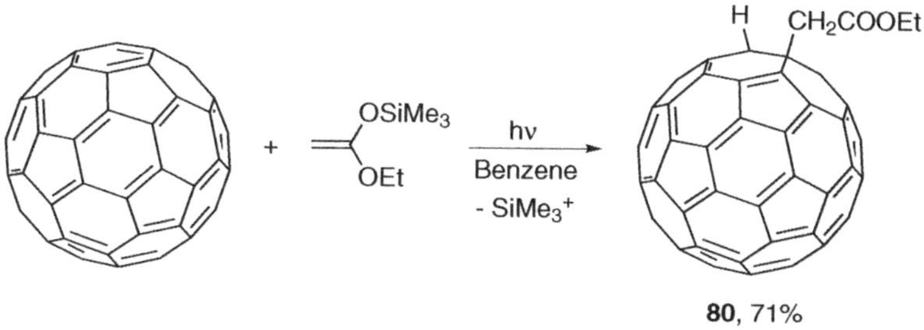

Scheme 5.35.

As for α-amino radicals, these are in principle available in the same way by PET oxidation and deprotonation of amines [34]. However, the first step is fast, given the low oxidation potential of these compounds, but the latter one has to compete with fast BET. With aromatic ketones as photocatalysts, as an example, ET quenching by amines was fast, but the production of trappable radicals was inefficient and was accompanied by extensive decomposition of the ketone. This was thus used in a stoichiometric amount, making the workup cumbersome. However, ketones bearing electron-donating substituents performed better [142]. Thus, the radical addition of tertiary amines and of silylated secondary amines to electrophilic alkenes by using either 4,4′-dimethoxybenzophenone (**84**) or Michler's ketone (**85**, both 10% mol equiv) as the photocatalysts was successful, as shown in Scheme 5.37.

When using a chiral enone such as (5*R*)-5-menthyloxy-2,5-dihydrofuran-2-one **87**, a complete facial stereoselectivity was observed in the radical addition step [143]. This reaction was exploited for the enantioselective synthesis of some necine bases, namely (+)-laburnine and (−)-isoretronecanol [143]. Furthermore, *N*,*N*-dialkylanilines (**86**) were used in the synthesis of 1,2,3,4-tetrahydroquinoline derivatives (**88**). The process was based on the stereoselective radical-tandem reaction of aniline derivatives with either the (5*R*)- or the (5*S*)-isomer of butenolide **87**, where the initial radical attack onto the enone was followed by the intramolecular attack of the radical adduct onto the aromatic ring [144, 145]. This gave the respective stereoisomers **88** with good selectively [146] and chiral induction [147].

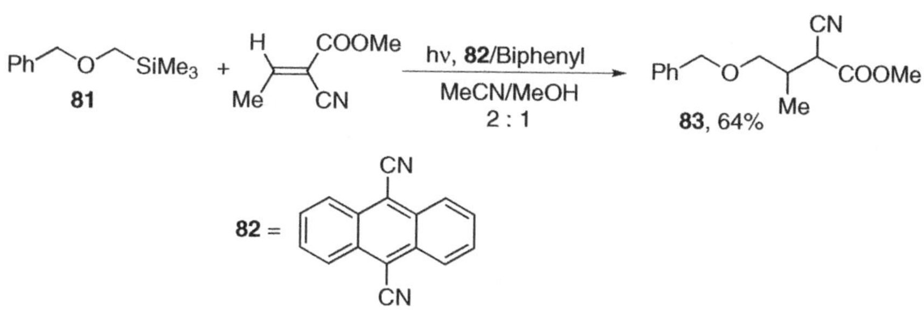

Scheme 5.36.

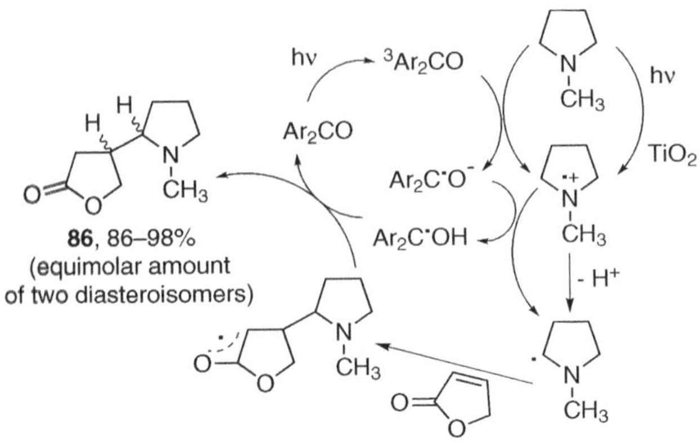

84, Ar = 4-MeO-C_6H_4,
85, Ar = 4-Me_2N-C_6H_4

Scheme 5.37.

Reduction of the butenolide was a side process, which was completely suppressed, however, in the presence of acetone (~6% vol., Scheme 5.38).

With anthraquinone (or the corresponding 2-sodium sulfonate derivative) as the photocatalyst, the alkylation of electrophilic alkenes was successful with tertiary, secondary, and primary amines [148–150]. When triethylamine was used, the addition of α-amino radicals onto methacrylates was followed by a 1,5-hydrogen abstraction reaction allowing addition of a further alkene molecule, so that multiple-olefin addition products were isolated at the end of the reaction [148]. Furthermore, when the amine used had a free N−H bond, a lactam was obtained by intramolecular reaction of the first-formed δ-amino ester [149]. In such a case, however, the thermal aza-Michael addition of the starting amine in the ground state competed. This drawback was mitigated by carrying out the irradiation at a lower temperature, thus slowing down the thermal path. Lactams were smoothly obtained also from *N*-allylamines [150].

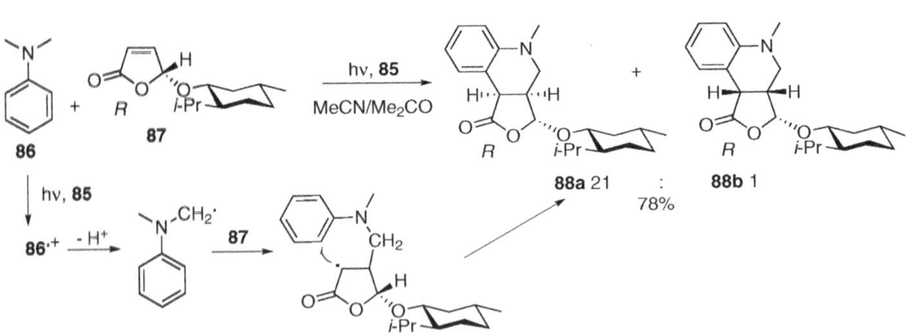

Scheme 5.38.

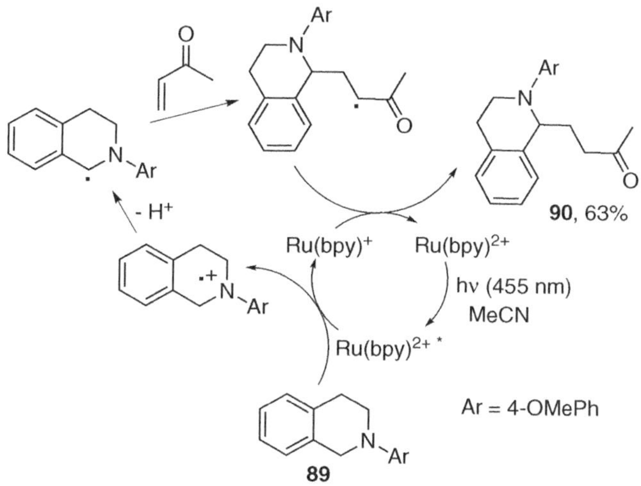

Scheme 5.39.

Visible light photocatalysis [28–34] was suitable for the generation of α-amino radicals by using Ru(II)-based complexes. As an example, the redox potential of the excited $Ru(bpy)^{2+}$ was sufficient to oxidize electron-rich amines, such as **89**, to the corresponding radical cations [34]. Deprotonation and addition of the resulting radical onto electron-withdrawing substituted olefins gave substituted *N*-aryltetrahydroisoquinolines **90** in a decent yield, while the Ru complex was regenerated (Scheme 5.39) [151].

Heterogeneous photocatalysis has been likewise applied by using either TiO_2 or ZnS (Scheme 5.37) [152, 153]. In this case a large excess of the amine (preferably used as the solvent) was required in order to obtain a high product yield, presumably because only the molecules adsorbed at the surface were involved in the ET step and, since escape of radical cations was slow, BET predominated unless the amine was in such a high concentration that its action as a base deprotonated efficiently the first formed intermediate [152, 153]. It is noteworthy that when *N*-methylpiperidine was used in the place of the analogue pyrrolidine, deprotonation from the methyl group followed by oxidation and loss of the group to yield piperidine became competitive with deprotonation from a ring methylene [152]. Acyclic amines gave no alkylation under the above conditions, but this was obtained in the presence of some thiocarbonyl or xanthate derivatives (e.g., *S*-methyl-*N*,*N*-diethyldithiocarbamate, 5 mol equiv) [154], apparently because the reversible radical trapping of the amino radical by the thiocarbonyl species stabilized the radicals and facilitated the addition to alkenes.

α-Amino radicals were likewise generated from α-silyl amines. In some instances, an unsaturated ketone played both the role of acceptor and that of reagent to be alkylated. Thus, the reaction between an enone and a (trimethylsilyl)amine was initiated by a PET reaction between the triplet enone and the amine and photoaddition onto the enone took place by coupling of the resulting radical ion pair [155, 156]. The intermolecular addition of a silyl amine **92** to enone **91** gave a good yield

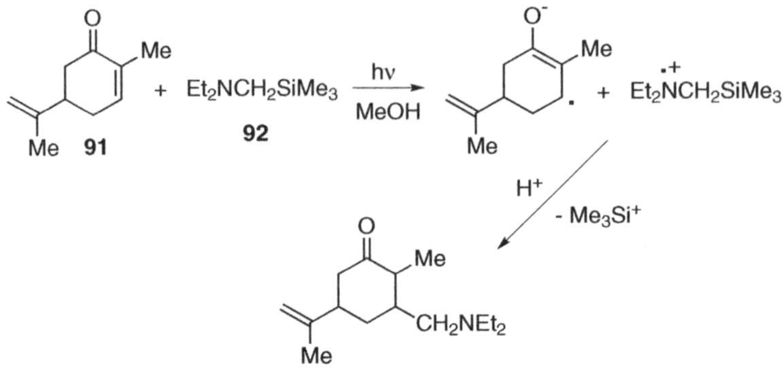

93, 86% (25% **91** conv.)

Scheme 5.40.

of cyclohexanone derivative **93**, although only a limited conversion of the starting enone was achieved (Scheme 5.40).

Better results were reported for the intramolecular version of the reaction, where *N*-(trimethylsilyl)methyl substituents were tethered to the β-position of the enone [157, 158].

The trimethylsilyl (TMS) group likewise lowered the oxidation potential of modest donors such as α-silyl carbamates (e.g., **94**). This made an aromatic nitrile such as 9,10-dicyanoanthracene (**82**) able to cause photoinduced ET [159] from the carbamates in the presence of a co-catalyst (biphenyl, **95**; see Scheme 5.41). The presence of a better electrofugal group made the generation of the radical more efficient. As an example, desilylation of the radical cation (in a protic medium), intramolecular cyclization onto a double bond, reduction by $\mathbf{82}^{\bullet -}$, and protonation gave carbamate **96** in a decent yield and in high diastereomeric excess. Related cyclizations were used for inducing structural changes in oligopeptides by introducing the structural motif present in compound **96** into the peptide chain.

An α-silyl carbamate (**97**) [159, 160] has also been irradiated in the presence of the **82**/biphenyl couple or pyrilium salts for the alkylation of electron-poor alkenes (Scheme 5.42) [160]. The intramolecular version of this alkylation was applied in peptide chemistry when *N*-trimethylsilylmethyl aminoacid derivatives were employed in the place of carbamates. Accordingly, cyclization took place forming a proline ring, and this caused a structural change in the α-helix secondary structure with respect to the starting peptide [160].

Photofixation of carbon dioxide was also investigated in the frame of artificial photosynthesis. Yanagida and co-workers found that visible-light-absorbing compounds such as poly(*p*-phenylene) and, even better, CdS-DMF were efficient photocatalysts for this reaction [161–163]. Initially, CO_2 was introduced into a DMF solution containing CdS-DMF and pre-irradiated for 1 h. Then triethylamine (TEA, as a sacrificial donor) and aromatic ketones or benzyl halides were added as

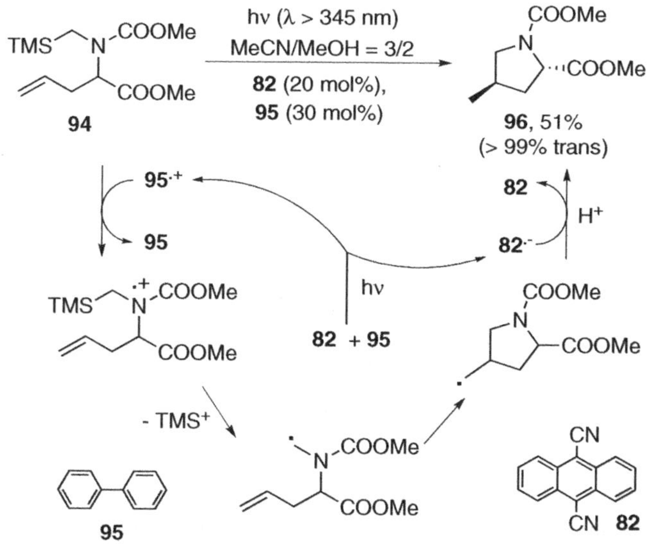

Scheme 5.41.

photofixation substrates. Photolysis took place in a sealed tube. The proposed mechanism is shown in Scheme 5.43. The catalyst was able to reduce both carbon dioxide and the organic additive. A benzylic acid (**98**) was finally formed in a modest yield by radical–radical anion coupling along with a roughly equimolar amount of benzopinacol.

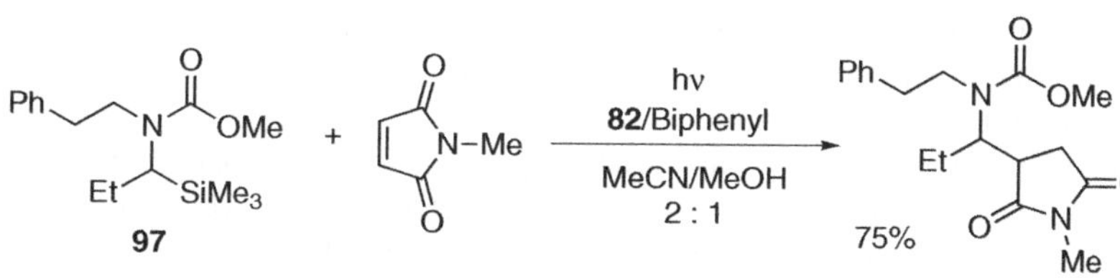

Scheme 5.42.

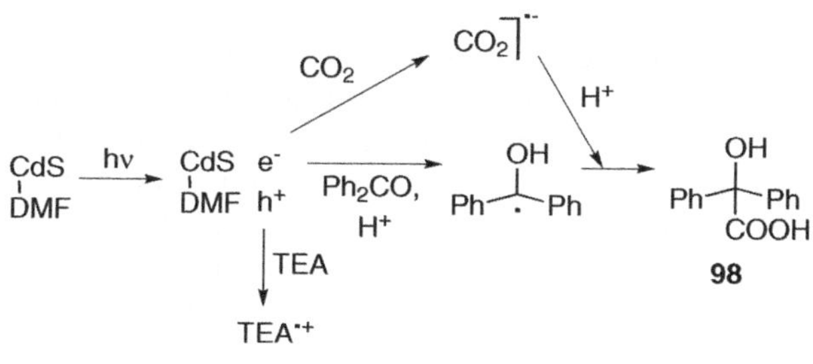

Scheme 5.43.

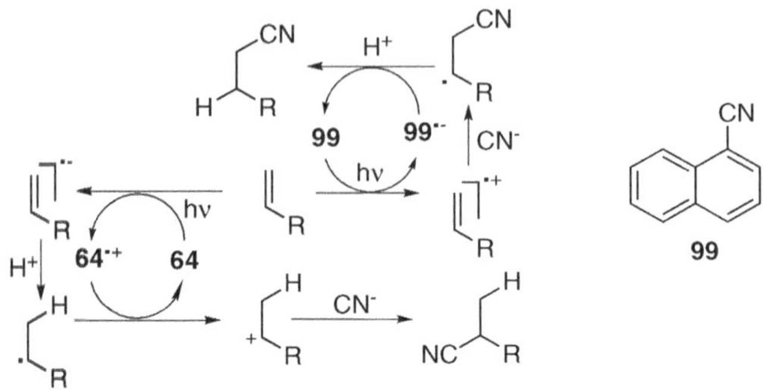

Scheme 5.44.

5.3.1.3 Addition of Other Groups Photochemical addition of cyanide ion onto a double bond could be accomplished both by monoelectronic oxidation and by monoelectronic reduction of the olefin [164, 165]. It is noteworthy that a different regioselectivity was observed according to the mechanism involved. Thus, 1-cyanonaphthalene (**99** in Scheme 5.44) induced oxidation of a phenylated olefin, (e.g., 2-phenylnorbornene), and gave the anti-Markovnikov cyanated product [164], whereas 1,4-dimethoxynaphthalene (**64**) induced reduction of 1,1-diphenylethylene and afforded the Markovnikov adduct [165]. As shown in Scheme 5.44, in both cases the most stable radical was formed, but the order of addition of proton and cyano anion was reversed.

A class of reactions that is being increasingly studied involves the use of a visible-light-absorbing photocatalyst [28–34] and concerns the generation of electrophilic radicals, prone to be trapped by electron-rich olefins. Some representative examples will be reported in the following.

Thus, excited [Cu(dap)$_2$Cl] (dap = 2,9-bis(*para*-anisyl)-1,10-phenanthroline) is a slightly stronger reducing agent (−1.43 V) with respect to the well-known Ru and Ir complexes and is thus able to reduce both diethylbromomalonate (**5**) and α-bromoacetophenone (**100**) to the corresponding radical anions. Bromide loss and trapping of the resulting radical by protected allylamine ensued and gave as end products alkylated compounds **101** and **102** via an atom-transfer radical addition (ATRA) to the olefin (Scheme 5.45) [166].

Enamides could likewise be alkylated in a three-component reaction via radicals and cations, where the double bond was functionalized by the formation of a C–C and a C–O bond [167]. The schematic representation of the reaction is shown in Scheme 5.46. Here the process was initiated by absorption of visible light by an Ir-based complex. PET generated malonyl radicals that added onto enamide **103**. The resulting radical adduct was oxidized to the corresponding iminium ion **104**$^+$. In turn, this was trapped by the nucleophilic solvent (EtOH) to form β-alkylated α-carbamido ethers (**105a,b**, actually a stable imine equivalent) under very mild conditions and in good to excellent yields [167].

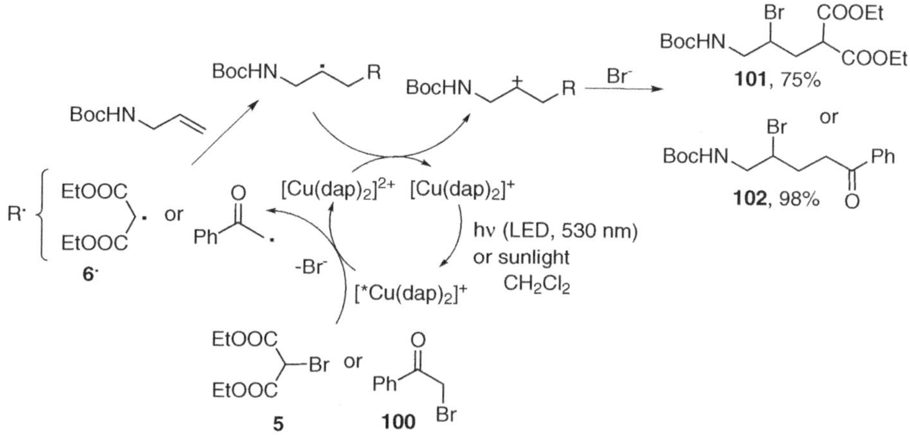

Scheme 5.45.

Indeed, the reductive power of the excited state of a photocatalyst was exploited in the generation of α-ketoradicals by ring opening of α-ketoepoxides (**106**). Both Ru- and Ir-based photocatalysts have been used along with 4-methyl Hantzsch ester (**110**) in the double role of quencher and hydrogen donor. Thus, reduction of **106** led to the radical anion **107**$^{\cdot-}$ that upon reaction with allyl sulfone **108** gave β-hydroxyketone **109** through a highly diastereoselective radical ring-opening/allylation tandem process (Scheme 5.47) [168].

Another approach involved the photoaddition of electron-rich alkenes (e.g., silyl ketene acetals **112**) to acetylpyridines **111** to produce β-hydroxyesters **113** in high yields. The end products resulted from intermolecular photoinduced ET between the two reagents followed by radical–radical coupling (Scheme 5.48) [169]. Oxetanes arising via [2 + 2]-cycloaddition were formed in trace amounts, but became the main products when using electron-poor alkenes.

The efficient **50** photocatalyzed one-electron oxidative dissociation of the selenium–selenium bond [170] in PhSeSePh could be exploited for promoting the generation of an electrophilic species ($PhSe^{+}$) that induced a cyclization step

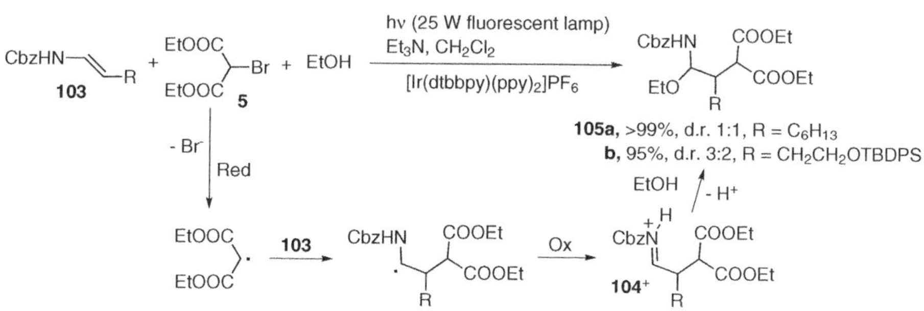

Scheme 5.46.

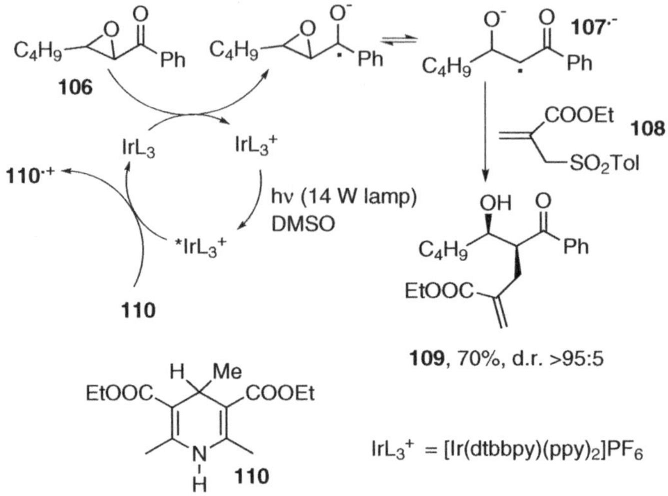

Scheme 5.47.

after addition onto various substituted C=C bonds [171–174]. As an example, both allyl propargyl ethers [172, 174] and ω-hydroxyalkenes **114** [173] were used as starting materials for this reaction and yielded phenyl-selenyl-substituted oxygen-based heterocycles **115** (Scheme 5.49). Thus, the phenyl selenide group was in every case incorporated in the end products, but was expediently converted into a methyl ether function by a further photocatalyzed oxidation step in MeOH. This involved alcohol-assisted fragmentation of the carbon–selenium bond in the radical cation $R'SeCH_2R^{\bullet+}$. The finally isolated ether RCH_2OMe **116** resulted from a one-pot process involving C–C bond formation as well as selenylation and deselenylation reactions [171]. Alkyl phenyl selenides could alternatively be reduced by excited **64**. In this case, loss of the $PhSe^-$ group followed by a radical intramolecular addition onto a C=C bond occurred [175].

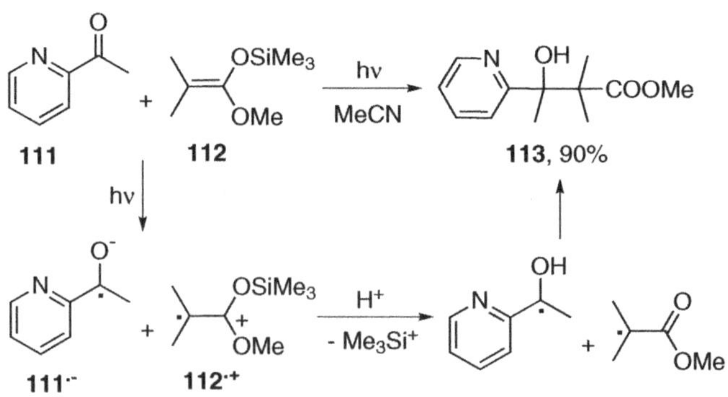

Scheme 5.48.

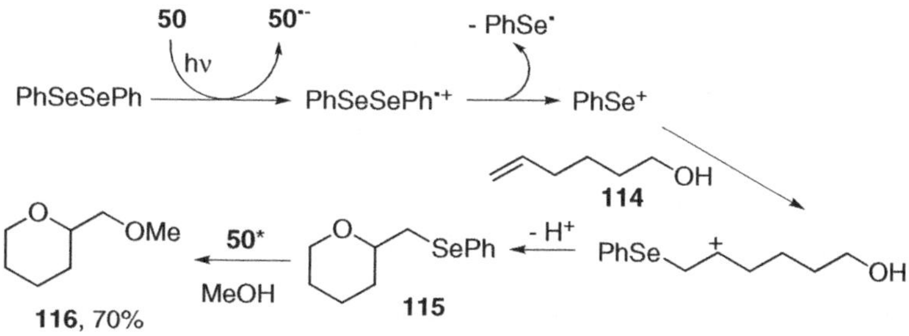

Scheme 5.49.

5.3.2 Intramolecular Addition onto C=C Bonds

Intramolecular attack of a radical ion to a tethered alkene proved to be a convenient method for cyclization reactions [176]. The different pathways accessible for achieving this target are briefly presented in the following.

*5.3.2.1 **Addition via Silyl Enol Ethers*** The **82**-photocatalyzed ($\lambda_{irr} = 410$ nm) cyclization of ω-pentenyl silyl ether **117** was observed in MeCN. The processes observed involved the formation of either a six-membered ring (**119**, by attack of the radical cation onto the alkene and ensuing desilylation, Scheme 5.50, path *a*) or a five-membered ring (**118**). In the presence of alcohols, desilylation was faster and

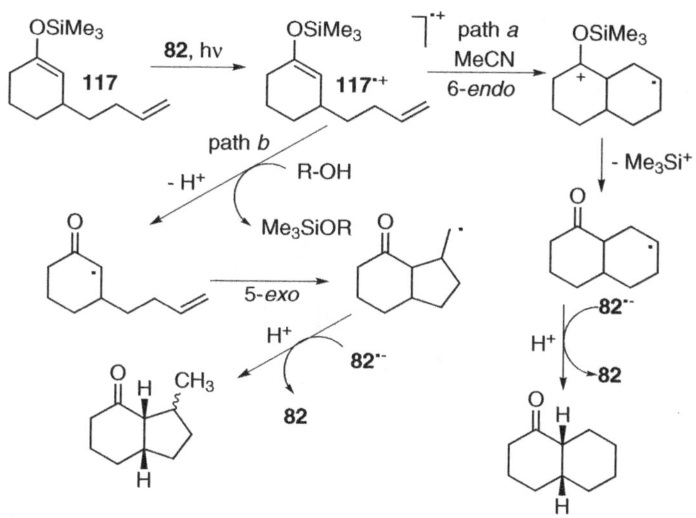

MeCN	118	0	119 25%
MeCN/*i*-PrOH 17/3		20	30%

Scheme 5.50.

thus the cyclization step involved a neutral radical, path b [177, 178]. Compound **50** was an alternative choice in the role of the photocatalyst; but due to the less favorable absorption characteristics, it had to be used in a larger amount and required the adoption of a lamp emitting at shorter wavelengths ($\lambda_{irr} = 350\,nm$), causing some photodecomposition of the ketone product.

The presence of substituents on the attacking double bond could affect both the regiochemistry of the cyclization step and the overall yield. Thus, when using highly substituted double bonds and cyclic enol ethers, polymerization (assumed to be the major side reaction) was hampered. Interestingly, a *cis* ring juncture was always obtained, a fact that was presumably due to a preferred reaction from the chair conformer bringing the substituents pseudoaxially arranged.

Adding phenanthrene as a co-catalyst made the reaction more efficient, with a reduction of the irradiation time required to two-fifths [178, 179]. The *endo/exo* regiochemistry of the ring closure could be directed by changing the silyl group or by adding alcohols to the reaction mixture. When the silyl enol ether moiety was embedded in a carbocyclic ring with a stereogenic center adjacent to the double bond (e.g., in compound **120**), this controlled the stereochemical outcome of the reaction (see Scheme 5.51), as predicted also by DFT calculations [179]. The PET method was also exploited for a cascade polycyclization reaction of cyclic enol ethers bearing a suitable cyclohexenylethyl or a cyclohexenylhexenyl side chain. This approach led to a (poly) cyclization representing a new strategy for the biomimetic stereoselective synthesis of a (quasi-steroidal) polycyclic carbon backbone [180].

An aromatic moiety was another convenient trap. Starting again from a silyl enol ether, a cyclization was again obtained. Methyl ketones **121** were initially converted into ethers **122** and **124**. Then, **50** catalyzed the PET oxidation of the last compounds to yield carboannelated products, namely, methyl ketones **123** and cyclic ketones **125**, respectively (Scheme 5.52) [181, 182]. This strategy was applied to the synthesis of phenanthridone alkaloids (e.g., 2,7-dideoxypancratistatin), again based on the above PET carbocyclization [183].

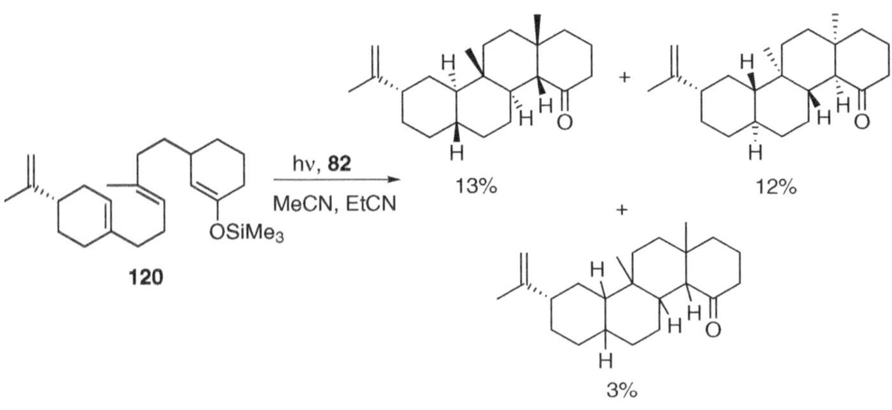

Scheme 5.51.

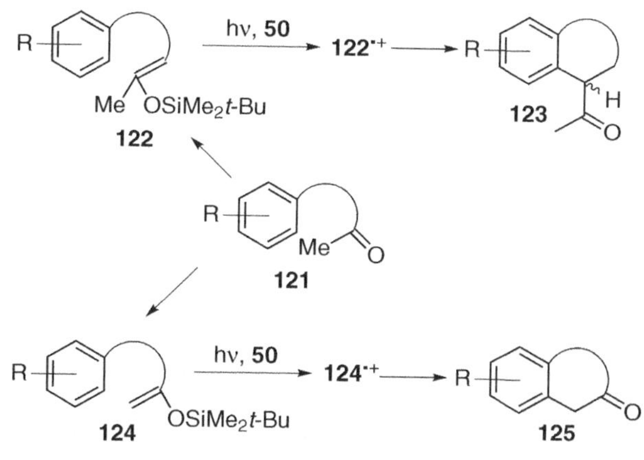

Scheme 5.52.

An alternative approach started from related vinyl ether bearing a selenyl group (**126a,b**). The last moiety was oxidized preferentially under **50** photocatalysis and underwent attack by the nucleophilic vinyl ether (see Scheme 5.53) to yield cyclopentyl ketones **127a,b** in high yields [184]. Interestingly, the intermolecular version of the reaction likewise gave satisfactory yields (60–70%) [184].

As mentioned in Section 5.3.1.1, another way to obtain alkyl radicals by PET-induced cleavage started from cyclopropanes rather than from olefins. This principle was applied to bicyclic silyloxycyclopropanes that underwent cleavage of the *endo* C–C bond and desilylation to give monocyclic β-keto radicals. When a double bond was present in the ring or in a pendant chain, cyclization reactions took place [185, 186]. Thus, some cyclic cyclopropyl silyl ethers bearing an olefinic or acetylenic side chain gave polycyclic (bicyclic to tetracyclic) products [185]. The initial oxidation gave a β-carbonyl radical cation that underwent nucleophile-assisted desilylation. Polycyclic compounds were formed via cyclopropane ring opening and radical addition with well-defined stereochemistry at the ring juncture. An example is shown in Scheme 5.54 with the dicyanoanthracene (**82**)-photocatalyzed reaction of 2-trimethylsilyloxy-tricyclo[6.3.0.0^{2,4}]undec-1(8)-ene (**128**) in MeCN to form tricyclic ketone **129** (66% yield).

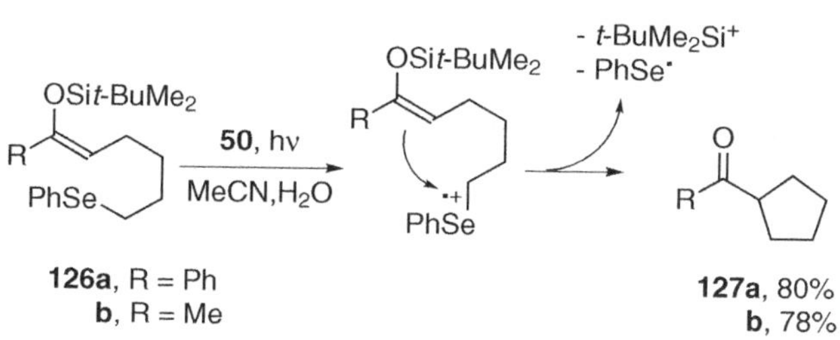

Scheme 5.53.

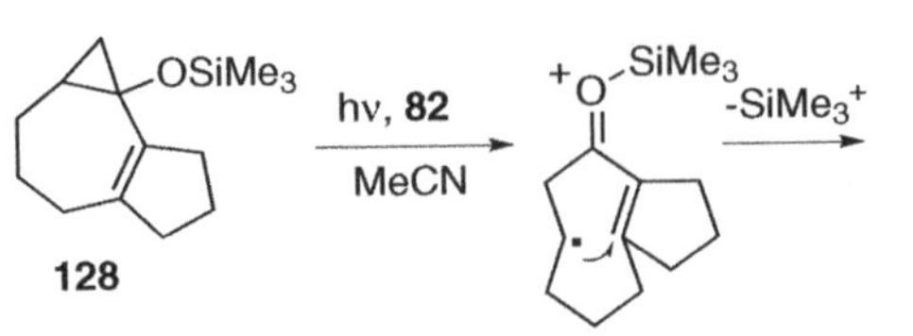

Scheme 5.54.

The final step of this reaction was either hydrogen transfer from the solvent (acetonitrile) or a stepwise electron transfer/protonation reaction involving traces of moisture present in solution. The experimentally determined stereoselectivity of the products was supported by the predictions from quantum chemical calculations [185].

5.3.2.2 ***Addition via Olefins*** The application of polyene cyclizations via PET in order to mimic sequential cyclizations that form polycyclic terpenoids in vivo is an ambitious target that has been extensively investigated. This approach requires that the subsequent chemical steps are sufficiently fast—in other words, that the starting polyene is correctly folded to prefigure the structure of the product so that BET is slowed down and more time is left to the chemically productive steps. Indeed, a useful move was the adoption of a microheterogeneous medium (generated by using a surfactant anion such as sodium dodecyl sulfate, SDS) that was initially identified as the only effective condition [187]. Another approach was the use of a sterically hindered electron acceptor such as 1,4-dicyano-2,3,5,6-tetramethylbenzene (**131**) that caused an enhanced separation of the radical ions. When this was used in combination with a secondary donor (biphenyl, **95**), cascade cyclizations could actually be performed also in homogeneous media (MeCN–water or MeCN–MeOH mixtures as the solvent). The cyclization was reasonably initiated by single-electron transfer from the ω-alkene moiety to the excited catalyst. The radical cation thus obtained (**130**$^{\bullet +}$) was trapped by *anti-Markovnikov* addition of water, and the resulting β-hydroxy radical **132**$^{\bullet}$ initiated the cyclization cascade. As a matter of fact, the key step in the cyclization reaction involved attack to an olefin by a radical, not by a radical cation, while it is usually thought that the latter intermediates are involved in enzymatic processes [188–190].

A simple model demonstrating a structure-dependent cyclization is reported in Scheme 5.55, where attack to an allylic acetate moiety led to a six-membered ring, while attack onto an α,β-unsaturated nitrile led to a five-membered cycle. The stability of the dicyano-substituted radical formed may be determining in the latter case, but further factors may have a role—for example, the termination mechanism [191].

A biomimetic synthesis of a steroid was accomplished in MeCN–H_2O 10:1 mixed solvent at −25°C starting from all-*trans* geranylgeranyl acetate [191–193]. In order to induce chirality in such PET-triggered cyclizations, a chiral auxiliary was linked to the starting polyunsaturated compound via a ketal function (see **133** in Scheme 5.56). The degree of asymmetric induction associated with this cyclization was remarkable, when taking into account that eight stereogenic centers were generated in the reaction and

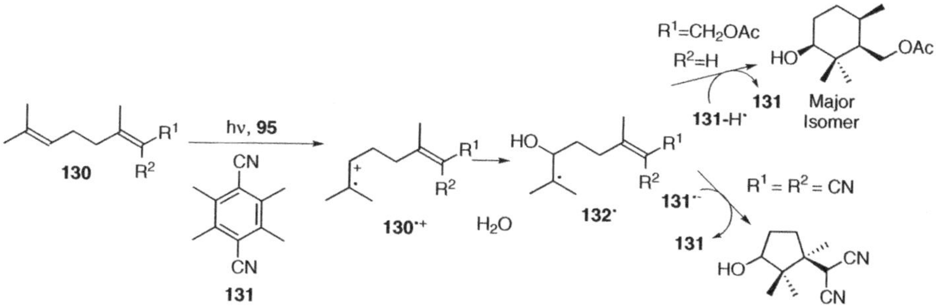

Scheme 5.55.

only 2 out of 256 possible isomers were formed, namely **134** and **135**, despite the fact that the chiral moiety was quite far from the initiation site (Scheme 5.56).

This strategy was successfully applied in the formal synthesis of the antitumoral (±)-stypoldione [194, 195] as well as of oxygenated tetracyclic diterpenes, such as

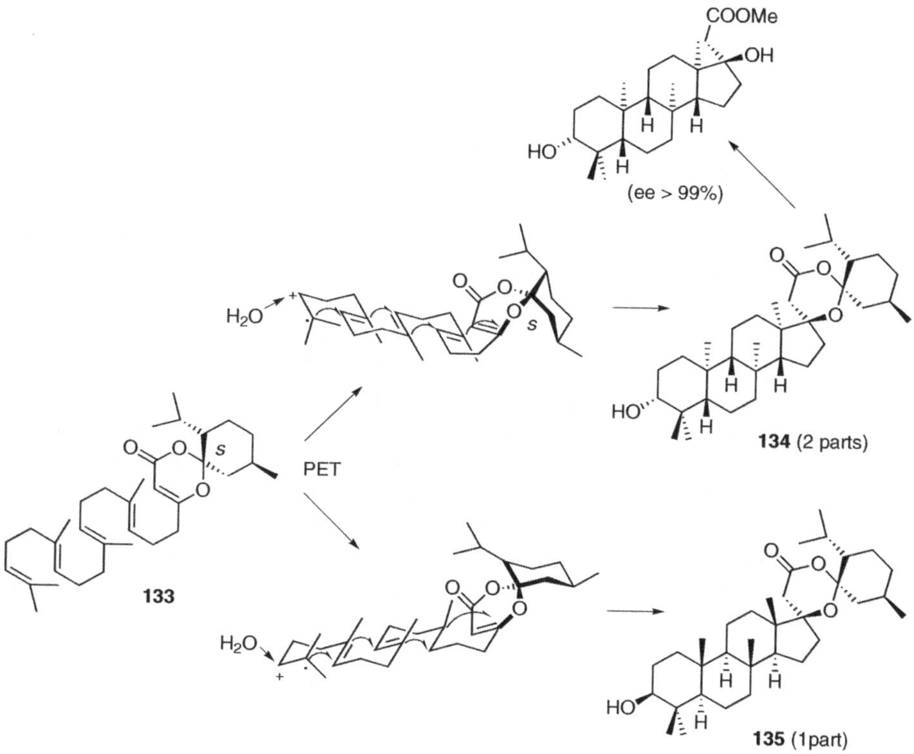

Scheme 5.56.

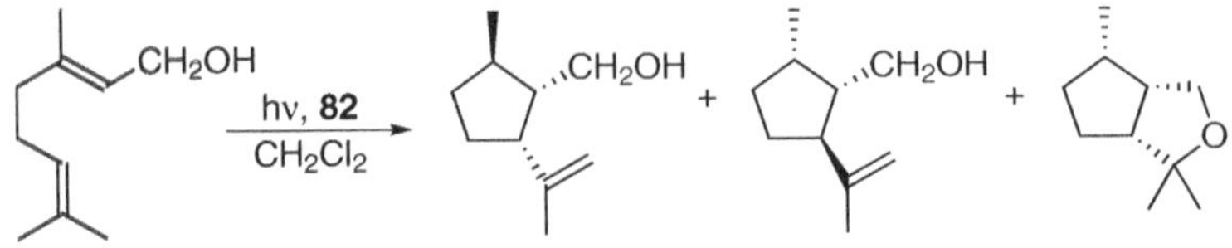

136 **137**, 70% **138**, 6% **139**, 6%

Scheme 5.57.

(±)-3-hydroxy-spongian-16-one [196]. Cyclization reactions could be carried out also in a less polar solvent such as dichloromethane or in CH_2Cl_2/MeOH mixtures, but not with dicyanotetramethylbenzene as the photocatalyst, since BET could not be avoided in this case. However, when passing to a cationic photocatalyst such as *N*-methylquinolinium hexafluorophosphate ($NMQ^+PF_6^-$), the coulombic attraction between opposite radical ions was eliminated and the reaction did take place [197].

In the case of geraniol (**136**), photolysis in the presence of **82** (~30% mol equiv) in a nonaqueous, low-polarity medium such as methylene chloride gave two isomeric five-membered ring derivatives (**137** and **138**), along with a small amount of the oxabicyclo[3.4.0]octane derivative **139** (Scheme 5.57) [198, 199].

Cyclic ethers such as **139** arose from tandem C–C and C–O bonds formation and predominated when using 1,4-dicyanobenzene (**55**) in the place of **82**, because with the latter photocatalyst fast BET from $\mathbf{82}^{\bullet -}$ to the radical cation prevented the intramolecular attack by the OH group [199].

5.3.2.3 Addition via Electron-Withdrawing Substituted Olefins Cyclization reactions were obtained also by monoelectronic reduction. Photocatalysts that have been considered above as oxidants may be used also as reducing agents. In fact, this was obtained by adding an auxiliary donor. This produced the radical anion of the catalyst that was an excellent reducing agent. Thus, photoinduced electron transfer to (visible-light-absorbing) excited 9,10-dicyanoanthracene (**82**) by 1-methoxynaphthalene (**64**) generated the radical anion of the nitrile, and this was able to reduce an unsaturated ester or a ketone (**140**). When a second such moiety was present in the reagent, intramolecular C–C bond formation led to the cyclization to five- or six-membered rings (Scheme 5.58) [200, 201]. In these reactions, the solvent was a mixture of DMF:*i*-PrOH:H_2O 88:10:2. A large amount of DMF was required in order to dissolve a sufficient amount of **82** so that light was efficiently absorbed and the process could take place. As an example, radical anion $\mathbf{82}^{\bullet -}$ initiated the reaction by reduction of a carbonyl (or carboxyl) moiety present in the precursor **141** (Scheme 5.59). The thus obtained enolate radical anion ($\mathbf{141}^{\bullet -}$) underwent C–C bond formation via radical cyclization with a strong preference for *trans*-1,2-stereochemistry (compound **142**) [202–205].

The generation of the key species $\mathbf{82}^{\bullet -}$ was accomplished through either of the two methods **A** or **B** (see Scheme 5.58), both of which involved a sacrificial electron donor for regenerating the photocatalyst. Systems that have been found convenient are those based on Ph_3P (the radical cation of which is trapped by water and gives the

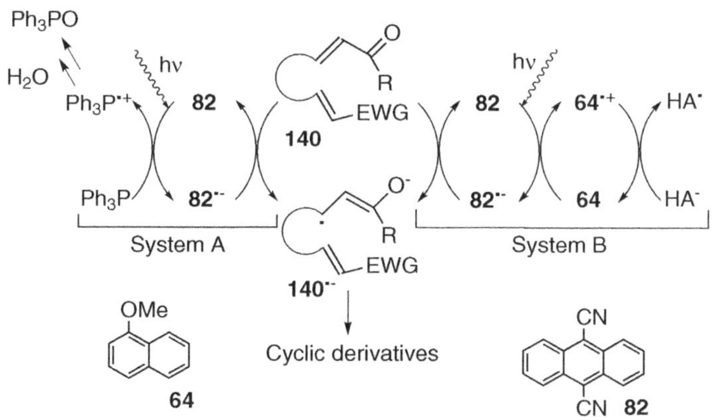

Scheme 5.58.

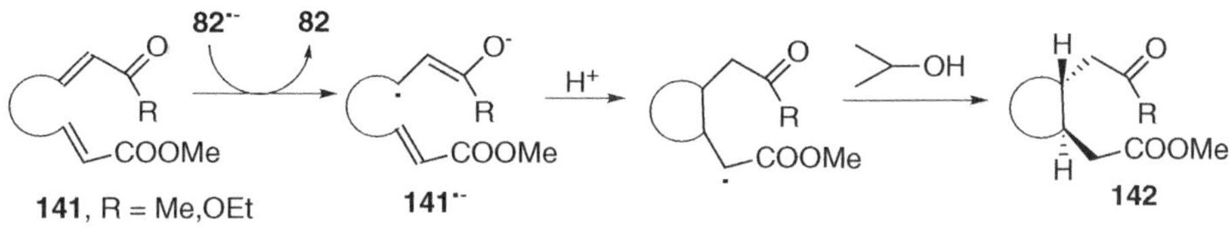

Scheme 5.59.

phosphine oxide) and the ascorbate anion HA^-, respectively. Suitable substrates for the reaction were α,β-unsaturated ketones (or esters, e.g., **140, 141**) tethered by an alkyl chain to an activated (electron-poor) olefin as illustrated in Scheme 5.59.

5.4 ADDITION TO CARBONYLS, IMINE, AND IMINIUM SALTS

5.4.1 Addition onto C=O

Photoaddition reactions at the carbonyl (or carboxyl) group are feasible when this moiety participates into the PET process as the acceptor. Typical are the reactions involving phthalimides (or maleimides, **143**) owing to their high singlet and triplet excited-state reduction potentials [206–210]. Thus, the initial PET reaction between the excited imide and a donor D-X led to the radical ion pair and the ensuing addition of D• (formed by X^+ loss) afforded the corresponding hydroxyphthalimidines **144** (Scheme 5.60). Both the intermolecular and (mainly) the intramolecular version of the process have been investigated in detail [206–209].

The salts of carboxylic acids were often used as the donors. Irradiation led to their photodecarboxylative addition to phthalimides in what was considered a versatile alternative to Grignard reactions [211–214]. As an example, the potassium salts of several carboxylic acids were used for the alkylation (including

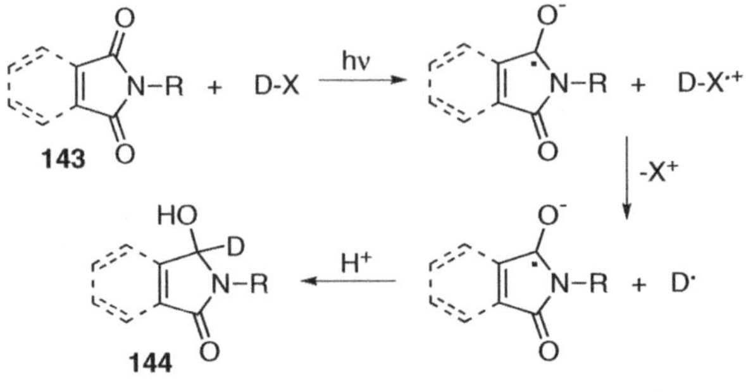

Scheme 5.60.

benzylation (**146a–c**) [211]) or in favorable cases acylation (**147a–c**) [214] (when using α-keto carboxylates) of *N*-methylphthalimide **145** in a satisfying yield (Scheme 5.61).

When starting from ω-arylpropanoic acids **148a–c**, a series of 3-(alkyl and aryl) methyleneisoindolin-1-one derivatives (**150a–c**, a class of compounds having potential biological activity) were conveniently accessed by dehydration/deprotection of the resulting photoproducts **149a–c** (Scheme 5.62) [213]. The intramolecular version allowed for the easy synthesis of medium- and large-ring compounds. The irradiation of various N-phthaloyl-ω-aminocarboxylic acids (e.g., **151a–c**) in aqueous acetone in the presence of a base induced a macrocyclization process (products containing up to a 13-membered ring **152a–c** were formed) in a high yield (Scheme 5.63) [215, 216].

When the tether to the nitrogen atom contained two electron-donating moieties such as a carboxylic group and a SMe group, the course of the reaction depended on the conditions used. As an example, the irradiation of *N*-phthaloyl-*S*-methyl-L-cystein (**153**) in acetone gave a mixture of the annulation product **154** and the thiazino isoindole **155**, with the latter one as the major component (Scheme 5.64) [217]. This arose from a process where the SMe group was the donor and dehydration occurred after the intramolecular radical–radical anion coupling. As for compound **154**, this was formed by two subsequent electron transfer reactions that induced decarboxylation.

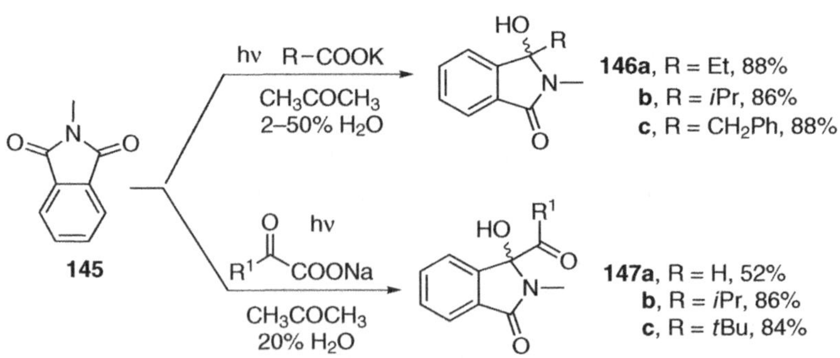

Scheme 5.61.

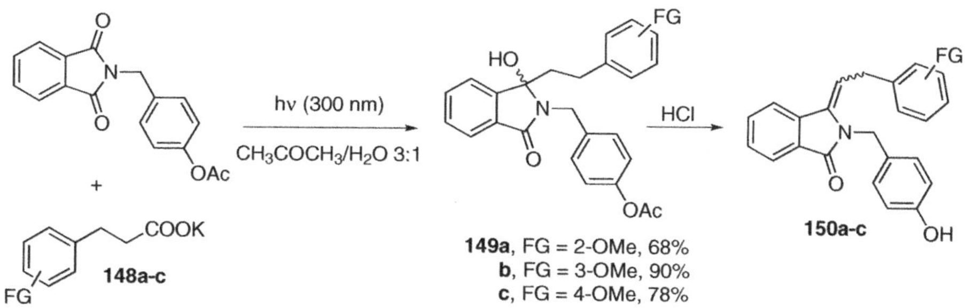

Scheme 5.62.

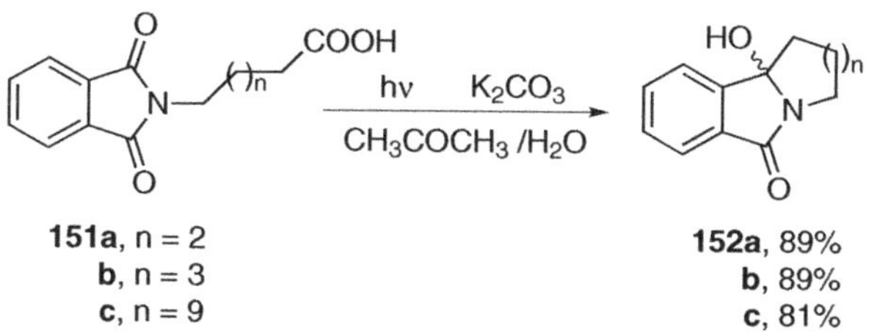

Scheme 5.63.

The presence of a base, however, activated the COOH group and prevented the acid-catalyzed dehydration of the hydroxy-lactam.

The reaction was extended to macrocyclic compounds of structure **157a–c**. These were prepared by irradiation of phthalimides **156a–c** containing *N*-linked ω-trimethylsilylmethyl-substituted polysulfonamide chains. In this case, the trimethylsilyl group acted as the donor moiety (Scheme 5.65) [218]. In some of these reactions, the electron transfer reaction path was accessed only when favored by a suitable reaction medium. Thus, the photolysis of *N*-silylpropyl-maleimide **158** in aprotic solvent (MeCN) gave the 2 + 2-dimer **159** in an almost quantitative yield [219]. However, the reaction was shifted to a PET process when adding water to the reaction mixture, and pyrrolizidine **160** was then the exclusive product (Scheme 5.66).

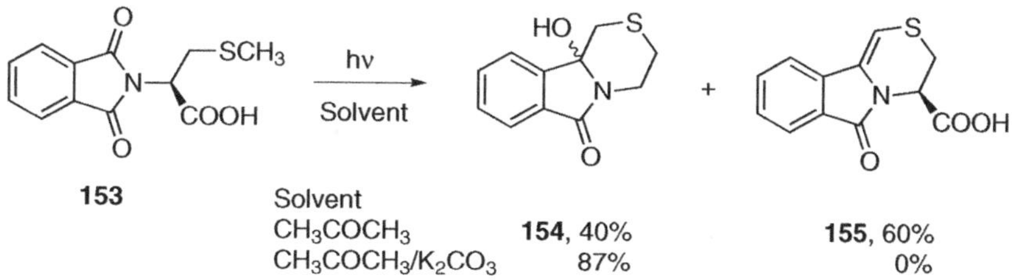

Scheme 5.64.

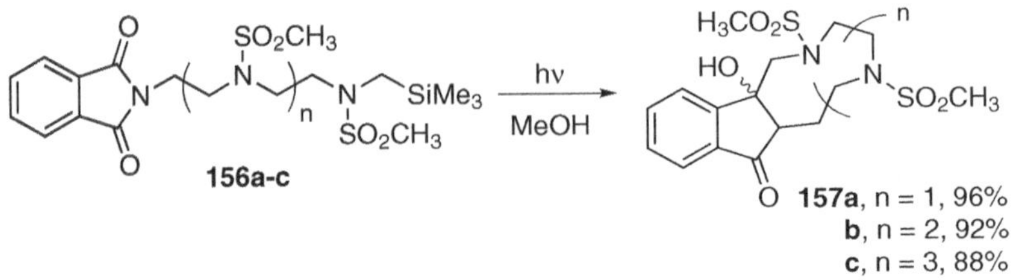

Scheme 5.65.

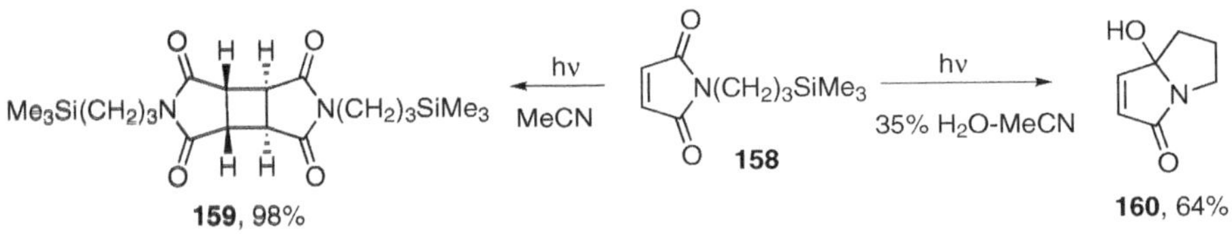

Scheme 5.66.

5.4.2 Addition onto C=N

In Section 5.3.1.2, the convenient generation and synthetic application of α-amino radicals (from the corresponding radical cation) was discussed [34]. However, the low oxidation potential of these radicals makes a further oxidation step, in general involving the *ground-state* photocatalyst, a common occurrence (Scheme 5.2, path *i*). In such a case, iminium cations, formed via an overall two-electron process, were the reactive intermediates in the place of the primarily formed aminoalkyl radicals [220–222]. Cyanoarenes (e.g., 1,4-dicyanonaphthalene, **50**) were again suitable for the regioselective generation of iminium ions by oxidation of unsymmetrical tertiary amines. The cation was formed by oxidation of the α-amino radical by the ground-state cyanoarene. Substituted pyrrolidines (e.g., **161**) were prepared and employed for the completely regioselective synthesis of bicyclic heterocyclic compounds, such as **162** (Scheme 5.67) [222].

Moreover, **50** photocatalysis in MeCN-water 4:1 v/v mixture caused the oxidation of amine **163** to the iminium cation through the following sequence: (i) SET to excited **50**, (ii) deprotonation of the radical cation (assisted by water present in the solvent), and (iii) SET from the radical to ground-state **50**. The cation was finally

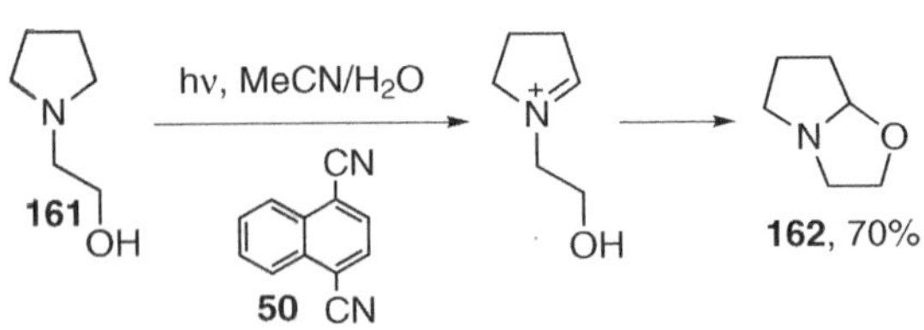

Scheme 5.67.

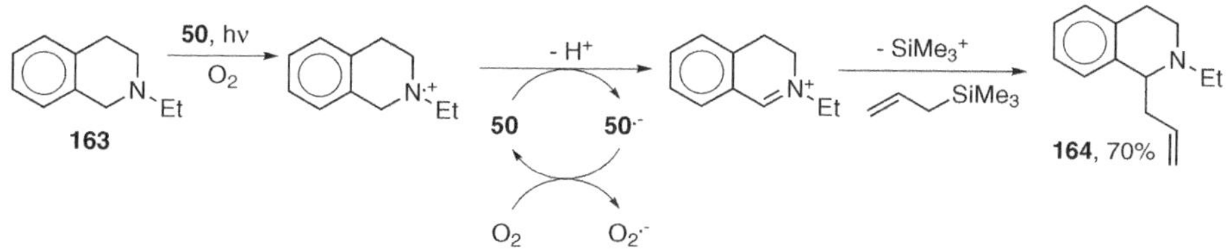

Scheme 5.68.

trapped (iv) by allyltrimethylsilane and afforded the allylated derivative **164** in 70% yield (Scheme 5.68) [221].

Trapping of the photogenerated iminium ions by cyanide ions or by trimethylsilylcyanide (TMSCN) was likewise effective. In this way, α-cyanoamines [223–228], in turn valuable intermediates for the synthesis of indole alkaloids, were synthesized. A typical example is shown in Scheme 5.69, where cyanoamine **166** was formed upon **82**-catalyzed photolysis of amine **165** and was converted into (±)-eburnamonine **167** in two steps from **166** [223].

Both 6-cyano-1,2,3,6-tetrahydropyridines [224] and indoloquinolizidine alkaloids [225] were obtained analogously. Visible light could be conveniently adopted when using photocatalysts such as *N,N′*-dimethyl-2,7-diazapyrenium bis(tetrafluoroborate) [$DAP^{2+}(BF_4^-)_2$] (**170**) or dyes such as methylene blue and Eosin [226, 227]. Interestingly, the reaction took place both in the presence and in the absence of oxygen. In the former case, singlet oxygen was envisaged as the first intermediate involved in the electron transfer steps (Scheme 5.70). This was supported by the occurring of the reaction also when $^1O_2$ was thermally generated from 1,4-dimethylnaphthalene endoperoxide [228, 229].

An alternative approach to iminium cations involved the oxidation of the amine as the first step followed by homolytic hydrogen abstraction from the amine radical

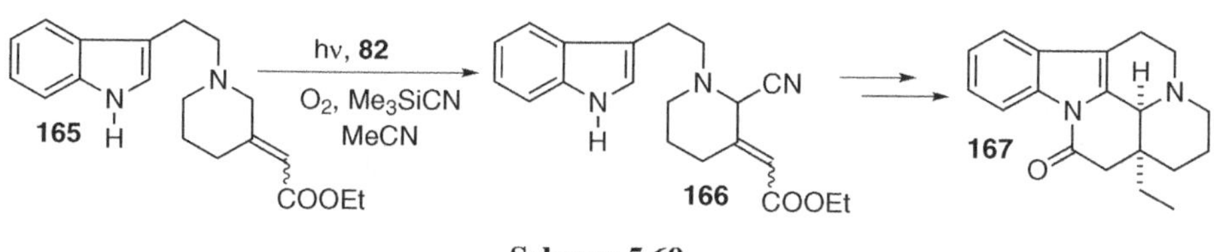

Scheme 5.69.

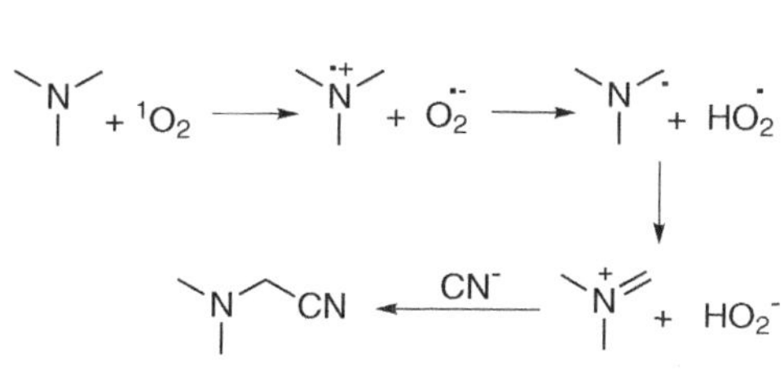

Scheme 5.70.

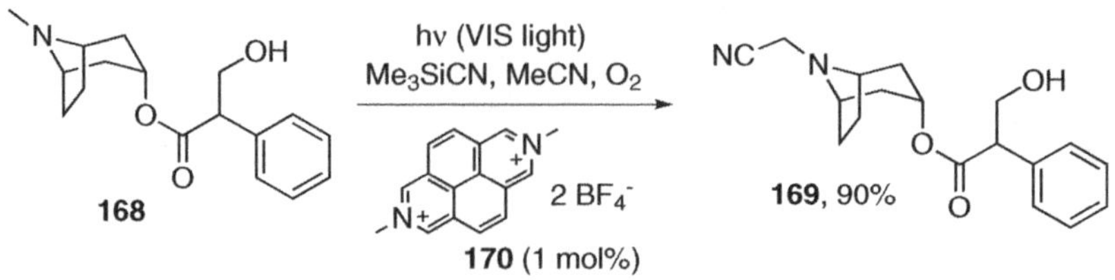

Scheme 5.71.

cation. Again the end result was the formation of an electrophilic iminium cation that was trapped by carbon nucleophiles. Oxygen was usually required as hydrogen atom scavenger. A series of alkaloids was photocyanated by using a visible-light-absorbing sensitizer [$DAP^{2+}(BF_4^-)_2$, (**170**); see above, 1 mol%] and Me_3SiCN as the cyanide source [230]. Under these conditions, atropine **168** gave α-aminonitrile **169** in 90% yield. The success of the reaction was in part attributed to the stability of the radical cation generated by monoelectronic reduction of the diazapyrenium salt (Scheme 5.71).

Recently, substituted *N*-aryl tetrahydroisoquinolines were used as starting materials for the preparation of isoquino[2,1-*a*]pyrimidines via the photocatalyzed formation of a C−N bond again by using Eosin Y disodium salt (**172**, 0.5 mol%) as the catalyst [37, 38, 231]. The strategy was in this case based on the intramolecular addition of a pendant *N*-tosyl moiety onto the iminium ion formed. Thus, 4-methyl-*N*-(2-(7-methyl-3,4-dihydroisoquinolin-2(1H)-yl)benzyl)benzenesulfonamide (**171**) was functionalized under mild conditions to give 3-methyl-5-tosyl-5,6,12,13-tetrahydro-4*bH*-isoquinolino[2,1-*a*]quinazoline (**173**) in 85% yield upon 25 h of irradiation by a 36-W fluorescent lamp (Scheme 5.72) [231]. On the other hand, the attempted extension of Eosin Y sensitization to the alkylation of diethyl ethylidenemalonate (with methyldiphenylamine as the radical precursor) resulted in a poor yield (9%), to be compared with 90% when an Ir complex was chosen (90%) [232].

Rose Bengal photocatalyzed the cyanation of the *N*-acetyl-2,16-dihydro derivative of the alkaloid tabersonine (33% yield, 2 h of irradiation) [233], as well as that of *N*-aryl-tetrahydroisoquinolines (a 5 mol% amount of the catalyst was used here in combination with graphene oxide) [234]. The mild conditions involved allowed for the C−H functionalization of these tertiary amines by using Me_3SiCN as

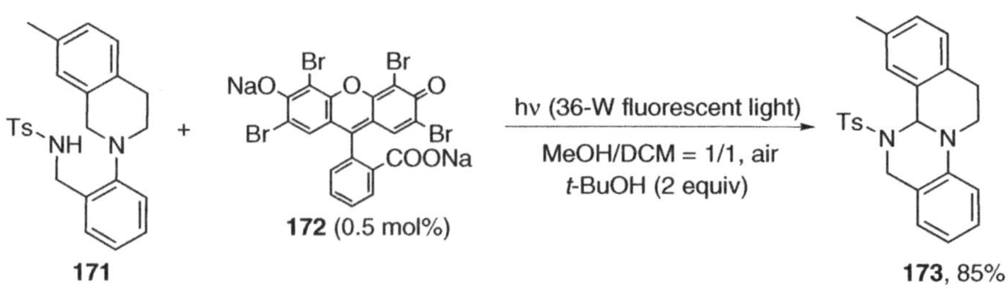

Scheme 5.72.

the nucleophile. Using graphene oxide improved the rate and the efficiency of the reaction. The process was likewise applied for the trifluoromethylation of the isoquinoline derivative by using $TMSCF_3$ as the nucleophile [234]. *N*-Aryl-tetrahydroisoquinolines underwent a smooth dehydrogenative coupling reaction in the presence of dyes (Rose Bengal and Eosin Y gave the best performance), thus offering a facile metal-free way of building sp^3–sp^3 C–C bonds [235–237]. The α-functionalization of tertiary amines via the Henry and Mannich reactions was thus obtained. As an example, irradiation with green light LEDs of a mixture of the isoquinoline and nitromethane (or a different nitroalkane) gave the corresponding Henry-coupled product in a high yield (at least with benzylamines) [236], which was further improved when solar light was used. Henry adducts from *N*-aryl tetrahydroisoquinolines were likewise reported by using Eosin Y bis-(tetrabutylammonium salt) (2 mol%) as the photocatalyst [237]. In this case, the role of superoxide radical anion ($O_2^{\bullet -}$) was documented (see further below, Scheme 5.74) [237]. Dehydrogenative-Mannich reactions catalysed by Rose Bengal (**174**) [37, 38] were likewise reported as illustrated in Scheme 5.73. Here the nucleophile was an enamine generated in situ from acetone (the solvent of the reaction) and an organocatalyst (pyrrolidine), a process promoted by TFA [236]. Under these conditions, the substituted isoquinoline **89** smoothly gave aminoketone **175** after 20 h of irradiation in almost quantitative yield (Scheme 5.73). When using cyclohexanone, substituting (L)-proline for pyrrolidine improved the yield and imparted a good diastereoselectivity [236]. The functionalization of *N*-aryl tetrahydroisoquinolines could took

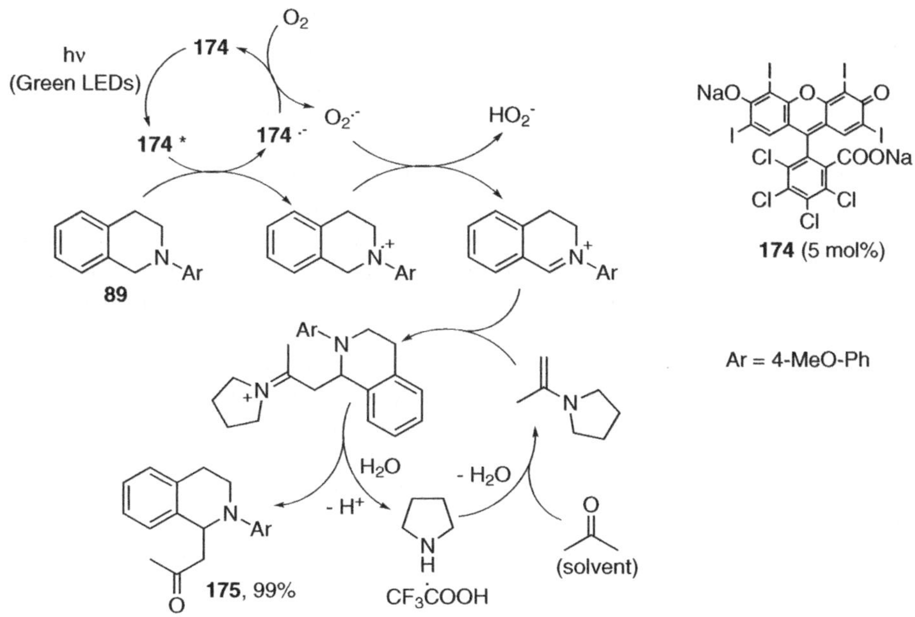

Scheme 5.73.

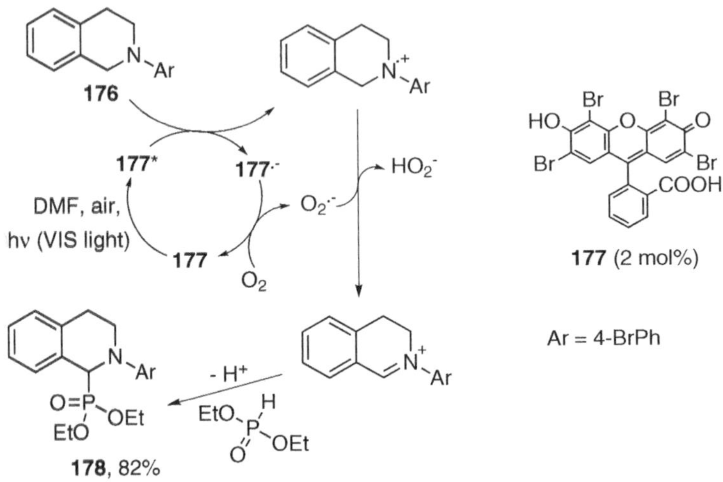

Scheme 5.74.

place even under anaerobic conditions by using bromotrichloromethane as the stoichiometric oxidant [238].

C–P bond formation by photocatalysis was recently reported and involved the reaction of a number of *N*-aryl tetrahydroisoquinolines (e.g., **176**) in the presence of an excess of diethyl phosphite. This resulted in the oxidative synthesis of α-amino phosphonates (**178**, Scheme 5.74) [37, 38, 235]. Green LEDs were used as the light source and Eosin Y (**177**, 2 mol%) was used as the photocatalyst, regenerated by the dissolved oxygen. In common with other photocatalytic reactions of the easily oxidized *N*-aryl tetrahydroisoquinolines, the mechanism involved oxidation by the excited sensitizer and hydrogen atom transfer from the radical cation to $O_2^{\bullet -}$, in this case followed by trapping of the resulting iminium ion by the phosphorus nucleophile (the yield of the phosphonate exceeded 80%, Scheme 5.74) [235].

The oxidative coupling of tetrahydroisoquinolines with dialkyl malonates (used as the solvent) was similarly obtained by irradiation at 530 nm in the presence of 2 mol% Eosin Y. The products were isolated in excellent yields since the excess of dialkyl malonate was easily eliminated by Kugelrohr distillation [235].

A further path that can be followed by photogenerated radical ion pairs is proton transfer between the oppositely charged ions followed by coupling of the neutral radicals [239]. Although different paths competed, chemoselective heterocoupling may predominate, as in the case of the photocatalyzed alkylation of some phenylcyano-*N*-benzoylimines (e.g., **179**) by using cycloalkenes and CdS supported on ZnS (Scheme 5.75) [240, 241]. Adduct **180** was formed in 81% yield. This reaction did not occur with silica-supported CdS, where a thermal reaction occurred leading to the hydrocyanation of the imine.

Another pathway involved the intramolecular addition of a radical ion onto a C=N bond, as observed in the PET induced 1-aza-di-π-methane rearrangements for

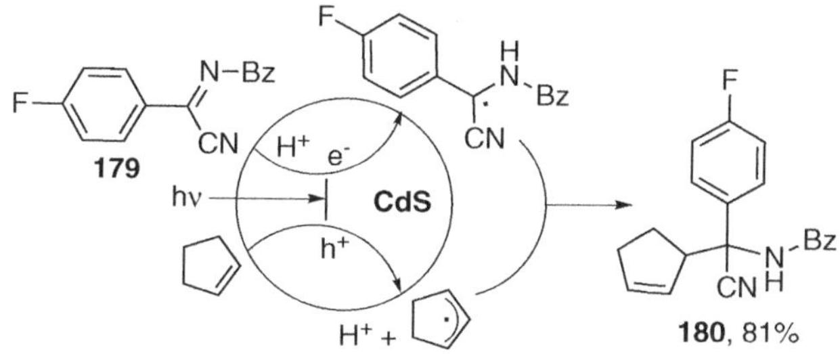

Scheme 5.75.

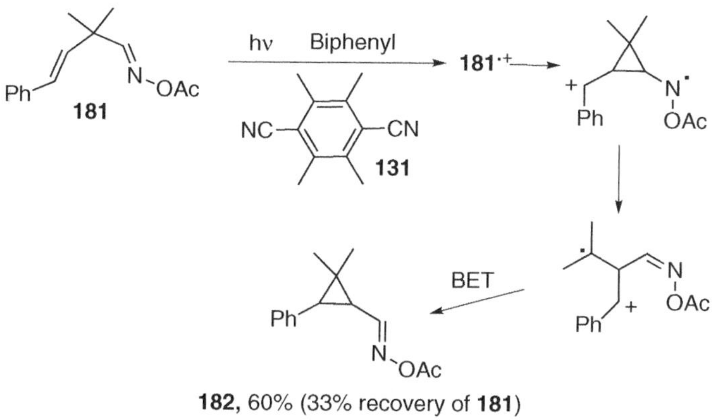

Scheme 5.76.

the construction of three membered rings. Thus, the styrene double bond in β,γ-unsaturated oxime acetate **181** underwent single electron oxidation by using 1,4-dicyano-2,3,5,6-tetramethylbenzene **131** (more effective than **82**) as the photocatalyst. This initiated a process leading to cyclopropane derivatives (e.g., **182**) that were obtained in modest yields (Scheme 5.76) [242, 243].

5.5 CYCLOADDITION

The Umpolung caused by ET in π systems has been exploited in cycloaddition reactions, similarly to the addition reactions reported above. Photochemical initiation, generally oxidative, has greatly contributed to enlarging the scope of this all-important class of reactions [244–247].

5.5.1 Formation of a Four-Membered Ring

The synthesis of four-membered rings by head-to-head [2 + 2] cycloaddition of alkenes (**183**) is conveniently obtained by photocatalyzed oxidation as illustrated in

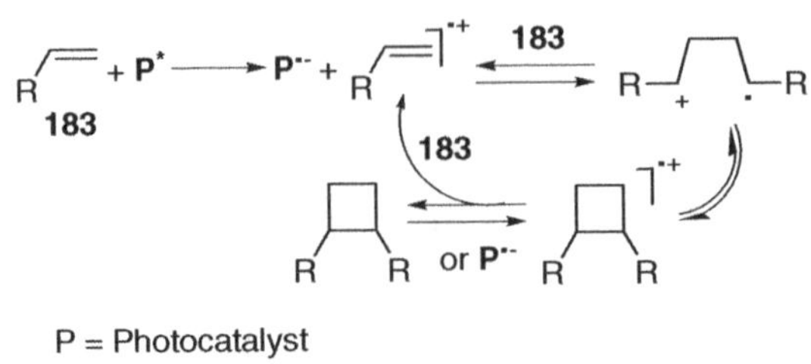

Scheme 5.77.

Scheme 5.77. The process is terminated by BET to the adduct (a distonic radical cation) either from the radical anion of the photocatalyst (and in this case the quantum yield of the cyclobutane formation was <1), or from the alkene. In the latter case a (short) chain process was initiated and the cyclobutane quantum yield was actually >1 for some derivatives.

The reaction occurred conveniently with radical cations stabilized by a heteroatom-bonded conjugated π moiety. The most extensively studied class is the [2 + 2] cyclodimerization of phenyl(aryl) vinyl ethers [248–254]. Thus, parent phenyl vinyl ether (**184**) gave the corresponding cyclobutane as a diastereoisomeric mixture (30% yield, isomer ratio 3:4) upon 1,4-dimethyl terephthalate (**185**, 40% mol equiv) photocatalysis in MeCN [248]. The reaction has been considered to involve an excited complex rather than free radical ions. The mechanism was not univocal, as suggested by the fact that the reaction quantum yield decreased when decreasing the **184** concentration, while formation of the *cis* isomer was largely preferred (~9:1) at [**184**] below 0.01 M, but not above [250]. When the alkene and the photocatalyst were included within NaX zeolites [255], the dimer ratios differed from those observed in solution (e.g., in the case of *trans*-anethole the *cis*/syn cyclobutanes were formed preferentially). Interestingly, commercial poly(ethylene terephthalate) could be used as photocatalyst and gave comparable amounts of the dimer with respect to monomeric **185**. The polymer was filtered out after the reaction, washed, and re-used, although its effectiveness decreased strongly after each run [249]. Inorganic semiconductors (ZnO, CdS) were likewise used as photocatalysts [252, 253], with the caveat that some samples of CdS caused the dimerization to occur in the dark as well [253], a fact that illustrated how difficult it was preparing samples of semiconductor sulfides that gave reproducible results.

[2 + 2] Cyclodimerization was also applied to aliphatic vinyl ethers. Thus, ethyl vinyl ether (100 mmol) in a benzene solution afforded the expected cyclobutanes in 39% yield upon photocatalysis by **50** (5 mol%) [256]. A mechanism via triplex (i.e., an excited complex involving the photocatalyst and two molecules of the ether), rather than the intermediacy of the free radical cations, was invoked in this case. Other photocatalytic systems such as phenanthrene/**55** or pyrilium salts could be conveniently used in the dimerization of α-methylstyrene [257] and of (dialkyl)indene [258], whereas *p*-methoxystyrene gave only the *trans*-head-to-head cyclodimer in a low yield (13%) when irradiated in the presence of **55** [259]. Aromatic enamines were reported to form cyclodimers in the presence of a wide range of photocatalysts (e.g., aromatic

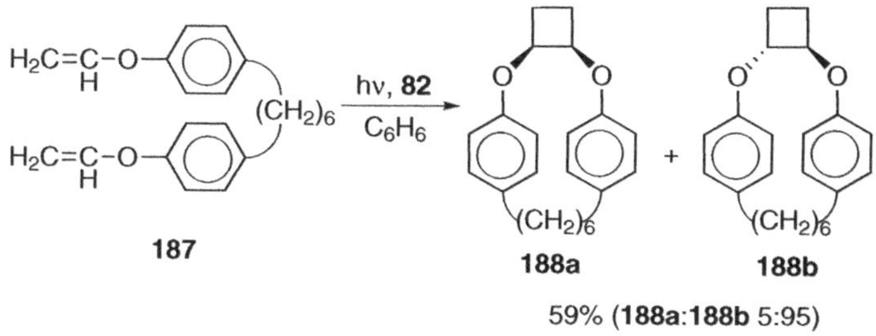

Scheme 5.78.

ketones, pyrilium salts), provided that oxygen was present in solution and prevented undesired polymerization [260]. The photocyclodimerization of *N*-vinylcarbazole has been early reported [261] and was later obtained by using **55** [262] as the photocatalyst. In this case the yields were significantly improved under aerated conditions by adding a cocatalyst such as perylene. In such a way, polymerization of *N*-vinylcarbazole was prevented and the overall turnover number of the photocatalyst was over 8000. Furthermore, the dimerization occurred efficiently under heterogeneous conditions by using a CdS dispersion [263]. Chloranil and fluorenone were likewise used with some success as photocatalysts [264].

An intramolecular example is the [2 + 2] cycloaddition of electron-rich styrenes tethered by an aliphatic link that procedeed conveniently upon visible light irradiation of ruthenium tris(bipyridine) salts (15 mol%) in the presence of methyl viologen (**186**$^{2+}$) as secondary acceptor. The reaction was successful with both symmetric and asymmetric derivatives with good to excellent yields [265]. Interestingly, the intramolecular version of the reaction was used for synthesizing macrocyclic 2,ω-dioxabicyclo[*n*.2.0] ring systems **188a,b** as shown in Scheme 5.78 (9,10-dicyanoanthracene **82** as the photocatalyst) [266].

The stereoselectivity of the process largely depended on the solvent employed. In most cases, the *cis* and *trans* isomers were predominant in MeCN and benzene, respectively. The **50**-photocatalyzed intramolecular [2 + 2] cycloaddition of di- and tetra-allylsilanes was also carried out and gave silabicyclo[3.2.0]heptanes in good yields [267]. It is worth noting that when an open-chain allyltrimethylsilane was subjected to the same irradiation conditions, no cycloadduct was formed. Analogously, 2-silacyclobuta[2.3]cyclophanes were synthesized starting from dimethyl-bis(4-vinylphenylmethyl)silanes [268]. As for the latter reaction, the same photocatalytic conditions were able to induce cycloreversion to give back the starting material.

The cross cycloaddition between two different olefins is of obvious importance and requires a difference in the redox properties of the partners. Having a conjugated and a nonconjugated alkene may seem an appropriate choice. However, when a mixture of an alkyl and a phenyl vinyl ether was irradiated in MeCN in the presence of 1,4-dycyanobenzene (**55**), a significant amount of cross adducts was obtained only by using a large excess of one of the ethers [269]. The same holds for dienes, often chosen as one of the partners. Thus, a [2 + 2] cycloadduct between 1,1′-dicyclopentenyl and ethyl vinyl ether was formed in a high yield upon **55** (35 mol%)

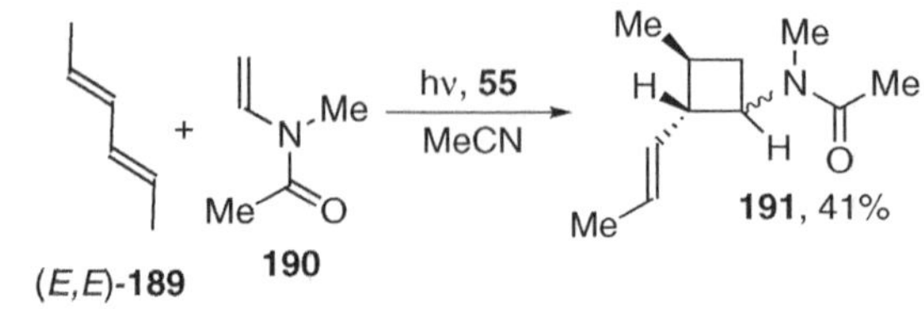

Scheme 5.79.

photocatalysis, but the ether was used in a 12-fold amount with respect to the diene [270]. However, an unusually clean reaction took place between diene **189** and *N*-methyl-*N*-vinylacetamide **190** (Scheme 5.79). This implied not only that **190** was not itself liable to dimerization, but also that the radical cation of the more easily oxidized term (the diene) reacted preferentially with **190** rather than with neutral **189** [271]. This behavior was attributed to the powerful stabilization by the amide function of the positive charge at an adjacent carbocationic center.

[2 + 2] Cycloaddition could be obtained also via radical anions. Thus, bis enones **192a–c** could be converted to cyclobutanes **193a–c** (Scheme 5.80). The reaction apparently involved the radical anion formed by reduction (a key role here had the stabilization by the lithium cation, the reaction did not proceed otherwise). This added onto the other moiety, functioning as Michael acceptor [272]. Yields and

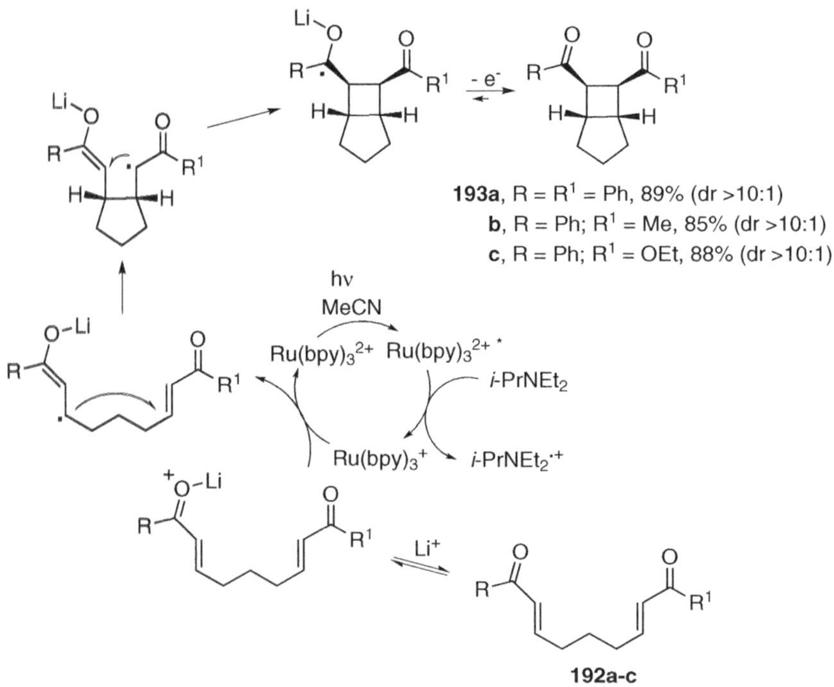

Scheme 5.80.

diastereoselctivity were much higher than when the same process was carried out at the cathode in the presence of a Co catalyst [273]. In that case the *trans* cyclobutane isomer and other products resulting from [4 + 2] cycloaddition and reductive dimerization were also formed.

Indeed, it was found that these side products were formed also when the *cis* cyclobutane was submitted to photosensitization for an *extended* time. This result was rationalized as involving the over-reduction of **193a–c** and subsequent reactions of the radical anion (Scheme 5.80). Thus, this would be a case differentiating the course of the electrochemical reaction, occurring at the cathode surface where the local concentration of electrons is much higher than when are donated by the photocatalyst in solution (see also Chapter 8) [9]. The reaction proceeded only with aryl substituted ketones, aliphatic derivatives apparently not being reduced under these conditions. However, aliphatic ketones served as Michael acceptors; this made possible cross cycloadditions, both intra- and intermolecularly. The reaction was hindered by bulky groups in β position, however [274].

5.5.2 Formation of a Five-Membered Ring

The PET-induced ring opening of three-membered ring derivatives (oxiranes, azirines or cyclopropanes) followed by reaction with a double bond offers a convenient access to five-membered ring compounds. Azirines (e.g., **194**) were extensively studied in [3 + 2] cycloaddition reactions under PET conditions [275–279]. A 2-azaallenyl radical cation was suggested as the intermediate (Scheme 5.81). Addition to acrylonitrile, ring closure

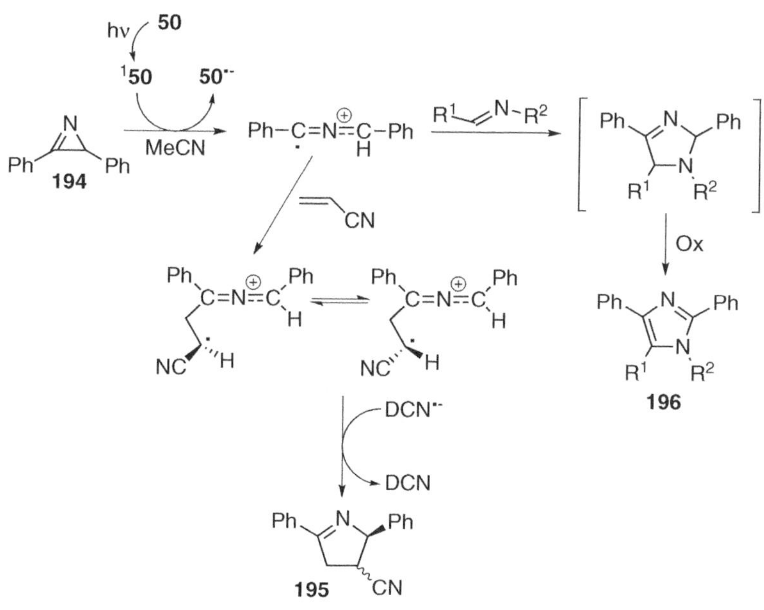

Scheme 5.81.

and BET led to didhydropyrroles **195**. Since the reaction occurred in two steps and involved an odd-electron species, it differed from the thermal concerted 1,3-dipolar cycloaddition. Moreover, the 2-azaallenyl radical cation was able to add to the C—N double bond of imines leading to 1-substituted imidazoles **196** via the spontaneous rearomatization of first-formed imidazolines (Scheme 5.81) [275, 276].

When bicyclic azirines were used in the reaction with imines, [*n*](2,4)imidazolophanes were obtained. The polymethylene bridge had to contain at least five units for making the reaction successful [277]. Further expanding this strategy, four azirine moieties were built on a cyclododecane ring and ET photocatalysis was exploited for opening the three-membered rings and add alkynes. In this way, a new synthesis of porphyrin systems was accomplished [278].

C_{60} fullerene was a suitable trap for the 2-azaallenyl cation. 1,9-(3,4-Dihydro-2-5-diphenyl-2*H*-pyrrolo)fullerene-60 was isolated from the reaction with diphenylazirine, although the concomitant formation of oligomeric adducts was detected [279].

2,3-Diphenyloxiranes likewise gave tetrahydrofurans upon **50** photocatalysis in the presence of α,β-unsaturated nitriles and esters [280]. The stereochemistry of the products depended on that of the starting oxiranes, in accordance with mechanistic studies showing that the *exo,endo* and *exo,exo* forms of carbonyl ylides were produced in significantly different ratios from *cis*- and *trans*-oxides [281]. This indicated the intermediacy of distinct, nonequilibrating, ring-opened radical cations in the two cases. The reaction was extended to other epoxides bearing one to three aryl groups, although tetraphenyloxirane gave no adducts, probably due to steric hindrance [282–284].

Triarylcyclopropanes **197** likewise underwent [3 + 2] cycloadditions after PET reaction with **55** in the presence of vinyl ethers. A cyclopentane derivative **198** was formed in 62% yield along with two byproducts (**199, 200**) in nonnegligible amounts (Scheme 5.82) [285]. The reaction required the use of an equimolar amount of the photocatalyst, but this was >80% recovered at the end of the reaction.

α,α′-Bis-(trimethylsilyl)-amines underwent two sequential desilylation processes following two single-electron transfer steps (**50** was the photocatalyst). In this way, a nonstabilized azomethine ylide was formed [286]. These ylides were the key intermediates in 1,3-dipolar cycloaddition reactions. Dipolarophiles such as aromatic ketones, α,β-unsaturated esters and amides were used, obtaining in every case the corresponding five-membered cyclic adducts [286].

A variation is the intramolecular [3 + 2] cycloaddition between phenylcyclopropylketones and multiple bonds. Irradiation of a Ru complex/amine system caused

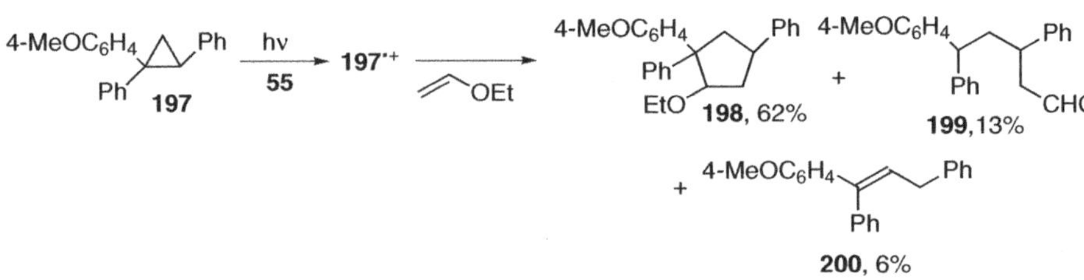

Scheme 5.82.

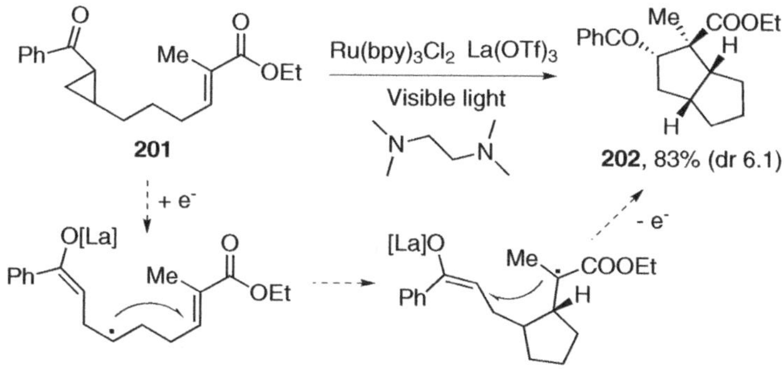

Scheme 5.83.

reduction of aromatic ketone and opening of the α,β-cyclopropane ring to give a distonic radical ion, in this case stabilized by lantanium salts, and intramolecular addition. Indeed, the reaction applied not only to electrophilic alkenes, but also to aliphatic alkenes and alkynes, giving access to a range of highly elaborated bicyclo [3.3.0]octanes with stereochemical control. Indeed, in a case it was checked that starting from the cyclopropyl ketone **201** as a 1:1 diastereochemical mixture, a single cyclopentane diastereomer **202** was obtained, in accord with the intermediacy of a nonstereogenic, ring-opened radical anion as indicated in Scheme 5.83 [287, 288]. Possibly, the distonic character of the radical anion from the opening of cyclopropyl ketones increased the chance of being trapped with respect to delocalized radical anions from enones.

N-Aryl tetrahydroisoquinolines are a convenient substrate for the photochemical generation of azomethine ylides and, in turn, a useful intermediate for the construction of five-membered ring by a dipolar cycloaddition reaction. The approach involved the photochemical oxidation of *N*-substituted derivatives **203a–c** to iminium ion followed by deprotonation by the HO_2^- formed and trapping of the ylide by a maleimide derivative as dipolarophile. The process resulted in the formation of a new fused pyrrolidine ring (compounds **204a–c**, Scheme 5.84) [289].

5.5.3 Formation of a Six-Membered Ring

As is well known, the Diels–Alder (DA) reaction does not occur when both diene and dienophile components are electron-rich compounds. However, photocatalysis offers a way for the *Umpolung* of one of the reagents, thus allowing the reaction to occur. Several examples were reported in the literature concerning the photocatalyzed DA reactions. The mechanism was similar to that proposed for the photocyclodimerization, and actually in some cases the two reactions occurred competitively (*vide supra*). Photocatalysis was generally preferable to ET thermally induced reactions, such as those initiated by ground-state acceptors such as aminylium salts (e.g., $Ar_3N^{\bullet +}SbF_6^-$), since in the latter case a large excess of dienophile was required and underwent competitively isomerization or polymerization under these conditions.

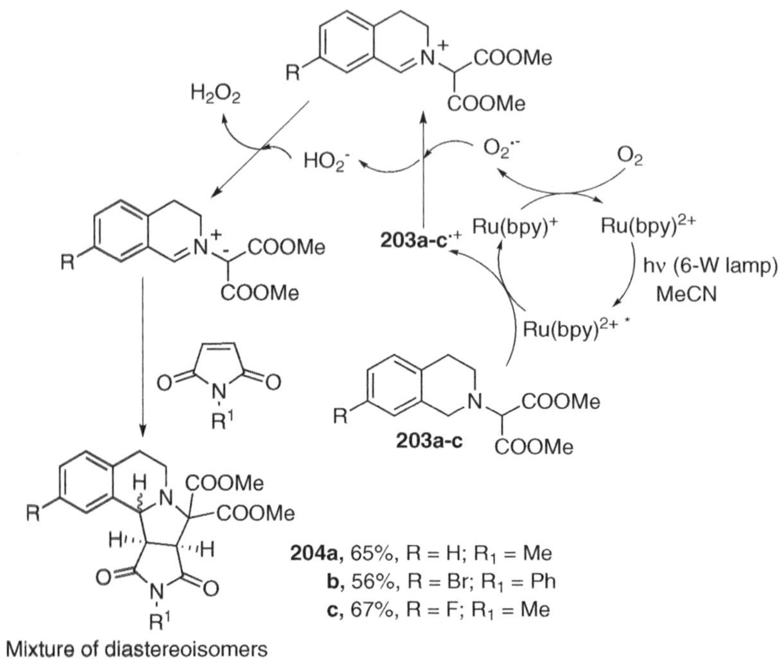

Scheme 5.84.

Thus, upon oxidative photocatalysis by an aromatic nitrile, 1,1-diphenylethylene led to the formation of a tetrahydronaphthalene (**205**) via attack by the radical cation onto a neutral molecule [290–292]. BET from the photocatalyst radical anion and rearomatization of the thus formed triene completed the sequence. It was also possible to trap the intermediate via ene reaction—for example, by using acrylonitrile as shown in the formation of **206** (Scheme 5.85) [292].

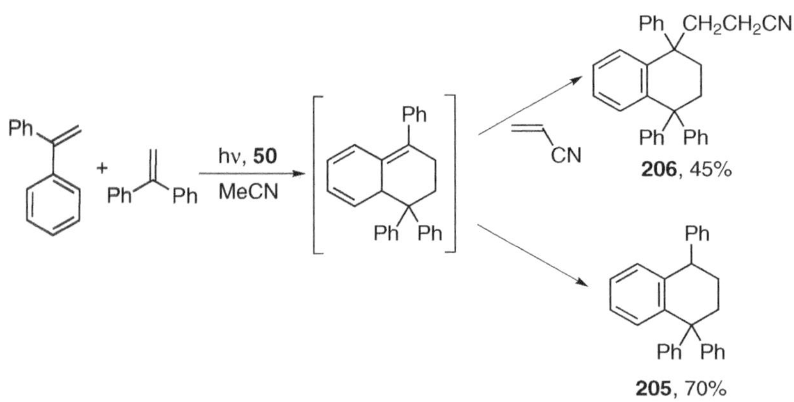

Scheme 5.85.

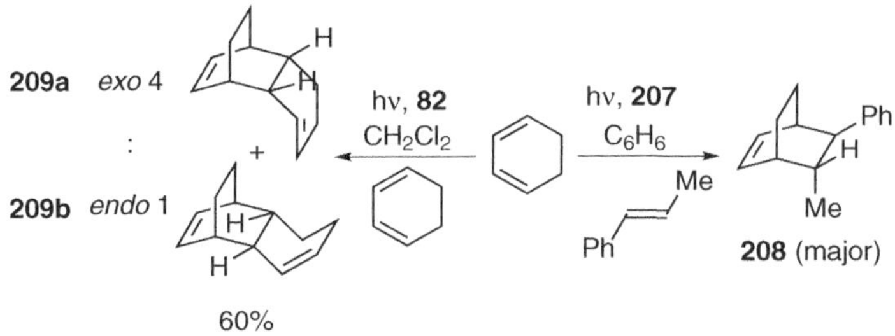

Scheme 5.86.

Dienes blocked in a *syn*-arrangement, such as cyclohexadienes, act as effective dienophiles under photocatalytic conditions. A triplex mechanism was invoked in this case. Thus, the exciplex formed by the excited singlet state of the photocatalyst (usually a cyanoarene, e.g., 2,6,9,10-tetracyanoanthracene, **207**) and a diene was trapped by a second molecule of diene to give a triplex that finally evolved to the cycloadduct [293–296]. β-Methylstyrene [293–295] and indene [296] were also used as dienophiles (see Scheme 5.86, right side). In the former case the *endo trans* isomer **208** was largely preferred and the selectivity was not affected when using a chiral catalyst such as (−)-1,1′-bis(2,4-dicyanonaphthalene) [295]. Cyclohexadiene was also used in the **82**-photocatalyzed [4 + 2] cyclodimerization in CH_2Cl_2 (Scheme 5.86, left side), which afforded a mixture of *endo*- and *exo*-dimers (**209a,b**) in more than 60% isolated yield [297]. Interestingly, the same reaction carried out under thermal conditions gave poor yields, whereas the concomitant formation of [2 + 2] cycloadducts was observed by changing the photocatalyst [297, 298].

2,4,6-Triphenylpyrylium tetrafluoroborate in CH_2Cl_2 was found to be particularly effective for inducing the Diels–Alder reaction of cyclohexadiene derivatives since only 0.1–1.5 mol% of the photocatalyst was required, light with wavelength >345 nm could be used, and the irradiation time was dramatically reduced with respect to the reaction catalyzed by cyanoarenes [299, 300]. Triphenylpyrylium salts could also be incorporated inside zeolites. Under these conditions, the 1,3-cyclohexadiene dimerization proceeded slower than in homogeneous solution and the *endo*[4 + 2] adduct was the major product [301]. 1-Methoxy-1,3-cyclohexadiene likewise underwent a cycloaddition reaction (70% yield) when irradiated in MeCN in the presence of 25 mol% **55** [302], while in the thermal $Ar_3N^{\bullet+}$-catalyzed reaction polymerization predominated.

A precursor (**211**) of the sesquiterpene (−)-β-selinene (**212**) was obtained in 60% yield by photochemical reaction of diene **210** and phenyl vinyl sulfide (Scheme 5.87), [303] in one of the few examples of the synthesis of a natural product by photocatalyzed DA reaction.

An excellent method for the cycloaddition of styrenes and butadiene derivatives was found in the visible light photocatalysis field by using ruthenium complexes as the catalysts [28–34]. While the system $Ru(bpy)_3^{2+}$/**186**$^{2+}$ mentioned above found limited application in this case, various experiments showed that the best results were obtained by using ruthenium tris-(bipyrazinyl) tetrakis(3,5-bis(trifluoromethyl)

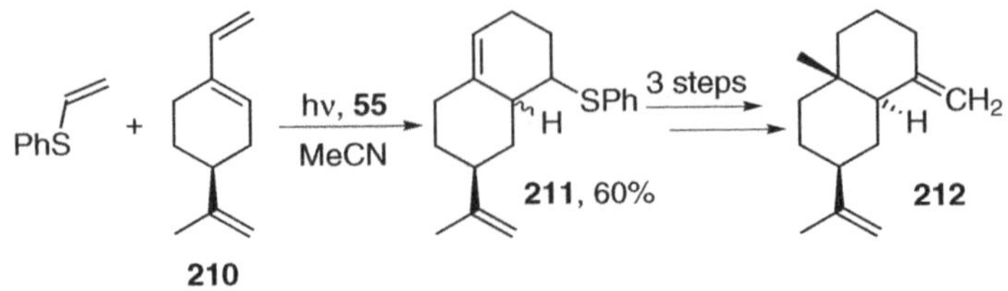

Scheme 5.87.

phenyl)borate (0.5 mol%) in air-equilibrated dichloromethane. The reaction was quite general, with a remarkable tolerance for steric and electronic factors, at least when moderately electron-rich styrenes were used [304]. An intramolecular version of the reaction was recently disclosed and applied to the gram scale preparation of adduct **214** in a good yield by using either fluorescent bulbs or simply ambient sunlight as the light source (Scheme 5.88) [305]. Useful applications of [4 + 2]-photocatalyzed cycloadditions were developed starting from indole derivatives [306–314]. The 2,3 double bond of these substrates can be considered an electron-rich 2π partner, but reaction in this role would cause the loss of aromaticity and actually only a few reactions between indoles and electron-poor heterodienes have been reported [315]. On the other hand, the monoelectronic oxidation of (substituted) indoles was conveniently obtained by irradiation in the presence of 5 mol% triarylpyrylium salts (**215a,b**$^{+}$, a cationic catalyst that was reduced to the neutral radical **215a,b**$^{\bullet}$) and the radical cation thus formed behaved as an electron-poor dienophile, smoothly reacting with dienes in a cycloaddition process (Scheme 5.89).

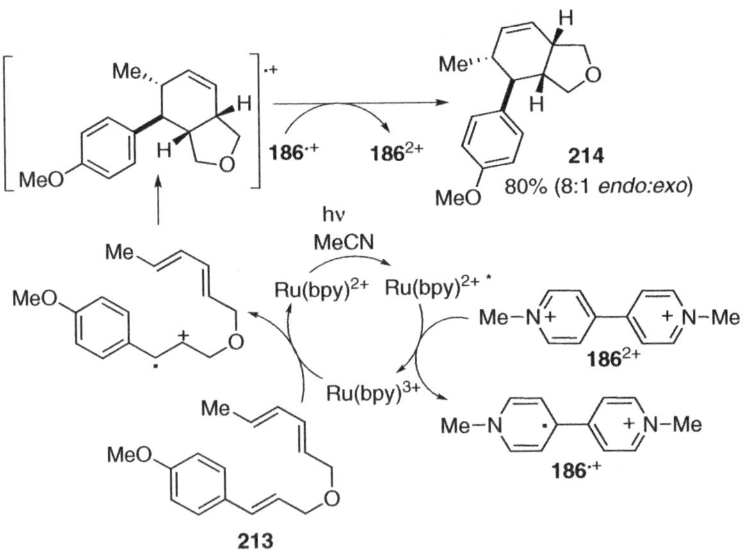

Scheme 5.88.

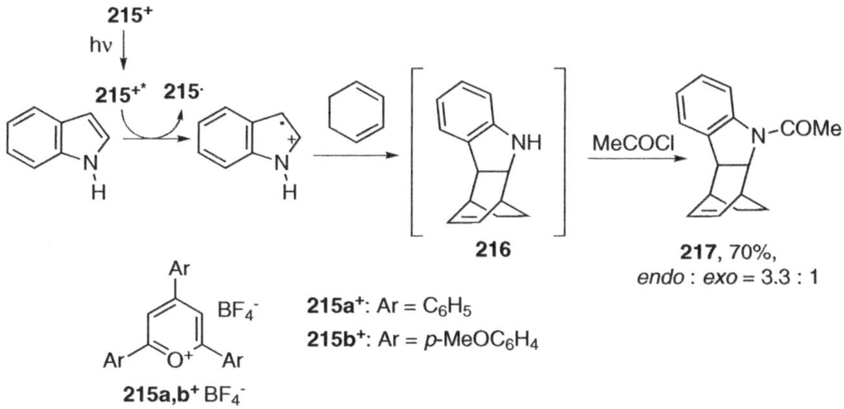

Scheme 5.89.

As a matter of fact, the photochemical path broadened the synthetic scope of the DA reaction and was applied for the synthesis of elaborated indole derivatives having biological activity. As an example, the indole radical cation underwent cycloaddition with cyclohexadienes. However, product **216** was more easily oxidized than starting indole and thus reacted further to give a complex mixture. This shortcoming was overcome, and amide **217** was prepared in a satisfactory yield when the reaction was performed in the presence of acetyl chloride, which functionalized the indoline as soon as it was formed and precluded further transformation [310]. A triplex was again invoked as the intermediate; indeed calculations supported a nonsynchronous nonconcerted reaction, involving a relatively stable intermediate [310–314]. This evidence could explain the complete regioselectivity observed; that is, a substituent in the 1-position (2-position) of the cyclohexadiene moiety was found in the 1-position (3-position) of the carbazole formed. The reaction was inefficient with indoles carrying a substituent on the 2,3 double bond (probably for steric reasons) and when open-chain dienes were employed.

Furthermore, exocyclic dienes could be used in some cases, as was shown in the 2,4,6-tris(4-methoxyphenyl)pyrylium tetrafluoroborate (**215b**$^+$) photocatalyzed reaction with (1,2)oxazinan-2-yl derivatives as dienes [309, 310].

Another class of suitable substrates for [4 + 2] cycloaddition reactions was shown to be that of 2-vinylindoles [311–314]. In this case, the indole derivatives reacted as the diene, and the course of the reaction was different. The first step was the oxidation of the indole **218**, followed by addition of the radical cation onto cyclohexadiene, cyclization, [1, 3] H shift, and BET. This finally yielded carbazole **219** (Scheme 5.90). It is worth noting that the use of chiral dienophiles led to a single isomer.

One of the limitations of these cycloaddition reactions was that these required two starting materials with oxidation potentials not differing by more than 0.5 V [311]. As an example, substituted enamines were suitable dienophiles for this reaction, since they had exactly the same oxidation potential of the above-mentioned

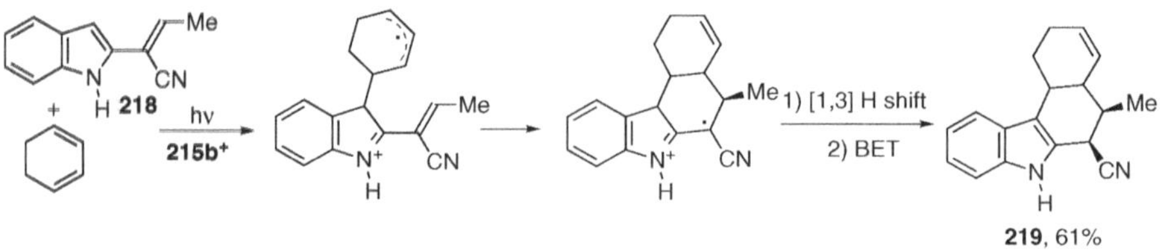

Scheme 5.90.

vinylindoles [313]. Moreover, vinylindoles gave pyrido[*1,2a*]indoles in the reaction with β-enamidoesters or -nitriles by using 2,4,6-tri-(*p*-anisyl)pyrylium tetrafluoroborate (**215b**$^+$) as the photocatalyst [314]. Unlike vinylindoles, 2-vinylbenzofurans mainly reacted as dienophiles and their reaction with cyclohexadiene showed a predominant *endo*-selectivity [316].

Furan behaved as a diene in the **50**-photocatalyzed reaction with indene. The reaction gave both the photocycloadduct and a substituted indene. The former product was the preferred one only when the furan ring had no substituent; a methyl group in the 2-position dramatically enhanced the yield of the substitution product [317].

The intramolecular version of the PET Diels–Alder reaction was rarely employed, although this path removed the requirement of a high dienophile concentration in order to prevent diene dimerization. As an example, the reaction of cyclohexadienyl derivatives bearing a flexible alkyl chain containing a double bond was investigated [318]. [4 + 2] Cycloadditions yielding the *endo-trans*-tricyclic compounds occurred to a variable extent depending on the photocatalyst used, since competing [2 + 2] reactions could also take place. The largest [4 + 2]/[2 + 2] ratio (~7) was observed in **82**-catalyzed experiments.

As a matter of fact, [2 + 2] cycloadducts or open chain dimers were often formed as side products in photocatalyzed Diels–Alder reactions [270, 319–324]. A representative example is shown in Scheme 5.91. Thus, irradiation of a benzene solution of **82** (2×10^{-2} M) containing cyclohexadiene (0.21 M) and 1-(trimethylsilyloxy)-1-methoxyethene (**220**, 0.86 M) gave a mixture of *endo* and *exo* [2 + 2] (**221a,b**) and

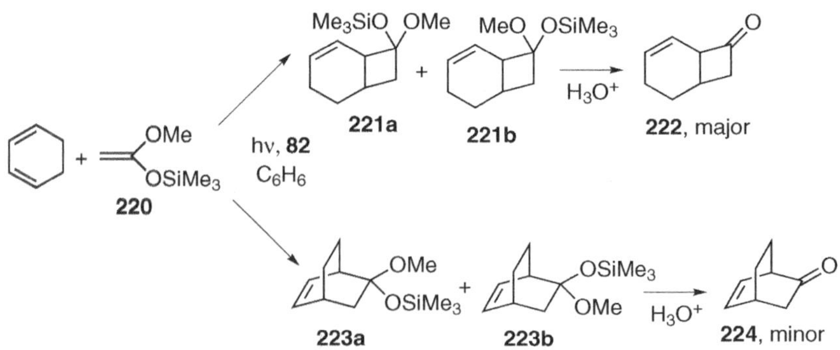

Scheme 5.91.

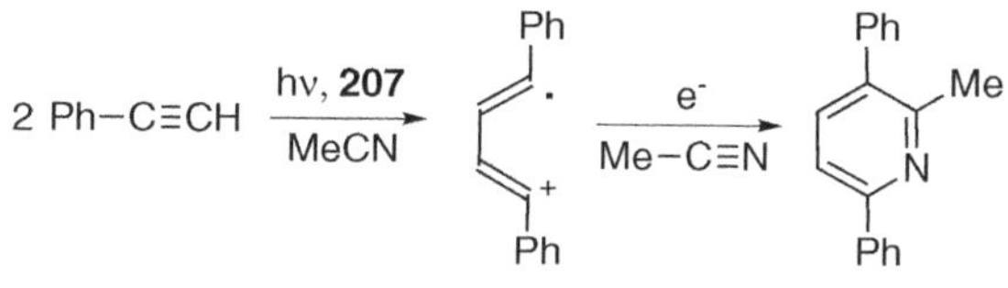

225

Scheme 5.92.

[4 + 2] cross-coupling cyclized products (**223a,b**) along with the cyclohexadiene dimers [324]. Acidic hydrolysis of the silicon-containing products yielded ketones **222** and **224** in a 3:1 ratio.

Finally, in the attempted cyclodimerization of phenylacetylene in MeCN, a diphenylpyridine **225** was isolated as the main product. This compound was envisaged as arising from trapping by the solvent of the dimeric radical cation intermediate (Scheme 5.92) [325].

In some cases, a Diels–Alder cycloaddition was induced by the initial formation of a radical anion rather than a cation. This is exemplified by the photocatalytic intramolecular hetero-Diels–Alder reaction of tethered bis(enones) shown in Scheme 5.93. The complexation by a Lewis acid [$Mg(ClO_4)_2$] of one of the enone moieties in compound **226** allowed for its monoelectronic reduction induced by a photogenerated Ru(I) complex. Coupling of the resulting enone radical anion (a formal electron-deficient heterodiene) with an electronically mismatched enone

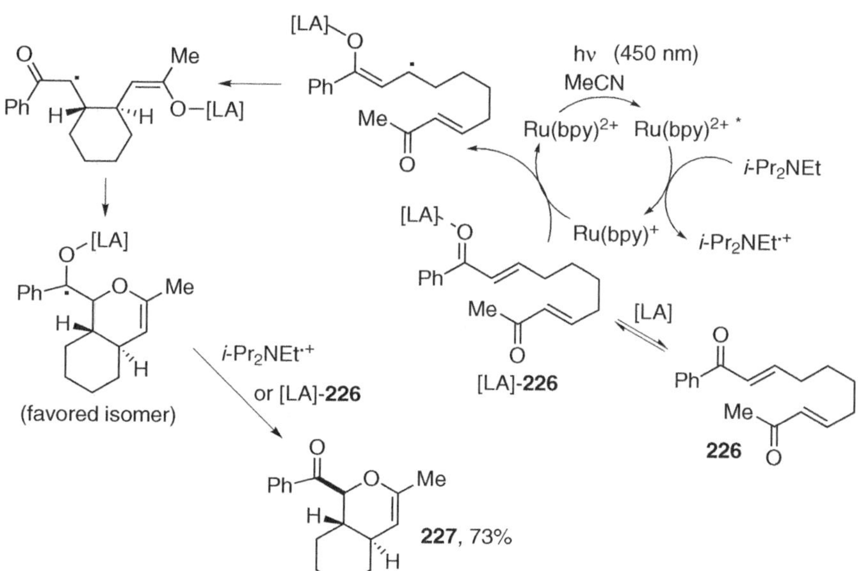

227, 73%

Scheme 5.93.

dienophile gave the desired dihydropyran **227** in a highly diastereoselective fashion (Scheme 5.93) [326].

5.6 FRAGMENTATION

As previously discussed, one of the possible fates of a radical ion is fragmentation to give a radical and a cation/anion. This reaction was indeed exploited to some extent for the generation in solution of other reactive species. Such a case was recently reported for quinone methides generation from a suitable precursor tethered to a water-soluble 1,8-naphthalenediimide derivative. The reactive species was obtained through an intramolecular PET reaction [327]. In other cases, the process had mainly a mechanistic interest as in the fragmentation of the radical cations of sulfides [328] or sulfoxides [329].

However, several examples of synthetically significant fragmentation reactions have been reported. Thus, the photochemical decarboxylation of free carboxylic acids can be accomplished by using this strategy, thus providing an alternative method to the Barton decarboxylation procedure. The reaction was particularly effective with *N*-Boc amino acids (**228**), where it was obtained via oxidation by the phenanthrene radical cation **229**$^{\bullet+}$, in turn generated by PET with 1,4-dicyanobenzene **55** (Scheme 5.94) [330].

Various examples involve ring cleavage. As an example, a retrocycloaddition could be photochemically induced in the case of 2,2-diarylthietanes (e.g., **231**, Scheme 5.95) [331]. In fact, the oxidation potential of heterocycle **231** was low enough to be oxidized by excited **82**, then causing ring opening and liberation of 2-cyano-1,1-diphenylethene (**232**) in a modest yield.

Other heterocyclic compounds were likewise cleaved upon photoinduced ET, as in the case of aryloxiranes that underwent C–C bond fragmentation [280] or of

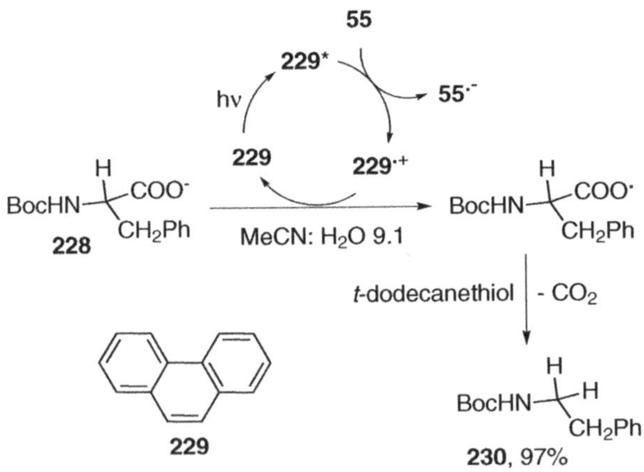

Scheme 5.94.

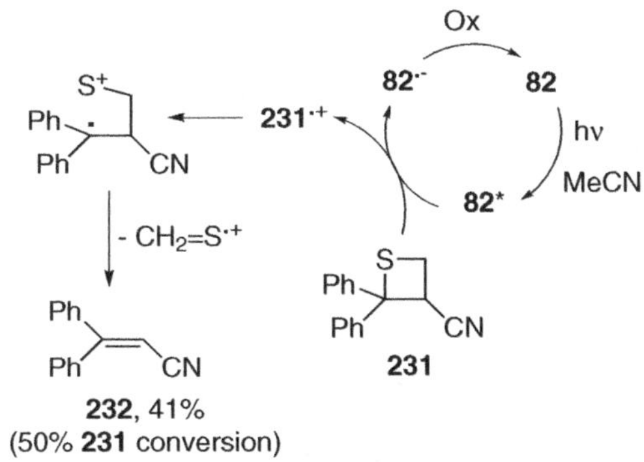

Scheme 5.95.

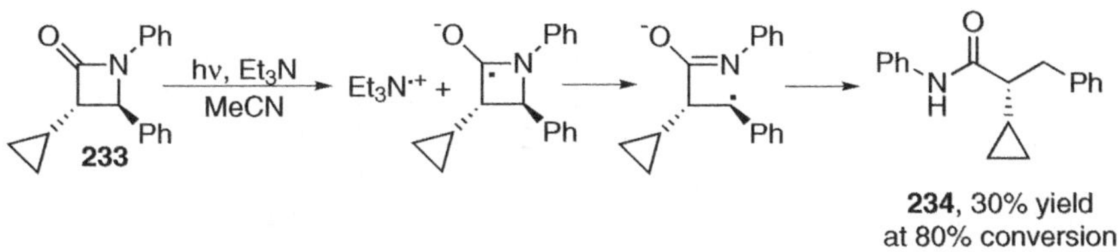

Scheme 5.96.

azetidin-2-ones that gave open chain amides [332]. In the latter case, the process was caused by a PET reaction between the starting lactam **233** and a tertiary amine (Scheme 5.96). Diaryl cyclopropanes (**235a,b**) were likewise opened upon ET photocatalysis [333]. The distonic radical cation formed was easily trapped by nitrogen oxide, allowing the preparation of substituted 2-isoxazolines (**236a,b**, Scheme 5.97), as well as by oxygen (see also Section 5.7).

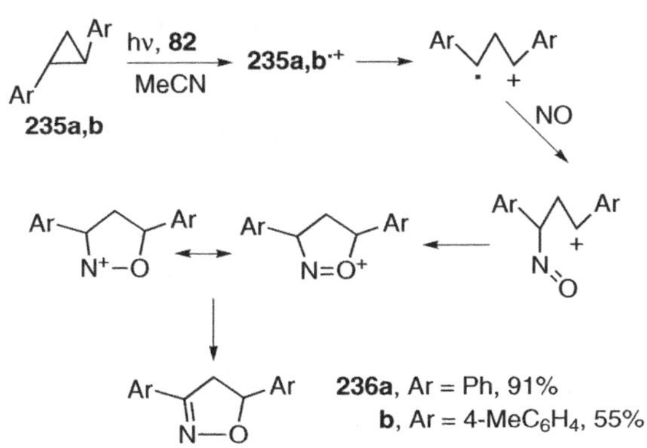

Scheme 5.97.

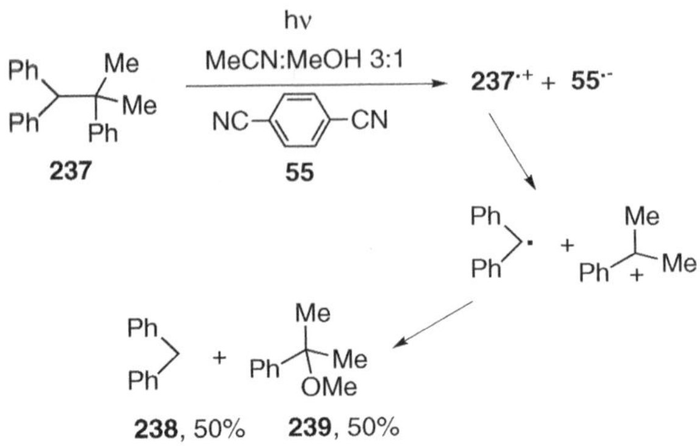

Scheme 5.98.

Fragmentation is likewise effective on acyclic compounds, such as 1,1,2,2-tetraphenylethane or related compounds, as indicated in Scheme 5.98. Thus, the **55** photomediated oxidation of compound **237** induced the C–C cleavage at the corresponding radical cation stage, thus forming a benzyl radical and a benzyl cation and finally leading to diphenylmethane **238** and to ether **239**, respectively, in a comparable yield [334, 335].

As for the last example, the carbocation formed in the carbon–carbon bond cleavage can be trapped intramolecularly by an OH group, thus forming oxygen heterocycles [336–345]. This electron-transfer-initiated cyclization (ETIC) reaction relies on the easy liberation of a benzyl radical upon radical cation fragmentation. A typical example involving *N*-methylquinolinium hexafluorophosphate ($NMQPF_6$, **240**) as the photocatalyst is shown in Scheme 5.99. Thus, a PET reaction between **240*** and homobenzylic ether **241** occurred and the stereochemistry of the resulting furanoside **243** (or pyranoside) system depended on whether the cyclization step arose directly from radical cation **241**$^{\bullet+}$ or rather from

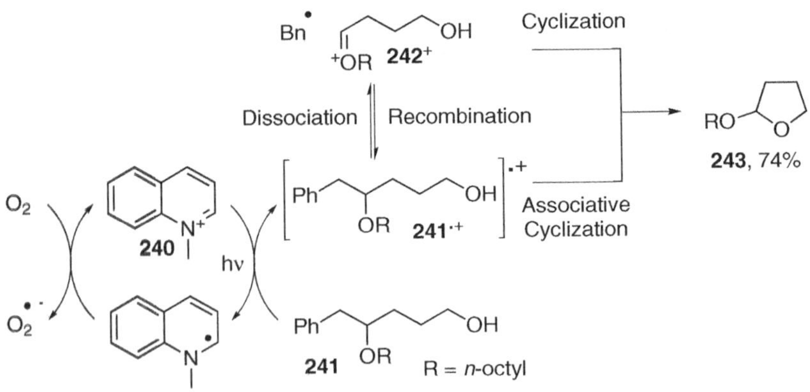

Scheme 5.99.

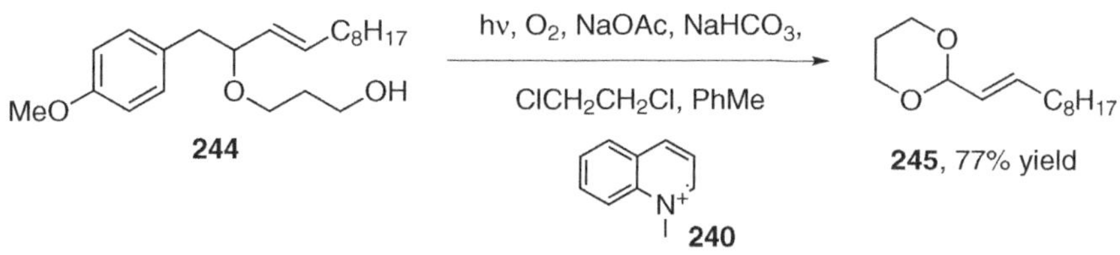

Scheme 5.100.

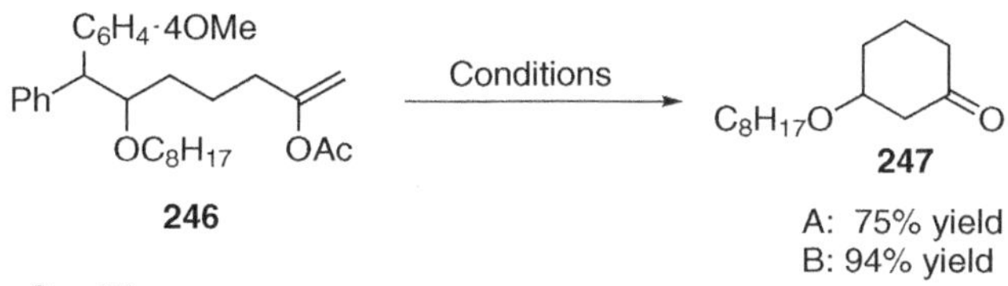

Conditions:
A: hv, **240** (2.5%), O_2, NaOAc, DCE, PhMe (3.5:1)
B: $(NH_4)_2Ce(NO_3)_6$ (**248**, 2.2 eqv.), $NaHCO_3$, DCE, MeCN (4:1)

Scheme 5.101.

oxonium cation **242**$^+$ [336, 337]. Interestingly, the oxidation potentials of homobenzylic ethers was appropriately tuned by inserting electron-donating substituents (e.g., a methoxy group) on the aromatic ring in order to make the compounds easier to oxidize (Scheme 5.100) [338, 339]. Acetal **245** was obtained in this way by oxidative fragmentation of the anisole derivative **244** [338]. Diphenylmethanes and other compounds bearing two aromatic rings at the same carbon were often oxidized also by thermal inorganic oxidants such as ceric (IV) ammonium nitrate (**248**). As an example, the oxidation of enol acetate **246** to cyclohexanone **247** gave a better yield (94%) under thermal than under photochemical conditions, but **248** had to be used in excess (2.2 equiv, Scheme 5.101) [339].

In an effort to make the process greener, the benzyl moiety was tethered to a soluble oligonorbornene-based polymeric support. After the reaction and the photorelease of the product, the support was precipitated by adding methanol and removed by filtration [340]. Moreover, the oxocarbenium ions could be employed as effective promoters of polyepoxide cascade cyclization reactions to form polyethers [341].

Electron transfer initiated cyclization (ETIC) was recently applied to the synthesis of lactodehydrothyrsiferol (a protein phosphatase 2A inhibitor) through a cascade reaction initiated by an epoxide alkylation through a PET-generated oxocarbenium ion [342]. Accordingly, a key intermediate for the preparation of this inhibitor was obtained by irradiation of **249** that gave the *spiro* compound **250** in 45% yield (75% based on recovered **249**, Scheme 5.102).

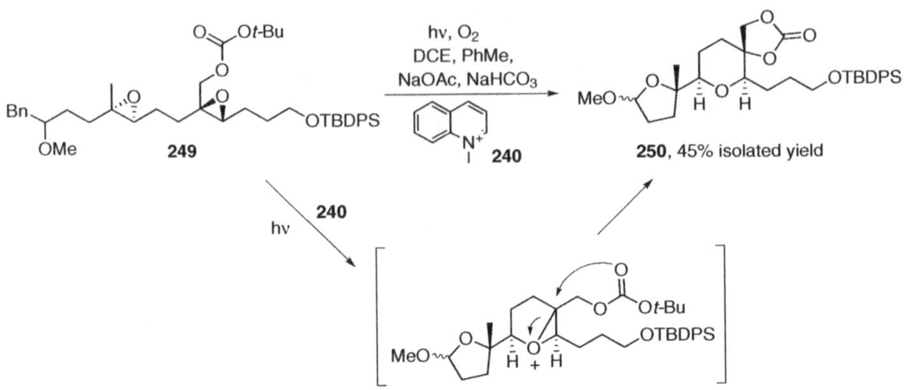

Scheme 5.102.

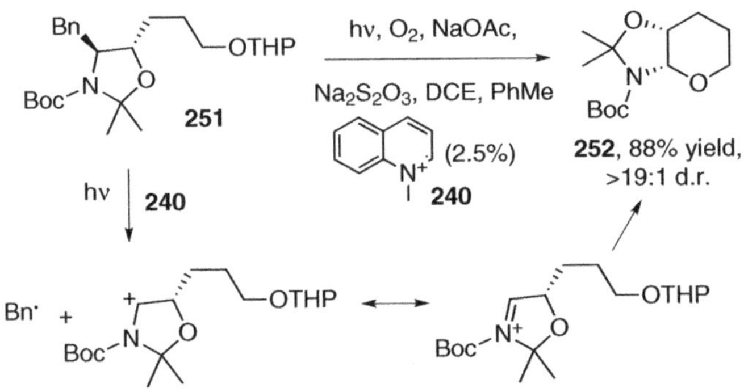

Scheme 5.103.

Acyliminium ions could likewise be generated by single-electron oxidation of homobenzylic amides and carbamates (**251** or **253**) providing (*N*,*O*)-acylaminals (**252** or **254a,b**) by intramolecular cyclization onto a (protected) OH group (Schemes 5.103, 5.104) [343, 344]. Starting from more eleborated structures allowed for the synthesis of the amido trioxadecalin core (compound **256**) by oxidative diastereoselective cyclization of **255** (Scheme 5.105) [345].

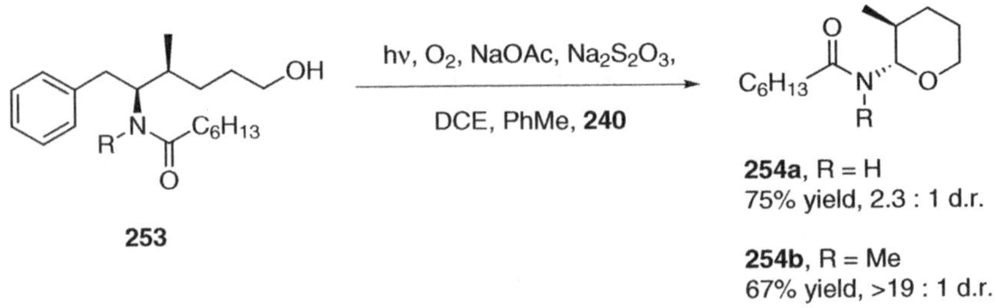

Scheme 5.104.

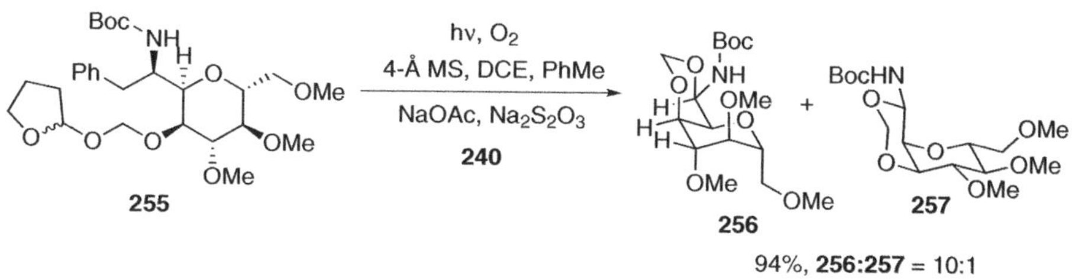

Scheme 5.105.

5.7 OXIDATION

Oxygenation upon photocatalysis via electron transfer has a rather general scope for various classes of organic compounds. The general scheme involves electron transfer from the donor to the excited photocatalyst and from the radical anion of the latter to molecular oxygen. The combination between the thus formed radical cation and either superoxide anion or molecular oxygen leads to the chemical reaction. Many (hetero)aromatic molecules act as photocatalysts in this way, but others sensitize oxygen from the ground triplet state to the reactive singlet one. Also note that some of them operate through either of the two mechanisms, electron transfer or singlet oxygen, depending on the nature of the donor and on the reaction conditions. Distinguishing the two mechanisms is no trivial matter and has been the subject of in-depth investigation, but synthetic applications are not frequent. Most of the reported examples involves sulfides, alkenes, and three-membered cyclic compounds.

The oxidation of sulfides to sulfoxides, particularly of aromatic derivatives that react poorly with singlet oxygen, has been the subject of much mechanistic work [346–348]. The process may be accompanied by some fragmentation of the radical cation when stable radicals or ions are formed (e.g., when a cumyl group is present) [328]. Also note that in aqueous solution and by using an appropriate photocatalyst, sulfoxides are formed through a different mechanism, namely the hydration of the sulfide radical cation [349].

Alkenes, and in particular arylalkenes, are susceptible to ET photocatalyzed oxidative cleavage, e.g. tetraphenylethylene **259** gave benzophenone, reasonably via a dioxetane **260** that cleaved (Scheme 5.106) [350–353].

Upon ET photocatalysis, three-membered rings cleaved and added oxygen to form dioxolanes from cyclopropanes, ozonides from epoxides, and dioxazolidines from aziridines [354–356]. In most cases, these reactions proceeded via a radical chain process and the quantum yield was found to be much larger than 1, provided that back electron transfer was minimized by adding a cocatalyst. This was an aromatic hydrocarbon, typically biphenyl or a triphenyl, that participated to the process and increased the generation of free radical cations by secondary electron transfer (Scheme 5.107) [357]. Furthermore, 1,1-diarylalkenes added a neutral molecule forming, as mentioned above, dimeric distonic radical cations. These were likewise trapped by molecular oxygen and gave 1,2-dioxolanes

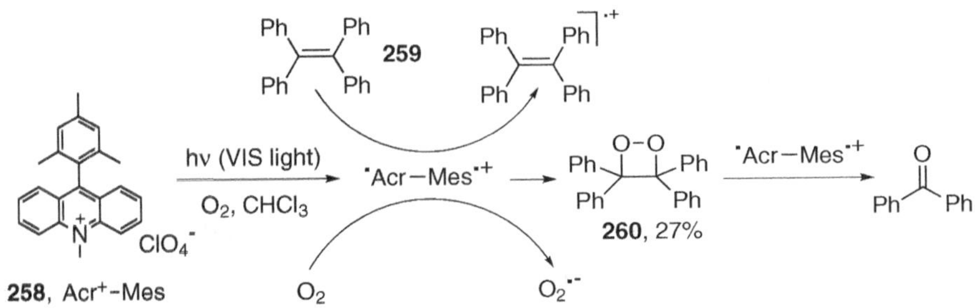

Scheme 5.106.

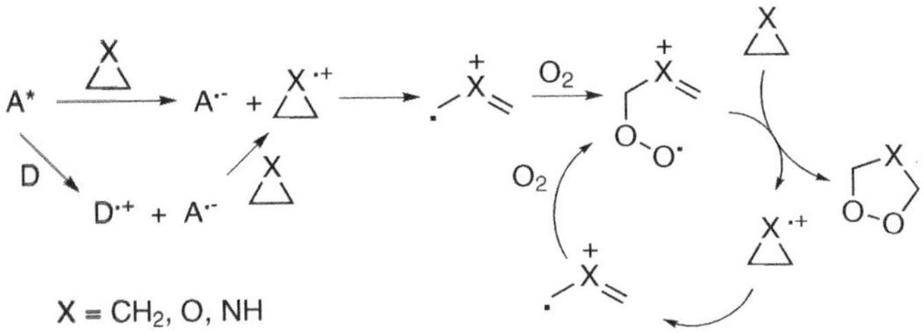

Scheme 5.107.

through a chain mechanism in a good yield, at least for electron-rich derivatives [358].

A few other oxidation processes should be mentioned. The epoxidation of alkenes by using water as both electron and oxygen donor has been obtained under photocatalysis by a ruthenium complex with hexachlorplatinate as a sacrificial donor (see Chapter 1 for details of the mechanism) [359].

Moreover, graphite-like carbon nitride (g-C_3N_4) has been used as the photocatalyst for the generation of phthalimidoyl radicals that were responsible for the oxidation of α-pinene (**59**) to the α,β-unsaturated ketone **261** (Scheme 5.108) [360].

Finally, the dehydrogenation of a 2,5-dihydrothiophene (**262**) to the corresponding aromatic compound (**264a–c**) has been carried out by ET photocatalysis by a platinum complex **263**$^{+}$ followed by deprotonation (Scheme 5.109) [361].

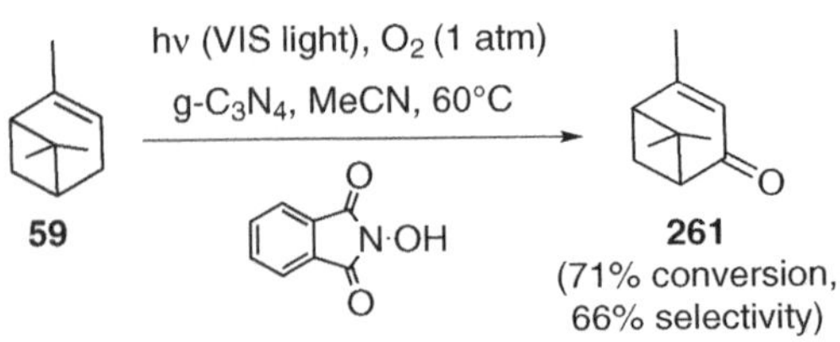

Scheme 5.108.

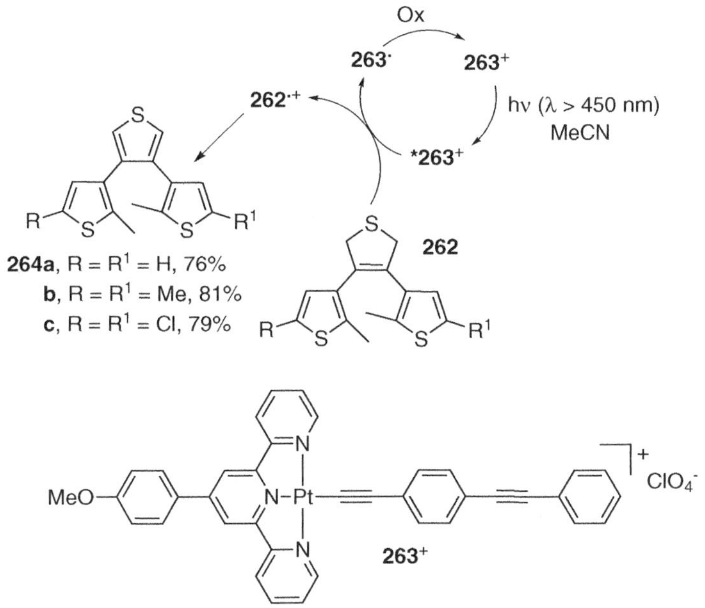

Scheme 5.109.

REFERENCES

1. Forbes, M. D. E., Ed. *Carbon-Centered Free Radicals and Radical Cations. Structure, Reactivity, and Dynamics*, John Wiley & Sons, Hoboken, NJ, 2010.
2. Albini, A., Fasani, E., Mella, M. (1993). PET-reactions of aromatic compounds. *Topics in Current Chemistry*, **168**, 143–173.
3. Schmittel, M., Burghart, A. (1997). Understanding reactivity patterns of radical cations. *Angewandte Chemie International Edition in English*, **36**, 2550–2589.
4. Mella, M., Fagnoni, M., Freccero, M., Fasani, E., Albini. A. (1998). New synthetic method via cation fragmentation. *Chemical Society Reviews*, **27**, 81–89.
5. Albini, A., Fagnoni, M., Mella, M. In Fabbrizzi, L., Poggi, A., Eds. *Photoinduced Electron Transfer: Perspectives in Organic Synthesis. Chemistry at the Beginning of the Third Millennium: Molecular Design, Supramolecules, Nanotechnology and Beyond*, Springer, Berlin, 2000, pp. 83–101.
6. Pincock, J. A. (2003). The 30 year anniversary of a seminal paper on radical ions in solution (radical ions in photochemistry. I. The 1,1-diphenylethylene cation radical). *Canadian Journal of Chemistry*, **81**, 413–416.
7. Hoffmann, N. (2008). Efficient photochemical electron transfer sensitization of homogeneous organic reactions. *Journal of Photochemistry and Photobiology, C: Photochemistry*, **9**, 43–60.
8. Poniatowski, A. J., Floreancig, P. E. Radical cation fragmentation reactions in organic synthesis. In Forbes, M. D. E., Ed. *Wiley Series on Reactive Intermediates in Chemistry*

and Biology, 3: *Carbon-Centered Free Radicals and Radical Cations: Structure, Reactivity, and Dynamics*, John Wiley & Sons, Hoboken, NJ, 2010, pp. 43–60.

9. Ischay, M. A., Yoon, T. P. (2012). Accessing the synthetic chemistry of radical ions. *European Journal of Organic Chemistry*, 3359–3372.
10. Rossi, R. A., Peñéñory, A. B. (2006). Strategies in synthetic radical organic chemistry. Recent advances on cyclization and $S_{RN}1$ reactions. *Current Organic Synthesis*, **3**, 121–158.
11. Rossi, R. A., Peñéñory, A. B. The photostimulated $S_{RN}1$ process: Reaction of haloarenes with carbanions. In Horspool, W., Lenci, F., Eds. *Handbook of Organic Photochemistry and Photobiology*, second edition, CRC Press, Boca Raton, FL, 2004.
12. (a) Rossi, R. A., Photoinduced aromatic nucleophilic substitution reactions. In *Molecular and Supramolecular Photochemistry. Synthetic Organic Photochemistry*, Vol. 12, Griesbeck, A. G., Mattay, J., Eds. Marcel Dekker, New York, 2005, pp. 495–527.
13. Budén, M. E., Martín, S. E., Rossi, R. A. Recent advances in the photoinduced radical nucleophilic substitution reactions. In Griesbeck, A., Oelgemoeller, M., Ghetti, F., Eds. *CRC Handbook of Organic Photochemistry and Photobiology*, third edition, CRC Press, Boca Raton, FL, 2012, pp. 347–367.
14. Peñeñory, A. B., Argüello, J. E. Aromatic and heteroaromatic substitution by $S_{RN}1$ and S_N1 reactions. In Albini, A., Fagnoni, M., Eds. *Handbook of Synthetic Photochemistry*, Wiley-VCH, Weinheim, 2010, pp. 319–352.
15. Mangion, D., Arnold, D. R. (2002). Photochemical nucleophile–olefin combination, aromatic substitution reaction. Its synthetic development and mechanistic exploration. *Accounts of Chemical Research*, **35**, 297–304.
16. Fagnoni, M., Albini, A. Photonucleophilic substitution reactions. In Ramamurthy, V., Schanze, K., Eds. *Molecular and Supramolecular Photochemistry*, Vol. 14: *Organic Photochemistry and Photophysics*, Marcel Dekker, New York, 2006, pp. 131–178.
17. Mangion, D., Arnold, D. R. The photochemical nucleophile–olefin combination, aromatic substitution (photo-NOCAS) reaction. In Lenci, F., Horspool, W., Eds. *CRC Handbook of Organic Photochemistry and Photobiology*, second edition, CRC Press, Boca Raton, FL, 2004, pp. 40-1–40-17.
18. Ishii, H., Imai, Y., Hirano, T., Maki, S., Niwa, H., Ohashi, M. (2000). Isoquinoline formation by a novel photochemical intermolecular cycloaddition of 1,1-diphenylethene derivatives to 1,4-dicyanobenzene. *Tetrahedron Letters*, **41**, 6467–6471.
19. Chow, Y. L., Wang, S.-S., Johansson, C. I., Liu, Z.-L. (1996). Mechanistic studies on the dual reaction pathways of singlet excited dibenzoyl(methanato)boron difluoride ($DBMBF_2$): Reactions of the excimer and exciplexes. *Journal of the American Chemical Society*, **118**, 11725–11732.
20. Kim, B. D., Kim, G. S., Tu, B., Shim, S. C. (2001). Photoaddition of 1,4-diphenyl-1, 3-butadiyne to dimethyl fumarate: A novel exciplex mechanism. *Tetrahedron Letters*, **42**, 2369–2372.
21. Mattay, J., Vondenhof, M. (1991). Contact and solvent-separated radical ions in organic photochemistry. *Topics in Current Chemistry*, **159**, 219–255.
22. Russell, G. A., Danen, W. C. (1966). Coupling reactions of the 2-nitro-2-propyl anion. *Journal of the American Chemical Society*, **88**, 5663–5665.
23. Kornblum, N., Michael, R. E., Kerber, R. C. (1966). Chain reactions in substitution processes which proceed via radical–anion intermediates. *Journal of the American Chemical Society*, **88**, 5662–5663.

24. Guastavino, J. F., Rossi, R. A. (2009). Synthesis of 1,1-bis(trimethylstannyl)cyclopropanes by the $S_{RN}1$ mechanism. *Organometallics*, **28**, 2646–2649.

25. Szabo, R., Crozet, M. D., Vanelle, P. (2008). Original $S_{RN}1$ reactions on new non-nitrated heterocyclic system. *Synlett*, **18**, 2836–2840.

26. Maillard-Boyer, M., Castera-Ducros, C., Verhaeghe, P., Sifredi, F., Rathelot, P., Vanelle, P. (2010). Access to original vinylic chlorides in the quinazoline series via a monoelectronic transfer reaction approach. *Molecules*, **15**, 2719–2729.

27. Zink, L., Crozet, M. D., Terme, T., Vanelle, P. (2011). Long distance-$S_{RN}1$ in nitroimidazole series favored by temperature. *Tetrahedron Letters*, **52**, 6991–6996.

28. Yoon, T. P., Ischay, M. A., Du, J. (2010). Visible light photocatalysis as a greener approach to photochemical synthesis. *Nature Chemistry*, **2**, 527–532.

29. Narayanam, J. M. R., Stephenson, C. R. J. (2011). Visible light photoredox catalysis: Applications in organic synthesis. *Chemical Society Reviews*, **40**, 102–113.

30. Teply, F. (2011). Photoredox catalysis by $[Ru(bpy)_3]^{2+}$ to trigger transformations of organic molecules. Organic synthesis using visible-light photocatalysis and its 20th century roots. *Collection of Czechoslovak Chemical Communications*, **76**, 859–917.

31. Tucker, J. W., Stephenson, C. R. J. (2012). Shining light on photoredox catalysis: Theory and synthetic applications. *The Journal of Organic Chemistry*, **77**, 1617–1622.

32. Xuan, J., Xiao, W.-J. (2012). Visible-light photoredox catalysis. *Angewandte Chemie International Edition*, **51**, 6828–6838.

33. Maity, S., Zheng, N. (2012). A photo touch on amines: New synthetic adventures of nitrogen radical cations. *Synlett*, **23**, 1851–1856.

34. Shi, L., Xia, W. (2012). Photoredox functionalization of C–H bonds adjacent to a nitrogen atom. *Chemical Society Reviews*, **41**, 7687–7697.

35. Nicewicz, D. A., MacMillan, D. W. C. (2008). Merging photoredox catalysis with organocatalysis: The direct asymmetric alkylation of aldehydes. *Science*, **322**, 77–80.

36. Nagib, D. A., Scott, M. E., MacMillan, D. W. C. (2009). Enantioselective α-trifluoromethylation of aldehydes via photoredox organocatalysis. *Journal of the American Chemical Society*, **131**, 10875–10877.

37. Ravelli, D., Fagnoni, M. (2012). Dyes as visible light photoredox organocatalysts. *ChemCatChem*, **4**, 169–171.

38. Ravelli, D., Fagnoni, M., Albini, A. (2013). Photoorganocatalysis. What for? *Chemical Society Reviews*, **42**, 97–113.

39. Neumann, M., Fueldner, S., Koenig, B., Zeitler, K. (2011). Metal-free, cooperative asymmetric organophotoredox catalysis with visible light. *Angewandte Chemie International Edition*, **50**, 951–954.

40. Zhou, D.-Y., Dou, H.-Y., Zhao, C.-X., Chen, Q.-Y. (2006). A novel example of vinylic $S_{RN}1$: Reactions of 3,3-difluoro-1-iodo-2-phenylcyclopropene with thiolate ions. *Journal of Fluorine Chemistry*, **127**, 740–745.

41. Roche, M., Terme, T., Vanelle, P. (2012). First long-distance $S_{RN}1$ on a propargylic chloride. *Tetrahedron Letters*, **53**, 4184–4187.

42. Rossi, R. A., Penenory, A. B. (2006). Strategies in synthetic radical organic chemistry. Recent advances on cyclization and $S_{RN}1$ reactions. *Current Organic Synthesis*, **3**, 121–158.

43. Pierini, A. B., Baumgartner, M. T., Rossi, R. A. (1987). Photostimulated reactions of haloarenes with 2-naphtylamide ions. A facile synthesis of 1-aryl-2-naphthylamines. *Tetrahedron Letters*, **28**, 4653–4656.

44. Bardagi, J. I., Rossi, R. A. (2010). Short access to 6-substituted pyrimidine derivatives by the $S_{RN}1$ mechanism. Synthesis of 6-substituted uracils through a one-pot procedure. *The Journal of Organic Chemistry*, **75**, 5271–5277.

45. Corsico, E. F., Rossi, R. A. (2000). Reactions of trimethylstannyl ions with mono-, di- and trichloro-substituted aromatic substrates by the $S_{RN}1$ mechanism. *Synlett*, 227–229.

46. Silbestri, G. F., Lo Fiego, M. J., Lockhart, M. T., Chopa, A. B. (2010). Selective synthesis of mono- and distannylpyridines from chloropyridinols via an $S_{RN}1$ mechanism. *Journal of Organometallic Chemistry*, **695**, 2578–2585.

47. Santiago, A. N., Basso, S. M., Montanez, J. P., Rossi, R. A. (2006). Trimethylstannylation of mono- and dichloroarenes by the $S_{RN}1$ mechanism in liquid ammonia. *Journal of Physical Organic Chemistry*, **19**, 829–835.

48. Layman, W. J., Jr., Greenwood, T. D., Downey, A. L., Wolfe, J. F. (2005). Synthesis of 2*H*-1,2-benzothiazine 1,1-dioxides via heteroannulation reactions of 2-iodobenzenesulfonamide with ketone enolates under $S_{RN}1$ conditions. *The Journal of Organic Chemistry*, **70**, 9147–9155.

49. Szabo, R., Crozet, M. D., Vanelle, P. (2008). Synthesis of new 3-nitroimidazo[1,2-*a*] pyridine derivatives by $S_{RN}1$ reactions. *Heterocycles*, **75**, 2263–2274.

50. Baumgartner, M. T., Pierini, A. B., Rossi, R. A. (1992). The reactivity of oxygen nucleophiles with aryl radicals in the $S_{RN}1$ mechanism. *Tetrahedron Letters*, **33**, 2323–2326.

51. Tempesti, T. C., Pierini, A. B., Baumgartner, M. T. (2005). A different route to the synthesis of 9,10-disubstituted phenanthrenes. *The Journal of Organic Chemistry*, **70**, 6508–6511.

52. D'Auria, M., De Luca, E., Mauriello, G., Racioppi, R. (1998). Photochemical synthesis of 4(5)-nitro-2-arylimidazoles. *Journal of the Chemical Society, Perkin Transactions 1*, 271–274.

53. D'Auria, M., De Mico, A., D'Onofrio, F., Piancatelli, G. (1987). Photochemical synthesis of bithienyl derivatives. *Journal of the Chemical Society, Perkin Transations 1*, 1777–1780.

54. Elisei, F., Latterini, L., Aloisi, G. G., D'Auria, M. (1995). Photoinduced substitution reactions in halothiophene derivatives. Steady state and laser flash photolytic studies. *Journal of Physical Chemistry*, **99**, 5365–5372.

55. Bardagì, J. I., Vaillard, S. E., Rossi, R. A. (2007). Synthesis of 4 substituted chromanes and 1 substituted benzo[*f*]chromanes by tandem 6-*exo trig* cyclization-$S_{RN}1$ reactions. *ARKIVOC*, **4**, 73–83.

56. Vaillard, S. E., Postigo, A., Rossi, R. A. (2004). Fast tin-free hydrodehalogenation and reductive radical cyclization reactions: A new reduction process. *The Journal of Organic Chemistry*, **69**, 2037–2041.

57. Laha, J. K., Barolo, S. M., Rossi, R. A., Cuny, G. D. (2011). Synthesis of carbolines by photostimulated cyclization of anilinohalopyridines. *The Journal of Organic Chemistry*, **76**, 6421–6425.

58. Vaillard, V. A., Guastavino, J. F., Budén, M. E., Bardagí, J. I., Barolo, S. M., Rossi, R. A. (2012). Synthesis of 6-substituted 2-pyrrolyl and indolyl benzoxazoles by intramolecular

O-arylation in photostimulated reactions. *The Journal of Organic Chemistry*, **77**, 1507–1519.

59. Javier, F., Guastavino, J. F., Rossi, R. A. (2012). Synthesis of benzo-fused heterocycles by intramolecular α-arylation of ketone enolate anions. *The Journal of Organic Chemistry*, **77**, 460–472.
60. Buden, M. E., Rossi, R. A. (2007). Syntheses of phenanthridines and benzophenanthridines by intramolecular ortho-arylation of aryl amide ions with aryl halides via $S_{RN}1$ reactions. *Tetrahedron Letters*, **48**, 8739–8742.
61. Barolo, S. M., Teng, X., Cuny, G. D., Rossi, R. A. (2006). Syntheses of aporphine and homoaporphine alkaloids by intramolecular ortho-arylation of phenols with aryl halides via $S_{RN}1$ reactions in liquid ammonia. *The Journal of Organic Chemistry*, **71**, 8493–8499.
62. Chen, Q. Y., Li, Z. T. (1993). Pentafluorophenylation of aromatics with pentafluorophenyl perfluoro- and polyfluoroalkanesulfonates. A Photoinduced electron-transfer cation diradical coupling process. *The Journal of Organic Chemistry*, **58**, 2599–2604.
63. Fagnoni, M. (2003). Photoinduced electron transfer reactions in heterocyclic chemistry. *Heterocycles*, **60**, 1921–1958.
64. Freccero, M., Pratt, A., Albini, A., Long, C. (1998). A kinetic evaluation of carbon–hydrogen, carbon–carbon, and carbon–silicon bond activation in benzylic radical cations. *Journal of the American Chemical Society*, **120**, 284–297.
65. Kyushin, S., Masuda, Y., Matsushita, K., Nakadaira, Y., Ohashi, M. (1990). Novel alkylation of aromatic nitriles via photo-induced electron transfer of group 14 metal–carbon σ donors. *Tetrahedron Letters*, **31**, 6395–6398.
66. Fagnoni, M., Mella, M., Albini, A. (1994). Scope and mechanism of the electron transfer photoinduced alkylation of an aromatic nitrile. *Tetrahedron*, **50**, 6401–6410.
67. Vanossi, M., Mella, M., Albini. A. (1994). 2 + 2 + 2 Cycloaddition vs radical ion chemistry in the photoreactions of 1,2,4,5-benzenetetracarbonitrile with alkenes in acetonitrile. *Journal of the American Chemical Society*, **116**, 10070–10075.
68. Sulpizio, A., Albini, A., d'Alessandro, N., Fasani, E., Pietra, S. (1989). Photochemical reaction between 1,4-naphthalenedicarbonitrile and α-substituted benzylic derivatives. *Journal of the American Chemical Society*, **111**, 5773–5777.
69. Albini, A., Fasani, E., Freccero, M. (1996). The photochemical reaction between arenenitriles and benzylic donors. *Advances in Electron Transfer Chemistry*, **5**, 103–140.
70. Tsujimoto, K., Nakao, N., Ohashi, M. (1992). Electron-donating behaviour of aliphatic carboxylic acids in the photoreaction with 1,2,4,5-tetracyanobenzene. *Journal Chemical Society Chemical Communications*, 366–367.
71. Fasani, E., Peverali, D., Albini, A. (1994). Alkyl radicals from *t*-butyl esters through photoinduced electron transfer. *Tetrahedron Letters*, **35**, 9275–9278.
72. Fasani, E., Mella, M., Albini, A. (1995). Aliphatic radicals from ethers via photoinduced electron transfer: Selective formation and chemistry. *Journal of the Chemical Society, Perkin Transations 2*, 449–452.
73. Mella, M., Freccero, M., Albini, A. (1994). Photochemical reaction of benzene-1,2,4,5-tetracarbonitrile with the ketals of cyclic and bicyclic ketones. *The Journal of Organic Chemistry*, **59**, 1047–1052.
74. Lewis, F. D., Zebrowski, B. E., Correa, P. E. (1984). Photochemical reactions of arenecarbonitriles with aliphatic amines. 1. Effect of arene structure on aminyl vs.

α-aminoalkyl radical formation. *Journal of the American Chemical Society*, **106**, 187–193.

75. Mella, M., d'Alessandro, N., Freccero, M., Albini. A. (1993). Photochemical reaction of arenecarbonitriles in the presence of alkylsilanes, silyl ethers and silyl amines. *Journal of the Chemical Society, Perkin Transations 2*, 515–519.
76. Mizuno, K., Nishiyama, T., Takahashi, N., Inoue. H. (1996). Photoinduced C3-unit introduction to electron-deficient molecules by use of cyclopropanone silyl acetals via electron transfer. *Tetrahedron Letters*, **37**, 2975–2978.
77. Lan, J. Y., Schuster. G. B. (1985). Photoalkylation of dicyanoarenes with alkyltriphenylborate salts. *Journal of the American Chemical Society*, **107**, 6710–6711.
78. Mizuno, K., Ikeda, M., Otsuji, Y. (1985). A novel photosubstitution of dicyanobenzenes by allylic and benzylic silanes. *Tetrahedron Letters*, **26**, 461–464.
79. Mizuno, K., Nakanishi, K., Otsuji, Y. (1988). Photosubstitution of dicyanobenzenes by group 14 organometallic compounds via photoinduced electron-transfer. Additive and medium effects on photoinduced electron transfer reaction. *Chemistry Letters*, **17**, 1833–1836.
80. Mella, M., Freccero, M., Albini, A. (1996). The photochemical approach to the functionalization of open-chain and cyclic alkanes: 1. Single electron transfer oxidation. *Tetrahedron*, **52**, 5533–5548.
81. Mella, M., Freccero, M., Soldi, T., Fasani, E., Albini, A. (1996). Oxidative functionalization of adamantane and some of its derivatives in solution. *The Journal of Organic Chemistry*, **61**, 1413–1422.
82. Tsuji, M., Higashiyama, K., Yamauchi, T., Kubo, H., Ohmiya, S. (2001). Photosubstitution reaction of cyanoaromatics with aliphatic amides. *Heterocycles*, **54**, 1027–1032.
83. Fagnoni, M., Vanossi, M., Mella, M., Albini, A. (1996). Limits to oxidation of organic substrates. SET oxidative processes of commonly used solvents as revealed through the photochemical reaction with 1,2,4,5-benzenetetracarbonitrile. *Tetrahedron*, **52**, 1785–1796.
84. Torriani, R., Mella, M., Fasani, E., Albini, A. (1993). On the mechanism of the photochemical reaction between 1,4-dicyanobenzene and 2,3-dimethylbutene in the presence of nucleophiles. *Tetrahedron*, **53**, 2573–2580.
85. Nishiyama, T., Mizuno, K., Otsuji, Y., Inoue, H. (1995). Photoinduced electron transfer reaction of mono- and dicyanonaphthalenes with allyltrimethylsilane: Synthesis of benzotricyclo[4.2.1.$0^{3,8}$]nonenes by reductive photoallylation and intramolecular ($2\pi + 2\pi$) photocycloaddition. *Tetrahedron*, **51**, 6695–6706.
86. Tamai, T., Mizuno, K., Hashida, I., Otsuji, Y. (1993). Photoinduced electron-transfer reactions of arylmethyl-substituted 14 group compounds: Photoarylmethylation and photooxygenation. *Bulletin of the Chemical Society of Japan*, **66**, 3747–3754.
87. Lan, J. Y., Schuster, G. B. (1986). Free radical formation in the photooxidative alkylations of dicyanonaphthalene with alkyltriphenylborate salts. *Tetrahedron Letters*, **27**, 4261–4264.
88. Fukuzumi, S., Ohkubo, K., Suenobu, T., Kato, K., Fujitsuka, M., Ito, O. (2001). Photoalkylation of 10-alkylacridinium ion via a charge-shift type of photoinduced electron transfer controlled by solvent polarity. *Journal of the American Chemical Society*, **123**, 8459–8467.

89. Mella, M., Fagnoni, M., Albini, A. (1994). Radicals through photoinduced electron transfer. Addition to olefin and addition to olefin–aromatic substitution reactions. *The Journal of Organic Chemistry*, **59**, 5614–5622.

90. Fagnoni, M., Mella, M., Albini, A. (1995). Radical addition to alkenes via electron transfer photosensitization. *Journal of the American Chemical Society*, **117**, 7877–7881.

91. Mangion, D., Arnold, D. R., Cameron, T. S., Robertson, K. N. (2001). The electron transfer photochemistry of allenes with cyanoarenes. Photochemical nucleophile–olefin combination, aromatic substitution (photo-NOCAS) and related reactions. *Journal of the Chemical Society, Perkin Transations 2*, 48–60.

92. McManus, K. A., Arnold, D. R. (1995). The photochemical nucleophile–olefin combination, aromatic substitution (photo-NOCAS) reaction. Part 10: intramolecular reactions involving alk-4-enols and 1,4-dicyanobenzene. *Canadian Journal of Chemistry*, **73**, 2158–2169.

93. Lu, Z.-F., Yue, J.-J., Zhu, Y., Hu, H.-W., Xu, J.-H. (2009). Photo-multicomponent reactions leading to the construction of isocoumarins and large ring lactone precursors. *Photochemical & Photobiological Sciences*, **8**, 217–223.

94. Arnold, D. R., Snow, M. S. (1988). The photochemical nucleophile–olefin combination, aromatic substitution reaction (Part 2): methanol–cyclic olefins, 1,4-dicyanobenzene. *Canadian Journal of Chemistry*, **66**, 3012–3026.

95. Arnold, D. R., Chan, M. S. W., McManus. K. A. (1996). Photochemical nucleophile–olefin combination, aromatic substitution (photo–NOCAS) reaction, Part 12. Factors controlling the regiochemistry of the reaction with alcohol as the nucleophile. *Canadian Journal of Chemistry*, **74**, 2143–2166.

96. de Lijser, H. J. P., Arnold, D. R. (1997). Radical ions in photochemistry. 44. The photo-NOCAS reaction with acetonitrile as the nucleophile. *The Journal of Organic Chemistry*, **62**, 8432–8438.

97. Chan, M. S. W., Arnold, D. R. (1997). Photochemical nucleophile–olefin combination, aromatic substitution (photo-NOCAS) reaction, Part 15. Investigations involving fluoride anion as the nucleophile and the effect of fluorine substitution on the relative stability of the reaction intermediates. *Canadian Journal of Chemistry*, **75**, 1810–1819.

98. Arnold, D. R., McManus, K. A., Chan. M. S. W. (1997). Photochemical nucleophile–olefin combination, aromatic substitution (photo-NOCAS) reaction, Part 13. The scope and limitations of the reaction with cyanide anion as the nucleophile. *Canadian Journal of Chemistry*, **75**, 1055–1075.

99. Arnold, D. R., Du, X. (1994). The photochemical nucleophile–olefin combination, aromatic substitution (photo-NOCAS) reaction. Part 5: Methanol–monoterpenes (α- and β-pinene, tricyclene, and nopol), 1,4-dicyanobenzene. *Canadian Journal of Chemistry*, **72**, 403–414.

100. Arnold, D. R., Du, X., de Lijser, H. J. P. (1995). The photochemical nucleophile–olefin combination, aromatic substitution (photo-NOCAS) reaction. Part 8: Methanol–2-carene, and 1,4-dicyanobenzene. *Canadian Journal of Chemistry*, **73**, 522–530.

101. Yasuda, M., Shiragami, T., Matsumoto, J., Yamashita, T., Shima, K. Photoamination with ammonia and amines. In Ramamurthy, V., Schanze, K., Eds. *Molecular and Supramolecular Photochemistry*, Vol. 14: *Organic Photochemistry and Photophysics*, Marcel Dekker, New York, 2006, pp. 207–253.

102. Yasuda, M., Kojima, R., Tsutsui, H., Utsunomiya, D., Ishii, K., Jinnouchi, K., Shiragami, T., Yamashita, T. (2003). Redox-photosensitized aminations of 1,2-benzo-1,3-cycloalkadienes, arylcyclopropanes, and quadricyclane with ammonia. *The Journal of Organic Chemistry*, **68**, 7618–7624.

103. Yamashita, T., Itagawa, J., Sakamoto, D., Nakagawa, Y., Matsumoto, J., Shiragami, T., Yasuda, M. (2007). Redox-photosensitized amination of alkenes and alkadienes with ammonia and alkylamines. *Tetrahedron*, **63**, 374–380.

104. Ohashi, M., Maeda, H., Mizuno, K. (2010). Photochemical formal alkadiene insertion into an aromatic C–CN bond using cyanide ion as a catalyst. *Chemistry Letters*, **39**, 462–463.

105. Ohashi, M., Nakatani, K., Maeda, H., Mizuno, K. (2008). Photoinduced tandem three-component coupling of propanedinitrile, 2,5-dimethylhexa-2,4-diene, and cyanoarenes. *The Journal of Organic Chemistry*, **73**, 8348–8351.

106. Matsumoto, J., Yoshinaga, Y., Hamasaki, A., Kawasaki, T., Yamashita, T., Shiragami, T., Yasuda, M. (2011). Temperature dependence of photosensitized nucleophilic addition to alkenes. *Bulletin of the Chemical Society of Japan*, **84**, 1130–1132.

107. Lu, Z.-F., Shen, Y.-M., Yue, J.-J., Hu, H.-W., Xu, J.-H. (2008). Photoinduced three-component reactions of tetracyanobenzene with alkenes in the presence of 1,3-dicarbonyl compounds as nucleophiles. *The Journal of Organic Chemistry*, **73**, 8010–8015.

108. Zhang, M., Lu, Z.-F., Liu, Y., Grampp, G., Hu, H.-W., Xu, J.-H. (2006). Photoreactions of 1,2,4,5-benzenetetracarbonitrile with arylethenes—Photo-olefin dimerization–aromatic substitution reactions. *Tetrahedron*, **62**, 5663–5674.

109. Ishii, H., Imai, Y., Hirano, T., Maki, S., Niwa, H., Ohashi, M. (2000). Isoquinoline formation by a novel photochemical intermolecular cycloaddition of 1,1-diphenylethene derivatives to 1,4-dicyanobenzene. *Tetrahedron Letters*, **41**, 6467–6471.

110. Mangion, D., Frizzle, M., Arnold, D. R., Cameron, T. S. (2001). The photochemistry of acrylonitrile with methoxylated naphthalenes: Introducing the photochemical electrophile–olefin combination, aromatic substitution (photo-EOCAS) reaction. *Synthesis*, 1215–1222.

111. Kitagawa, F., Murase, M., Kitamura, N. (2002). A mechanistic study of photocyanation of pyrene in oil-in-water emulsion systems. *The Journal of the Organic Chemistry*, **67**, 2524–2531.

112. Ohkubo, K., Kobayashi, T., Fukuzumi, S. (2012). Photocatalytic alkoxylation of benzene with 3-cyano-1-methylquinolinium ion. *Optics Express*, **20** (S2), A360–A365.

113. Caronna, T., Gambarotti, C., Palmisano, L., Punta, C., Recupero, F. (2003). Sunlight induced functionalisation of some heterocyclic bases in the presence of polycrystalline TiO_2. *Chemical Communications*, 2350–2351.

114. Xuhui Ju, X., Liang, Y., Jia, P., Li, W., Yu, W. (2012). Synthesis of oxindoles *via* visible light photoredox catalysis. *Organic and Biomolecular Chemistry*, **10**, 498–501.

115. Rao, C. N., Hoz, S. (2012). Photostimulated reduction of nitriles by SmI_2. *The Journal of Organic Chemistry*, **77**, 4029–4034.

116. Hintz, S., Heidbreder, A., Mattay, J. (1996). Radical ion cyclizations. *Topics in Current Chemistry*, **177**, 77–124.

117. Pandey, G. (1993). Photoinduced electron transfer (PET) in organic synthesis. *Topics in Current Chemistry*, **168**, 175–221.

118. Cermenati, L., Dondi, D., Fagnoni, M., Albini, A. (2003). Titanium dioxide photocatalysis of adamantane. *Tetrahedron*, **59**, 6409–6414.
119. Fagnoni, M., Mella, M., Albini. A. (1998). Electron-transfer-photosensitized conjugate alkylation. *The Journal of Organic Chemistry*, **63**, 4026–4033.
120. Ellison, D. K., Trulove, P. C., Iwamoto, R. T. (1986). Photoelectrochemical synthesis with Pt-coated TiO_2 powder: Alkylation of 1-alkylpyridinium ion. *Tetrahedron*, **42**, 6405–6410.
121. Fagnoni, M., Mella, M., Albini, A. (1995). Alkylation of alkenes by radicals generated through photoinduced single electron transfer. *Tetrahedron*, **51**, 859–864.
122. Fagnoni, M., Mella, M., Albini, A. (1997). Tandem energy transfer-electron transfer in the photosensitized alkylation of α,β-unsaturated ketones. *Journal of Physical Organic Chemistry*, **10**, 777–780.
123. Mizuno, K., Nishiyama, T., Takahashi, N., Inoue, H. (1996). Photoinduced C3-unit introduction to electron-deficient molecules by use of cyclopropanone silyl acetals via electron transfer. *Tetrahedron Letters*, **37**, 2975–2978.
124. Abe, M., Nojima, M., Oku, A. (1996). Photoinduced electron transfer reactions of cyclopropanone acetal with conjugated enones in the presence of a redox-type photosensitizer. *Tetrahedron Letters*, **37**, 1833–1836.
125. Oku, A., Miki, T., Abe, M., Ohira, M., Kamada, T. (1999). Donor–acceptor binary photocatalysis system illustrated by the intramolecular cyclization of a cyclopropanone acetal tethered to cyclohexen-2-one. *Bulletin of the Chemical Society of Japan*, **72**, 511–517.
126. Rinderhagen, H., Waske, P. A., Mattay, J. (2006). Facile ring opening of siloxycyclopropanes by photoinduced electron transfer. A new way to β-keto radicals. *Tetrahedron*, **62**, 6589–6593.
127. Mella, M., Fagnoni, M., Albini, A. (1999). Benzyl radicals from toluene by phosensitization with naphthalene-1,4-dicarbonitrile—Benzylation and hydroxymethylation of unsaturated compounds. *European Journal of Organic Chemistry*, 2137–2142.
128. Montanaro, S., Ravelli, D., Merli, D., Fagnoni, M., Albini, A. (2012). Decatungstate as photoredox catalyst: Benzylation of electron-poor olefins. *Organic Letters*, **14**, 4218–4221.
129. Cermenati, L., Mella, M., Albini, A. (1998). Titanium dioxide photocatalysed alkylation of maleic acid derivatives. *Tetrahedron*, **54**, 2575–2582.
130. Cermenati, L., Albini, A. (2002). Titanium dioxide photocatalysis for radical alkylation. *Journal of Advanced Oxidation Technologies*, **5**, 58–66.
131. Cermenati, L., Fagnoni, M., Albini, A. (2003). TiO_2-photocatalyzed reactions of some benzylic donors. *Canadian Journal of Chemistry*, **81**, 560–566.
132. Cermenati, L., Richtie, C., Albini, A. (1998). Solar light induced carbon–carbon bond formation via TiO_2 photocatalysis. *Chemical Communications*, 805–806.
133. Mizuno, K., Hayamizu, T., Maeda, H. (2003). Regio- and stereoselective functionalization of electron-deficient alkenes by organosilicon compounds via photoinduced electron transfer. *Pure and Applied Chemistry*, **75**, 1049–1054.
134. Hayamizu, T., Maeda, H., Ikeda, M., Mizuno, K. (2001). Highly stereoselective carbon-functionalization of electron-deficient arylalkenes by use of organosilicon compounds via photoinduced electron transfer. *Tetrahedron Letters*, **42**, 2361–2364.
135. Hayamizu, T., Ikeda, M., Maeda, H., Mizuno, K. (2001). Photoallylation and photoreduction of cyclohexylidenepropanedinitrile by use of allyltrimethylsilane via

photoinduced electron transfer: Control of the product ratio depending on pK_a values of additives. *Organic Letters*, **3**, 1277–1280.

136. Hayamizu, T., Maeda, H., Mizuno, K. (2004). Diastereoselective protonation on radical anions of electron-deficient alkenes via photoinduced electron transfer. *The Journal of Organic Chemistry*, **69**, 4997–5004.
137. Maity, S., Zhu, M., Shinabery, R. S., Zheng, N. (2012). Intermolecular [3 + 2] cycloaddition of cyclopropylamines with olefins by visible-light photocatalysis. *Angewandte Chemie International Edition*, **51**, 222–226.
138. Mikami, K., Matsumoto, S., Ishida, A., Takamuku, S., Suenobu, T., Fukuzumi, S. (1995). Addition of ketene silyl acetals to the triplet excited state of C_{60} via photoinduced electron transfer leading to the fullerene acetates. *Journal of the American Chemical Society*, **117**, 11134–11141.
139. Fukuzumi, S., Suenobu, T., Fujitsuka, M., Ito, O., Tonoi, T., Matsumoto, S., Mikami, K. (1999). Addition of group 14 organometallic compounds to C_{60} via photoinduced electron transfer. Direct detection of radical ion pair intermediates. *Journal of Organometallic Chemistry*, **574**, 32–39.
140. Gutenberger, G., Blechert, S., Steckan, E. (2000). α-Cyano-γ-lactones by photoinduced electron transfer-catalyzed oxidation of α-hydroxyalkylsilanes in the presence of α-cyano acrylates. *Tetrahedron Letters*, **41**, 461–464.
141. Gutenberger, G., Steckhan, E., Blechert, S. (1998). α-Silyl ethers as hydroxymethyl anion equivalents in photoinduced radical electron transfer additions. *Angewandte Chemie International Edition*, **37**, 660–662.
142. Bertrand, S., Glapski, C., Hoffmann, N., Pete, J.-P. (1999). Highly efficient photochemical addition of tertiary amines to electron deficient alkenes. Diastereoselective addition to (5*R*)-5-menthyloxy-2[5*H*]-furanone. *Tetrahedron Letters*, **40**, 3169–3174.
143. Bertrand, S., Hoffmann, N., Pete, J.-P. (2000). Highly efficient and stereoselective radical addition of tertiary amines to electron-deficient alkenes—application to the enantioselective synthesis of necine bases. *European Journal of Organic Chemistry*, 2227.
144. Hoffmann, N., Bertrand, S., Marinković, S., Pesch, J. (2006). Efficient radical addition of tertiary amines to alkenes using photochemical electron transfer. *Pure and Applied Chemistry*, **78**, 2227–2246.
145. Bertrand, S., Hoffmann, N., Pete, J.-P., Bulach, V. (1999). Stereoselective radical-tandem reaction of aniline derivatives with (5*R*)-5-menthyloxy-2,5-dihydrofuran-2-one initiated by photochemical induced electron transfer. *Chemical Communications*, 2291–2292.
146. Bertrand, S., Hoffmann, N., Humbel, S., Pete, J.-P. (2000). Diastereoselective tandem addition-cyclization reactions of unsaturated tertiary amines initiated by photochemical electron transfer (PET). *The Journal of Organic Chemistry*, **65**, 8690–8703.
147. Marinković, S., Brulè, C., Hoffmann, N., Prost, E., Nuzillard, J.-M., Bulach, V. (2004). Origin of chiral induction in radical reactions with the diastereoisomers (5*R*)- and (5*S*)-5-*l*-menthyloxyfuran-2[5*H*]-one. *The Journal of Organic Chemistry*, **69**, 1646–1651.
148. Das, S., Kumar, J. S. D., Thomas, K. G., Shivaramayya, K., George, M. V. (1994). Photocatalyzed multiple additions of amines to α,β-unsaturated esters and nitriles. *The Journal of Organic Chemistry*, **59**, 628–634.
149. Das. S., Kumar, J. S. D., Shivaramayya, K., George, M. V. (1995). Anthraquinone-photocatalysed addition of amines to α,β-unsaturated esters: A novel route to

indolizidone, pyrrolizidone and related ring systems. *Journal of the Chemical Society, Perkin Transations* **1**, 1797–1799.

150. Das, S., Kumar, J. S. D., Shivaramayya, K., George, M. V. (1996). Formation of lactams via photoelectron-transfer catalyzed reactions of *N*-allylamines with α,β-unsaturated esters. *Tetrahedron*, **52**, 3425–3434.
151. Kohls, P., Jadhav, D., Pandey, G., Reiser, O. (2012). Visible light photoredox catalysis: Generation and addition of *N*-aryltetrahydroisoquinoline-derived α-amino radicals to Michael acceptors. *Organic Letters*, **14**, 672–675.
152. Marinković, S., Hoffmann, N. (2001). Efficient radical addition of tertiary amines to electron-deficient alkenes using semiconductors as photochemical sensitisers. *Chemical Communications*, 1576–1578.
153. Marinković, S., Hoffmann, N. (2004). Diastereoselective radical tandem addition-cyclization reactions of aromatic tertiary amines by semiconductor-sensitized photochemical electron transfer. *European Journal of Organic Chemistry*, 3102–3107.
154. Harakat, D., Pesch, J., Marinković, S., Hoffmann, N. (2006). Thiocarbonyl compounds as regulating reagent in the radical addition of tertiary amines with alkenes using photoelectron transfer conditions. *Organic and Biomolecular Chemistry*, **4**, 1202–1205.
155. Yoon, U. C., Mariano, P. S. (1992). Mechanistic and synthetic aspects of amine-enone single electron transfer photochemistry. *Account of Chemical Research*, **25**, 233–240.
156. Hasegawa, E., Xu, W., Mariano, P. S., Yoon, U.-C., Kim, J.-U. (1988). Electron-transfer-induced photoadditions of the silyl amine Et_2NCH_2TMS to α,β-unsaturated cyclohexenones. Dual reaction pathways based on ion-pair-selective cation-radical chemistry. *Journal of the American Chemical Society*, **110**, 8099–8111.
157. Xu, W., Zhang, X.-M., Mariano, P. S. (1991). Single electron transfer promoted photocyclization reactions of (aminoalkyl)cyclohexenones. Mechanistic and synthetic features of processes involving the generation and reactions of amine cation and α-amino radicals. *Journal of the American Chemical Society*, **113**, 8863–8878.
158. Jeon, T. J., Lee, C.-P., Mariano, P. S. (1991). Radical cyclization reactions of α-silyl amine α,β-unsaturated ketone and ester systems promoted by single electron transfer photosensitization. *Journal of the American Chemical Society*, **113**, 8847–8863.
159. Jonas, M., Blechert, S., Steckhan, E. (2001). Photochemically induced electron transfer (PET) catalyzed radical cyclization: A practical method for inducing structural changes in peptides by formation of cyclic amino acid derivatives. *The Journal of Organic Chemistry*, **66**, 6896–6904.
160. Meggers, E., Steckhan, E., Blechert, S. (1995). Radical C–C bond formation by photoinduced electron transfer addition of α-silyl carbamates to acceptor-substituted alkenes. *Angewandte Chemie International Edition in English*, **34**, 2137–2139.
161. Kanemoto, M., Ankyu, H., Wada, Y., Yanagida, S. (1992). Visible-light induced photofixation of CO_2 into benzophenone catalyzed by colloidal CdS microcrystallites. *Chemistry Letters*, 2113–2114.
162. Ogata, T., Hiranaga, K., Matsuoka, S., Wada, Y., Yanagida, S. (1993). Visible-light induced photocatalytic fixation of CO_2 into benzophenone using poly(*p*-phenylene) as a photocatalyst. *Chemistry Letters*, **22**, 983–984.
163. Fujiwara, H., Kanemoto, M., Ankyu, H., Murakoshi, K., Wada, Y., Yanagida, S. (1997). Visible-light induced photofixation of carbon dioxide into aromatic ketones and benzyl

halides catalysed by CdS nanocrystallites. *Journal of Chemical Society Perkin Transations 2*, 317–322.

164. Maroulis, A. J., Shigemitsu, Y., Arnold, D. R. (1978). Radical ions in photochemistry. 5. Photosensitized (electron transfer) cyanation of olefins. *Journal of the American Chemical Society*, **100**, 535–541.
165. Arnold, D. R., Maroulis, A. J. (1977). Radical ions in photochemistry. 4. The 1,1-diphenylethylene anion radical by photosensitization (electron transfer). *Journal of the American Chemical Society*, **99**, 7355–7356.
166. Pirtsch, M., Paria, S., Matsuno, T., Isobe, H., Reiser, O. (2012). [Cu(dap)$_2$Cl] As an efficient visible-light-driven photoredox catalyst in carbon–carbon bond-forming reactions. *Chemistry—A European Journal*, **18**, 7336–7340.
167. Courant, T., Masson, G. (2012). Photoredox-initiated α-alkylation of imines through a three-component radical/cationic reaction. *Chemistry—A European Journal*, **18**, 423–427.
168. Larraufie, M.-H., Pellet, R., Fensterbank, L., Goddard, J.-P., Lacôte, E., Malacria, M., Ollivier, C. (2011). Visible-light-induced photoreductive generation of radicals from epoxides and aziridines. *Angewandte Chemie International Edition*, **50**, 4463–4466.
169. Park, H. J., Yoon, U. C., Lee, H.-Y., Cho, D. W., Mariano, P. S. (2012). Photoaddition reactions of acetylpyridines with silyl ketene acetals: SET vs [2 + 2]-cycloaddition pathways. *The Journal of Organic Chemistry*, **77**, 10304–10313.
170. Pandey, G., Gadre, S. R. (2004). Generation and mesolytic dynamics of organoselenane and selenosilane radical ions: Development of mechanistically interesting and synthetically useful chemistry. *Accounts of Chemical Research*, **37**, 201–210.
171. Pandey, G., Sekhar, B. B. V. S., Bhalerao, U. T. (1990). Photoinduced single electron transfer initiated heterolytic carbon–selenium bond dissociation. Sequential one-pot selenenylation and deselenenylation reaction. *Journal of the American Chemical Society*, **112**, 5650–5651.
172. Pandey, G., Sekhar, B. B. V. S. (1993). Enyne cyclization via photoinduced electron transfer (PET) generated electrophilic selenium species: A new carbon–carbon bond formation strategy. *Journal of the Chemical Society, Chemical Communications*, 780–782.
173. Pandey, G., Sekhar, B. B. V. S. (1994). Photoinduced electron transfer initiated activation of organoselenium substrates as carbocation equivalents: Sequential one-pot selenylation and deselenylation reaction. *The Journal of Organic Chemistry*, **59**, 7367–7372.
174. Pandey, G., Sekhar, B. B. V. S. (1995). Enyne cyclisation by photoinduced electron transfer (PET) promoted in situ generated electrophilic selenium species: A new carbocyclisation strategy. *Tetrahedron*, **51**, 1483–1494.
175. Pandey, G., Rao, K. S. S. P., Rao, K. V. N. (1996). Photosensitized electron transfer promoted reductive activation of carbon–selenium bonds to generate carbon-centered radicals: Application for unimolecular group transfer radical reactions. *The Journal of Organic Chemistry*, **61**, 6799–6804.
176. Oelgemoeller, M., Bunte, J. O., Mattay, J. Photoinduced electron transfer cyclizations via radical ions. In Griesbeck, A. G., Mattay, J., Eds. *Molecular and Supramolecular Photochemistry*, Vol. 12: *Synthetic Organic Photochemistry*, Marcel Dekker, New York, 2005, pp. 269–297.
177. Heidbreder, A., Mattay, J. (1992). PET-oxidative cyclization of unsaturated enol silyl ethers. *Tetrahedron Letters*, **33**, 1973–1976.

178. Hintz, S., Mattay, J., van Eldik, R., Fu, W.-F. (1998). Regio- and stereoselective cyclization reactions of unsaturated silyl enol ethers by photoinduced electron transfer—Mechanistic aspects and synthetic approach. *European Journal of Organic Chemistry*, 1583–1596.

179. Bunte, J. O., Heilmann, E. K., Hein, B., Mattay, J. (2004). Cyclizations of silyl enol ether radical cations—The cause of the stereoselectivity. *European Journal of Organic Chemistry*, 3535–3550.

180. Bunte, J. O., Rinne, S., Schäfer, C., Neumann, B., Stammler, H.-G., Mattay, J. (2003). Synthesis and PET oxidative cyclization of silyl enol ethers: Build-up of quasi-steroidal carbocycles. *Tetrahedron Letters*, **44**, 45–48.

181. Pandey, G., Krishna, A., Girjia, K., Karthikeyan, M. (1993). Intramolecular nucleophilic addition of silylenol ether to photosensitized electron transfer (PET) generated arene radical cations: A novel non-reagent based carboannulation reaction. *Tetrahedron Letters*, **34**, 6631–6634.

182. Pandey, G., Karthikeyan, M., Murugan, A. (1998). New intramolecular α-arylation strategy of ketones by the reaction of silyl enol ethers to photosensitized electron transfer generated arene radical cations: Construction of benzannulated and benzospiroannulated compounds. *The Journal of Organic Chemistry*, **63**, 2867–2872.

183. Pandey, G., Murugan, A., Balakrishnan, M. (2002). A new strategy towards the total synthesis of phenanthridone alkaloids: Synthesis of (+)-2,7-dideoxypancratistatin as a model study. *Chemical Communications*, 624–625.

184. Pandey, G., Sochanchingwung, R. (1994). Photoinduced electron transfer (PET) promoted cross-coupling of organoselenium and organosilicon compounds: A new carbon–carbon bond formation strategy. *Journal of the Chemical Society, Chemical Communications*, 1945–1946.

185. Rinderhagen, H., Mattay, J. (2004). Synthetic applications in radical/radical cationic cascade reactions. *Chemistry—A European Journal*, **10**, 851–874.

186. Waske, A. P., Mattay, J. (2005). Synthesis of cyclopropyl silyl ethers and their facile ring opening by photoinduced electron transfer as key step in radical/radical cationic cascade reactions. *Tetrahedron*, **61**, 10321–10330.

187. Hoffmann, U., Gao, Y., Pandey, B., Klinge, S., Warzecha, K.-D., Krüger, C., Roth, H. D., Demuth, M. (1993). Light-induced polyene cyclizations via radical cations in micellar medium. *Journal of the American Chemical Society*, **115**, 10358–10359.

188. Warzecha, K.-D., Xing, X., Demuth, M. (1997). Cyclization of terpenoid polyalkenes via photoinduced electron transfer-versatile single-step syntheses of mono- and polycycles. *Pure and Applied Chemistry*, **69**, 109–112.

189. Warzecha, K.-D., Xing, X., Demuth, M., Goddard, R., Kessler, M., Krüger, C. (1995). Cascade cyclizations of terpenoid polyalkenes triggered by photoelectron transfer—biomimetics with photons. *Helvetica Chimica Acta*, **78**, 2065–2076.

190. Görner, H., Warzecha, K.-D., Demuth, M. (1997). Cyclization of terpenoid dicarbonitrile polyalkenes upon photoinduced electron transfer to 1,4-dicyano-2,3,5,6-tetramethylbenzene and other cyanoarenes. *Journal of the Physical Chemistry A*, **101**, 9964–9973.

191. Heinemann, C., Xing, X., Warzecha, K.-D., Ritterskamp, P., Görner, H., Demuth, M. (1998). An asymmetric induction principle and biomimetics with photons via electron transfer. *Pure and Applied Chemistry*, **70**, 2167–2176.

192. Heinemann, C., Demuth, M. (1997). Biomimetic cascade cyclizations of terpenoid polyalkenes *via* photoinduced electron transfer. Long-distance asymmetric induction by a chiral auxiliary. *Journal of the American Chemical Society*, **119**, 1129–1130.
193. Heinemann, C., Demuth, M. (1999). Short biomimetic synthesis of a steroid by photoinduced electron transfer and remote asymmetric induction. *Journal of the American Chemical Society*, **121**, 4894–4895.
194. Xing, X., Demuth, M. (1999). An efficient formal total synthesis of (±)-stypoldione via photochemically triggered biomimetic cyclizations of terpenoid polyalkenes. *Synlett*, 987–990.
195. Xing, X., Demuth, M. (2001). Application of photoinduced biomimetic cascade cyclizations of terpenoid polyalkenes for the synthesis of (±)-stypoldione. *European Journal of Organic Chemistry*, 537–544.
196. Goeller, F., Heinemann, C., Demuth, M. (2001). Investigations of cascade cyclizations of terpenoid polyalkenes via radical cations. A biomimetic-type synthesis of (±)-3-Hydroxy-spongian-16-one. *Synthesis*, 1114–1116.
197. Ozser, M. E., Icil, H., Makhynya, Y., Demuth, M. (2004). Electron-transfer-initiated cascade cyclizations of terpenoid polyalkenes in a low-polarity solvent: One-step synthesis of mono- and polycylic terpenoids with various functionalities. *European Journal of Organic Chemistry*, 3686–3692.
198. Weng, H., Scarlata, C., Roth, H. D. (1996). Electron transfer photochemistry of geraniol and (*E,E*)-farnesol. A novel "tandem," 1,5-cyclization, intramolecular capture. *Journal of the American Chemical Society*, **118**, 10947–10953.
199. Roth, H. D., Herbertz, T., Sauers, R. R., Weng, H. (2006). Intramolecular nucleophilic capture of radical cations by tethered hydroxy functions. *Tetrahedron*, **62**, 6471–6489.
200. Pandey, G., Hajra, S. (1994). A novel photosystem for harvesting visible light to drive photoinduced electron transfer (PET) reductions: β-Activation of α,β-unsaturated ketones for radical cyclizations. *Angewandte Chemie International Edition*, **33**, 1169–1171.
201. Pandey, G., Hajra, S., Ghorai, M. K. (1994). Visible light initiated photosensitised electron transfer (PET) reductive β-activation of α,β-unsaturated ketones for radical cyclisation: A new concept in promoting radical reactions. *Tetrahedron Letters*, **35**, 7837–7840.
202. Pandey, G., Hajra, S., Ghorai, M. K., Kumar, K. R. (1997). Designing photosystems for harvesting photons into electrons by sequential electron-transfer processes: Reversing the reactivity profiles of α,β-unsaturated ketones as carbon radical precursor by one electron reductive β-activation. *Journal of the American Chemical Society*, **119**, 8777–8787.
203. Pandey, G., Hajra, S., Ghorai, M. K. (1997). Visible light initiated photosensitized electron transfer cyclizations of aldehydes and ketones to tethered α,β-unsaturated esters: Stereoselective synthesis of optically pure *C*-furanosides. *The Journal of Organic Chemistry*, **62**, 5966–5973.
204. Pandey, G., Ghorai, M. K., Hajra, S. (1998). A new strategy for the construction of carbo- and oxycycles by intramolecular reductive coupling of α,β-unsaturated esters. *Tetrahedron Letters*, **39**, 1831–1834.
205. Pandey, G., Gaikwad, A. L., Gadre, S. R. (2006). Chiral 6-phenyl-2,3-bismethylene-methoxycarbonyl-[1,4]-dioxane as a designer synthon for an efficient and short synthesis

of optically pure 2,6-dioxabicyclo[3.3.0]octane-3,7-dione. *Tetrahedron Letters*, **47**, 701–703.

206. Griesbeck, A. G., Kramer, W., Oelgemöller, M. (1999). Synthetic applications of photoinduced electron transfer decarboxylation reactions. *Synlett*, 1169–1178.
207. Yoon, U. C., Mariano, P. S. (2001). The synthetic potential of phthalimide SET photochemistry. *Accounts of Chemical Research*, **34**, 523–533.
208. Oelgemöller, M., Griesbeck, A. G. (2002). Photoinduced electron transfer chemistry of phthalimdes: An efficient tool for C–C-bond formation. *Journal of Photochemistry and Photobiology C: Photochemistry Reviews*, **3**, 109–127.
209. McDermott, G., Yoo, D. J., Oelgemöller, M. (2005). Photochemical addition reactions involving phthalimides. *Heterocycles*, **65**, 2221–2257.
210. Yoon, U. C., Mariano, P. S. Mechanistic and synthetic aspects of SET-promoted photocyclization reactions of silicon substituted phthalimides. In Ramamurthy, V., Schanze, K., Eds. *Molecular and Supramolecular Photochemistry*, Vol. 14: *Organic Photochemistry and Photophysics*, Marcel Dekker, New York, 2006, pp. 179–206.
211. Hatoum, F., Gallagher, S., Baragwanath, L., Lex, J., Oelgemöller, M. (2009). Photodecarboxylative benzylations of phthalimides. *Tetrahedron Letters*, **50**, 6335–6338.
212. Griesbeck, A. G., Oelgemöller, M. (1999). Photodecarboxylative addition of carboxylates and α-keto carboxylates to phthalimides. *Synlett*, 492–494.
213. Hatoum, F., Engler, J., Zelmer, C., Wißen, J., Motti, C. A., Lex, J., Oelgemöller, M. (2012). Photodecarboxylative addition of carboxylates to phthalimides: A concise access to biologically active 3-(alkyl and aryl)methylene-1*H*-isoindolin-1-ones. *Tetrahedron Letters*, **53**, 5573–5577.
214. Griesbeck, A. G., Oelgemöller, M., Lex, J. (2000). Photodecarboxylative additions of α-keto carboxylates to phthalimides alkylation, acylation and ring expansion. *Synlett*, 1455–1457.
215. Griesbeck, A. G., Henz, A., Peters, K., Peters, E.-M., von Schnering, H. G. (1995). Photo electron transfer induced macrocyclization of *N*-phthaloyl-ω-aminocarboxylic acids. *Angewandte Chemie International Edition*, **34**, 474–476.
216. Griesbeck, A. G., Henz, A., Kramer, W., Lex, J., Nerowski, F., Oelgemöller, M., Peters, K., Peters, E.-M. (1997). Synthesis of medium- and large-ring compounds initiated by photochemical decarboxylation of o-phthalimidoalkanoates. *Helvetica Chimica Acta*, **80**, 912–933.
217. Griesbeck, A. G., Hirt, J., Kramer, W., Dallakian, P. (1998). Photochemistry of phthaloylcysteine, its methyl ester and C-unprotected S-alkyl derivatives. *Tetrahedron*, **54**, 3169–3180.
218. Yoon, U. C., Oh, S. W., Lee, J. H., Park, J. H., Kang, K. T., Mariano, P. S. (2001). Applications of phthalimide photochemistry to macrocyclic polyether, polythioether, and polyamide synthesis. *The Journal of Organic Chemistry*, **66**, 939–943.
219. Yoon, U. C., Oh, S. W., Lee, S. M., Cho, S. J., Gamlin, J., Mariano, P. S. (1999). A solvent effect that influences the preparative utility of *N*-(silylalkyl) phthalimide and *N*-(silylalkyl) maleimide photochemistry. *The Journal of Organic Chemistry*, **64**, 4411–4418.
220. Pandey, G., Rani, K. S., Lakshmaiah, G. (1992). Direct carbon–carbon bond formation strategy at α-position of tertiary amines by photoinduced electron transfer (PET) processes. *Tetrahedron Letters*, **33**, 5107–5110.

221. Pandey, G., Kumaraswamy, G., Reddy, P. Y. (1992). Regioselective generation of iminium cation by PET processes: Its in situ trapping by intramolecular nucleophiles and synthesis of some biologically active heterocycles. *Tetrahedron*, **48**, 8295–8308.

222. Pandey, G., Gadre, S. R. (2003). Sequential two electron photooxidation of *t*-amines: Generation of a regiospecific iminium cation and its application in organic synthesis. *ARKIVOC*, **iii**, 45–54.

223. Da Silva Goes, A., Ferroud, C., Santamaria, J. (1995). Regioselective single electron transfer photocatalytic synthesis of 2-cyano-3-ethylidenepiperidines. New route to the total synthesis of (±) *cis*-eburnamonine. *Tetrahedron Letters*, **36**, 2235–2238.

224. Ferroud, C., Cavalcanti de Amorim, E. L., Dallery, L., Santamaria, J. (1994). Single electron transfer regioselective photocatalytic synthesis of 2-cyano-1,2,5,6-tetrahydropyridines. *Synthesis*, 291–294.

225. Santamaria, J., Kaddachi, M. T., Ferroud, C. (1992). Trimethylsilyl cyanide as a superior trapping agent for iminium ion intermediates in the DCA-sensitized photooxygenation of indole derivatives and analogues. Application to the total synthesis of some indoloquinolizidine alkaloids. *Tetrahedron Letters*, **33**, 781–784.

226. Santamaria, J., Kaddachi, M. T., Rigaudy, J. (1990). Electron-transfer activation. Photocyanation of tertiary amines. *Tetrahedron Letters*, **31**, 4735–4738.

227. Santamaria, J., Khuong-Huu, F. (1978). Oxydation photochimique d'amines tertiaires et d'alcaloides—IX: Oxydation photochimique d'alcaloides quinolizidiniques et indolizidiniques tetracycliques: Sparteine et α-isosparteine, lupanine et α-isolupanine, camoensidine et tetrahydroleontidine. *Tetrahedron*, **34**, 1523–1528.

228. Ferroud, C., Rool, P., Santamaria, J. (1998). Singlet oxygen mediated alkaloid tertiary amines oxidation by single electron transfer. *Tetrahedron Letters*, **39**, 9423–9426.

229. Baciocchi, E., Del Giacco, T., Lapi, A. (2006). Dual pathways for the desilylation of silylamines by singlet oxygen. *Organic Letters*, **8**, 1783–1786.

230. Santamaria, J., Kaddachi, M. T., Rigaudy, J. (1990). Electron-transfer activation. Photocyanation of tertiary amines. *Tetrahedron Letters*, **31**, 4735–4738.

231. Xuan, J., Feng, Z.-J., Duan, S.-W., Xiao, W.-J. (2012). Room temperature synthesis of isoquino[2,1-*a*][3,1]oxazine and isoquino[2,1-*a*]pyrimidine derivatives via visible light photoredox catalysis. *RSC Advances*, **2**, 4065–4068.

232. Miyake, Y., Nakajima, K., Nishibayashi, Y. (2012). Visible-light-mediated utilization of α-aminoalkyl radicals: Addition to electron-deficient alkenes using photoredox catalysts. *Journal of the American Chemical Society*, **134**, 3338–3341.

233. Santamaria, J., Herlem, D., Khuong-Huu, F. (1977). Oxydation photochimique d'amines tertiaires et d'alcaloïdes-VIII: Oxydation photochimique d'alcaloïdes indolinoindolizidiniques, vincadifformine, tabersonine, $N_{(a)}$-acétyl dihydro-2,16 vincadifformine et $N_{(a)}$-acétyl dihydro-2,16 tabersonine. *Tetrahedron*, **33**, 2389–2392.

234. Pan, Y., Wang, S., Kee, C. W., Dubuisson, E., Yang, Y., Loh, K. P., Tan, C.-H. (2011). Graphene oxide and Rose Bengal: Oxidative C–H functionalisation of tertiary amines using visible light. *Green Chemistry*, **13**, 3341–3344.

235. Hari, D. P., König, B. (2011). Eosin Y catalyzed visible light oxidative C–C and C–P bond formation. *Organic Letters*, **13**, 3852–3855.

236. Pan, Y., Kee, C. W., Chen, L., Tan, C.-H. (2011). Dehydrogenative coupling reactions catalysed by Rose Bengal using visible light irradiation. *Green Chemistry*, **13**, 2682–2685.

237. Liu, Q., Li, Y.-N., Zhang, H.-H., Chen, B., Tung, C.-H., Wu, L.-Z. (2012). Reactivity and mechanistic insight into visible-light-induced aerobic cross-dehydrogenative coupling reaction by organophotocatalysts. *Chemistry—An European Journal*, **18**, 620–627.

238. Freeman, D. B., Furst, L., Condie, A. G., Stephenson, C. R. J. (2012). Functionally diverse nucleophilic trapping of iminium intermediates generated utilizing visible light. *Organic Letters*, **14**, 94–97.

239. Fagnoni, M., Dondi, D., Ravelli, D., Albini, A. (2007). Photocatalysis for the formation of the C–C bond. *Chemical Reviews*, **107**, 2725–2756.

240. Gärtner, M., Ballmann, J., Damm, C., Heinemann, F. W., Kisch, H. (2007). Support-controlled chemoselective olefin–imine addition photocatalyzed by cadmium sulfide on a zinc sulfide carrier. *Photochemical & Photobiological Sciences*, **6**, 159–164.

241. Kisch, H., Mitoraj, D. Preparative and mechanistic semiconductor photocatalysis. In Griesbeck, A., Oelgemoeller, M., Ghetti, F., Eds. *CRC Handbook of Organic Photochemistry and Photobiology*, third edition, CRC Press, Boca Raton, FL, 2012, pp. 293–328.

242. Armesto, D., Ortiz, M. J., Agarrabeitia, A. R., Aparicio-Lara, S., Martin-Fontecha, M., Liras, M., Martinez-Alcazar, M. P. (2002). Photochemical reactivity of 1-substituted-1-aza-1,4-dienes promoted by electron-acceptor sensitizers. Di-π-methane rearrangements and alternative reactions via radical-cation intermediates. *The Journal of Organic Chemistry*, **67**, 9397–9405.

243. Armesto, D., Caballero, O., Ortiz, M. J., Agarrabeitia, A. R., Martin-Fontecha, M., Torres, M. R. (2003). Novel photoreactions of 2-aza-1,4-dienes in the triplet excited state and via radical-cation intermediates. 2-Aza-di-π-methane rearrangements yielding cyclopropylimines and *N*-vinylaziridines. *The Journal of Organic Chemistry*, **68**, 6661–6671.

244. Mattay, J. (1987). Charge transfer and radical ions in photochemistry. *Angewandte Chemie International Edition*, **26**, 825–845.

245. Bauld, N. L. (1989). Cation radical cycloadditions and related sigmatropic reactions. *Tetrahedron*, **45**, 5307–5363.

246. Mattay, J., Trampe, G., Runsink, J. (1988). Radical ions and photochemical charge transfer phenomena. Part 21. Olefin radical cation cycloadditions. *Chemische Bericte*, **121**, 1991–2005.

247. Nathan, L., Bauld, N. L., Bellville, D. J., Harirchian, B., Lorenz, K. T., Pabon, R. A., Jr., Reynolds, D. W., Wirth, D. D., Chiou, H. S., Marsh, B. K. (1987). Cation radical pericyclic reactions. *Accounts of Chemical Research*, **20**, 371–378.

248. Kuwata, S., Shigemitsu, Y., Odaira, Y. (1973). Photosensitized cyclodimerization of phenyl vinyl ethers. *The Journal of Organic Chemistry*, **38**, 3803–3805.

249. Albini, A., Spreti, S. (1986). Poly(ethylene terephthalate) as an insoluble electron transfer photosensitizer. *Journal of the Chemical Society, Chemical Communications*, 1426–1427.

250. Mizuno, K., Kagano, H., Kasuga, T., Otsuji, Y. (1983). Stereoselectivity in the photocyclodimerization of phenyl vinyl ether via electron-transfer. *Chemistry Letters*, 133–136.

251. Mattes, S. L., Luss, H. R., Farid, S. (1983). Photochemical electron-transfer reactions: Effect of ring opening of the cyclobutane radical cation on the dimerization of phenyl vinyl ether. *Journal of Physical Chemistry*, **87**, 4779–4781.

252. Draper, A. M., Ilyas, M., De Mayo, P., Ramamurthy, V. (1984). Surface photochemistry: Semiconductor-photoinduced dimerization of phenyl vinyl ether. *Journal of the American Chemical Society*, **106**, 6222–6230.

253. Ilyas, M., De Mayo, P. (1985). Surface photochemistry: Cadmium sulfide-mediated dimerization of phenyl vinyl ether. The dark reaction. *Journal of the American Chemical Society*, **107**, 5093–5099.

254. Lewis, F. D., Kojima, M. (1988). Electron-transfer-induced photoisomerization, dimerization, and oxygenation of *trans*- and *cis*-anethole. The role of monomer and dimer cation radicals. *Journal of the American Chemical Society*, **110**, 8664–8670.

255. Brancaleon, L., Brousmiche, D., Rao, V. J., Johnston, L. J., Ramamurthy, V. (1998). Photoinduced electron transfer reactions within zeolites: Detection of radical cations and dimerization of arylalkenes. *Journal of the American Chemical Society*, **120**, 4926–4933.

256. Mizuno, K., Hashizume, T., Otsuji, Y. (1983). Photocyclodimerization of alkyl, benzyl, and trimethylsilyl vinyl ethers in the presence of 1,4-dicyanonaphthalene. *Journal of the Chemical Society, Chemical Communications*, 772–773.

257. Farid, S., Shealer, S. E. (1973). Radical cations: Photochemical and ferric ion-induced formation and reactions of indene radical cation. *Journal of the Chemical Society, Chemical Communications*, 677–678.

258. Asanuma, T., Gotoh, T., Tsuchida, A., Yamamoto, M., Nishijima, Y. (1977). Photosensitized dimerization of styrene derivatives by a cation radical transfer mechanism. *Journal of the Chemical Society, Chemical Communications*, 485–486.

259. Yamamoto, M., Asanuma, T., Nishijima, Y. (1975). Photocyclodimerization of *p*-methoxystyrene in the presence of 1,4-dicyanobenzene. *Journal of the Chemical Society, Chemical Communications*, 53–54.

260. Carruthers, R. A., Crellin, R. A., Ledwith, A. (1969). Cation-radicals: Oxygen catalysis in the photosensitised cyclodimerisation of aromatic enamines. *Journal of the Chemical Society D: Chemical Communications*, 252–253.

261. Ellinger, L. P. (1964). The polymerization of vinylcarbazole by electron acceptors I. *Polymer*, **5**, 559–578.

262. Tazuke, S., Kitamura, N. (1977). Photoinduced ternary electron transfer reactions: A novel application to electron transfer sensitization. *Journal of the Chemical Society, Chemical Communications*, 515–516.

263. Al-Ekabi, H., De Mayo, P. (1986). Surface photochemistry: The CdS photoinduced dimerization of N-vinylcarbazole. *Tetrahedron*, **42**, 6277–6284.

264. Crellin, R. A., Lambert, M. C., Ledwith, A. (1970). Photochemical 2 + 2 cycloaddition via a cation–radical chain reaction. *Journal of the Chemical Society D: Chemical Communications*, 682–683.

265. Ishay, M. A., Lu, Z., Yoon, T. P. (2010). [2 + 2] Cycloadditions by oxidative visible light photocatalysis. *Journal of the American Chemical Society*, **132**, 8572–8574.

266. Mizuno, K., Kagano, H., Otsuji, Y. (1983). Regio- and stereoselective intramolecular photocycloaddition: Synthesis of macrocyclic 2,ω-dioxabicyclo[n.2.0] ring system. *Tetrahedron Letters*, **24**, 3849–3850.

267. Nakanishi, K., Mizuno, K., Otsuji, Y. (1991). A novel intramolecular [2π + 2π] photocycloaddition of di- and tetra-allylsilanes. *Journal of the Chemical Society, Chemical Communications*, 90–92.

268. Mizuno, K., Nakanishi, K., Otsuji, Y., Hayamizu, T., Maeda, H., Adachi, T., Ishida, A., Takamuku, S. (2003). Formation and cycloreversion of 2-silacyclobuta[2.3]cyclophanes via photoinduced electron transfer. *Journal of Photoscience*, **10**, 121–126.

269. Mizuno, K., Ueda, H., Otsuji, Y. (1981). Photoinduced [2 + 2] cycloaddition between aryl vinyl ether and alkyl vinyl ether via electron transfer. *Chemistry Letters*, 1237–1240.

270. Pabon, R. A., Bellville, D. J., Bauld, N. L. (1984). Selective cyclobutane adduct formation in competition with Diels–Alder addition in cation radical cycloadditions. *Journal of the American Chemical Society*, **106**, 2730–2731.

271. Bauld, N. L., Harirchian, B., Reynolds, D. W., White, J. C. (1988). A strategy for intermolecular Diels-Alder cycloaddition to enamides. *Journal of the Americal Chemical Society*, **110**, 8111–8117.

272. Ischay, M. A., Anzovino, M. E., Du, J., Yoon, T. P. (2008). Efficient visible light photocatalysis of [2 + 2] enone cycloadditions. *Journal of the American Chemical Society*, **130**, 12886–12887.

273. Wang, L.-C., Jang, H.-Y., Roh, Y., Lynch, V., Schultz, A. J., Wang, X., Krische, M. J. (2002). Diastereoselective cycloreductions and cycloadditions catalyzed by Co(dpm)$_2$-silane (dpm) 2,2,6,6-tetramethylheptane-3,5-dionate): Mechanism and partitioning of hydrometallative versus anion radical pathways. *Journal of the American Chemical Society*, **124**, 9448–9453.

274. Du, J., Yoon, T. P. (2009). Crossed intermolecular [2 + 2] cycloadditions of acyclic enones via visible light photocatalysis. *Journal of the American Chemical Society*, **131**, 14604–14605.

275. Albrecht, E., Averdung, J., Bischof, E. W., Heidbreder, A., Kirschberg, T., Müller, F., Mattay, J. (1994). Photoinduced electron transfer (PET) in organic synthesis. [3 + 2]-type cycloaddition, cyclization and C–C bond cleavage reactions. *Journal of Photochemistry and Photobiology, A: Chemistry*, **82**, 219–232.

276. Müller, F., Mattay, J. (1993). [3 + 2] Cycloadditions with azirines under the conditions of photoinduced electron transfer: A new method for the synthesis of imidazoles and heterophanes. *Chemische Berichte*, **126**, 543–549.

277. Müller, F., Mattay, J. (1992). A new synthesis for imidazolo- and pyrrolophanes by [3 + 2] cycloaddition with azaallenyl radical cations. *Angewandte Chemie, International Edition in English*, **31**, 209–210.

278. Müller, F., Karwe, A., Mattay, J. (1992). A new synthesis of porphyrin systems by four sequential [3 + 2] cycloadditions of an alkyne with azaallenyl radical cations. *The Journal of Organic Chemistry*, **57**, 6080–6082.

279. Averdung, J., Albrecht, E., Lauterwein, J., Luftmann, H., Mattay, J., Mohn, H., Müller, W. H., Meer, H. U. (1994). Photoreaktionen mit fulleren-C_{60} [3 + 2]-photocycloaddition von 2,3-diphenyl-2*H*-azirin. *Chemische Berichte*, **127**, 787–789.

280. Albini, A., Arnold, D. R. (1978). Radical ions in photochemistry. 6. The photosensitized (electron transfer) ring opening of aryloxiranes. *Canadian Journal of Chemistry*, **56**, 2985–2993.

281. Bhattacharyya, K., Das, P. K. (1999). Photosensitized ring opening in phenyl oxiranes. *Research on Chemical Intermediates*, **25**, 645–665.

282. Clawson, P., Lunn, P. M., Whiting, D. A. (1984). The photochemistry of 2,3-bis (*p*-methoxyphenyl)oxirane: Trapping of a C–C cleaved intermediate in an electron-

transfer sensitised process. *Journal of the Chemical Society, Chemical Communications*, 134–135.

283. Clawson, P., Lunn, P. M., Whiting, D. A. (1990). Synthetic studies on *O*-heterocycles via cycloadditions. Part 1. Photochemical (electron transfer sensitised) C–C cleavage of diaryloxiranes. *Journal of the Chemical Society, Perkin Transations 1*, 153–157.
284. Das, P. K., Muller, A. J., Griffin, G. W. (1984). Photoinduced electron-transfer processes involving substituted stilbene oxides. *The Journal of Organic Chemistry*, **49**, 1977–1985.
285. Tomioka, H., Kobayashi, D., Hashimoto, A., Murata, S. (1989). [3 + 2] Cycloaddition of cyclopropane with vinyl ether via photoinduced electron transfer. *Tetrahedron Letters*, **30**, 4685–4688.
286. Pandey, G., Lakshmaiah, G., Kumaraswamy, G. (1992). A new and efficient strategy for nonstabilized azomethine ylide via photoinduced electron transfer (PET) initiated sequential double desilylation. *Journal of the Chemical Society, Chemical Communications*, 1313–1314.
287. Yoon, T. P. (2013). Visible light photocatalysis: The development of photocatalytic radical ion cycloadditions. *ACS Catalysis*, **3**, 895–902.
288. Z. Lu, Z., Shen, M., Yoon, T. P. (2011). [3 + 2] Cycloadditions of aryl cyclopropyl ketones by visible light photocatalysis. *Journal of the American Chemical Society*, **133**, 1162–1164.
289. Rueping, M., Leonori, D., Poisson, T. (2011). Visible light mediated azomethine ylide formation-photoredox catalyzed [3 + 2] cycloadditions. *Chemical Communications*, **47**, 9615–9617.
290. Neunteufel, R. A., Arnold, D. R. (1973). Radical ions in photochemistry. I. 1,1-diphenylethylene cation radical. *Journal of the American Chemical Society*, **95**, 4080–4081.
291. Maroulis, A. J., Arnold, D. R. (1979). Photosensitized (electron transfer)[2*e* + 4*e*] cross-cycloaddition of alkylated olefins to phenylated olefins. *Journal of the Chemical Society, Chemical Communications*, 351–352.
292. Arnold, D. R., Borg, R. M., Albini, A. (1981). Photosensitized (electron transfer)[2*e* + 4*e*] dimerization and cross-cycloaddition of phenylated olefins: Trapping the intermediate. *Journal of the Chemical Society, Chemical Communications*, 138–139.
293. Hartsough, D., Schuster, G. B. (1989). The triplex Diels–Alder reaction: Stereospecific addition of methylstyrenes to 1,3-cyclohexadiene. *The Journal of Organic Chemistry*, **54**, 3–4.
294. Akbulut, N., Hartsough, D., Kim, J. I., Schuster, G. B. (1989). The triplex Diels–Alder reaction of 1,3-dienes with enol, alkene, and acetylenic dienophiles: Scope and utility. *The Journal of Organic Chemistry*, **54**, 2549–2556.
295. Kim, J. I., Schuster, G. B. (1990). Enantioselective catalysis of the triplex Diels–Alder reaction: Addition of trans-β-methylstyrene to 1,3-cyclohexadiene photosensitized with (−)-1,1′-bis(2,4-dicyanonaphthalene). *Journal of the American Chemical Society*, **112**, 9635–9637.
296. Calhoun, G. C., Schuster, G. B. (1986). Triplex catalyzed Diels–Alder reactions: Addition of indene to 1,3-cyclohexadiene. *Tetrahedron Letters*, **27**, 911–914.
297. Jones, C. R., Allman, B. J., Mooring, A., Spahic, B. (1983). Photosensitized [4 + 2] cyclodimerization of 1,3-cyclohexadiene. *Journal of the American Chemical Society*, **105**, 652–654.

298. Mattay, J., Gersdorf, J., Mertes, J. (1985). A new type of electron acceptor for Diels–Alder reactions via radical cations. *Journal of the Chemical Society, Chemical Communications*, 1088–1090.

299. Mlcoch, J., Steckhan, E. (1985). Photochemically initiated, electron-transfer-catalyzed Diels–Alder reactions of electron-rich dienophiles. *Angewandte Chemie International Edition in English*, **24**, 412–414.

300. Mattay, J., Vondenhof, M., Denig, R. (1989). Pyrylium salts as photosensitizers in homogeneous and heterogeneous electron-transfer catalysis.—A comparison with cyano arenes. *Chemische Berichte*, **122**, 951–958.

301. Fornés, V., Garcia, H., Miranda, M. A., Mojarrad, F., Sabater, M.-J., Suliman, N. N. E. (1996). Zeolite-based heterogeneous photosensitizers containing triphenylpyrylium and dibenzotropylium cations. Modifications of the product selectivity in the cyclodimerization of 1,3-cyclohexadiene. *Tetrahedron*, **52**, 7755–7760.

302. Pabon, R. A., Bellville, D. J., Bauld, N. L. (1983). Cation radical Diels–Alder reactions of electron-rich dienophiles. *Journal of the American Chemical Society*, **105**, 5158–5159.

303. Harirchian, B., Bauld, N. L. (1989). Cation radical Diels–Alder cycloadditions in organic synthesis. A formal total synthesis of (−)-β-selinene. *Journal of the American Chemical Society*, **111**, 1826–1828.

304. Shishi, L., Ischay, M. A., Fry, C. G., Yoon, T. P. (2011). Radical cation Diels–Alder cycloadditions by visible light photocatalysis. *Journal of the American Chemical Society*, **133**, 19350–19353.

305. Lin, S., Padilla, C. E., Ischay, M. A., Yoon, T. P. (2012). Visible light photocatalysis of intramolecular radical cation Diels–Alder cycloadditions. *Tetrahedron Letters*, **53**, 3073–3076.

306. Gieseler, A., Steckhan, E., Wiest, O. (1990). Photoinduced, electron-transfer-catalyzed Diels–Alder reaction between indole and 1,3-cyclohexadienes. *Synlett*, 275–276.

307. Gieseler, A., Steckhan, E., Wiest, O., Knoch, F. (1991). Photochemically induced radical-cation Diels–Alder reaction of indole and electron-rich dienes. *The Journal of Organic Chemistry*, **56**, 1405–1411.

308. Wiest, O., Steckhan, E., Grein, F. (1992). Selectivity in radical cation Diels–Alder reactions of indole and electron-rich dienes—A semiempirical approach. *The Journal of Organic Chemistry*, **57**, 4034–4037.

309. Haberl, U., Steckhan, E., Blechert, S., Wiest, O. (1999). Electron-transfer-induced Diels–Alder reactions of indole and exocyclic dienes: Synthesis and quantum-chemical studies. *Chemistry—An European Journal*, **5**, 2859–2865.

310. Peglow, T., Blechert, S., Steckhan, E. (1999). Radical cation Diels–Alder reaction between indoles and exocyclic 1,3-dienes with incorporated intentional cleaving points. *Chemical Communications*, 433–434.

311. Wiest, O., Steckhan, E. (1993). Electron transfer-catalyzed Diels–Alder reactions with 2-vinylindoles. *Angewandte Chemie, International Edition in English*, **32**, 901–903.

312. Wiest, O., Steckhan, E. (1993). Radical cation Diels–Alder reaction of indoles and exocyclic dienes. *Tetrahedron Letters*, **34**, 6391–6394.

313. Gürtler, C. F., Steckhan, E., Blechert, S. (1996). [4 + 2]-Cycloaddition reactions between β-acceptor-substituted enamines and 2-vinylindole radical cations acting as heterodienes. *The Journal of Organic Chemistry*, **61**, 4136–4143.

314. Gürtler, C. F., Blechert, S., Steckhan, E. (1994). Formation of pyrido[1,2*a*]indoles by electron transfer catalyzed Diels–Alder reaction between 2-vinylindoles and β-enamidoesters. *Synlett*, 141–142.

315. Benson, S. C., Palabrica, C. A., Snyder, J. K. (1987). Indole as a dienophile in inverse electron demand Diels–Alder reactions. *5H*-Pyridazino[4,5-*b*]indoles as cycloadducts with 3,6-dicarbomethoxy-1,2,4,5-tetrazine. *The Journal of Organic Chemistry*, **52**, 4610–4614.

316. Botzem, J., Haberl, U., Steckhan, E., Blechert, S. (1998). Radical cation cycloaddition reactions of 2-vinylbenzofurans and 2-vinylfurans by photoinduced electron transfer. *Acta Chemica Scandinava*, **52**, 175–193.

317. Mizuno, K., Kaji, R., Okada, H., Otsuji, Y. (1978). Addition of furans to electron-rich olefins through photo-induced electron-transfer in the presence of electron acceptors. *Journal of the Chemical Society, Chemical Communications*, 594–595.

318. Wölfle, I., Chan, S., Schuster, G. B. (1991). The triplex Diels–Alder reaction: Intramolecular cycloaddition of phenyl-substituted alkenes to 1,3-dienes. *The Journal of Organic Chemistry*, **56**, 7313–7319.

319. Majima, T., Pac, C., Nakasone, A., Sakurai, H. (1981). Redox-photosensitized reactions. 7. Aromatic hydrocarbon-photosensitized electron-transfer reactions of furan, methylated furans, 1,1-diphenylethylene, and indene with *p*-dicyanobenzene. *Journal of the American Chemical Society*, **103**, 4499–4508.

320. Mattes, S. L., Farid, S. (1983). Photosensitized electron-transfer reactions. Interception of the geminate radical ion pair. *Journal of the American Chemical Society*, **105**, 1386–1387.

321. Calhoun, G. C., Schuster, G. B. (1984). Radical cation and triplex Diels–Alder reactions of 1,3-cyclohexadiene. *Journal of the American Chemical Society*, **106**, 6870–6871.

322. Calhoun, G. C., Schuster, G. B. (1986). The triplex Diels–Alder reaction of indene and cyclic dienes. *Journal of the American Chemical Society*, **108**, 8021–8027.

323. Al-Ekabi, H., De Mayo, P. (1987). Surface photochemistry. The cadmium sulfide mediated reactions of 1,1-di-*p*-anisylethylene. *The Journal of Organic Chemistry*, **52**, 4756–4759.

324. Akbulut, N., Schuster, G. B. (1988). Triplex-catalyzed Diels–Alder and [2 + 2] cycloaddition reactions of enol ethers and ketene acetals. *Tetrahedron Letters*, **29**, 5125–5128.

325. Mattes, S. L., Farid, S. (1980). Photosensitized electron-transfer reactions of phenylacetylene. *Journal of the Chemical Society, Chemical Communications*, 126–128.

326. Hurtley, A. E., Cismesia, M. A., Ischay, M. A., Yoon, T. P. (2011). Visible light photocatalysis of radical anion hetero-Diels–Alder cycloadditions. *Tetrahedron*, **67**, 4442–4448.

327. Percivalle, C., La Rosa, A., Verga, D., Doria, F., Mella, M., Palumbo, M., Di Antonio, M., Freccero, M. (2011). Quinone methide generation via photoinduced electron transfer. *The Journal of Organic Chemistry*, **76**, 3096–3106.

328. Bonesi, S. M., Fagnoni, M., Albini, A. (2008). Electron transfer oxidation of sulfides: Structure and medium effect. *Journal of Sulfur Chemistry*, **29**, 367–376.

329. Baciocchi, E., Del Giacco, T., Lanzalunga, O., Mencarelli, P., Procacci, B. (2008). Photosensitized oxidation of alkyl phenyl sulfoxides. C–S bond cleavage in alkyl phenyl sulfoxide radical cations. *The Journal of Organic Chemistry*, **73**, 5675–5682.

330. Yoshimi, Y., Itou, T., Hatanaka, M. (2007). Decarboxylative reduction of free aliphatic carboxylic acids by photogenerated cation radical. *Chemical Communications*, 5244–5246.

331. Kensuke Shima, K., Sazaki, A., Nakabayashi, K., Yasuda, M. (1992). Ring-splitting reaction of 2,2-diarylthietane by electron-transfer photosensitization of 9,10-dicyanoanthracene. *Bulletin of the Chemical Society of Japan*, **65**, 1472–1474.

332. Pérez-Ruiz, R., Sáez, J. A., Domingo, L. R., Jiménez, M. C., Miranda, M. A. (2012). Ring splitting of azetidin-2-ones via radical anions. *Organic and Biomolecular Chemistry*, **10**, 7928–7932.

333. Mizuno, K., Ichinose, N., Tamai, T., Otsuji, Y. (1992). Insertion of nitrogen oxide and nitrosonium ion into the cyclopropane ring: A new route to 2-isoxazolines and its mechanistic studies. *The Journal of Organic Chemistry*, **57**, 4669–4675.

334. Popielarz, R., Arnold, D. R. (1990). Radical ions in photochemistry. Carbon–carbon bond cleavage of radical cations in solution: Theory and application. *Journal of the American Chemical Society*, **112**, 3068–3082.

335. Ishiguro, K., Osaki, T., Sawaki, Y. (1992). A novel role of methanol in in-cage reaction of radical cations during photosensitized C–C bond cleavage of 1,1,2,2-tetraphenylethane. *Chemistry Letters*, 743–746.

336. Kumar, V. S., Floreancig, P. E. (2001). Electron transfer initiated cyclizations: Cyclic acetal synthesis through carbon–carbon σ-bond activation. *Journal of the American Chemical Society*, **123**, 3842–3843.

337. Kumar, V. S., Aubele, D. L., Floreancig, P. E. (2001). Aerobic organocatalytic photoinitiated arene oxidations: Application to electron transfer initiated cyclization reactions. *Organic Letters*, **3**, 4123–4125.

338. Seiders II, J. R., Wang, L., Floreancig, P. E. (2003). Tuning reactivity and chemoselectivity in electron transfer initiated cyclization reactions: Applications to carbon–carbon bond formation. *Journal of the American Chemical Society*, **125**, 2406–2407.

339. Wang, L., Seiders II, J. R., Floreancig, P. E. (2004). Structure–reactivity relationships in oxidative carbon–carbon bond forming reactions: A mild and efficient approach to stereoselective syntheses of 2,6-disubstituted tetrahydropyrones. *Journal of the American Chemical Society*, **126**, 12596–12603.

340. Liu, H., Wan, S., Floreancig, P. E. (2005). Oxidative cyclorelease from soluble polymeric supports. *The Journal of Organic Chemistry*, **70**, 3814–3818.

341. Kumar, V. S., Wan, S., Aubele, D. L., Floreancig, P. E. (2006). Oxidatively generated electrophiles as initiators of epoxide cascade cyclization processes. *Tetrahedron: Asymmetry*, **16**, 3570–3578.

342. Clausen, D. J., Wan, S., Floreancig, P. E. (2011). Total synthesis of the protein phosphatase 2A inhibitor lactodehydrothyrsiferol. *Angewandte Chemie International Edition*, **50**, 5178–5181.

343. Aubele, D. L., Floreancig, P. E. (2002). Oxidative synthesis of cyclic acyl aminals through carbon–carbon σ-bond activation. *Organic Letters*, **4**, 3443–3446.

344. Aubele, D. L., Rech, J. C., Floreancig, P. E. (2004). Catalytic aerobic generation of acyliminium ions through electron-transfer-mediated carbon–carbon bond activation. *Advanced Synthesis and Catalysis*, **346**, 359–366.

345. Rech, J. C., Floreancig, P. E. (2003). An oxidative entry into the amido trioxadecalin ring system. *Organic Letters*, **5**, 1495–1498.

346. Clennan, E. L., Liao, C. (2008). The role of sulfide radical cations in electron transfer promoted molecular oxygenations at sulfur. *Journal of the American Chemical Society*, **130**, 4057–4068.

347. Bonesi, S. M., Manet, I., Freccero, M., Fagnoni, M., Albini, A. (2006). Photosensitized oxidation of sulfides: Discriminating between the singlet-oxygen mechanism and electron transfer involving superoxide anion or molecular oxygen. *Chemistry—A European Journal*, **12**, 4844–4857.

348. Baciocchi, E., Del Giacco, T., Elisei, F., Gerini, Maria F., Guerra, M., Lapi, A., Liberali, P. (2003). Electron transfer and singlet oxygen mechanisms in the photooxygenation of dibutyl sulfide and thioanisole in MeCN sensitized by *N*-methylquinolinium tetrafluoroborate and 9,10-dicyanoanthracene. The probable involvement of a thiadioxirane intermediate in electron transfer photooxygenations. *Journal of the American Chemical Society*, **125**, 16444–16454.

349. Gillo, P., Hamelin, O., Batat, P., Jonusauskas, G., McClenaghan, N. D., Menage, S. (2012). Photocatalyzed sulfide oxygenation with water as the unique oxygen atom source. *Inorganic Chemistry*, **51**, 2222–2230.

350. Suga, K., Ohkubo, K., Fukuzumi, S. (2003). Oxygenation of α-methylstyrene with molecular oxygen, catalyzed by 10-methylacridinium ion via photoinduced electron transfer. *Journal of Physical Chemistry A*, **107**, 4339–4346.

351. Mori, T., Takamoto, M., Tate, Y., Shinkuma, J., Wada, T., Inoue, Y. (2001). Photoinduced electron transfer oxidation of α-methylstyrene with molecular oxygen sensitized by dimethoxybenzenes: A non-singlet-oxygen mechanism. *Tetrahedron Letters*, **42**, 2505–2508.

352. Ohkubo, K., Nanjo, T., Fukuzumi, S. (2005). Efficient photocatalytic oxygenation of aromatic alkene to 1,2-dioxetane with oxygen via electron transfer. *Organic Letters*, **7**, 4265–4268.

353. Kotani, H., Ohkubo, K., Fukuzumi, S. (2004). Photocatalytic oxygenation of anthracenes and olefins with dioxygen via selective radical coupling using 9-mesityl-10-methylacridinium ion as an effective electron-transfer photocatalyst. *Journal of the American Chemical Society*, **126**, 15999–16006.

354. Lopez, L. (1990). Photoinduced electron transfer oxygenations. *Topics in Current Chemistry*, **156**, 117–166.

355. Schaap, A. P., Siddiqui, S., Prasad, G., Palomino, E., Lopez, L. (1984). Cosensitized electron transfer photo-oxygenation: The photochemical preparation of 1,2,4-trioxolanes, 1,2-dioxolanes and 1,2,4-dioxazolidines. *Journal of Photochemistry*, **25**, 167–181.

356. Ishii, K., Sone, T., Shigeyama, T., Noji, M., Sugiyama, S. (2006). Photoreactions of β-aziridinylacrylonitriles and acrylates with alkenes: The substituent effects on the formation of [3 + 2] cycloadducts. *Tetrahedron*, **62**, 10865–10878.

357. Tamai, T., Ichinose, N., Tanaka, T., Sasuga, T., Hashida, I., Mizuno, K. (1998). Generation of polyphenylene radical cations and their cosensitization ability in the 9,10-dicyanoanthracene-sensitized photochemical chain reactions of 1,2-bis(4-methoxyphenyl)cyclopropane. *The Journal of Organic Chemistry*, **63**, 3204–3212.

358. Gollnick, K., Schnatterer, A. (1984). Formation of 1,2-dioxanes by electron transfer photooxygenations of 1,1-disubstituted ethylenes. *Tetrahedron Letters*, 2735–2738.

359. Shigeaki, F., Isobe, T., Takagi, S., Tryk, D. A., Inoue, H. (2003). Highly efficient and selective epoxidation of alkenes by photochemical oxygenation sensitized by a

ruthenium(II) porphyrin with water as both electron and oxygen donor. *Journal of the American Chemical Society*, **125**, 5734–5740.

360. Pengfei, Z., Yong, W., Jia, Y., Congmin, W., Chao, Y., Markus, A., Haoran, L. (2011). Visible-light-induced metal-free allylic oxidation utilizing a coupled photocatalytic system of g-C_3N_4 and *N*-hydroxy compounds. *Advanced Synthesis and Catalysis*, **353**, 1447–1451.

361. Chen, Y.-Z., Wang, D.-H., Chen, B., Zhong, J.-J., Tung, C.-H., Wu, L.-Z. (2012). Visible light-induced synthesis of 3,4-diarylthiophenes from 3,4-diaryl-2,5-dihydrothiophenes. *The Journal of Organic Chemistry*, **77**, 6773–677.

6

PHOTOGENERATION OF CARBOCATIONS AND CARBANIONS

6.1 INTRODUCTION

The key role of carbocations and carbanions in organic synthesis has been made familiar to us starting from first-level textbooks, as has been the effect that structure and environment have on their stability. As an example, the heterolysis of the C–halogen (or C–O) bond in alkyl halides (or sulfonates) offers an entry for the thermal generation of carbocations, albeit such a (pseudo) unimolecular cleavage is sufficiently fast to have a role only when a stable (e.g., tertiary or benzylic) cation is involved, at least under the commonly used preparative conditions. As for carbanions and related anions such as enolates, these are the most important synthetic intermediates in organic chemistry; their thermal generation, however, depends on (a) the C–H bond acidity of a suitable moiety (or moieties) in the organic precursor or (b) the use of a strong base.

The scenario is quite different when using the photochemical approach (consideration is explicitly restricted to the relatively weak light flux conditions characteristic of preparative works, where only monophotonic processes occur, to the exclusion of the high flux conditions, such as those involved upon absorption of a high-energy, short laser pulse, where multiphotonic processes may have a role). Here the involvement of positive versus negative species is in reversed order with respect to thermal chemistry. As far as it is presently known, the photogeneration of anions occurs only in rare instances and has a limited synthetic interest. On the contrary, the photoheterolysis of suitable C–X bonds to give carbocations appears to

Photochemically-Generated Intermediates in Synthesis, First Edition.
Angelo Albini and Maurizio Fagnoni.

have a much larger scope than its thermal counterpart and to actually exploit the high energy associated with the radiation absorbed [1, 2]. In this case, even the thermally inaccessible cleavage of strong C–X bonds such as vinyl–X or aryl–X bonds can occur. On the other hand, as mentioned in Chapter 2, the photocleavage of a C–halogen bond is subjected to competitive homolytic fragmentation to give a radical pair [3, 4]. In some cases, the formation of an ion couple has been envisaged to arise from an electron transfer process within the radical pair [Eq. (6.1)].

$$R{-}X + h\nu \rightarrow (R^{\bullet}X^{\bullet}) \rightarrow R^{+}\,X^{-} \qquad (6.1)$$

Properly, the process depicted in Eq. (6.1) does not fit exactly in the scope of the present chapter, since the ions are formed in a secondary process; however, it was deemed appropriate to include them here, at least for the cases where this led to a clean reaction. As a matter of fact, the two mechanisms, direct and secondary ion generation, have not always been clearly distinguished, nor has the alternative been easily determined in some cases.

6.2 APPLICATION OF PHOTOGENERATED ANIONS

As mentioned above, the photogeneration of anions for synthetic purposes has a limited scope and is essentially limited to a number of photodecarboxylation reactions (mainly from arylacetic acids) [5]. One of the earliest examples dealt with the photodecarboxylation in water of 2-, 3-, and 4-pyridylacetic acids to form the corresponding methyl pyridines [6]. The quantum yield of such processes was found to be pH-dependent and reached the maximum value at the isoelectric point of the pyridylacetic acids, suggesting that the excited state of the zwitterion was involved in the process [6]. In the case of phenylacetic acids, the course of the reaction depended on whether the free acid or the corresponding carboxylate anion was present [5]. In the former case, a competitive mechanism operated and led to benzyl radicals in the place of benzyl anions [7, 8].

Benzannelated carboxylic acids, such as 5*H*-dibenzo[*a,d*]cyclohepten-5-carboxylic acid (**1**), were particularly prone to photoinduced carbon dioxide loss to form the corresponding benzyl anions as shown in Scheme 6.1 [9, 10]. The mechanism was confirmed by irradiation in the presence of D_2O, since deuterium incorporation in the end hydrocarbon (**2**) supported the heterolytic cleavage of the carboxylate anion [9, 10].

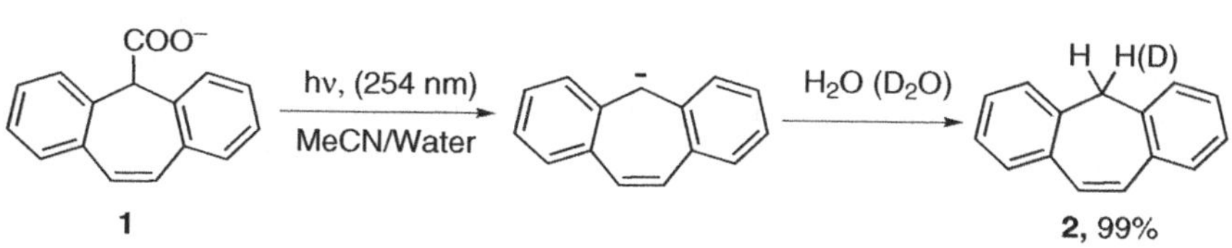

Scheme 6.1.

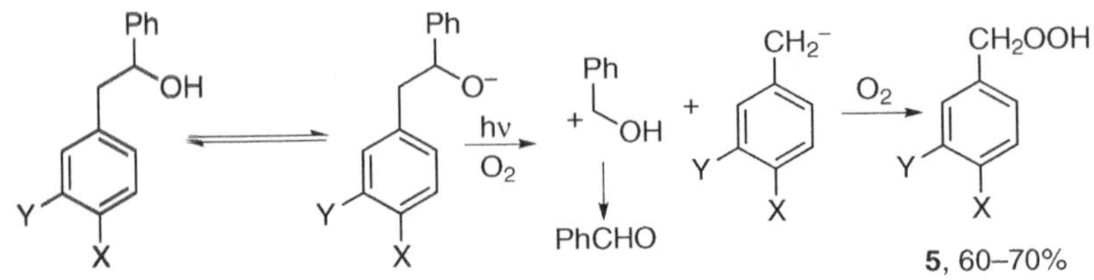

3, X = H, Y = NO_2
4, X = NO_2, Y = H

Scheme 6.2.

Interestingly, biologically active carboxylic acids can undergo a related photo-decarboxylation, thus acting as convenient sources of carbanions [11–13]. In the photolysis of the nonsteroidal anti-inflammatory drug ketoprofen (2-(3-benzoyl-phenyl)-propionic acid), it was found that the lifetime of the intermediate could be extended to some hours. Thus, under carefully controlled conditions the corresponding benzyl anion was formed in THF via the singlet excited state [14]. Another series of carboxylic acids prone to photodecarboxylation are those bearing the nitro-phenylmethyl group as in 3- or 4-nitrophenylacetates [15, 16].

Moreover, the nitro group has been used to develop a different elimination reaction leading to benzyl carbanions, in this case via a photo-retro-aldol-type reaction [17]. The proposed mechanism is illustrated in Scheme 6.2. Thus, photolysis of compound **3** (or **4**) caused the heterolytic cleavage of the benzylic C–C bond from the triplet excited state to form a nitrobenzyl carbanion. In turn, these anions were readily trapped by molecular oxygen to give isolable hydroperoxides (**5**) at pH < 12 in a satisfactory yield.

An analogous photo-retro-aldol process has been observed upon irradiation of ketals such as α-(*p*-nitrophenyl)acetophenone ethylene glycol ketal by photolysis at pH ≥ 7 [17].

6.3 APPLICATION OF PHOTOGENERATED CATIONS

As mentioned in the introduction, the photochemical approach has a significant role for the generation of different kinds of alkyl, vinyl, and aryl cations under mild conditions. Although in many cases mixtures were formed, in various instances alkylated or arylated products have been isolated in a good yield. The photo-heterolytic fragmentation of a C–X bond allows for the generation of even unstabilized cations whereas the photoprotonation of a double bond is likewise used in the formation of alkyl cations. As mentioned in Section 6.1, the competitive C–X homolytic fragmentation is a serious limitation in order to have a clean formation of carbocations. Stabilized cations can be easily obtained by the hetero-lytic bond fission of a C–X bond, and solvolyzed products were mainly formed [18]. However, the photogeneration of unstabilized cations is more appealing; various examples will be collected here, encompassing alkyl, vinyl, and aryl cations.

6.3.1 Alkyl Cations

The main pathways for the photogeneration of alkyl cations are the photoheterolysis of alkyl-halogen bonds and the photoinduced addition of a proton onto an alkene. As for the former process, the mechanism is not always unambiguous and quite often photohomolysis is a strong competitor so that a carbon radical is formed along with (or rather than) a carbocation [3]. Selected examples are presented below, where the generation of a cationic intermediate and its role in the formation of the end products were actually proven.

6.3.1.1 Unsubstituted Alkyl Cations A classic example deals with the photolysis of haloadamantanes (Scheme 6.3) [19]. Thus, irradiation of 1-iodoadamantane **6** in DMSO caused the formation of the corresponding adamantyl cation that upon solvent addition gave 1-adamantanol **8** in 90% yield. In a similar way the photolysis of diiodoadamantane **7** in MeOH afforded the disubstituted product **9** in a high yield along with a small amount of adamantane. However, starting from the dibromo analogue a mixture resulted with 1-methoxyadamantane as the main product, showing that in this case homolysis (resulting in the end in reductive elimination) was competitive [19].

The competition between radical- and cation-derived products was also observed for bicyclic halides (e.g., 1-halonorbornanes or 1-halomethylnorbornanes), where the iodides gave predominantly (or exclusively) the product arising from the nucleophilic trapping of a cationic intermediate, whereas bromides shifted the photochemical reaction toward reduction [20]. The high-energy cubyl cation was easily accessible by irradiation of 1,4-diiodocubane in wet acetonitrile where both 4-iodoacetamidocubane (12%) and 1,4-diacetamidocubane (the main product, 50%) were isolated [21].

In other cases the photoheterolysis could be assisted, as was the case for the irradiation of 2-phenylethyl iodide where the intermediacy of a phenonium ion allowed for the formation of ionic deriving products [22]. This was confirmed by the different behavior observed in the irradiation of the homologue 4-phenyl-1-butyl

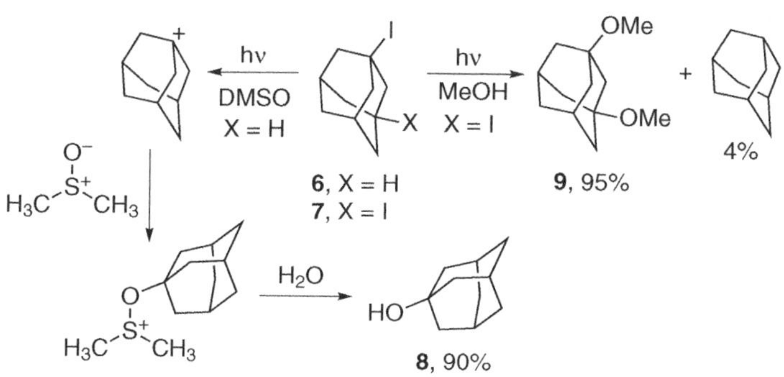

Scheme 6.3.

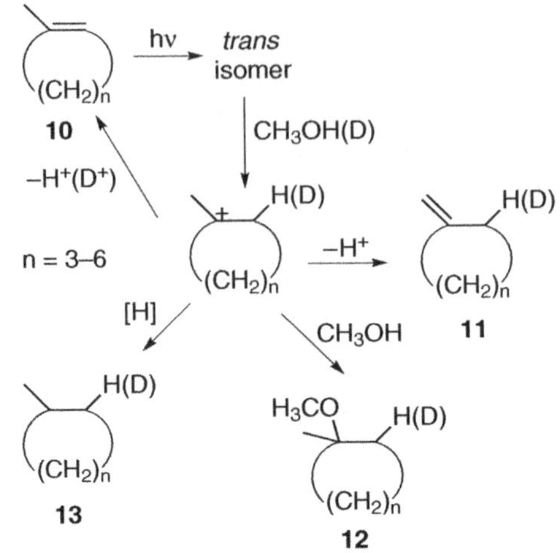

Scheme 6.4.

iodide. An alternative route to phenonium ion is the addition of a photogenerated triplet phenyl cation onto a double bond (see Section 6.3.3).

The protonation of an alkene (or alkyne, *vide infra*) is an alternative way for the photogeneration of cations that has been sparsely documented [23–26]. The role of the cation has not always been unambiguously clarified. The most representative case is probably that of medium-sized cycloalkenes, where irradiation led to strained *trans* isomers prone to accept a proton from a hydroxylic medium. Scheme 6.4 collects the pathways observed. Thus, irradiation of a methanolic solution of a methyl cycloalkene (from cyclopentene to cyclooctene, **10**) led to a carbocation intermediate that deprotonated to afford either the starting alkene or an isomer (**11**), or alternatively added to the solvent to give ether **12** or finally formed alkane **13** through a reducing pathway. The mechanism was confirmed by deuterium labeling experiments and in a few cases the process led to a clean reaction. As an example, 1-methylcycloheptene gave the methoxylated derivative as the exclusive end product [24].

Other strained olefins containing the phenylallene moiety gave interesting results in the photoaddition of alcohols via the intermediacy of an allyl cation [27]. Thus, (*E*)-bicyclo[3.3.l]non-1-ene underwent a photomethanolysis reaction to give 1-methoxy bicyclo[3.3.l]nonane after 10 h of irradiation with a 450-W Hanovia lamp [28].

In one case, the intermediacy of an α-iodocation was postulated—that is, in the photolysis of CH_2I_2 in the presence of alkenes. As a result, a cyclopropanation took place, as demonstrated in the case of 1-methylcyclohexene (68% yield) [29]. It is noteworthy that this approach gave comparable or better results than the thermal Simmons–Smith reaction and indeed led to higher yields with highly substituted alkenes, while the thermal reaction typically suffered from steric effects [29].

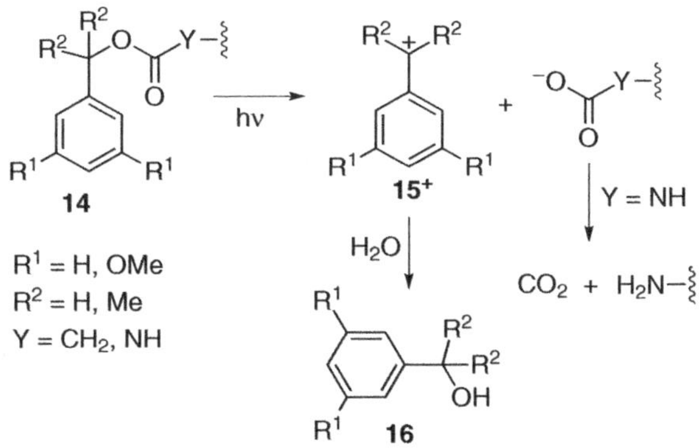

Scheme 6.5.

6.3.1.2 Benzyl Cations An input for developing the photogeneration of benzyl cations came indirectly with the target of developing a photoremovable protecting group and of releasing biologically important anions (e.g., a carbamate) by exploiting the heterolysis of a benzylic C–O bond [30–33]. About 50 years ago, glycine was protected with a *N*-benzyloxycarbonyl (Cbz) group (**14**, $R^1 = R^2 = H$, Y = NH in Scheme 6.5) and upon irradiation was found to give cation **15**$^+$ along with a carbamate anion [30]. Glycine, resulting from the usual carbon dioxide loss from the carbamate, and benzyl alcohols **16** were the end products, although the process was quite inefficient (Scheme 6.5). The presence of two methoxy groups in the meta position (R^1 = OMe, R^2 = H), however, made the process appealing and the photodeprotection of 3,5-dimethoxybenzyloxycarbonylglycine took place in 85% yield [32]. Moreover, the cleavage of the C–O bond was facilitated by introducing methyl groups in the benzylic position (R^2 = Me).

Recently, the design of trityl-based photolabile hydroxyl protecting groups was again based on the photogeneration of benzyl cations [34, 35]. The presence of a *m*-dimethylamino group is mandatory to reach a high photochemical deprotection yield. Noteworthy, this protecting group has the advantage to absorb significantly in the region of natural and artificial light, as demonstrated in the photodeprotection of monosaccharide **17** (Scheme 6.6). This compound is thermally stable, but is easily photocleaved to alcohol **18** under indoor lighting illumination (4–5 h of exposure, 84% yield) [34, 35].

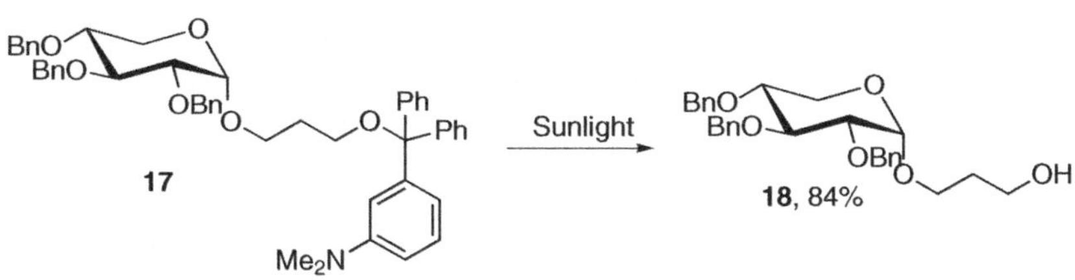

Scheme 6.6.

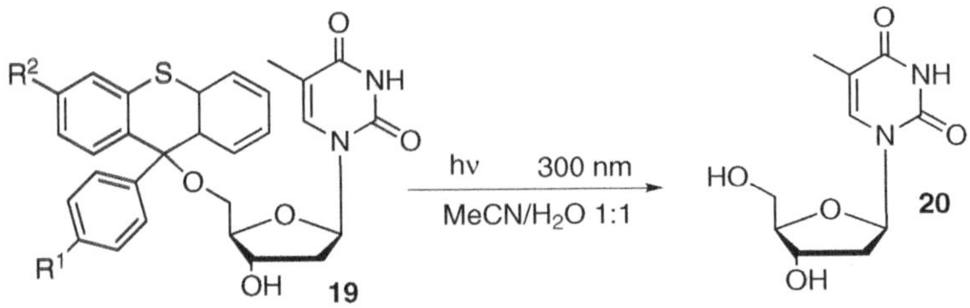

97%, R^1 = H, R^2 = H
93%, R^1 = MeO, R^2 = H
95%, R^1 = Me, R^2 = H
97%, R^1 = H, R^2 = MeO

Scheme 6.7.

Pixyl (9-phenylxanthen-9-yl) derivatives were likewise used for the photodeprotection of alcohols again via a benzyl cation intermediate and applied for the combinatorial synthesis of oligoribonucleotides [36, 37]. Later, 9-arylthioxanthyl derivatives (**19**) have been introduced and proven effective in the photodeprotection of the 5'-hydroxyl functionality of thymidine, as illustrated in Scheme 6.7. The presence of electron-donating methoxy and methyl substituents (R^1 and R^2 in Scheme 6.7) imparted to these derivatives a higher reactivity in comparison with the unsubstituted parent S-pixyl analogues, and quantitative deprotection was achieved in a shorter time while producing negligible amounts of byproducts. The improved cation stabilization resulting in the presence of a 10% v/v amount of 2,2,2-trifluoroethanol in the reaction medium further improved the performance of these thymidine precursors [38].

Photoinduced heterolytic solvolysis in methanol has been likewise reported for the case of both 3,5-dimethoxybenzyl iodide and diethyl (3,5-dimethoxybenzyl) phosphate, where the corresponding methyl ethers were isolated in a high yield [39].

Alternative electrofugal groups for the photogeneration of benzyl cations were the trialkylammonium, as reported for the photolysis of 4-methoxy-l-naphthylmethylammonium chloride in MeOH [40], and phosphonium groups, the latter used for the generation of the α,α-dimethylbenzyl (cumyl) cation [41]. A particular case is that of benzhydryl cations photocatalytically generated in various ways (e.g., from benzhydryl triarylphosphonium salts), but only for mechanistic purposes [42].

In rare instances, the benzyl cation formed was used for a synthetically valuable target, such as the formation of a carbon–heteroatom bond. A recent example is illustrated in Scheme 6.8 and is related to a photo-Ritter reaction [43]. Accordingly, irradiation of arylmethyl bromides **21a** and **21b** in MeCN gave 2-(acetamidomethyl) naphthalene and 2-(acetamidomethyl)anthraquinone (**22a,b**) in 63% and 72% yield, respectively.

At any rate, benzyl phosphates appear to be the best benzyl cation precursors. In fact, the reaction mechanism has been proven to involve a singlet-state heterolytic process leading to an ion pair [44]. As a result, the irradiation of benzyl phosphates **23** in neat benzene gave the benzylated derivative **24** in a satisfactory amount

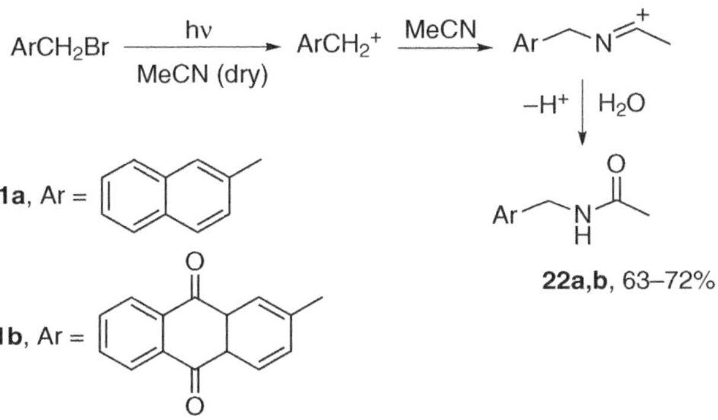

Scheme 6.8.

(65% yield), while the formation of a C–O (by reaction with methanol or butanol to give compounds **25**) or of a C–N bond (by a Ritter reaction, to give compound **26**) took place when the reaction was carried out in alcohols or in a MeCN/H_2O mixture, respectively (Scheme 6.9) [45].

6.3.1.3 α-Ketoalkyl Cations A mechanistically different case is that of the photochemical generation of α-keto carbocation **28**$^+$, since this is formed in the triplet state. A benzoin diethyl phosphate (**27**) was used as the precursor, and the photoreactivity was mainly investigated with attention to the phosphate photorelease [46]. It is interesting to note that solvolysis (to give **29**) was the almost exclusive process when compound **27** was irradiated in 2,2,2-trifluoroethanol (TFE), a cation-stabilizing solvent. It appears that the triplet cation **28**$^+$ is the first-formed intermediate and undergoes ISC to the singlet before solvent addition (Scheme 6.10, see below for analogies). No synthetic applications of this carbocation are known as yet.

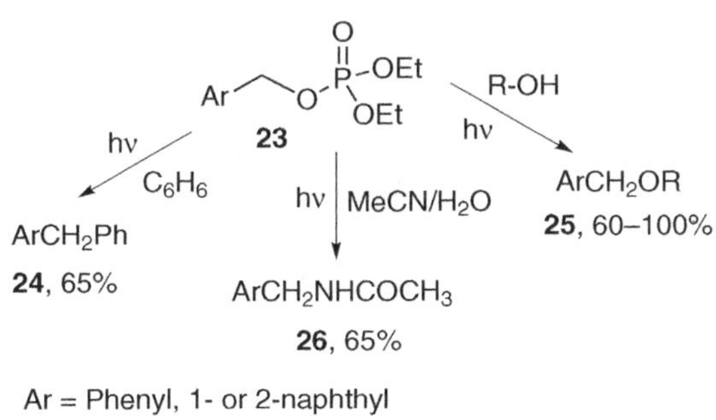

Scheme 6.9.

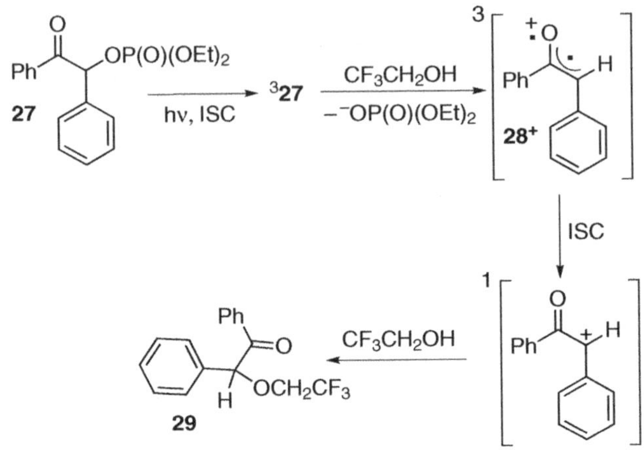

Scheme 6.10.

6.3.2 Vinyl Cations

Vinyl cations [47–49] are known since the 1970s and are mainly obtained by thermal solvolysis of the corresponding vinyl bromides or vinyl triflates [50–53]. However, several reports appeared later in the literature dealing with the photochemical generation of these important cations. The earliest examples were concerned with the irradiation of (di)phenylacetylene in hydrophilic solvents. The initial step here was photoprotonation of the triple bond to give a vinyl cation that then added to the reaction medium to give the corresponding methyl vinyl ether (in MeOH) or an aromatic ketone (by tautomerization of the enol, in water-containing solvents) [54, 55]. Thus, methoxystilbene was the major product in the irradiation of diphenylacetylene [54] in MeOH, and similarly the photohydration of monosubstituted acetylenes was observed in aqueous acetonitrile [55].

However, the most common way for arriving at vinyl cations by photochemical means involves the photoheterolytic cleavage of the C−X bond in vinyl halides [56–61]. Unfortunately, the photolysis of the latter halides is rarely clean; two kinds of reactive intermediates, vinyl radicals and vinyl cations, are formed under these conditions. Further competitive photochemical reactions complicated the matter, namely, photoisomerization of the double bond or cyclization to phenanthrenes (when starting from stilbenes) (Scheme 6.11, path *a*) [56–61]. Homolysis mainly led to reduction by abstraction of a hydrogen atom from the reaction medium (path *b*). As for the vinyl cation, this resulted either via direct heterolytic fragmentation (path *c*) or via secondary electron transfer within the radical pair (path *d*). Many (mainly mechanistically oriented) reports described the competition of the two paths in model vinyl halides such as 2-halo-1,1-diphenylethylenes and 1,1-diaryl-2-halopropenes [62, 63]. As a general rule, homolysis was preferred, and vinyl radicals were formed unless charge stabilizing substituents (e.g., an electron-rich aromatic ring) or a bromine atom were present. Moreover, when

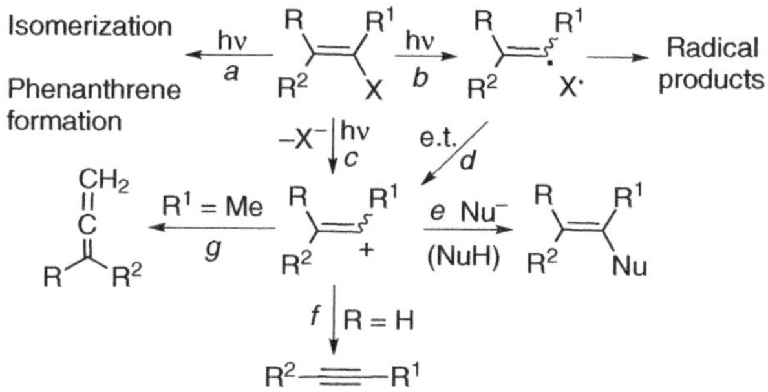

Scheme 6.11.

cation-deriving products were observed, it was not trivial to understand whether these arose through either of the two paths, *b*→*d* or *c* (Scheme 6.11). The reverse process may also occur, as reported in a couple of cases. Thus, the irradiation of (*E*)-bromostyrene appeared to cause heterolysis of the C−Br bond, but then a radical pair was formed via an electron transfer step within the initially formed ion pair [64]. A peculiar behavior was exhibited by olefins bearing a CF_3 group directly bound to the double bond. In this case, a 1,2-fluorine shift with deprotonation occurred as the main competitive process; as a result, variable amounts of substituted vinyl fluorides were formed [65].

Cation-deriving products can be increased with respect to those arising from the radical path in different ways. One is the addition of an oxidant (e.g., copper(II) salts) [66, 67] in solution to oxidize the photochemically formed vinyl radical. The operation of this principle has been demonstrated for the case of 1,1-diaryl-2-bromopropenes and related compounds [66], but has not been used in synthesis as yet. Another way is lowering the temperature. Thus, an increased ratio of ionic to radical products was observed in the irradiation of (iodomethylene)cyclohexane, 1-iodocyclopentene, or 1-iodocyclohexene when operating at a lower temperature [68]. Furthermore, it was found that when carrying out the reaction on *ortho*-substituted phenylvinyl halides, the cationic route took advantage from the convenient conformation of the aromatic ring in the radical pair, which assured a better electron transfer with the halogen atom [69].

From the synthetic point of view, the trouble is that even when cleanly formed, a vinyl cation finds different processes accessible and a mixture of products is expected. Competition at the cation level involves addition of a nucleophile (path *e*), deprotonation to give an alkyne (path *f*) or an allene (path *g*), and further intramolecular electrophilic additions. The latter reactions may have some synthetic interest (*vide infra*, Scheme 6.11). In most of the examples shown, the double bond is substituted with at least a phenyl group. The process occurring at the vinyl cation level depends mainly on the nature of the α- (or β)-aryl substituent(s) and on the choice of the leaving group (usually a bromide or a iodide) [70].

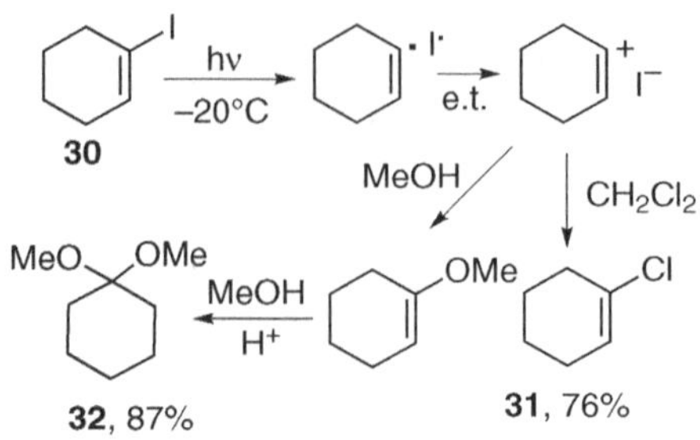

Scheme 6.12.

An important effect is obviously exerted by the nature of the nucleophile. Quite often, the solvent is incorporated in the end product. As an example, Scheme 6.12 illustrates the different behavior of a vinyl cation according to the reaction medium. Thus, irradiation at −20°C of 1-iodocyclohexene **30** gave the corresponding chloride **31** in dichloromethane (DCM) and the acetal **32** in MeOH in 76% and 87% yield, respectively [71].

The photolysis of vinyl bromides (e.g., β-arylvinyl bromides, **33**) with an aliphatic nitrile (e.g., MeCN) as the reaction medium caused the addition of the photogenerated vinyl cations to the solvent in a Ritter-type reaction with the intermediacy of nitrilium ions **34**$^+$. The end products were isoquinoline derivatives **35** [72] (Scheme 6.13), obtained in most cases in a modest yield, since hydrogen atom abstraction by the vinyl radical strongly competed. The same vinyl cations could be smoothly generated by a thermal silver assisted reaction where isoquinolines **35** were obtained in a satisfactory yield [72].

Better nucleophiles reacted selectively. Thus, photolysis of 1,2,2-triaryl-l-bromoethenes in acetonitrile in the presence of a 5 molar excess of PhSMe caused the substitution of the bromine atom by the phenylthio group in variable yields [73].

As one may expect, charged nucleophiles are excellent traps for vinyl cations. As for oxygen-centered anions, the use of an alkoxide anion was reported to cause attack to the phenyl ring besides vinylic substitution, as depicted in Scheme 6.14a [74, 75]. Thus, irradiation of triarylvinyl bromides **36** in alcohol ROH in the presence of the corresponding alkoxide (RO^-) gave variable amounts of 1-alkoxy-1-aryl-2,2-diphenylethenes (**37a–c**) and 3,3-dialkoxy-6-(2,2-diphenylvinylidene)-1,4-cyclohexadienes (**38a–c**). This behavior was ascribed to the great stabilization exerted by

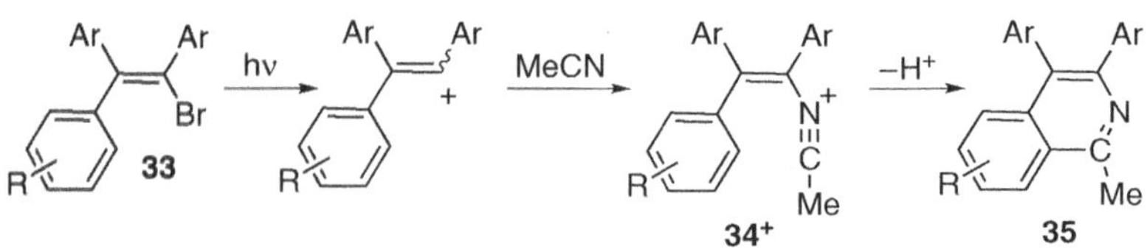

Scheme 6.13.

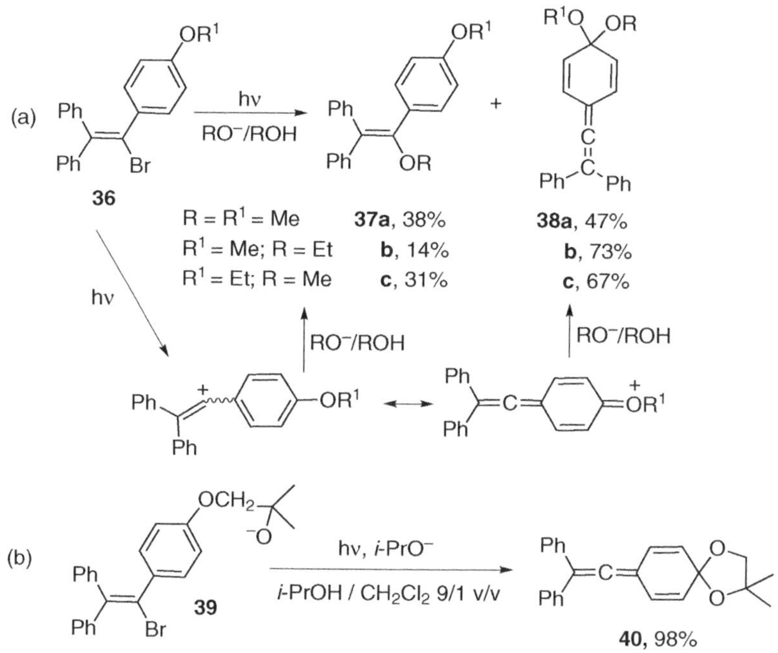

Scheme 6.14.

the OR^1 group onto the vinyl cation formed that allowed the delocalization of the positive charge (Scheme 6.14a). Interestingly, when the photolysis took place in ROH, avoiding the alkoxyde anion but in the presence of a weak base (e.g., TEA), compounds **37a–c** were exclusively formed in a high yield [74, 75]. The formation of a cross-conjugated cyclohexadienemethylene moiety occurred as a clean process upon irradiation of bromide **39**. In this case, crystalline 2,2-dimethyl-8-(2,2-diphenylvinylidene)-1,4-dioxaspiro[4.5]-deca-6,9-diene (**40**) was obtained in 98% yield by intramolecular cyclization of the alkoxide moiety onto the vinyl cation intermediate (Scheme 6.14b) [76].

The azide anion was successfully used in a photochemical cascade reaction starting with the vinyl azide initially formed [77, 78]. The reaction was performed in a two-phase system (water-DCM) by using the tetrabutylammonium salt of the azide as the phase-transfer catalyst (PTC). Scheme 6.15 illustrates the concept. Thus, azides **42** were photochemically cleaved under the reaction conditions to form azirines **43** that underwent a photoinduced ring-opening process to yield 1,3-dipoles prone to react with olefins, such as dimethyl fumarate. As a result, 1-pyrroline derivatives (**44a–d**) were formed in a good yield, particularly when the phenyl groups were substituted by at least a –Me or a –OMe group. The same process, when carried out on monoaryl vinyl halides, gave poor results [77, 78]. The involvement of azirine **43** and of the azomethinylide deriving from ring opening was further confirmed by carrying out

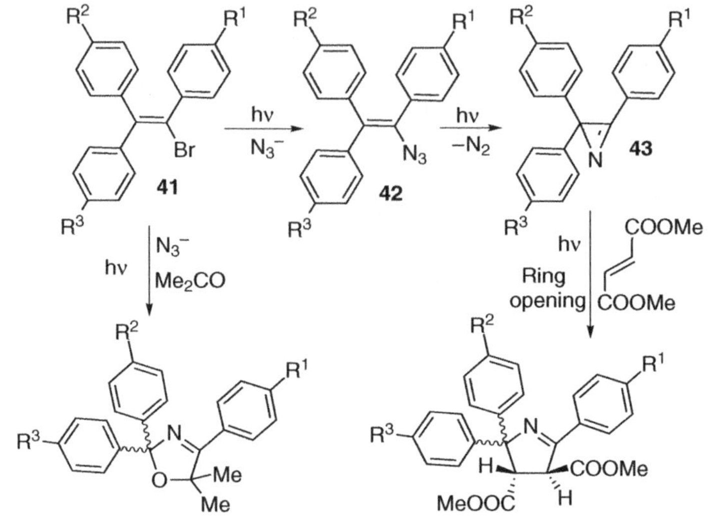

45a, $R^1 = R^2 = R^3 = OMe$ 90%
b, $R^1 = OMe$; $R^2 = R^3 = H$ 54%

44a, $R^1 = R^2 = R^3 = OMe$ 92%
b, $R^1 = OMe$; $R^2 = R^3 = H$ 86%
c, $R^1 = Me$; $R^2 = R^3 = H$ 75%
d, $R^1 = R^2 = R^3 = H$ 45%

Scheme 6.15.

the reaction in acetone, where the last intermediate was trapped by the solvent to give an oxazoline (**45a,b**, Scheme 6.15) [77, 78].

Ambident anions such as the cyanate and the thiocyanate anion were also tested. Attack at the *N*-site was exclusive with the former anion and the resulting vinyl isocyanates **47a–c** underwent a photocyclization process to yield isoquinolinones **48a–c** as the sole products (Scheme 6.16) [79, 80]. The reaction was conveniently carried out in a two-phase system (water-DCM) in the presence of tetrabutylammonium bromide as PTC. On the other hand, the thiocyanate anion exhibited its ambident nature that caused the formation of three different products, namely, both

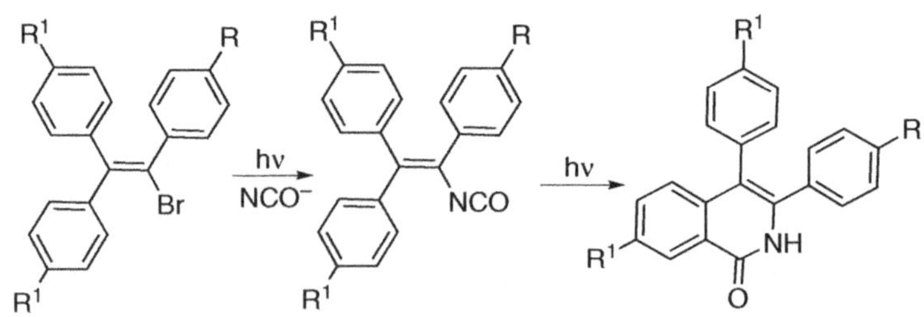

46a, $R^1 = R = OMe$
b, $R^1 = H$; $R = OMe$
c, $R^1 = H$; $R = Me$

47a–c

48a, $R^1 = R = OMe$, 93%
b, $R^1 = H$; $R = OMe$, 91%
c, $R^1 = H$; $R = Me$, 77%

Scheme 6.16.

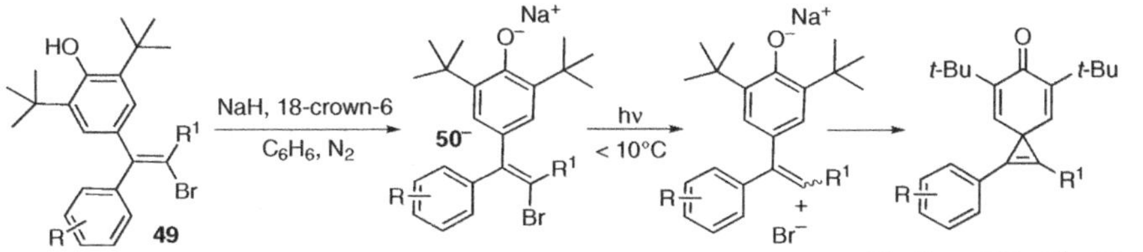

51a, R = *p*-OMe; R[1] = Ph, 74%
b, R = H; R[1] = Ph, 89%
c, R = *o*-Me; R[1] = Ph, 51%
d, R = *p*-Me; R[1] = Ph, 38%

Scheme 6.17.

isothioquinolinones and vinyl isothiocyanates (deriving from attack at the *N*-site) and vinyl thiocyanates (deriving from attack at the *S*-site). The last compounds were formed in a higher yield when the vinyl cation formed was stabilized (e.g., by the substitution with a *p*-methoxyphenyl group). This induced a preferential reaction with the softer S atom of the thiocyanate anion [79, 80].

Finally, reactions via intramolecular trapping will be considered. These have been documented, although in only a few cases, and involve both π-bond and *n*-nucleophiles. Thus, vinyl cations were intermediates for the preparation of *spiro*[2.5]octa-1,4,7-trien-6-ones (**51a–d**) starting from vinyl phenols **49** (Scheme 6.17). Phenols were converted into the corresponding sodium salts by treatment with NaH, and a crown ether was added to make the solution transparent. Irradiation of the dark red solution of anion **50**$^-$ at $< 10°C$ afforded the corresponding vinyl cation that added to the phenolate ring to give the above-mentioned spiro compounds in moderate to good yields (Scheme 6.17) [81].

In another example, a five-membered ring was formed. Thus, in the photolysis of (*o*-methoxyphenyl)vinyl bromides (**52a–c**), the vinyl cation added to the O atom of the methoxy group to give a benzofuran **53a–c** upon MeBr loss (Scheme 6.18). The reaction was not always clean, but in most cases the benzofuran was the exclusive (or the main) product and it was seldom accompanied by other (rearranged) cationic adducts [82].

Increasing again the size of the ring formed, another intramolecular arylation of vinyl cations with C–C bond formation allowed the smooth synthesis of dibenzo[*b*,*f*]

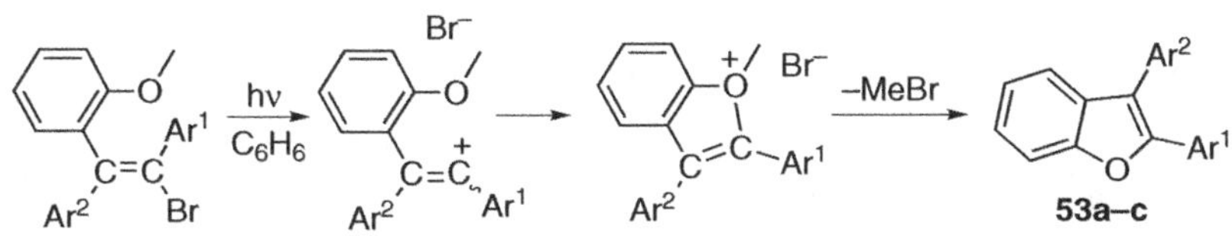

53a–c
Main product

52a, Ar[1] = *p*-$MeOC_6H_4$; Ar[2] = *o*-$MeOC_6H_4$
b, Ar[1] = Ph; Ar[2] = *o*-$MeOC_6H_4$
c, Ar[1] = Ph; Ar[2] = *p*-$MeOC_6H_4$

Scheme 6.18.

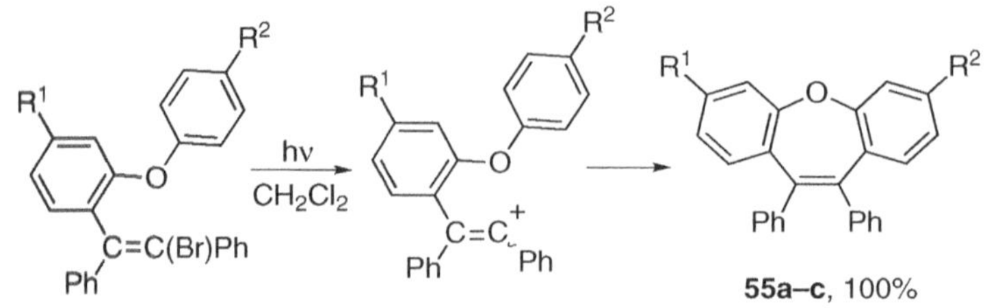

Scheme 6.19.

oxepins. The approach is outlined in Scheme 6.19 and involves the irradiation of a vinyl bromide having a diphenylether moiety, such as β-[*o*-(aryloxy)phenyl]vinyl bromides (**54a–c**) [83, 84]. The vinyl cation formed underwent intramolecular addition onto a phenyl ring and afforded the dibenzoxepins **55a–c** in almost quantitative yield. It is noteworthy that the same reaction was achieved thermally by a solvolysis reaction of (**54a–c**) in 60% EtOH, but a high temperature was required in that case (160°C) [83, 84]. Moreover, the competitive cyclization of the vinyl cation onto the oxygen atom forming a five-membered ring was ruled out by computational studies that demonstrated the endothermicity of such an alternative pathway, in contrast with the exothermicity of the formation of the seven-membered ring [84].

As indicated in Scheme 6.11 above, the vinyl cation can deprotonate to give an alkyne. In some cases this process was preceded by an aryl shift to form a more stable vinyl cation. This usually led to mixtures of limited synthetic interest [85]. However, in a few cases this route has been used for the preparation of tolans (e.g., **57a–c**) in modest to good yields by irradiation at 10°C of 2,2-di-*p*-, *m*-, and *o*-anisylvinyl bromides **56a–c** (Scheme 6.20) [86]. In the latter case, 1,2-migration of an anisyl group to give a more stable intermediate occurred before the final deprotonation.

Another interesting class of vinyl cation precursors is that of the corresponding phenyl iodonium salts [87–89]. Although the irradiation of these compounds always gives rise to a mixture of products, this approach has some interest since vinyl cations

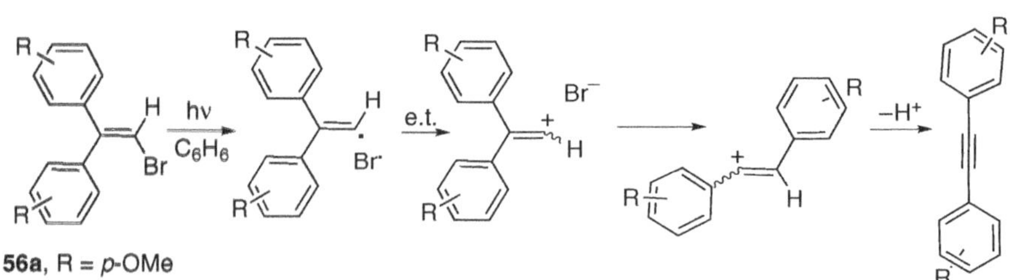

Scheme 6.20.

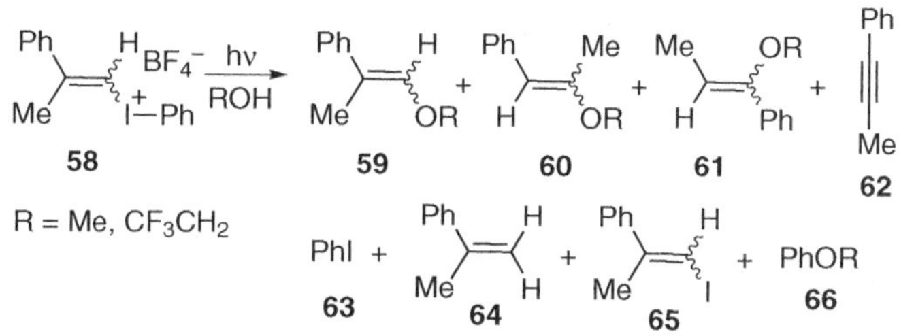

Scheme 6.21.

can be formed both thermally (by solvolysis) and photochemically, thus allowing to point out some differences. The varied chemistry of phenyl iodonium salts is exemplified in Scheme 6.21 for the case of 2-phenyl-1-propenyl(phenyl)-iodonium tetrafluoroborates **58** in alcoholic solvents. It was found that the thermal and photochemical solvolysis of **58** are complementary techniques for vinyl cation generation. Mechanistically, the cleavage of the C–I bond was assisted by the group in the *trans* position in the first case, whereas it occurred unassisted in the latter one [87, 88]. The primary vinyl cation formed gave in part (rearranged) solvolysis products (**59–61**) and the corresponding alkyne (**62**). Other byproducts deriving from the Ph–I detachment or from the concurrent vinyl radical formation were likewise observed in the reaction (**63–66**, Scheme 6.21). One should further notice that the mechanism followed was affected by the nucleophilicity of the alcohol used.

Iodonium salts were efficiently employed in the first unambiguous photogeneration of six- and five-membered cyclic vinyl cations starting from 4-*tert*-butylcyclohex-1-enyl(phenyl)iodonium tetrafluoroborate and cyclopent-1-enyl(phenyl) iodonium tetrafluoroborate [89].

A peculiar vinyl cation has been recently obtained from phosphonium-iodonium ylide **67** (Scheme 6.22). In fact, irradiation of **67** caused the liberation of PhI and an

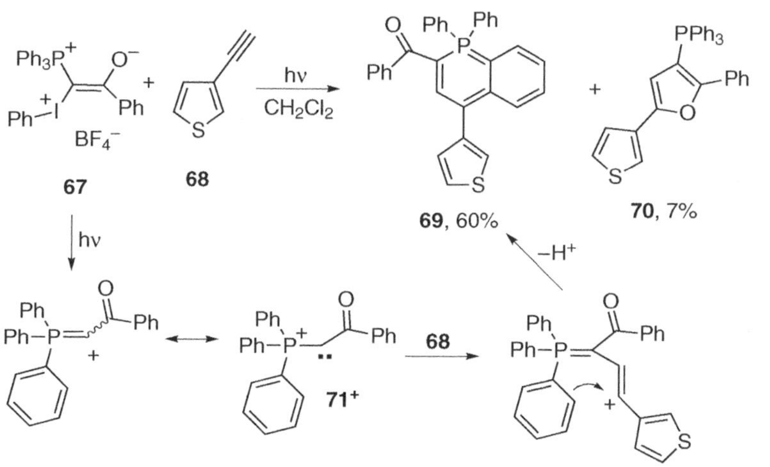

Scheme 6.22.

intermediate with the characteristics of both a vinyl cation and an electrophilic carbene **71**$^+$ was readily formed. This added onto a triple bond (e.g., in compound **68**) and gave a further vinyl cation. Intramolecular electrophilic addition onto a phenyl group ultimately led to phosphinoline **69** in 60% yield [90]. However, depending on the substituents present on the triple bond, a competitive process was observed from the same **71**$^+$ intermediate, namely a 1,3-dipolar cycloaddition with the alkyne to give substituted furan **70** in variable yields (Scheme 6.22).

6.3.3 Phenyl Cations

Cross-coupling reactions of aromatic compounds have become a main class of reactions in synthesis thanks to the dramatic progress of transition metal-catalyzed processes in recent decades [91]. However, the high price and toxicity of the catalysts used and their limited stability promoted a growing interest for the development of metal-free processes [92]. The metal-assisted process begins by oxidative addition of the metal across a carbon–heteroatom bond. Such a process may thus be mimicked by the generation of phenyl cations, the key intermediates of an ArS_N1 reaction [92]. With respect to vinyl and alkyl cations, here there is the advantage that the requirement that aromaticity is conserved inhibits most side paths operating in those cases. Thus, reactions via phenyl cation are expected to follow a clean course, and, as will be shown below, this is indeed the case. However, the thermal generation of these intermediates is no trivial issue and is actually limited to the solvolysis of some perfluoroalkylsulfonic aryl esters [93], to the solvolytic cyclization of (trifluoromethanesulfonyl)oxydienynes [94–96], and to the dissociation after β-decay of tritium substituted benzene. The thermal decomposition of aryldiazonium, diaryliodonium, and triarylsulfonium salts is no convenient method for this purpose, because the chain reductive processes initiated by single electron transfer and homolytic cleavage to afford a phenyl radical competes with the heterolytic route leading to phenyl cations. Further note that this competition depends on the structure of the salt and on the reaction medium [97–103].

However, in most cases the decomposition of diazonium salts was found to be accelerated by irradiation and under those conditions phenyl cations appeared to be the intermediates. This process was studied as the photochemical version of the Balz–Schiemann reaction for the synthesis of aryl fluorides and in some cases it was very efficient, as indicated in Scheme 6.23 [104–106]. The photochemical approach has the advantage of being carried out at a lower temperature than the thermal analogue and is generally cleaner. Thus, irradiation of diazonium tetrafluoroborate salts **72** and **74** gave the corresponding aryl fluorides in pyridine hydrofluoride or by using a $BF_3{\cdot}Et_2O$, HF solution [104–106].

This photochemical fluorination likewise took place in ionic liquids (e.g., 1-butyl-3-methylimidazolium tetrafluoroborate ([bmim][BF_4]), where fluoroimidazole derivatives were obtained by photolysis at 0°C of the corresponding diazonium salts [107].

A complication in the use of phenyl cations is their dual nature, since either of the two spin states, the singlet or the triplet, can be generated and reacts (Scheme 6.24) [108–116].

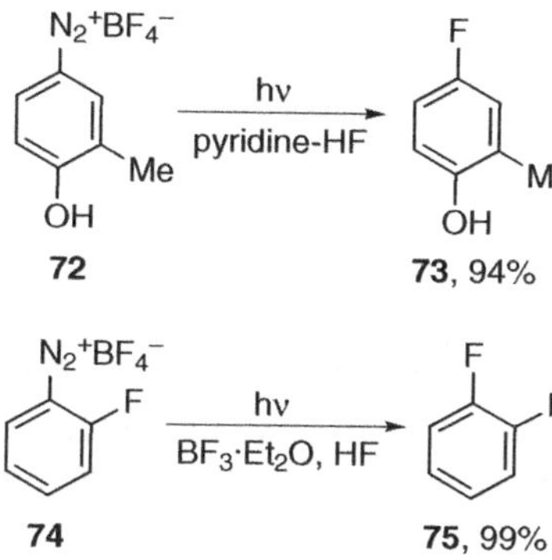

Scheme 6.23.

DFT calculations demonstrated that the former intermediate was a localized carbocation (with $\pi^6\sigma^0$ orbital occupance) (**76**, Scheme 6.24a), whereas in the triplet the charge was delocalized over the ring and the two unpaired electrons resided in orthogonal orbitals, the σ orbital at the divalent carbon and the π system ($\pi^5\sigma^1$ structure, **77**, Scheme 6.24b). The last intermediate thus had a diradical character that was reminiscent of triplet carbene, with the difference that one of the electrons was delocalized over the π system rather than residing on a *p* orbital [108–115]. The geometry of the two states was different. The singlet exhibited a puckered structure with a cumulene character at the C_2–C_1–C_6 moiety (**76a**). The deformation was largely affected both by the nature and by the position of the substituent(s), the largest change resulting when an electron-donating substituent was placed in *meta* or *para* positions with respect to the divalent carbon (formula **76b,** Scheme 6.24a) [108, 109]. On the contrary, a regular hexagon geometry was calculated for the corresponding triplet phenyl cations, rather independently from the position and the nature of the substituent(s) present (formula **77a**, Scheme 6.24b).

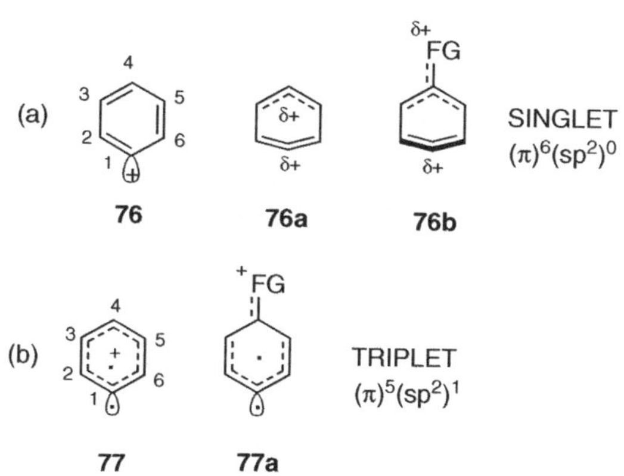

Scheme 6.24.

The differences between the electronic distributions of singlet and triplet phenyl cations are reflected in the chemoselectivity observed in their reactions. Accordingly, singlet aryl cations react unselectively, while triplets add efficiently to π-bond nucleophiles (olefins, alkynes, aromatics) but not to n-bond nucleophiles (water, alcohols), except when charged (e.g., the cyanide anions). As for the generation, a thermal reaction leads to the singlet, or it would lead if it happened. The singlet is also the lowest-lying spin state in many phenyl cations (except when strong π-donating substituents, such as F, OH, or NH_2, are present) [108–115]. Under photochemical conditions, on the other hand, the cations are generated in the same multiplicity as the excited state from which these arise. Subsequent ISC at the aryl cation level may lead to the lowest of the two states.

The most obvious precursors for aryl cations are aryldiazonium ions that split off such a good leaving group as molecular nitrogen (see Scheme 6.25, path *a*). This was found to be indeed the case, and the substituents present on the aromatic ring affected the result. As an example, irradiation of 4-*N,N*-diethylamino or 4-morpholino benzenediazonium salts gave products resulting from the singlet phenyl cation, while the products distribution observed for the 4-benzoyl derivative was ascribed to the triplet [117]. The unselective singlet ($^1Ar^+$) underwent solvolysis reactions, and the products formed were acetamides in acetonitrile (via a Ritter reaction), ethers in alcohols, and *N*-phenylacetamides in substituted acetamides (see Scheme 6.25) [119–121]. These reactions have limited synthetic value, with a few exceptions such as the synthesis of aryl trifluoroethyl ethers by photodecomposition of diazonium salts in 2,2,2-trifluoroethanol [117, 122]. On the other hand, an electron-withdrawing substituent caused ISC to the triplet state of the salt that in any case could be accessed by energy transfer sensitization—for example, by an aromatic ketone [118, 119].

A further class of phenyl cation precursors that has sometimes been used is that of iodonium salts. Worth mentioning is the formation of naphthyl cation from naphthalen-1-yl(phenyl)- and naphthalen-2-yl(phenyl)-iodonium tetrafluoroborate [123] that led to the corresponding ethers as the main products.

More appealing from the synthetic point of view is the use of phenyl cations for building aryl–carbon bonds by reaction with carbon-based nucleophiles [124, 125]. Diazonium salts are themselves strong electrophiles and thus are not suitable as

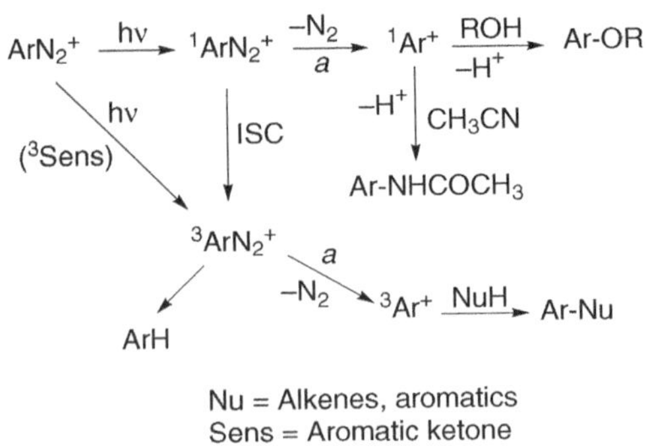

Scheme 6.25.

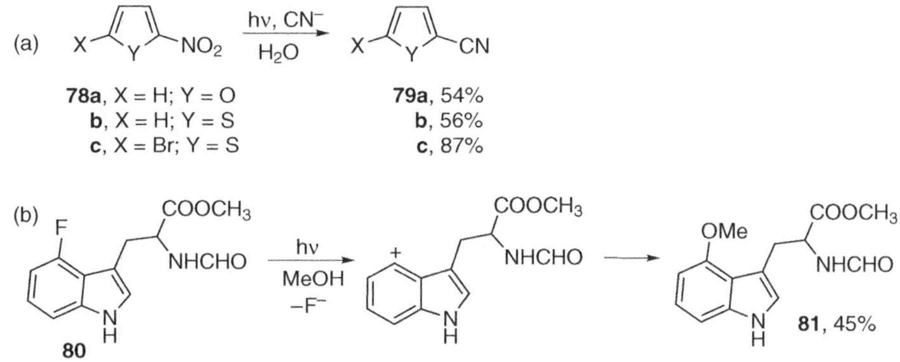

Scheme 6.26.

precursors in this case, due to the competitive thermal reaction forming azo derivatives. For this aim, uncharged precursors must be considered.

The intermediacy of a phenyl cation formed from a neutral precursors was claimed in a number of photosubstitution reactions [124]. Representative examples are collected in Scheme 6.26. Thus, the photocyanation of both nitrofuran **78a** and of substituted nitrothiophenes **78b,c** (Scheme 6.26a) appeared to be an ArS_N1 process [126]. It is noteworthy that the nitro group was selectively split off in preference to the generally highly photolabile bromine. Some haloindoles likewise underwent photoheterolytic cleavage, as was the case in the irradiation of (±)-*N*-formyl-4-fluorotryptophan methyl ester (**80**) in MeOH that afforded (±)-*N*-formyl-4-methoxytryptophan methyl ester (**81**) in 45% yield (Scheme 6.26b) [127].

A large number of new examples have been reported in recent years, however, and the scope of the process has been explored. It turned out that triplet phenyl cations can be smoothly generated by UV irradiation of electron-rich phenyl halides or esters [92, 128–131]. The reaction was effective with substituted anilines [132–137], phenols [138–141], anisoles [142, 143], thioanisoles [144], and even alkylbenzenes [116, 145, 146], provided that polar or protic solvents (e.g., ethyl acetate, acetonitrile, alcohols or water/acetonitrile mixtures) were used. 2,2,2-Trifluoroethanol is the elective solvent for explorative studies due to the low nucleophilicity and the high ion stabilization power. In fact, in this solvent even chlorobenzene underwent photoheterolytic cleavage, at least in part [147]. Actually, the triplet cation was formed due to the efficient intersystem crossing (ISC) in these aromatic derivatives, and cleavage of the triplet precursor gave the phenyl cation in the same multiplicity (Scheme 6.27) [92, 128–131].

In most cases, the triplet cation (that as mentioned has a biradicalic character) was reduced by the solvent. However, in the presence of a sufficient amount (0.05 to 2 M) of a π-nucleophile (such as an alkyne, an alkene, an aromatic, or a heteroaromatic), these cations were selectively trapped and arylation forming new Ar–C bonds was the main or the exclusive process observed [92, 128–131]. As an example, the addition of triplet phenyl cations $^3\mathbf{82}^+$ onto a double bond led to a distonic diradical

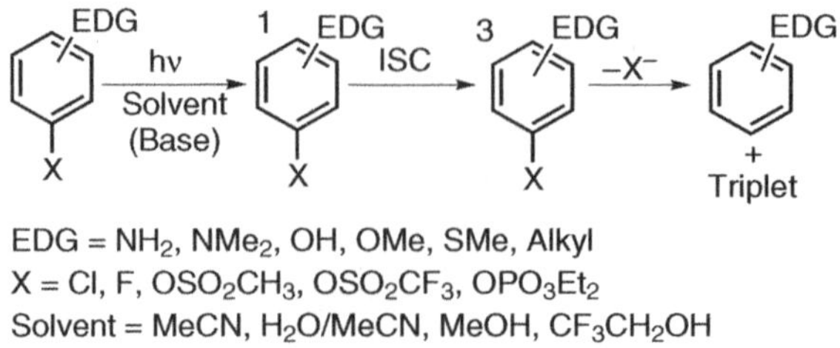

Scheme 6.27.

adduct ^{3}**83**$^{+}$ that upon ISC afforded a singlet phenethyl cation **84**$^{+}$ and/or a σ-bridged intermediate, termed "phenonium ion" (**85**$^{+}$, Scheme 6.28) [148, 149].

In turn, the phenethyl cation/phenonium ion intermediate gave the final product by adding a nucleophile. The course actually followed depended on the nature and the position of the substituent(s) present on the aromatic ring, on the structure of the alkene, and on the reaction conditions, leaving some margin for directing the process. A typical example is the competition between the counterion (the leaving group of the initial heterolytic step) and the solvent. Thus, when 4-*N,N*-dimethylchloroaniline (**86**) was irradiated in acetonitrile in the presence of 1-hexene, the phenyl cation formed added to the alkene according to the Markovnikov's rule and the chloride anion was incorporated in the final products (**87** and **88**, Scheme 6.29a) [136]. On the other hand, when the irradiation was performed in a nucleophilic medium (e.g., methanol or even TFE) and with a more substituted alkene, as in the reaction of 4-chlorophenol with 2,3-dimethyl-2-butene (Scheme 6.29b) [142], the main arylation product (**89**) was that arising from the addition of a solvent molecule onto the phenonium ion. In alcoholic solvents the use of a base (e.g., cesium carbonate) was often mandatory in order to buffer the acidity liberated in the process [150].

More synthetically interesting results were obtained when adding a second nucleophile to the reaction mixture, thus resulting in a three-component reaction. The reaction met some success when the second nucleophile did not compete with the alkene in the reaction with the phenyl cation. Thus, irradiation of 4-chloroaniline in acetonitrile in the presence of $NaBH_4$ gave moderate yields of 4-alkylanilines resulting from the trapping of the phenonium cation by the borohydride anion [151]. Additional three-component reactions have been reported. The success of such processes required that the adduct cation had available no easy competing path and combination with the third reagent could occur. Thus, the irradiation of a solution containing 4-chloroaniline, an alkene, and a secondary amine gave

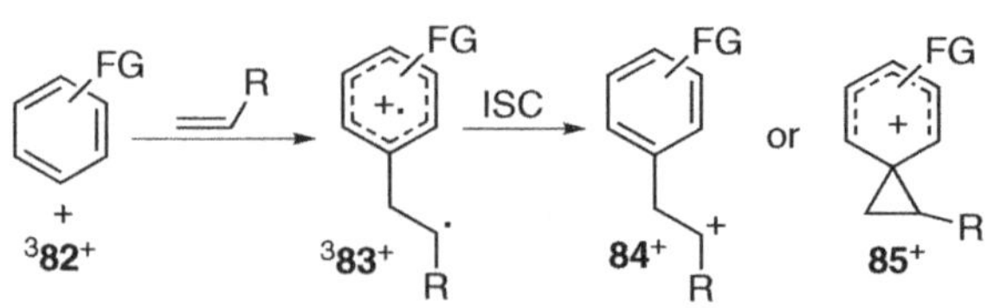

Scheme 6.28.

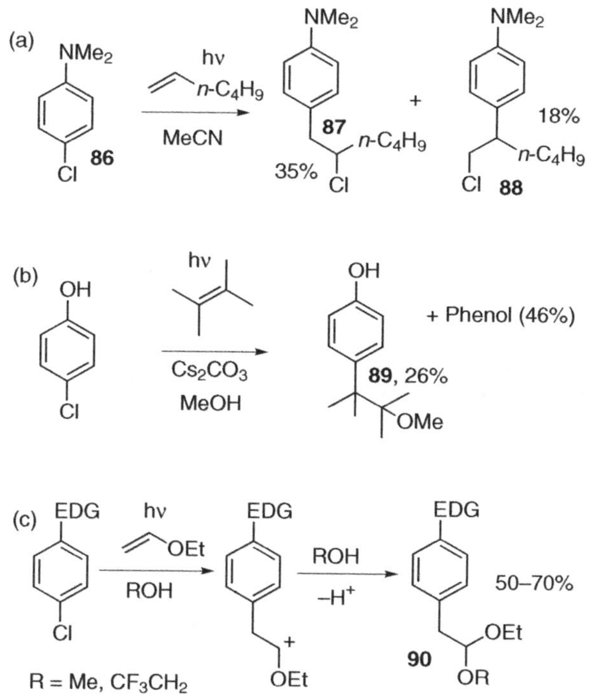

Scheme 6.29.

aminoalkylanilines [152]. Likewise, the addition of an aryl cation onto ethyl vinyl ether or related compounds resulted in a three-component process, which afforded the acetals of β-arylaldehydes or ketones **90** in good isolated yields (Scheme 6.29c) [146].

From the previous examples, it appears that the addition of aryl cations onto alkenes often led to mixtures of products, due to the role of competing processes either in the addition step or in the evolution of the adduct cation. However, selective arylations do occur in appropriate cases. As an example, an alkene bearing a tethered oxygen-based nucleophile was found to be arylated at the C–C double bond and then to undergo intramolecular attack resulting in the tandem formation of an aryl–carbon and a C–O bond. Thus, the carboxylic moiety in ω-alkenoic acids attacked the phenonium ion forming γ-benzyl lactones (**91**, in the case of 4-pentenoic acid; Scheme 6.30, path *a*) or β-phenyl lactones (**92**, when 3-butenoic acid was used, path *b*) [145]. When a cyclic alkene (such as 2-cyclopentenyl acetic acid, path *c*), was tested, the reaction was fully regio- and stereoselective, affording the corresponding furanones **93a,b** as a single diastereoisomer with the simultaneous generation of three stereogenic centers.

A similar reaction was observed with alkenols in the reaction with 4-hydroxyphenyl cation. When using 4-pentenol, benzyl tetrahydrofuran **94** was successfully

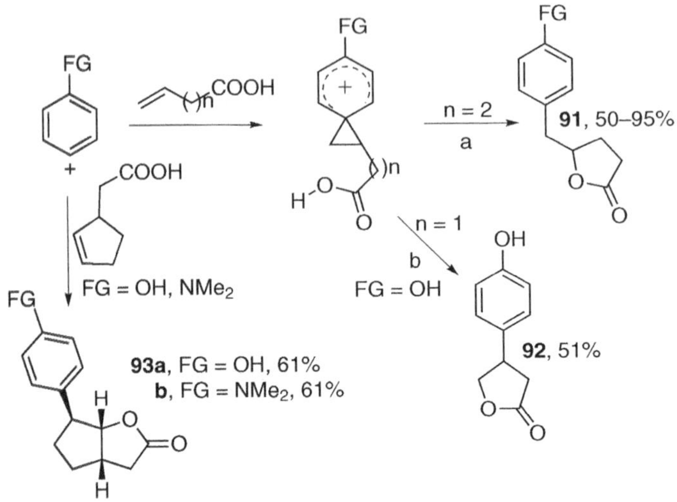

Scheme 6.30.

prepared, whereas the reaction with 3-butenol gave phenyl tetrahydrofuran (**95**, Scheme 6.31) [153].

On the other hand, the presence of a good electrofugal group in the adduct cation made elimination the dominant path. When allyltrimethylsilane (ATMS) was used as the nucleophile, loss of the trimethylsilyl cation led to the formation of an allyl aromatic (Scheme 6.32). This allylation reaction has been efficiently applied to the synthesis of many bioactive allyl phenols and allyl anisoles, like eugenol (**97c**) and safrole (**97d**) [154]. These derivatives, which are present in nature in the plants of the genus *Piper*, found several applications as drugs or cosmetic products. The mild conditions of the photochemical protocol allowed for skipping the protection of the phenolic group usually required in organometallic mediated procedures. Arene diazonium salts have been likewise adopted for the synthesis of allylbenzenes via irradiation in the presence of ATMS [118, 119].

Silyl enol ethers can likewise be employed as nucleophiles. The mechanism involves again addition of the cation onto the double bond and subsequent

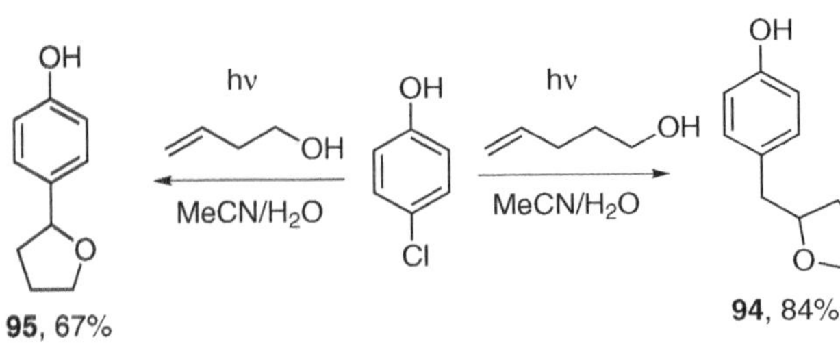

Scheme 6.31.

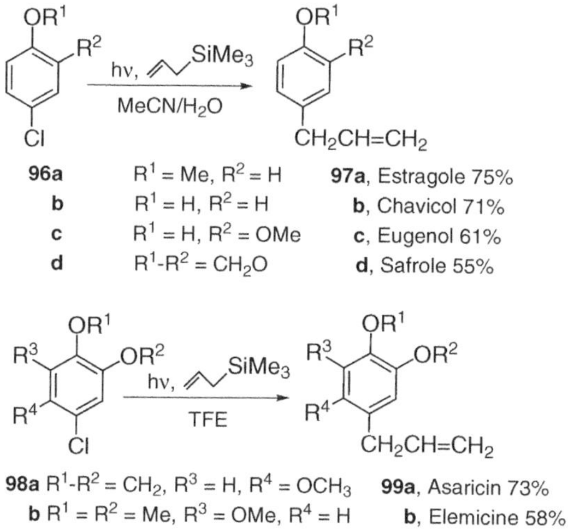

Scheme 6.32.

elimination of the trialkylsilyl cation, finally leading to α-aryl ketones in high yields. Analogously, using ketene silyl acetals as traps, the process afforded in good yield α-aryl propionic esters (e.g., **101**, Scheme 6.33). These products are synthetically useful as precursors of some important nonsteroidal anti-inflammatory drugs, such as alminoprofen (**102**) and indoprofen (**103**) [155].

When an acidic hydrogen was present in the α-position with respect to the positive charge in the adduct cation, either benzylic or allylic deprotonation occurred, giving a styrene or an allylbenzene, respectively. For example, in the reaction of 4-chloroanisole (**96a**) in the presence of 2,3-dimethyl-2-butene (Scheme 6.34a), allylic deprotonation led to the corresponding allylanisole **104** in good isolated yield. When

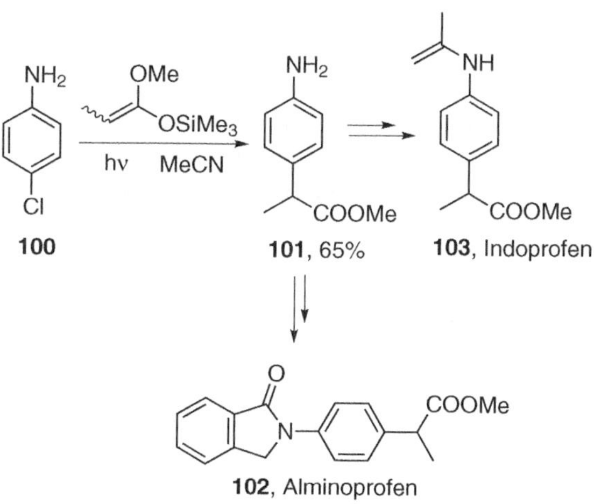

Scheme 6.33.

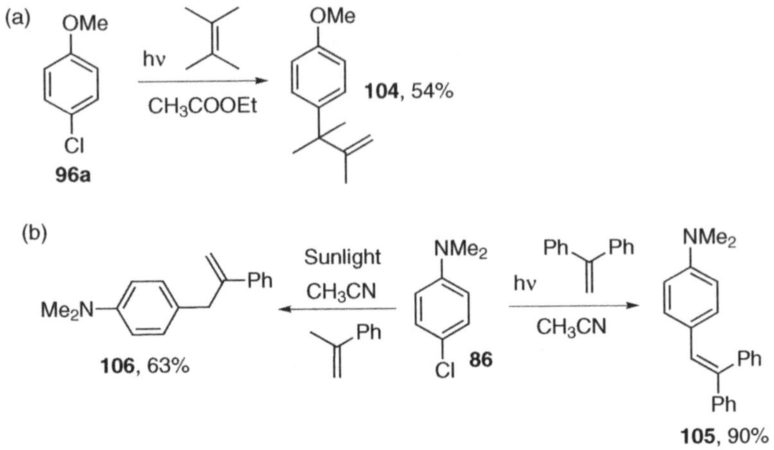

Scheme 6.34.

1,1-diphenylethylene was used as the olefin (Scheme 6.34b), deprotonation at the benzylic position gave the corresponding triphenyl ethylene **105** in almost quantitative yield [136]. Allyl moieties were likewise introduced by exposing the reaction mixture to sunlight. Under these conditions, the 4-*N,N*-dimethylaminophenyl cation was smoothly generated and used for the addition onto α-methylstyrene. This gave the allylation product **106** in 63% yield after 3 days of irradiation [156].

Both deprotonation and solvent addition have been observed using dienes as nucleophiles. In the presence of a conjugated system, addition of a phenyl cation onto the double bond was usually followed by deprotonation, affording the corresponding dienyl compound (e.g. **107**, Scheme 6.35) [157]. On the other hand, when using 1,4-octadiene, the intermediate cation underwent a second intramolecular electrophilic addition giving a mixture of bicyclic products.

When the irradiations were carried out in ion-stabilizing media, such as trifluoroethanol, isomerization of the cation often intervened and involved either hydride or alkyl shift. These reactions strongly support the carbocationic nature of the intermediates involved. A representative example is given by the reaction of phenyl cations with cyclohexene (Scheme 6.36). In this case, either a Wagner–Meerwein rearrangement or a ring contraction occurred, affording a 1-arylcyclohexyl (**108**) and a cyclopentylmethyl (**109**) ether, respectively [142].

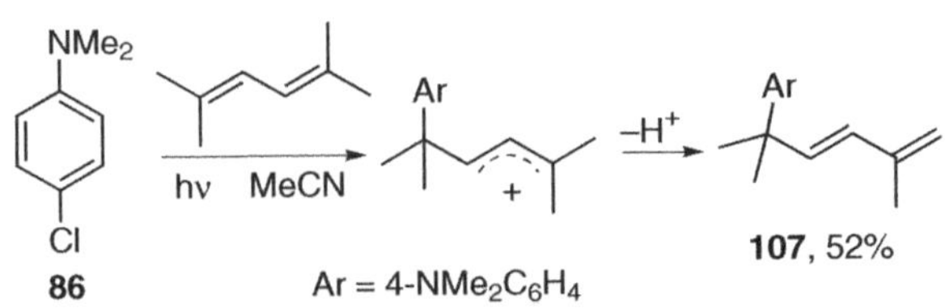

Scheme 6.35.

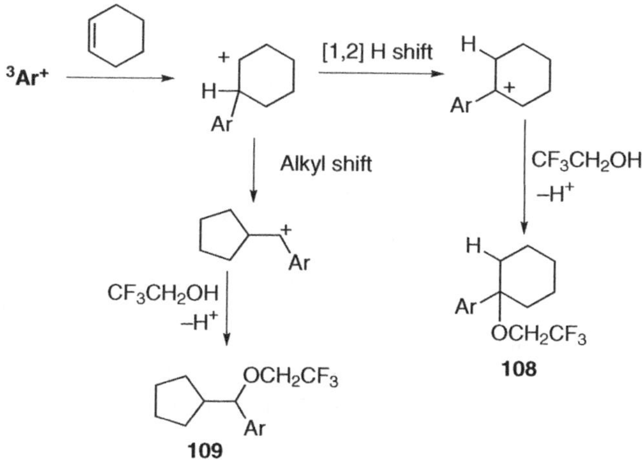

Scheme 6.36.

Similar processes have been detected from 3-aryl-2-norbornyl cations formed upon addition of a phenyl cation onto norbornene [158, 159].

Another class of π-bond nucleophiles that are efficiently trapped by phenyl cations is that of alkynes. The addition of a phenyl cation onto a terminal C–C triple bond led to a β-phenyl vinyl cation (**110**$^+$) that in turn gave the corresponding arylalkyne **111** after proton loss (Scheme 6.37, path *a*). No evidence for other conceivable processes, such as solvent addition or allene formation, was found. The preparation of arylalkynes is important, since these compounds are valuable

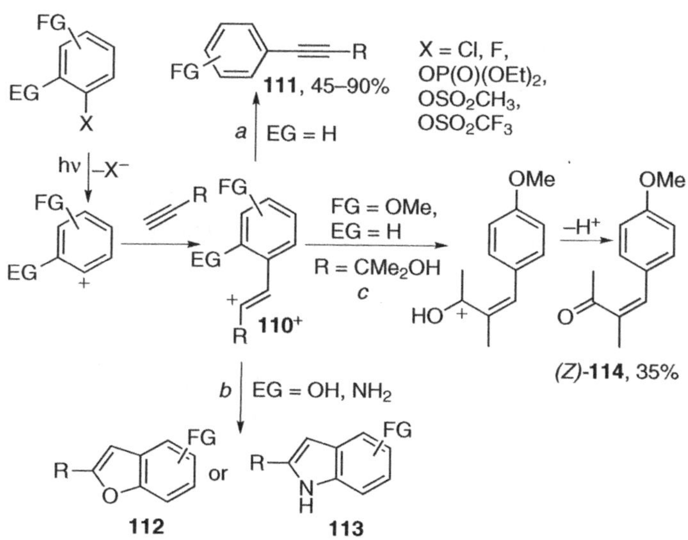

Scheme 6.37.

intermediates in organic synthesis, in view of the easy derivatization of their triple bond. This photochemical approach is up to now one of the rare examples of a metal-free Sonogashira-like reactions that can be applied to different substrates, always affording good yields of the end products [160].

When the starting aromatic bore a hydroxy or an amino substituent in the *ortho*-position with respect to the leaving group, the exclusive reaction was trapping of the cation by the alkyne, followed by an intramolecular cyclization to give the corresponding benzofurans (**112**) or indoles (**113**, Scheme 6.37, path *b*) [92, 161]. This is remarkable because benzofurans (and indoles) are largely diffuse among natural compounds and pharmaceutically active molecules. Moreover, this is one of the few examples of synthesis of benzofurans starting from chlorophenols, while thermal methods usually employ more expensive bromo- or iodophenols.

In a particular case, when using a propargylic alcohol, such as 2-methyl-3-butyn-2-ol, in the place of a simple alkyne, methyl group migration at the vinyl cation level occurred and gave α,β-unsaturated ketone (*Z*)-**114** as the exclusive product (Scheme 6.37, path *c*) [160].

Another class of π-traps is that of aromatics. Despite recent advances in organometallic chemistry, the synthesis of biaryls by cross-coupling at an unsubstituted position still remains a difficult challenge. Phenyl cations became useful here, and the arylation of benzene forming mono-substituted biphenyls was easily accomplished by this type of reactions. Thus, 4-aminobiphenyl (**115**) was prepared in quantitative yield by irradiation of the corresponding mesylate or phosphate in TFE in the presence of benzene (Scheme 6.38a) [162]. The same compound was obtained also starting from 4-chloroaniline in acetonitrile, but in a lower yield (42%). Interestingly, when changing the solvent to an imidazolium-based ionic liquid,

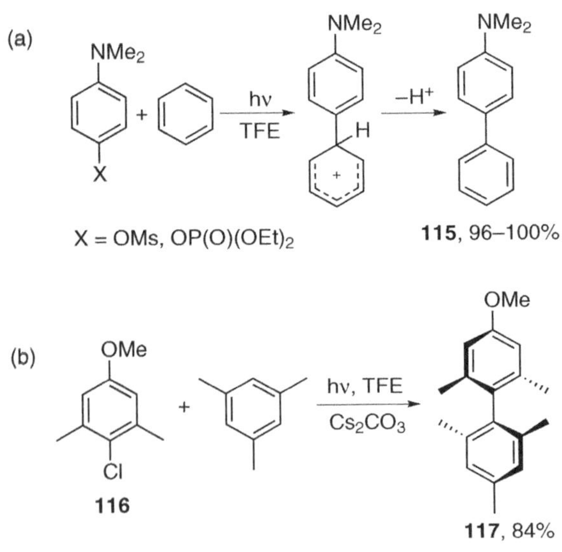

Scheme 6.38.

the product yield increased up to 70% [163]. Biphenyls were likewise formed from phenyl cations photogenerated from arene diazonium salts, although a larger excess of benzene had to be used [118, 119].

The synthesis of poly-substituted biaryls—in particular, of asymmetric or crowded derivatives, which are a key structural feature in many pharmaceutically and biologically active compounds, as well as in several BINOL ligands—was accomplished in a similar way. It was found that phenyl cations reacted smoothly with symmetric methyl benzenes in trifluoroethanol. Steric hindrance caused no major limitation to the process efficiency, and biphenyls with two, three, or even four substituents *ortho* to the axial bond were easily prepared. For example, irradiation of 3,5-dimethyl-4-chloroanisole (**116**) in the presence of mesitylene afforded the tetra-*ortho*-methyl-substituted biphenyl (**117**) in 84% yield (Scheme 6.38b). This reaction was fully chemoselective, consistently involving a clean mono-arylation, while no detectable competing functionalization of the benzylic hydrogens of mesitylene was observed [164].

The arylation of unsymmetric aromatics such as anisole gave a mixture of methoxy biphenyls arising from the attack of the phenyl cation onto the *o*-, *m*-, and *p*- positions. In most cases, a marked preference for the products formed via an *ortho* attack (at least two-thirds of the arylated products) was observed, rather independently on the nature of the phenyl cation used and on the nature of the leaving group photodetached in the reaction [165]. The different *o*/*m*/*p* ratio observed was used as a tool to fingerprint the singlet/triplet nature of phenyl cations.

Electron-rich heteroaromatics (e.g., furan, pyrrole, and thiophene) acted as efficient phenyl cation traps and were used for the formation of aryl–heteroaryl bonds. It is noteworthy that heterocycles were arylated with a high regioselectivity (α/β ratio > 98) confirming the general pattern of electrophilic substitution with these heterocycles. More precisely, thiophene was selectively arylated in the α-position (**118**, 54% yield) in the reaction with 4-*N,N*-dimethylphenyl cation (Scheme 6.39, path *a*). The same selectivity was found even when a single free α-position was available, as in the case of 2,4-dimethylpyrrole, where the corresponding heterobiaryl (**119**, Scheme 6.39, path *b*) was isolated in 75% yield. With 2,5-dimethylfuran, the arylation occurred with equal efficiency in the β- position and gave the 3-phenylfuran derivative in 70% yield [166].

Triplet phenyl cations are efficiently trapped not only by π-nucleophiles, but also by inorganic anions, such as iodide or cyanide. Benzonitriles, an important class of

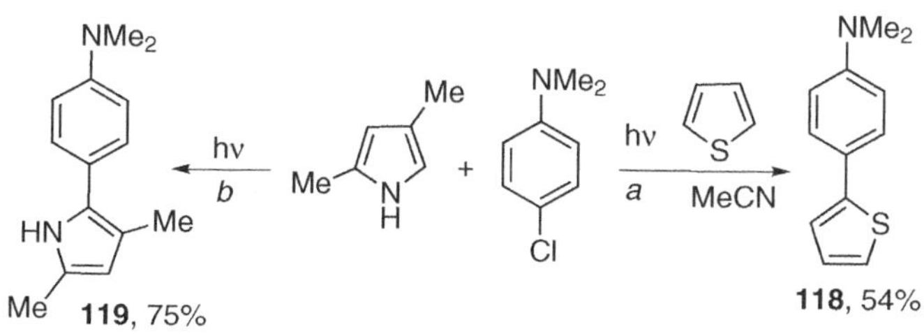

Scheme 6.39.

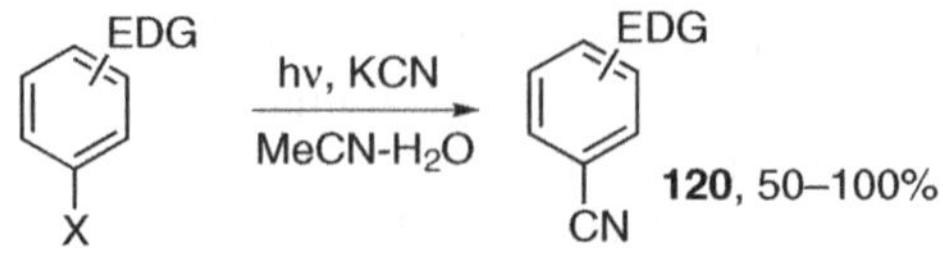

EDG = NH_2, NMe_2, OH, OMe, SMe
X = F, Cl, OSO_2CF_3, OSO_2CH_3, $OP(O)(OEt)_3$

Scheme 6.40.

compounds that find applications among agrochemicals and pharmaceuticals, were prepared by the reaction of phenyl cations with cyanide ion. Cyanation of arenes under uncatalyzed conditions is a demanding task and usually takes place by attack of a cyanide ion onto an electron-poor aromatic, while cyanation of electron-rich aromatics is more challenging. Once again, the irradiation of electron-donating phenyl halides or esters in a water/acetonitrile solution of KCN overcame this limitation and afforded several electron-rich benzonitriles **120** in satisfactory yields and under mild conditions (Scheme 6.40) [167].

The reductive dehalogenation of aromatics is often a key step both in organic synthesis and in the treatment of halogenated wastes, which are toxic and polluting compounds. Recently, the phenyl cation-mediated protocol has been applied to the photo(hydro)dehalogenation of aryl chlorides (e.g., **121**) and fluorides by using mild reducing agents (such as 2-propanol or hypophosphorous acid). This was considered a "green" photochemical protocol that avoided the use of neurotoxic organotin hydrides or of aggressive metal hydrides, as well as of expensive and labile metal catalysts, which are often required in thermal reduction (Scheme 6.41) [168].

As for chlorophenols, the photoreduction can be induced by adding various reducing agents. Since these processes can take place in an aqueous environment, this has important implications in the photoreactions of polychlorophenols (largely used as antimicrobials). Although the photochemistry of chlorophenols in solution strictly depends on the reaction medium, the irradiation of 4-chlorophenol in water gave the corresponding 4-hydroxyphenyl cation [138–141]. In deaerated solutions hydroquinone was obtained as the main product, but under aerated solutions *p*-benzoquinone was the exclusive product. The latter product was formed by oxygen trapping of the 4-oxocyclohexa-2,5-dienylidene carbene (**122**) formed by deprotonation of the initially formed cation (Scheme 6.42). On the other hand, the presence

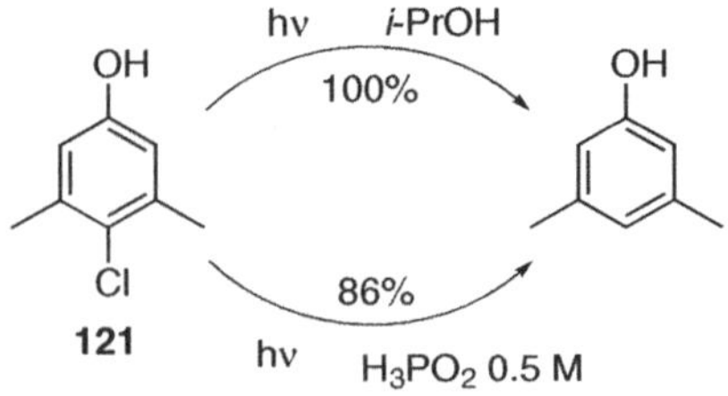

Scheme 6.41.

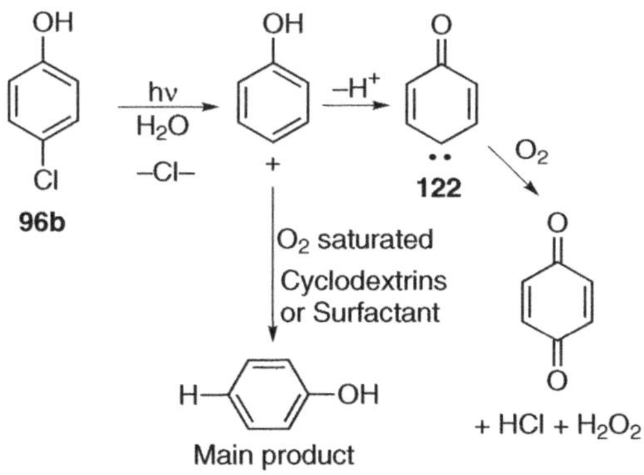

Scheme 6.42.

in solution of a complexing agent such as α-cyclodextrin (α-CD) caused a protection of the substrate from oxidation processes; as a consequence, a cleaner reaction occurred and led to phenol as the exclusive product [169]. A similar photochemistry took place in micelles (obtained by dissolving a surfactant in the aqueous mixture), where hydrogen abstraction from the surfactant predominated (Scheme 6.42) [170].

Intramolecular reactions between phenyl cations and a chain tethered to the aromatic ring have been likewise investigated. Photosolvolysis of 2-chloropropylbenzene (**123**) in TFE afforded a mixture of products containing propylbenzene, indane, and 2-trifluoroethoxypropylbenzene along with other solvolysis products (Scheme 6.43a) [116]. The intermediacy of a 2-propylphenyl cation was envisaged that either added to the solvent or underwent intramolecular hydrogen abstraction from the propyl chain. A useful variation was later tested when a nucleophile moiety was introduced in the chain, as in the case of *O*-allyloxy derivatives (e.g., **124**). The phenyl cation formed by irradiation of 1-(3-methylbut-2-enyloxy)-2-chlorobenzene in AcOEt gave 3-isopropenyl-2,3-dihydrobenzofuran (**125**) in 80% yield. It is

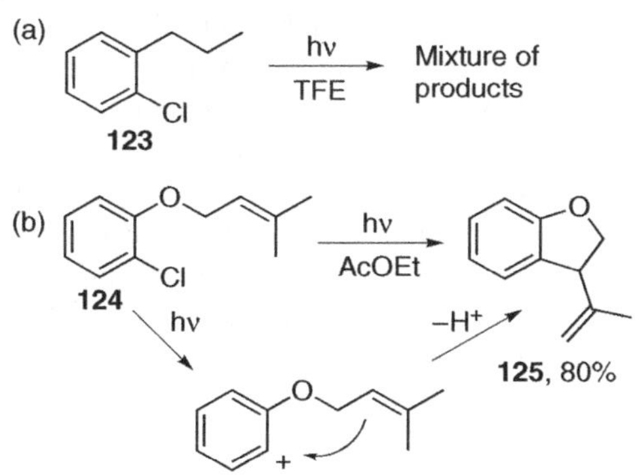

Scheme 6.43.

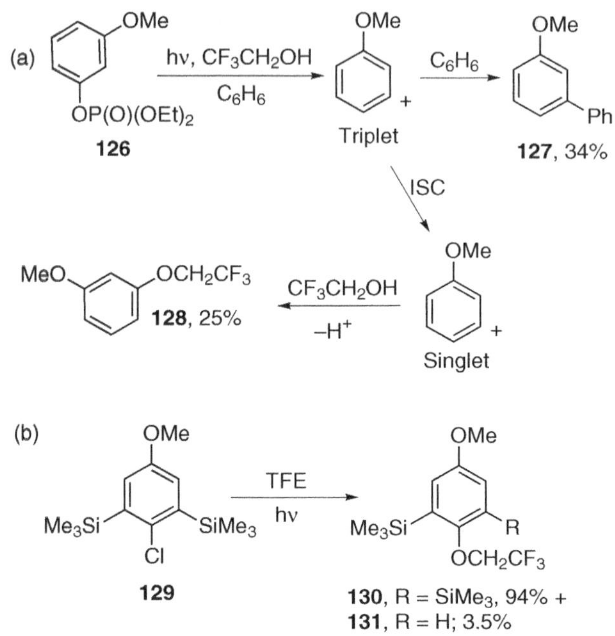

Scheme 6.44.

noteworthy that the phenyl cation addition onto the tethered double bond occurred exclusively by the *5-exo* mode (Scheme 6.43b) [171].

As mentioned above, singlet phenyl cations can be generated directly by photolysis of some diazonium salts. In particular cases, however, these cations were formed (at least in part) by intersystem crossing from the corresponding triplets and the related solvolityc chemistry was then observed. This path appears to be followed for cations bearing an electron-donating group in the *meta* position or a silicon-based substituent in the *ortho* position. As an example, irradiation of phosphate **126** in TFE in the presence of benzene led to the expected arylation product **127** in 34% yield, but the singlet cation had a role and gave the trifluoroethylether **128** in 25% yield by solvent addition (Scheme 6.44a) [172]. On the other hand, the stability imparted to the singlet cation by the two silyl groups in *bis*-silylated anisole **129** shifted the reaction toward solvolysis from the singlet as the exclusive process to form compounds **130** and **131** (Scheme 6.44b) [173].

The high stabilizing effect of the trimethylstannnyl group on triplet phenyl cations and the electrofugal ability of the $-SnMe_3^+$ moiety caused a further variation of the reaction course and the formation of a benzyne intermediate by two consecutive elimination (of chloride anion and of $SnMe_3^+$ cation). Thus, the irradiation of stannylated *N*,*N*-dimethylchloro aniline **132** in TFE gave, along with other arylation products, the isomeric trifluoroethyl ethers **133** and **134** in an overall 20% yield (Scheme 6.45) [174].

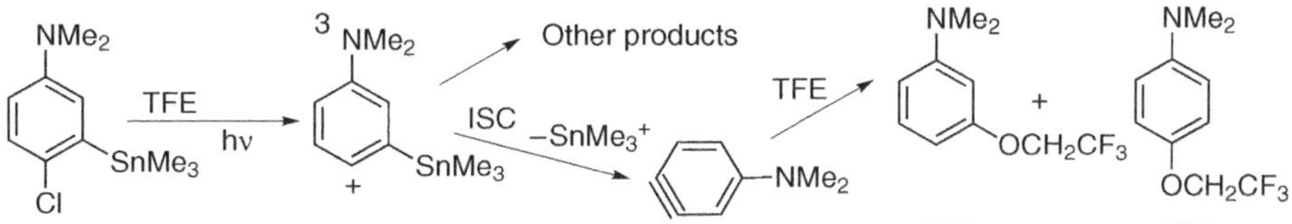

132

133 134

133 + **134** = 20%

Scheme 6.45.

It thus appears that the formation of Ar—C bonds via phenyl cations does offer a convenient alternative to metal catalyzed cross coupling reactions and is endowed by a peculiar "green" character [175, 176]. It may be mentioned that the photochemistry of aryl cations has been developed also in a nonsynthetic direction, since these intermediates play an important role in the photodegradation of fluoroquinolone [177] and oxazolidinone [178] antibiotics.

REFERENCES

1. McClelland, R. A. (1996). Flash photolysis generation and reactivities of carbenium ions and nitrenium ions. *Tetrahedron*, **52**, 6823–6858.
2. Das, P. K. (1993). Transient carbocations and carbanions generated by laser flash photolysis and pulse radiolysis. *Chemical Reviews*, **93**, 119–144.
3. Kropp, P. J. (1984). Photobehavior of alkyl halides in solution: radical, carbocation, and carbene intermediates. *Accounts of Chemical Research*, **17**, 131–137.
4. CRC Handbook of Organic Photochemistry and Photobiology, third edition, Griesbeck, A., Oelgemoeller, M., Ghetti, F., Eds. CRC Press, Boca Raton, FL, 2012.
5. Lukeman, M., Photodecarboxylation of arylacetic acids. In ref. 4, pp. 715–726.
6. Stermitz, F. R., Huang, W. H. (1971). Photochemistry of *N*-heterocycles. VII. Thermal and photodecarboxylation of 2-, 3-, and 4-pyridylacetic acid. *Journal of the American Chemical Society*, **93**, 3427–3431.
7. Epling, G. A., Lopes, A. (1977). Fragmentation pathways in the photolysis of phenylacetic acid. *Journal of the American Chemical Society*, **99**, 2700–2704.
8. Meiggs, T. O., Grosweiner, L. I., Miller, S. I. (1972) Extinction coefficient and recombination rate of benzyl radicals. I. Photolysis of sodium phenylacetate. *Journal of the American Chemical Society*, **94**, 7981–7986.
9. McAuley, I., Krogh, E., Wan, P. (1988). Carbanion intermediates in the photodecarboxylation of benzannelated acetic acids in aqueous solution. *Journal of the American Chemical Society*, **110**, 600–602.
10. Krogh, E., Wan, P. (1992). Photodecarboxylation of diarylacetic acids in aqueous solution: Enhanced photogeneration of cyclically conjugated eight π electron carbanions. *Journal of the American Chemical Society*, **114**, 705–712.
11. Borsarelli, C. D., Braslavsky, S. E., Sortino, S., Marconi, G., Monti, S. (2000). Photodecarboxylation of ketoprofen in aqueous solution. A time-resolved laser-induced optoacoustic study. *Photochemistry and Photobiology*, **72**, 163–171.

12. Xu, M., Wan, P. (2000). Efficient photodecarboxylation of aroyl-substituted phenylacetic acids in aqueous solution: A general photochemical reaction. *Chemical Communications*, 2147–2148.
13. Suzuki, T., Osanai, Y., Isozaki, T. (2012). Effect of basic amino acids on photoreaction of ketoprofen in phosphate buffer solution. *Photochemistry and Photobiology*, **88**, 884–888.
14. Laferrière, M., Sanramé, C. N., Scaiano, J. C. (2004). A remarkably long-lived benzyl carbanion. *Organic Letters*, **6**, 873–875.
15. Margerum, J. D., Petrusis, C. T. (1969). Photodecarboxylation of nitrophenylacetate ions. *Journal of the American Chemical Society*, **91**, 2467–2472.
16. Morrison, J., Wan, P., Corrie, J. E. T., Munasinghe, V. R. N. (2003). Chemistry of photogenerated α-hydroxy-*p*-nitrobenzyl carbanions in aqueous solution: Protonation vs. disproportionation. *Canadian Journal of Chemistry*, **81**, 586–597.
17. Wan, P., Muralidharan, S. (1988). Structure and mechanism in the photo-retro-aldol type reactions of nitrobenzyl derivatives. Photochemical heterolytic cleavage of carbon–carbon bonds. *Journal of the American Chemical Society*, **110**, 4336–4345.
18. Abraham, H.-W. Stabilized carbocations generated by photoheterolysis. In ref. 4, pp. 419–443.
19. Perkins, R. R., Pincock, R. E. (1975). Photochemical substitution reactions of adamantyl halides. *Tetrahedron Letters*, **16**, 943–946.
20. Kropp, P. J., Poindexter, G. S., Pienta, N. J., Hamilton, D. C. (1976). Photochemistry of alkyl halides. 4. 1-norbornyl, 1-norbornylmethyl, 1- and 2-adamantyl, 1-octyl bromides and iodides. *Journal of the American Chemical Society*, **98**, 8135–8144.
21. Reddy, D. S., Sollott, G. P. (1989). Photolysis of cubyl iodides: Access to the cubyl cation. *The Journal of Organic Chemistry*, **54**, 722–723.
22. Bhalerao, V. K., Nanjundiah, B. S., Sonawane, H. R., Nair, P. M. (1986). Photolysis of 2-phenylethyl and 4-phenyl-1-butyl halides in alcoholic solvents. *Tetrahedron*, **42**, 1487–1496.
23. Wan, P., Yates, K. (1984). Photogenerated carbonium ions and vinyl cations in water. *Reviews of Chemical Intermediates*, **5**, 157–181 and reference therein.
24. Kropp, P. J., Reardon, E. J., Gaibel, Z. L. F., Jr., Williard, K. F., Hattaway, J. H., Jr., (1973). Photochemistry of alkenes. Direct irradiation in hydroxylic media. *Journal of the American Chemical Society*, **95**, 7058–7067.
25. Kropp, P. J., Krauss, H. J. (1967). Photochemistry of cycloalkenes. III. Ionic behavior in protic media and isomerization in aromatic hydrocarbon media. *Journal of the American Chemical Society*, **89**, 5199–5208.
26. Kropp, P. J. (1969). Photochemistry of cycloalkenes. V. Effects of ring size and substitution. *Journal of the American Chemical Society*, **91**, 5783–5791.
27. Klett, M. W., Johnson, R. P. (1983). Cumulene photochemistry: Photoaddition of neutral methanol to phenylallenes. *Tetrahedron Letters*, **24**, 1107–1110.
28. Wiseman, J. R., Kipp, J. E. (1982). (*E*)-Bicyclo[3.3.l]non-1-ene. *Journal of the American Chemical Society*, **104**, 4688–4689.
29. Kropp, P. J., Pienta, N. J., Sawyer, J. A., Polniaszek, R. P. (1981). Photochemistry of alkyl halides—VII: Cyclopropanation of alkenes. *Tetrahedron*, **37**, 3229–3286.
30. Barltrop, J. A., Schofield, P. (1962). Photosensitive protecting groups. *Tetrahedron Letters*, **3**, 697–699.

31. Barltrop, J. A., Schofield, P. (1965). Organic photochemistry. Part II. Some photosensitive protecting groups. *Journal of the Chemical Society*, 4758–4765.
32. Chamberlin, J. W. (1966). Use of the 3, 5-dimethoxybenzyloxycarbonyl group as a photosensitive N-protecting group. *The Journal of Organic Chemistry*, **31**, 1658–1660.
33. Bochet, C. G., Blanc, A. Photolabile protecting groups in organic synthesis In ref. 4, pp. 73–93.
34. Wang, P., Zhou, L., Zhang, X., Liang, X. (2010). Facilitated photochemical cleavage of benzylic C—O bond. Application to photolabile hydroxyl-protecting group design. *Chemical Communications*, **46**, 1514–1516.
35. Zhou, L., Yang, H., Wang, P. (2011). Development of trityl-based photolabile hydroxyl protecting groups. *The Journal of Organic Chemistry*, **76**, 5873–5881.
36. Misetic, A., Boyd, M. K. (1998). The pixyl (Px) group: A Novel photocleavable protecting group for primary alcohols. *Tetrahedron Letters*, **39**, 1653–1656.
37. Wan, P., Yates, K., Boyd, M. K. (1985). Adiabatic photodehydroxylation of 9-phenylxanthen-9-ol. Observation of carbocation fluorescence in neutral aqueous solution. *The Journal of Organic Chemistry*, **50**, 2881–2886.
38. Coleman, M. P., Boyd, M. K. (2002). S-pixyl analogues as photocleavable protecting groups for nucleosides. *The Journal of Organic Chemistry*, **67**, 7641–7648.
39. DeCosta, D. P., Howell, N., Pincock, A. L., Pincock, J. A., Rifai, S. (2000). 1,3-Dimethoxy-5-methylene-1,3-cyclohexadiene compounds with leaving groups at C6: Generation, solvolytic reactivity, and their importance in the photochemistry of 3,5-dimethoxybenzyl derivatives. *The Journal of Organic Chemistry*, **65**, 4698–4705.
40. Foster, B., Gaillard, B., Pincock, A. L., Pincock. J. A., Sehmbey, C. (1987). Substituent effects on homolytic versus heterolytic photocleavage of (1-naphthylmethyl)trimethylammonium chlorides. *Canadian Journal of Chemistry*, **65**, 1599–1607.
41. Ammer, J., Mayr, H. (2010). Electrophilic reactivity of the α,α-dimethylbenzyl (cumyl) cation. *Macromolecules*, **43**, 1719–1723.
42. Ammer, J., Sailer, C. F., Riedle, E., Mayr, H. (2012). Photolytic generation of benzhydryl cations and radicals from quaternary phosphonium salts: How highly reactive carbocations survive their first nanoseconds. *Journal of the American Chemical Society*, **134**, 11481–11494 and references therein.
43. Bi, N.-M., Ren, M.-G., Song, Q.-H. (2010). Photo–Ritter reaction of arylmethyl bromides in acetonitrile. *Synthetic Communications*, **40**, 2617–2623.
44. Givens, R. S., Matuszewski, B., Athey, P. S., Stoner, M. R. (1990). Photochemistry of phosphate esters: An efficient method for the generation of electrophiles. *Journal of the American Chemical Society*, **112**, 6016–6021.
45. Givens, R. S., Matuszewski, B. (1984). Photochemistry of phosphate esters: An efficient method for the generation of electrophiles. *Journal of the American Chemical Society*, **106**, 6860–6861.
46. Rajesh, C. S., Givens, R. S., Wirz, J. (2000). Kinetics and Mechanism of Phosphate Photorelease from benzoin diethyl phosphate: Evidence for adiabatic fission to an α-Keto cation in the triplet state. *Journal of the American Chemical Society*, **122**, 611–618.
47. Stang, P. J., Rappoport, Z., Hanack, M., Subramanian, R. L. *Vinyl Cations*, Academic Press, New York, 1979.

48. Rappoport, Z. Vinyl cations. In Abramovitch, R. A., Ed. *Reactive Intermediates*, Volume **3**, Plenum press, New York, 1983, Chapter 7 and references cited therein.
49. Olah, G. A., Prakash, G. K. S., Molnar, A., Sommer, J. *Superacid Chemistry*, second edition, John Wiley & Sons, Hoboken, NJ, 2009.
50. Rappoport, Z., Noy, E., Houminer, Y. (1976). Vinylic cations from solvolysis. XXIII. Degenerate β-anisyl rearrangement during the solvolysis of 2-anisyl-1,2-diphenylvinyl bromides. Free ions as intermediates. *Journal of the American Chemical Society*, **98**, 2238–2244.
51. Jackel, K. P., Hanack, M. (1977). Vinylkationen, 24. Eine [1,2]-Hydridwanderung bei der Solvolyse von Phenylvinyl-trifluormethansulfonaten. *Chemische Berichte*, **10**, 199–207.
52. Stang, P. J., Dueber, T. E. (1977). Vinyl cations. 12. Mechanism of reaction of *cis*- and *trans*-3-phenyl-2-buten-2-yl triflates. Evidence for vinylidene phenonium ions. *Journal of the American Chemical Society*, **99**, 2602–2610.
53. Lee, C. C., Paine, A. J., Ko, E. C. F. (1977). Degenerate rearrangements in solvolytic studies with *cis*- and *trans*-2-phenyl-1,2-di-p-tolylvinyl-2-^{13}C bromides. *Journal of the American Chemical Society*, **99**, 7267–7272.
54. Roberts, T. D. (1971). *cis*-Stilbene as an intermediate in the photochemical cyclization of diphenylacetylene to phenanthrene. *Journal of the Chemical Society D: Chemical Communications*, 362–362.
55. Wooldridge, T., Roberts, T. D. (1973). Vinyl cations as intermediates in photohydration of acetylenes. *Tetrahedron Letters*, **14**, 4007–4008.
56. Kropp, P. J. Photobehavior of alkyl halides. In Horspool, W. M., Song, P. S. Eds., *CRC Handbook of Organic Photochemistry and Photobiology*, CRC Press, Boca Raton, FL, 1995, pp. 1142–1170.
57. Kitamura, T. C–X bond fission in alkene system. In Horspool, W. M., Song, P. S., Eds. *CRC Handbook of Organic Photochemistry and Photobiology*, CRC Press, Boca Raton, FL, 1995, pp. 1171–1180.
58. Lodder, G., Cornelisse, J. Recent advances in the photochemistry of the carbon–halogen bond. In Patai, S., Rappoport, Z., Eds. *The Chemistry of Functional Groups-Supplement D2: The Chemistry of Halides, Pseudohalides and Azides*, Part 2, John Wiley & Sons, Chichester, 1995, pp. 861–972.
59. Lodder, G. Photochemical generation and reactivity. In Rappoport, Z., Stang, P. J., Eds. *Dicoordinated Carbocations*, John Wiley & Sons, Chichester, 1997, pp. 377–431.
60. Kropp, P. J. Photobehavior of alkyl halides. In Lenci, F., Horspool, W., Eds. *CRC Handbook of Organic Photochemistry and Photobiology*, second edition, CRC Press, Boca Raton, FL, 2004, pp. 1-1–1-32.
61. Kitamura, T. C–X bond fission in alkene system. In Lenci, F., Horspool, W., Eds. *CRC Handbook of Organic Photochemistry and Photobiology*, second edition, CRC Press, Boca Raton, FL, 2004, pp. 11-1–11-10.
62. Sket, B., Zupan, M. (1979). Photochemistry of 2-halogeno-1,1-diphenylethylenes. The photo–Fritsch–Buttenberg–Wiechell rearrangement. *Journal of the Chemical Society, Perkin Transactions* **1**, 752–756.
63. Kitamura, T., Kobayashi, S., Taniguchi, H. (1982). Photochemistry of vinyl halides. Vinyl cation from photolysis of 1,1-diaryl-2-halopropenes. *The Journal of Organic Chemistry*, **47**, 2323–2328.

64. Gronheid, R., Zuilhof, H., Hellings, M. G., Cornelisse, J., Lodder, G. (2003). Photochemical generation of a primary vinyl cation from (*E*)-bromostyrene: Mechanisms of formation and reaction. *The Journal of Organic Chemistry*, **68**, 3205–3215.

65. van Alem, K., Belder, G., Lodder, G., Zuilhof, H. (2005). Photochemical generation of highly destabilized vinyl cations: The effects of α- and *β*-trifluoromethyl versus α- and *β*-methyl substituents. *The Journal of Organic Chemistry*, **70**, 179–190.

66. Kitamura, T., Kobayashi, S., Taniguchi, H. (1986). Photolysis of vinyl halides. Preferential formation of vinyl cations by copper(II) salts. *Journal of the American Chemical Society*, **108**, 2641–2645.

67. Zupancic, N., Sket, B. (1991). The influence of metal(II) salts and solid supports on the nature of the photochemical reaction of 1,1-diphenyl-2-haloethene. *Journal of Photochemistry and Photobiology A:Chemistry*, **60**, 361–367.

68. Kropp, P. J., McNeely S. A., Davis, R. D. (1983). Photochemistry of alkyl halides. 10. Vinyl halides and vinylidene dihalides. *Journal of the American Chemical Society*, **105**, 6907–6915.

69. Kitamura, T., Muta, T., Kobayashi, S., Taniguchi, H. (1982). A novel ortho-substituent effect on formation of vinyl cations in the photolysis of vinyl bromides. *Chemistry Letters*, **11**, 643–644.

70. van Dorp, J. W. J, Lodder, G. (2008). Substituent Effects on the photogeneration and selectivity of triaryl vinyl cations. *The Journal of Organic Chemistry*, **73**, 5416–5428.

71. McNeely, S. A., Kropp, P. J. (1976). Photochemistry of alkyl halides. 3. Generation of vinyl cations. *Journal of the American Chemical Society*, **98**, 4319–4320.

72. Kitamura, T., Kobayashi, S., Taniguchi, H. (1984). Isoquinoline derivatives from the Ritter-type reaction of vinyl cations *Chemistry Letters*, **13**, 1351–1354.

73. Kitamura, T., Kabashima, T., Taniguchi, H. (1991). Phenylthiolation of arylvinyl bromides by photolysis. *The Journal of Organic Chemistry*, **56**, 3739–3741.

74. Kitamura, T., Kabashima, T., Kobayashi, S., Taniguchi, H. (1988). Drastic base dependence of products in photolysis of 1-(*p*-methoxyphenyl)-2,2-diphenylvinyl bromide. Ipso substitution by alkoxide ions. *Chemistry Letters*, **17**, 1951–1954.

75. Kitamura, T., Kabashima, T., Nakamura, I., Fukuda, T., Taniguchi, H. (1991). Ipso substitution of triarylvinyl cations by alkoxide anions. *Journal of the American Chemical Society*, **113**, 7255–7261.

76. Kitamura, T., Nakamura, I., Kabashima, T., Kobayashi, S., Taniguchi, H. (1989). A novel spiro adduct from intramolecular *ipso* substitution in the photolysis of an α-[p(2-hydroxyalkoxy)phenyl]vinyl bromide. *Journal of the Chemical Society, Chemical Communications*, 1154–1155.

77. Kitamura, T., Kobayashi, S., Taniguchi, H. (1979). Photolysis of vinyl bromides in the presence of tetrabutylammonium azide: Trapping of a vinyl cation with azide ion. *Tetrahedron Letters*, **20**, 1619–1622.

78. Kitamura, T., Kobayashi, S., Taniguchi, H. (1984). Photochemistry of vinyl halides. Heterocycles from reaction of photogenerated vinyl cations with azide anion. *The Journal of Organic Chemistry*, **49**, 4755–4760.

79. Kitamura, T., Kobayashi, S., Taniguchi, H. (1984). Reaction of photogenerated vinyl cations with ambient anions. *Chemistry Letters*, **13**, 1523–1526.

80. Kitamura, T., Kobayashi, S., Taniguchi, H. (1990). Photolysis of vinyl halides. Reaction of photogenerated vinyl cations with cyanate and thiocyanate ions. *The Journal of Organic Chemistry*, **55**, 1801–1805.
81. Ikeda, T., Kobayashi, S., Taniguchi, H. (1982). A new route to spiro[2.5]octa-1,4,7-trien-6-ones. *Synthesis*, 393–394.
82. Suzuki, T., Kitamura, T., Sonoda, T., Kobayashi, S., Taniguchi, H. (1981). Photochemistry of vinyl halides. Formation of benzofurans by photolysis of p-(o-methoxyphenyl)vinyl bromides. *The Journal of Organic Chemistry*, **46**, 5324–5328.
83. Kitamura, T., Kobayashi, S., Taniguchi, H. (1984). Formation of dibenz[*b,f*]oxepins from β(-(*o*-aryloxyphenyl)vinyl bromides. *Chemisty Letters*, **13**, 547–548.
84. Kitamura, T., Kobayashi, S., Taniguchi, H., Hori, K. (1991). Photolytic and solvolytic reactions of β[o-(aryloxy)-phenyl]vinyl bromides. Intramolecular arylation of vinyl cations into dibenzoxepins. *Journal of the American Chemical Society*, **113**, 6240–6245.
85. Kitamura, T., Kobayashi, S., Taniguchi, H., Fiakpui, C. Y., Chuck Lee, C., Rappoport, Z. (1984). Degenerate β-aryl rearrangements in photochemically generated triarylvinyl cations. *The Journal of Organic Chemistry*, **49**, 3167–3170.
86. Suzuki, T., Sonoda, T., Kobayashi, S., Taniguchi, H. (1976). Photochemistry of vinyl bromides: A novel 1 2-aryl group migration. *Journal of the Chemical Society, Chemical Communications*, 180.
87. Gronheid, R., Lodder, G., Ochiai, M., Sueda, T., Okuyama, T. (2001). Thermal and photochemical solvolysis of (*E*)- and (*Z*)-2-phenyl-1-propenyl(phenyl)iodonium tetrafluoroborate: benzenium and primary vinylic cation intermediates. *Journal of the American Chemical Society*, **123**, 8760–8765.
88. Gronheid, R., Lodder, G., Okuyama, T. (2002). Photosolvolysis of (*E*)-styryl(phenyl) iodonium tetrafluoroborate. Generation and reactivity of a primary vinyl cation. *The Journal of Organic Chemistry*, **67**, 693–702.
89. Slegt, M., Gronheid, R., van der Vlugt, D., Ochiai, M., Okuyama, T., Zuilhof, H., Overkleeft, H. S., Lodder, G. (2006). Photochemical generation of six- and five-membered cyclic vinyl cations. *The Journal of Organic Chemistry*, **71**, 2227–2235.
90. Matveeva, E. D., Podrugina, T. A., Taranova, M. A., Borisenko, A. A., Mironov, A. V., Gleiter, R., Zefirov, N. S. (2011). Photochemical synthesis of phosphinolines from Phosphonium–iodonium ylides. *The Journal of Organic Chemistry*, **76**, 566–572.
91. Ackermann, L. *Ed Modern Arylation Methods* Wiley-VCH, Weinheim, 2009.
92. Dichiarante, V., Fagnoni, M. (2008). Aryl cation chemistry as an emerging versatile tool for metal-free arylations. *Synlett*, 787–800.
93. Himeshima, Y., Kobayashi, H., Sonoda, T. (1985). A first example of generating aryl cations in the solvolysis of aryl triflates in trifluoroethanol. *Journal of the American Chemical Society*, **107**, 5286–5288.
94. Holweger, W., Hanack, M. (1984). Phenylkationem als reaktive zwischenstufen bei der solvolyse von dien-in-yl-triflaten. *Chemische Berichte*, **117**, 3004–3020.
95. Bleckmann, W., Hanack, M. (1984). Synthese und solvolyse von (Z)/(E)-1-methyl-2-[2-(1-propin-1-yl)phenyl]vinyl-triflat und 1-methylen-2-[2-(1-propin-1-yl)-phenyl]ethyl-triflat. *Chemische Berichte*, **117**, 3021–3033.
96. Hanack, M., Rieth, R. (1987). Synthese und solvolyse cyclisher dieninyl-triflate. *Chemische Berichte*, **120**, 1659–1666.

97. Zollinger, H. (1973). Reactivity and stability of arenediazonium ions. *Accounts of Chemical Research*, **6**, 335–341.

98. Zollinger, H. (1978). Nitrogen as leaving group: Dediazoniations of aromatic diazonium ions. *Angewandte Chemie International Edition*, **17**, 141–150.

99. Hegarty A. F. Kinetics and mechanism of reactions involving diazonium and diazo groups. In Patai, S., Ed. *The Chemistry of Diazonium and Diazo Groups*, John Wiley & Sons, London, 1978, pp. 511–519.

100. Zollinger, H. Dediazoniations of arenediazonium ions and related compounds. In Patai, S., Rappoport, Z., Eds. *The Chemistry of Triple-Bonded Functional Groups*, John Wiley & Sons, London, 1983, Suppl. C, pp. 603–669.

101. Zollinger, H. (1983). Mechanistic investigations on reactions of molecoular nitrogen with organic species. *Pure and Applied Chemistry*, **55**, 401–408.

102. Ando, W. Photochemistry of the diazonium and diazo groups. In Patai, S., Ed. *The Chemistry of Diazonium and Diazo Groups*, Part 1, John Wiley & Sons, Chichester, 1978, pp. 341–487.

103. Pazo-Llorente, R., Bravo-Diaz, C., Gonzalez-Romero, E. (2003). Solvolysis of some arenediazonium salts in binary EtOH/H_2O mixtures under acidic conditions. *European Journal of Organic Chemistry*, 3421–3428 and references therein.

104. Sawaguchi, M., Fukuhara, T., Yoneda, N. (1999). Preparation of aromatic fluorides: Facile photo-induced fluorinative decomposition of arenediazonium salts and their related compounds using pyridine-HF. *Journal of Fluorine Chemistry*, **97**, 127–133.

105. Shinhama, K., Aki, S., Furuta, T., Minamikawa, J. (1993). Facile conversion of arenediazonium salts to the corresponding fluoroarenes using boron trifluoride diethyl ether complex. *Synthetic Communications*, **23**, 1577–1582.

106. Canning, P. S. J., Maskill, H., McCrudden, K., Sexton, B. (2002). A product analytical study of the thermal and photolytic decomposition of some arenediazonium salts in solution. *Bullettin of the Chemical Society of Japan*, **75**, 789–800.

107. Heredia-Moya, J., Kirk, K. L. (2007). Photochemical Schiemann reaction in ionic liquids. *Journal of Fluorine Chemistry*, **128**, 674–678.

108. Lazzaroni, S., Dondi, D., Fagnoni, M., Albini, A. (2008). Geometry and energy of substituted phenyl cations. *The Journal of Organic Chemistry*, **73**, 206–211.

109. Lazzaroni, S., Dondi, D., Fagnoni, M., Albini, A. (2010). Selectivity in the reaction of triplet phenyl cations. *The Journal of Organic Chemistry*, **75**, 315–323.

110. Dill, J. D., Schleyer, P. v. R., Pople, J. A. (1977). Molecular orbital theory of the electronic structure of molecules. 31. Substituent stabilization of the phenyl cation. *Journal of the American Chemical Society*, **99**, 1–8.

111. Laali, K. K., Rasul, G., Prakash, G. K. S., Olah, G. A. (2002). DFT Study of substituted and benzannelated aryl cations: Substituent dependency of singlet/triplet ratio. *The Journal of Organic Chemistry*, **67**, 2913–2918.

112. Aschi, M., Harvey, J. N. (1999). Spin isomerisation of *para*-substituted phenyl cations. *Journal of the Chemical Society, Perkin Transactions* **2**, 1059–1062.

113. Filippi, A., Lilla, G., Occhiucci, G., Sparapani, C., Ursini, O., Speranza, M. (1995). Structure and reactivity of free arylium ions. 2. Reactions of decay-formed isomeric $XC_6H_4^+$ ions with gaseous and liquid nucleophile. *The Journal of Organic Chemistry*, **60**, 1250–1264.

114. Nicolaides, A., Smith, D. M., Jensen, F., Radom, L. (1997). Phenyl radical, cation, and anion. The triplet–singlet gap and higher excited states of the phenyl cation. *Journal of the American Chemical Society*, **119**, 8083–8088.

115. Apeloig, Y., Arad, D. (1985). Stabilization of the phenyl cation by hyperconjugation. *Journal of the American Chemical Society*, **107**, 5285–5286.

116. Hori, K., Sonoda, T., Harada, M., Yamazaki-Nishida, S. (2000). Theoretical study on the reactivity of phenyl cation with a propyl group at *ortho*-position. *Tetrahedron*, **56**, 1429–1436.

117. Gasper, S. M., Devadoss, C., Shuster, G. B. (1995). Photolysis of substituted benzene-diazonium salts: Spin-selective reactivity of aryl cations. *Journal of the American Chemical Society*, **117**, 5206–5211.

118. Milanesi, S., Fagnoni, M., Albini A. (2003). Cationic arylation through photo(sensitised) decomposition of diazonium salts. Chemoselectivity of triplet phenyl cations. *Chemical Communications*, 216–217.

119. Milanesi, S., Fagnoni, M., Albini, A. (2005). (Sensitized) photolysis of diazonium salts as a mild general method for the generation of aryl cations. Chemoselectivity of the singlet and triplet 4-substituted phenyl cations. *The Journal of Organic Chemistry*, **70**, 603–610.

120. Romsted, L. S., Zhang, J., Zhuang, L. (1998). Mechanism of reaction of an arenediazonium ion in aqueous solutions of acetamide, *N*-methylacetamide, and *N,N*-dimethylacetamide. A potential method for chemically tagging peptide bonds at aggregate interfaces. *Journal of the American Chemical Society*, **120**, 10046–10054.

121. Steenken, S., Askokkuna, M., Maruthamuthu, P., McClelland, R. A. (1998). Making photochemically generated phenyl cations visible by addition to aromatics: Production of phenylcyclohexadienyl cations and their reactions with bases/nucleophiles. *Journal of the American Chemical Society*, **120**, 11925–11931.

122. Canning, P. S. J., McCrudden, K., Maskill, H., Sexton, B. (1998). Dediazoniation reactions of arenediazonium ions under solvolytic conditions: Fluoride anion abstraction from trifluoroethanol and α-hydrogen atom abstraction from ethanol. *Chemical Communications*, 1971–1972.

123. Slegt, M., Minne, F., Zuilhof, H., Overkleeft, H. S., Lodder, G. (2007). Photochemical generation and reactivity of naphthyl cations: *Cine* substitution. *European Journal of Organic Chemistry*, 5353–5363.

124. Fagnoni, M., Albini, A. Photonucleophilic substitution reactions. In Ramamurthy, V., Schanze, K., Eds. *Molecular and Supramolecular Photochemistry. Organic Photochemistry and Photophysics*, Vol. **14**, Marcel Dekker, New York, 2006, pp. 131–178.

125. Peñeñory, A. B., Argüello, J. E. Aromatic and heteroaromatic substitution by $S_{RN}1$ and S_N1 reactions. In Albini, A., Fagnoni, M., Eds. *Handbook of Synthetic Photochemistry* Wiley-VCH, Weinheim, 2010 pp. 319–352.

126. Groen, M. B., Havinga, E. (1974). Photoreactions of aromatic compounds. XXXIII. nucleophilic photosubstitutions of some heteroaromatic nitro compounds. *Molecular Photochemistry*, **6**, 9–21.

127. Yang, N. C., Huang, A., Yang, D. H. (1989). Photodehalogenation of 4-haloindoles. *Journal of the American Chemical Society*, **111**, 8060–8061.

128. Fagnoni, M., Albini, A. (2005). Arylation reactions: The photo-S_N1 path via phenyl cation as an alternative to metal catalysis. *Accounts of Chemical Research*, **38**, 713–721.

129. Fagnoni, M. (2006). The photo S_N1 reaction via phenyl cation. *Letters in Organic Chemistry*, **3**, 253–259.

130. Pretali, L., Albini, A. Photochemistry of aryl halides. In Ref. 4, pp. 369–391.

131. Ravelli, D., Fagnoni, M. Photochemistry of phosphate and sulfonate esters. In ref. 4, pp. 393–417.

132. Freccero, M., Fagnoni, M., Albini, A. (2003). Homolytic vs heterolytic paths in the photochemistry of haloanilines. *Journal of the American Chemical Society*, **125**, 13182–13190.

133. Mella, M., Coppo, P., Guizzardi, B., Fagnoni, M., Freccero, M., Albini, A. (2001). Photoinduced, ionic meerwein arylation of olefins. *The Journal of Organic Chemistry*, **66**, 6344–6352.

134. Guizzardi, B., Mella, M., Fagnoni, M., Freccero, M., Albini, A. (2001). Generation and reactivity of the 4-aminophenyl cation by photolysis of 4-chloroaniline. *The Journal of Organic Chemistry*, **66**, 6353–6363.

135. Othmen, K., Boule, P., Szczepanik, B., Rotkiewicz, K. (2000). Photochemistry of 4-chloroaniline in solution. Formation and kinetic properties of a new carbene, 4-iminocyclohexa-2,5-dienylidene. *The Journal of Physical Chemistry A*, **104**, 9525–9534.

136. Fagnoni, M., Mella, M., Albini, A. (1999). Smooth synthesis of aryl- and alkylanilines by photoheterolysis of haloanilines in the presence of aromatics and alkenes. *Organic Letters*, **1**, 1299–1301.

137. Othmen, K., Boule, P. (1997). Phototransformation of 2-chloroaniline in aqueous solution. *Bulletin of Environmental Contamination and Toxicology*, **59**, 924–927.

138. Manet, I., Monti, S., Fagnoni, M., Protti, S., Albini, A. (2005). Aryl cation and carbene intermediates in the photodehalogenation of chlorophenols. *Chemistry—A European Journal*, **11**, 140–151.

139. Bonnichon, F., Grabner, G., Guyot, G., Richard, C. (1999). Photochemistry of substituted 4-halogenophenols: Effect of a CN substituent. *Journal of the Chemical Society, Perkin Transactions* **2**, 1203–1210.

140. Durand, A. P., Brown, R. G., Worral, D., Wilkinson, F. (1998). Study of the aqueous photochemistry of 4-fluorophenol, 4-bromophenol and 4-iodophenol. *Journal of the Chemical Society, Perkin Transactions* **2**, 365–370.

141. Grabner, G., Richard, C., Koehler, G. (1994). Formation and reactivity of 4-oxocyclohexa-2 5-dienylidene in the photolysis of 4-chlorophenol in aqueous solution at ambient temperature. *Journal of the American Chemical Society*, **116**, 11470–11480.

142. Protti, S., Fagnoni, M., Mella, M., Albini, A. (2004). Aryl cations from aromatic halides. Photogeneration and reactivity of 4-hydroxy(methoxy)phenyl cation. *The Journal of Organic Chemistry*, **69**, 3465–3473.

143. Zhang, G., Wan, P. (1994). Photogeneration of HF from fluoromethoxybenzenes in aqueous solution. *Journal of the Chemical Society, Chemical Communications*, 19–20.

144. Lazzaroni, S., Dondi, D., Fagnoni, M., Albini, A. (2007). Photochemical arylation reactions by 4-chlorothioanisole. *European Journal of Organic Chemistry*, 4360–4365.

145. Protti, S., Fagnoni, M., Albini, A. (2006). Benzyl (phenyl) γ- and δ-lactones via photoinduced tandem Ar–C, C–O bond formation. *Journal of the American Chemical Society*, **128**, 10670–10671.

146. Lazzaroni, S., Protti, S., Fagnoni, M., Albini, A. (2009). Photoinduced three-component reaction: A convenient access to 3-arylacetals or 3-arylketals. *Organic Letters*, **11**, 349–352.

147. Lazzaroni, S., Protti, S., Fagnoni, M., Albini, A. (2010). Participation of a heterolytic path in the photochemistry of chlorobenzene. *Journal of Photochemistry and Photobiology A: Chemistry*, **210**, 140–144.

148. Manet, I., Monti, S., Grabner, G., Protti, S., Dondi, D., Dichiarante, V., Fagnoni, M., Albini, A. (2008). Revealing phenylium, phenonium, vinylenephenonium, and benzenium ions in solution. *Chemistry—A European Journal*, **14**, 1029–1039.

149. Protti, S., Dondi, D., Mella, M., Fagnoni, M., Albini, A. (2011). Looking for a paradigm for the reactivity of phenonium ions. *European Journal of Organic Chemistry*, 3229–3237.

150. Terpolilli, M., Merli, D., Protti, S., Dichiarante, V., Fagnoni, M., Albini, A (2011). Cationic and radical intermediates in the acid photorelease from aryl sulfonates and phosphates. *Photochemical & Photobiological Sciences*, **10**, 123–127.

151. Coppo, P., Fagnoni, M., Albini, A. (2001). Photochemical conversion of 4-chloroaniline into 4-alkylanilines. *Tetrahedron Letters*, **42**, 4271–4273.

152. Guizzardi, B., Mella, M., Fagnoni, M., Albini, A. (2003). Phenonium ions from the addition of phenyl cations to alkenes. Photochemical synthesis of (rearranged) aminoalkylanilines from haloanilines in the presence of alkenes and amines. *The Journal of Organic Chemistry*, **68**, 1067–1074.

153. Protti, S., Dondi, D., Fagnoni, M., Albini, A. (2008). Photochemical arylation of alkenols: Role of intermediates and synthetic significance. *European Journal of Organic Chemistry*, 2240–2247.

154. Protti, S., Fagnoni, M., Albini, A. (2005). Expeditious synthesis of bioactive allylphenol constituents of the genus *Piper* through a metal-free photoallylation procedure. *Organic & Biomolecular Chemistry*, **3**, 2868–2871.

155. Fraboni, A., Fagnoni, M., Albini, A. (2003). A novel α-arylation of ketones, aldehydes, and esters via a photoinduced S_N1 reaction through 4-aminophenyl cations. *The Journal of Organic Chemistry*, **68**, 4886–4893.

156. Dichiarante, V., Fagnoni, M., Albini, A. (2010). *Solar*ylations via 4-aminophenyl cations. *The Journal of Organic Chemistry*, **75**, 1271–1274.

157. Guizzardi, B., Mella, M., Fagnoni, M., Albini, A. (2003). Photochemical reaction of *N,N*-dimethyl-4-chloroaniline with dienes: new synthetic paths via a phenyl cation. *Chemistry—A European Journal*, **9**, 1549–1555.

158. Mella, M., Esposti, S., Fagnoni, M., Albini, A. (2003). A novel access to 3-aryl-2-norbornyl cation. *Chemical Communications*, 738–739.

159. Mella, M., Fagnoni, M., Albini, A. (2004). On the addition of 4-(*N,N*-dimethylamino) phenyl cation to norbornene. *Organic & Biomolecular Chemistry*, **2**, 3490–3495.

160. Protti, S., Fagnoni, M., Albini, A. (2005). Photocross-coupling reaction of electron-rich aryl chlorides and aryl esters with alkynes. A novel metal-free alkynylation. *Angewandte Chemie International Edition*, **44**, 5675–5678.

161. Protti, S., Fagnoni, M., Albini, A. (2012). A photochemical route to 2-substituted benzo [*b*]furans. *The Journal of Organic Chemistry*, **77**, 6473–6479.

162. De Carolis, M., Protti, S., Fagnoni, M., Albini, A. (2005). Metal-free cross-coupling reactions of aryl sulfonates and phosphates through photoheterolysis of aryl–oxygen bonds. *Angewandte Chemie International Edition*, **44**, 1232–1236.

163. Dichiarante, V., Betti, C., Fagnoni, M., Maia, A., Landini, D., Albini, A. (2007). Characterizing ionic liquids as reaction media through a chemical probe. *Chemistry—A European Journal*, **13**, 1834–1841.

164. Dichiarante, V., Fagnoni, M., Albini, A. (2007). Metal-free synthesis of sterically crowded biphenyls by direct Ar–H substitution in alkyl benzenes. *Angewandte Chemie International Edition*, **46**, 6495–6498.

165. Slegt, M., Overkleeft, H. S., Lodder, G. (2007). Fingerprints of singlet and triplet phenyl cations. *European Journal of Organic Chemistry*, 5364–5375.

166. Guizzardi, B., Mella, M., Fagnoni, M., Albini, A. (2000). Easy photochemical preparation of dimethylaminophenylfurans, -pyrroles and -thiophenes. *Tetrahedron*, **56**, 9383–9389.

167. Dichiarante, V., Fagnoni, M., Albini, A. (2006). Convenient synthesis of electron-donating substituted benzonitriles by photolysis of phenyl halides and esters. *Chemical Communications*, 3001–3003.

168. Dichiarante, V., Fagnoni, M., Albini, A. (2009). Eco-friendly hydrodehalogenation of electron-rich aryl chlorides and fluorides by photochemical reaction. *Green Chemistry*, **11**, 942–945.

169. Manet, I., Monti, S., Bortolus, P., Fagnoni, M., Albini, A. (2005). The photochemistry of 4-chlorophenol in water revisited: The effect of cyclodextrins on cation and carbene reactions. *Chemistry—A European Journal*, **11**, 4274–4282.

170. Canevari, V., Fagnoni, M., Bortolus, P., Albini, A. (2011). Environmental implications of the surfactants effect on the photochemistry of(substituted)chlorophenols in water. *ChemSusChem*, **4**, 98–103.

171. Dichiarante, V., Fagnoni, M., Mella, M., Albini, A. (2006). Intramolecular photoarylation of alkenes by phenyl cation. *Chemistry—A European Journal*, **12**, 3905–3915.

172. Dichiarante, V., Dondi, D., Protti, S., Fagnoni, M., Albini, A. (2007). A meta effect in organic photochemistry? The case of S_N1 reactions in methoxyphenyl derivatives. *Journal of the American Chemical Society*, **129**, 5605–5611.

173. Dichiarante, V., Salvaneschi, A., Protti, S., Dondi, D., Fagnoni, M., Albini, A. (2007). The β effect of silicon in phenyl cations. *Journal of the American Chemical Society*, **129**, 15919–15926.

174. Protti, S., Dichiarante, V., Dondi, D., Fagnoni, M., Albini, A. (2012). Singlet/triplet phenyl cations and benzyne from the photodehalogenation of some silylated and stannylated phenyl halides. *Chemical Science*, **3**, 1330–1337.

175. Albini, A., Fagnoni, M. Photochemistry as a green synthetic method. In Mordini, A., Faigl, F., Eds. *New Methodologies and Techniques for a Sustainable Organic Chemistry*, Springer Science + Business Media B. V., 2008, pp. 279–293.

176. Dichiarante, V., Fagnoni, M., Albini, A. Photochemical arylation reactions. In ref. 91, pp. 513–535.

177. Albini, A., Monti, S. (2003). Photophysics and photochemistry of fluoroquinolones. *Chemical Society Reviews*, **32**, 238–250.

178. Fasani, E., Tilocca, F., Albini, A. (2009). Photochemistry of oxazolidinone antibacterial drugs. *Photochemistry and Photobiology*, **85**, 879–885.

7

PHOTOGENERATION OF CARBENES AND NITRENES

Carbenes and nitrenes exist in two multiplicities, singlet and triplet, that exhibit a quite different chemistry. The mode of generation, thermal or photochemical, has a paramount role in determining which spin state will be arrived at.

7.1 CARBENES

7.1.1 Structure and Chemical Behavior

Carbenes continue to be a favorite topic for mechanistic studies by using photochemical techniques, from laser flash photolysis to matrix isolation. These have given an invaluable contribution to the understanding of the chemistry of such intermediates. Although a range of interesting preparations have been developed by this way, in particular with regard to highly strained compounds, the role of carbenes in synthesis is generally not as important as that, for example, of carbon-centered radicals. In other cases, mixtures are obtained and the rapid development of the chemistry of metal-bonded carbenes ("carbenoids"), which are not considered here, has overcome the role of free carbenes for preparative chemistry. Therefore, only a short overview is presented here; on the other hand, excellent reviews are available [1, 2].

As hinted above, carbenes are neutral intermediates and exist as either singlet (σ^2 configuration) or triplet (σ^1p^1) states. The triplet is the ground state of the (bent) parent species $:CH_2$, but changing structure and medium largely affects the energy gap between σ and p orbitals and may revert the order of the states. Mesomeric and hyperconjugative effects add here to steric effect, important because it may push toward a linear structure and change the nature of the orbitals involved. Roughly,

Photochemically-Generated Intermediates in Synthesis, First Edition.
Angelo Albini and Maurizio Fagnoni.

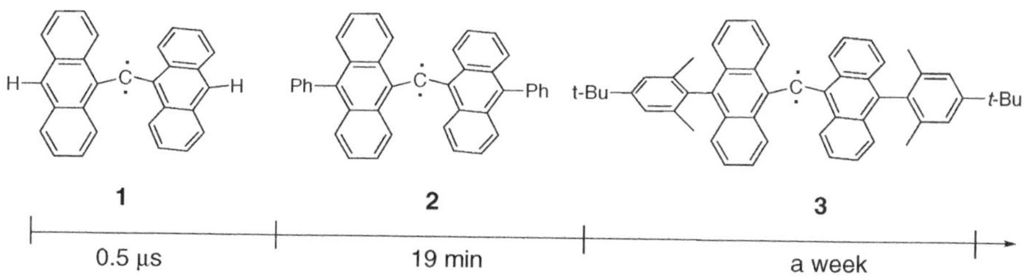

Scheme 7.1.

putting in position α to the carbene (or in *para* on a phenyl group bonded to that position) a heteroatom able to donate a lone pair makes the singlet the ground state, while a carbon-centered substituent or an electron-withdrawing group, such as NO_2, SO_2R, or COR, maintains the triplet as the ground state.

Furthermore, highly persistent carbenes are known which are borderline cases with respect to the concept of intermediate as has been used in this presentation. Indeed, persistent carbenes form a rapidly increasing field of research, or actually two, because the factors that stabilize each spin state are quite different from each other. As mentioned, triplets are stabilized by steric bulk, since this increases the bond angle at the carbon atom [3]. As an example, carbenes **1–3**, generated by photolysis of the corresponding diazo compounds survived in solution at room temperature for a time increasing from microseconds to weeks along the series. Indeed, also the typical carbene reactions were slowed down (Scheme 7.1). In the case of **3**, the reaction with oxygen occurred with a rate constant of $5.5 \times 10^5\,M^{-1}\,s^{-1}$ and hydrogen abstraction from 1,4-cyclohexadiene with a rate constant of $0.02\,M^{-1}\,s^{-1}$, which is certainly far from what expected from a reactive intermediate [4].

On the other hand, the singlet state is highly stabilized by lone-pair donating groups as in examples reported in Scheme 7.2 [5–7]. The solvent polarity, in turn, stabilizes preferentially the more polar singlet. Singlet carbenes stabilized by heteroatoms through the mesomeric effect include indefinitely persistent derivatives, some of which are crystalline solids (e.g., **4**, Scheme 7.2) [8].

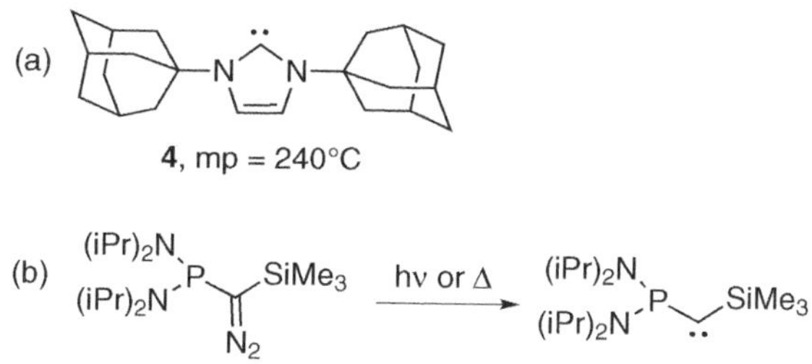

Scheme 7.2.

Both singlet and triplet persistent carbenes have progressed from sparse examples considered as a curiosity to a large number of cases, and predictive rules on their chemistry and stability have been recognized. Persistent singlet carbenes are useful ligands for metals [5–7], and triplet persistent carbenes have been considered for photosensitive magnetic materials [3].

Photochemistry offers a favorite access for the generation of such species from diazo compounds and from diazirines under sufficiently mild conditions allowing us to control subsequent reactions. As an example, di(1-adamantly)carbene **6** is photochemically generated in its triplet (ground) state and undergoes diagnostic reactions, such as oxygen addition to form a carbonyl oxide **7** or hydrogen abstraction (e.g., from the allylic position, compounds **10** and **11**), although the formation of the ether **8** in methanol supports that equilibration with the singlet occurs. In contrast, flash vacuum pyrolysis (FVP) of the same compound and rapid condensation yields a product of intramolecular insertion, the homoadamantene **12** (Scheme 7.3) [9, 10].

On the other hand, it has been suggested that excited states of the carbenes may have a role in photochemistry. Thus, diazirines bearing a tertiary alkyl group were found to give the same product distribution by thermolysis and by triplet photosensitization, but a different one by direct photolysis (Scheme 7.4) [11]. It was proposed that this is because direct irradiation gives rise to high-lying singlet states of carbenes.

From the practical point of view, the most convenient method for the generation of these intermediates in solution is photolysis of diazirines (**14**). Such precursors are, in turn, obtained by oxidative cyclization of imidines (**13**, path *a*) and nucleophilic substitution as appropriate (path b, Scheme 7.5) [12, 13].

Diazirines must be handled with care, but are more stable than the other common precursors—that is, diazo derivatives—historically more important but presently less often used at least for the (toxic and explosive) low-molecular-weight terms. Further photochemical precursors of carbenes are known. One that has been used to

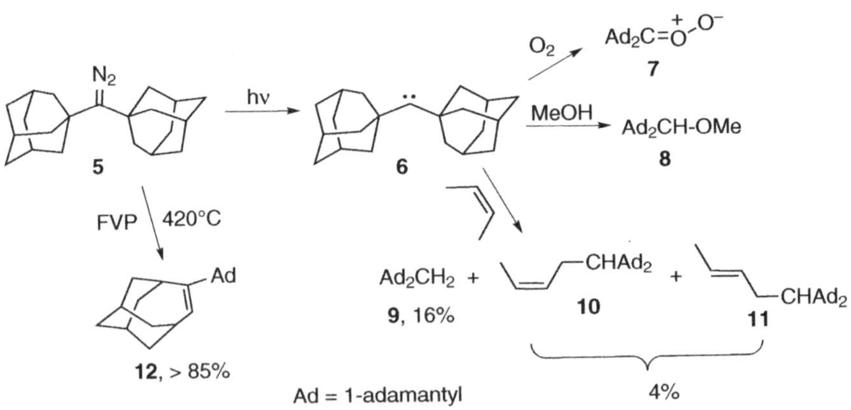

Scheme 7.3.

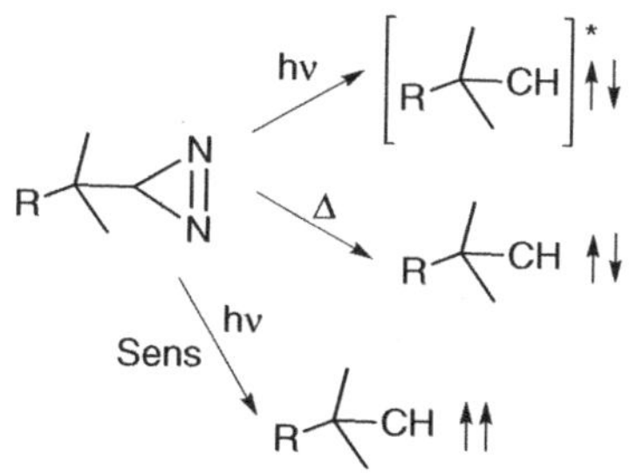

Scheme 7.4.

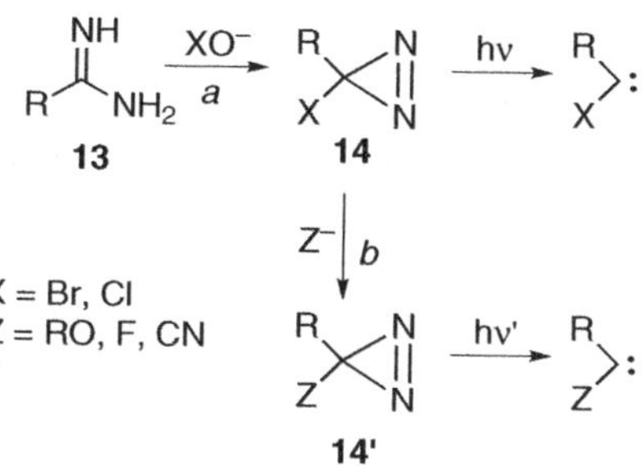

Scheme 7.5.

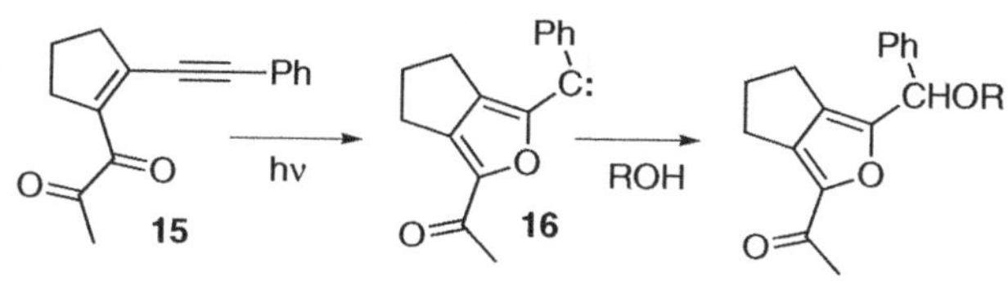

Scheme 7.6.

advantage in some cases is cyclopropaphenanthrene, via retro-cycloaddition [14]. Further approaches are possible. An interesting one involves α-diketones conjugated with an enyne system (**15**) that are photochemically equilibrated with a furylcarbene (**16**, Scheme 7.6). On this basis, more elaborated systems where an intramolecular trap is present have been tested with good results for the formation of polycyclic compounds [15].

7.1.2 Cycloaddition via Photogenerated Carbenes

Stereospecific cycloaddition is the most typical singlet carbene reaction. The rate of the process depends on whether the predominant interaction is that between the

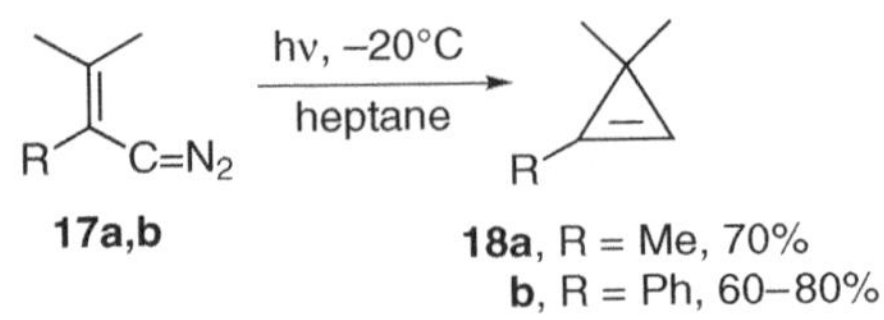

Scheme 7.7.

LUMO (p) of the carbene and the HOMO (π) of the alkene or vice versa between the HOMO (σ) of the carbene and LUMO (π^*) of the alkene. Indeed, direct measurements in flash photolysis demonstrated that some of these intermediates exhibit an electrophilic behavior (e.g., $:CCl_2$) with the relative rate of reaction with tetramethylethylene ~200 times greater than with acrylonitrile, while other ones are nucleophilic—for example, :CMe(OMe), for which the above rates ratio is 1:1000. With amphiphilic carbenes, such as :CCl(OMe), the two rates are in the same order of magnitude [16]. The large number of paths available generally causes the formation of mixtures from alkylcarbenes photochemically generated. This is one of the few cases where the thermal reaction may be cleaner. Much better results are obtained with conjugated carbenes, where generally fast insertion or cycloaddition of the singlet carbene funnels away the system before that competing processes intervene. A typical example is the synthesis of cyclopropenes **18a,b** from vinyldiazo compounds **17a,b**, occurring in 60–80% yield by irradiation in heptane at −20°C (Scheme 7.7) [17].

Unsaturated cyclic carbenes gave addition reactions. Polysubstituted 5-diazocyclopentadienes (**19**) added to olefins, both electrophilic and nucleophilic, to yield the corresponding spirocyclopropanes **20**, provided that the carbene had the matched electronic character (Scheme 7.8). These intermediates likewise added to benzene [18, 19].

7.1.3 C–H and C–C Insertion

Rearrangements during the photolysis of diazo compounds or diazirines have been observed, and the intricate mechanism has been studied in detail [20]. From the synthetic point of view, C–H insertion offers a favorite access to strained structures—for example, in the synthesis of the silacyclobutane **22** below, obtained as a single diastereoisomer (Scheme 7.9) [21].

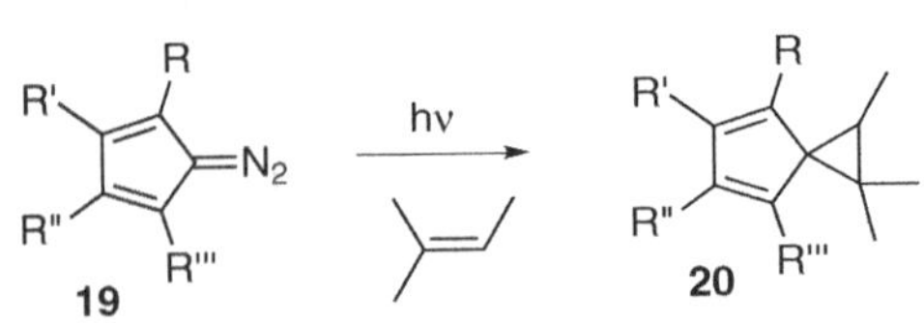

Scheme 7.8.

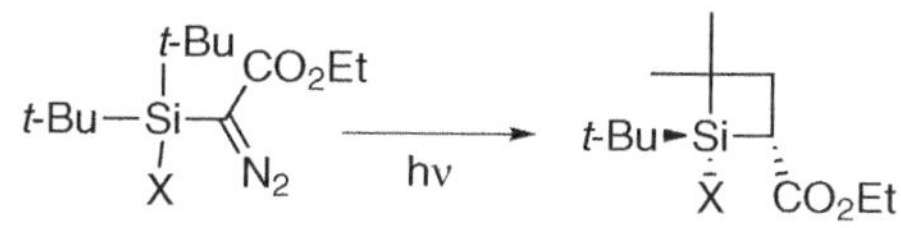

Scheme 7.9.

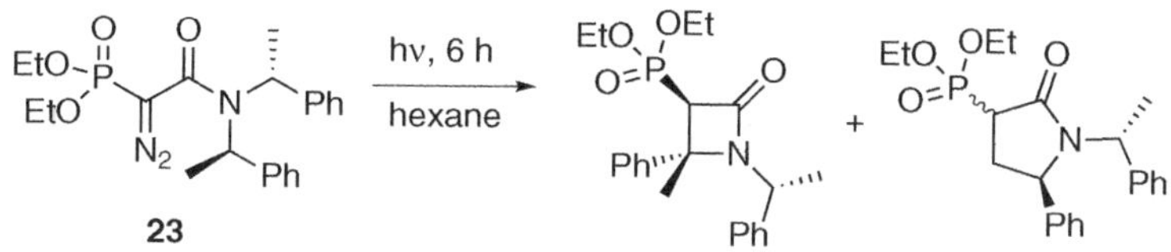

Scheme 7.10.

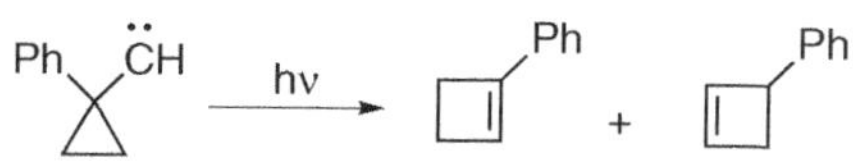

Scheme 7.11.

In the case of diazoacetamide **23**, enantiomerically pure lactam **24** was obtained by irradiation of either enantiomer in water-containing media (Scheme 7.10) [22]. The selectivity supports a reaction via the singlet carbene.

C–C insertion (1,2 shift) has been exploited for ring enlargement in cycloalkyl carbenes. As an example, cyclobutenes (**27**, **28**) have been obtained from cyclopropyl carbene **26** (Scheme 7.11) [23], and cubane **29** has been homologated to homocubane derivatives **32** and **33** via a highly strained olefin **31** (Scheme 7.12) [24].

7.1.4 Ketenes from the Wolff Rearrangement

Most interesting from the synthetic point of view is the Wolff rearrangement of α-diazoketones to give ketenes. As for the mechanism, it has been long debated whether it involves a stepwise process with a carbene as an actual intermediate or is instead a concerted process leading directly from the excited state of the precursor to the ketene. The variety of applications of ketenes makes the rearrangement, which occurs with a complete retention of configuration, useful in synthesis [25, 26]. This has been largely applied for reducing the ring dimension in cyclic diazoketones [27] as well as in [2 + 2] cycloaddition with alkenes. Thus, ketenes difficult to generate

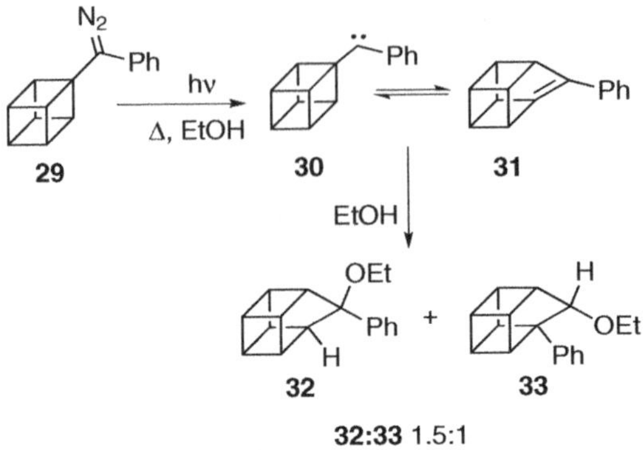

Scheme 7.12.

otherwise have been obtained photochemically. As an example, photodecomposition has been successful for preparing the carbene from the Meldrum's acid diazo derivative **34** and thus for the synthesis of cyclobutanones **35a–d** (Scheme 7.13) [28]. The mechanism is complex. In methanol the Wolff rearrangement occurred directly from the initially formed S_2 state, while competitive relaxation to the lowest singlet S_1 caused rearrangement to the diazirine and generation of the carbene that was finally trapped by the solvent.

Photochemistry is the method of choice for strained molecules such as those arising from intramolecular [2 + 2] cycloaddition of the ketene, as in the case of **37a,b** (Scheme 7.14) [29]. Further notice that in the case of diazo compound **39** photolysis led to the ketene **40** and the cycloadduct **41** from it, while the Rh(II) catalyzed reaction led to carbenoid attack to the C=O function and formation of an ylide (**42**, Scheme 7.15) [30].

The cycloaddition of **45** to alkynes **46** gave regiospecifically cyclobutenones (**47**) that upon further irradiation opened up to dienylketenes **48**. In turn, this underwent π^6 electrocyclization to give, after tautomerization, a highly functionalized phenol (**50**). The reaction was carried out photochemically, but heating was found useful for completing the process (Scheme 7.16) [31, 32].

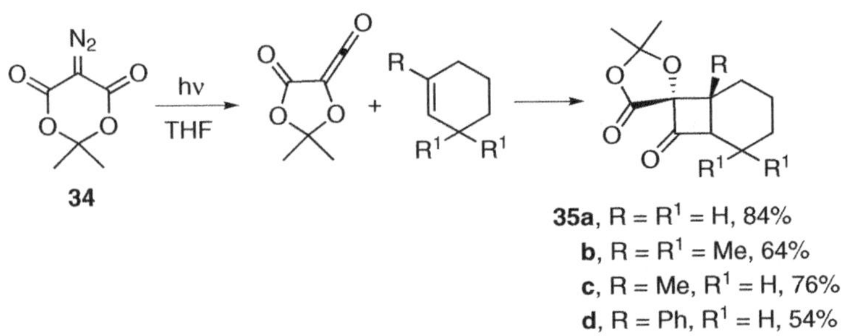

Scheme 7.13.

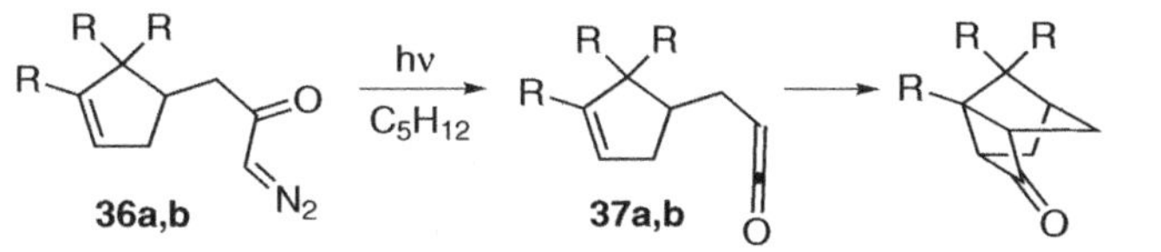

38a, R = H, 80%
b, R = Me, 79%

Scheme 7.14.

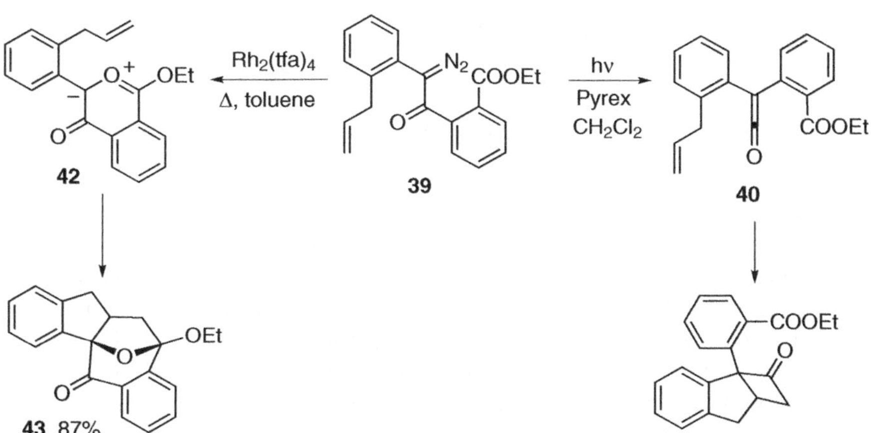

43, 87%

41, 80%

Scheme 7.15.

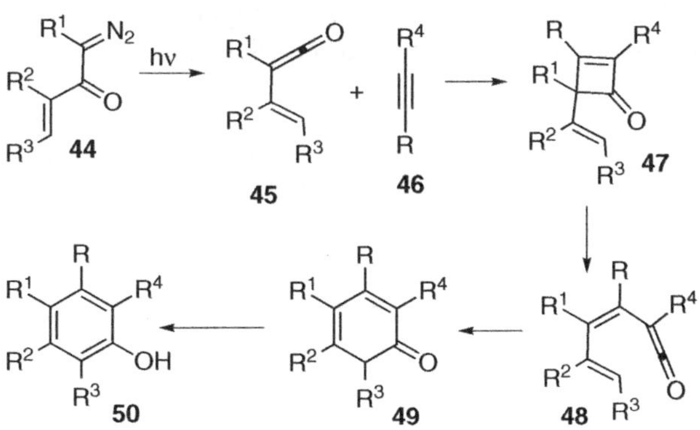

Scheme 7.16.

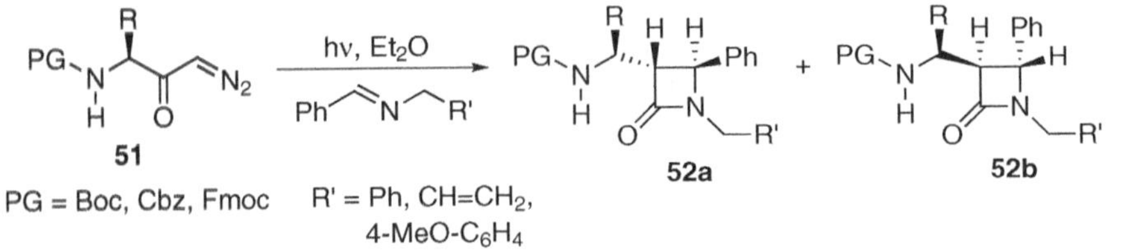

Scheme 7.17.

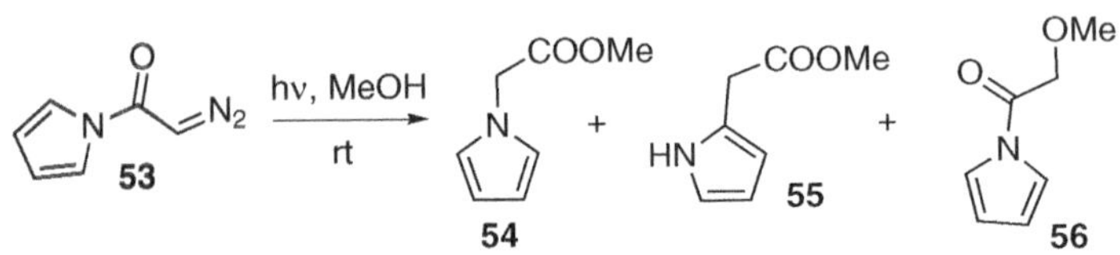

Scheme 7.18.

The analogous addition to imines offers an access to β-lactams. These products are interesting also because the incorporation of this feature in peptide strains stabilizes β-turns. Thus, photolysis or microwave pyrolysis of amino acid-derived diazo ketones **51** in the presence of imines provided the stereoisomeric β-lactams **52a,b**, both of them with *trans* configuration at the lactam ring (Scheme 7.17) [33, 34].

Competition between different paths is often occurring. As an example, *N*-diazoacetylpyrrole **53** yielded ester **54** via the Wolff rearrangement, ketoether **56** by OH insertion from the carbene, and ester **55** via a vinylogous Wolff rearrangement (Scheme 7.18) [35].

7.1.5 OH Insertion

OH Insertion is a diagnostic reaction of singlet carbene and has been used for generating ethers, although different processes such as the Wolff rearrangement usually compete [36]. In the case of the *N,N*-diethylacetamide derivative **57**, both OH, CH insertion and rearrangement occurred (Scheme 7.19) [37].

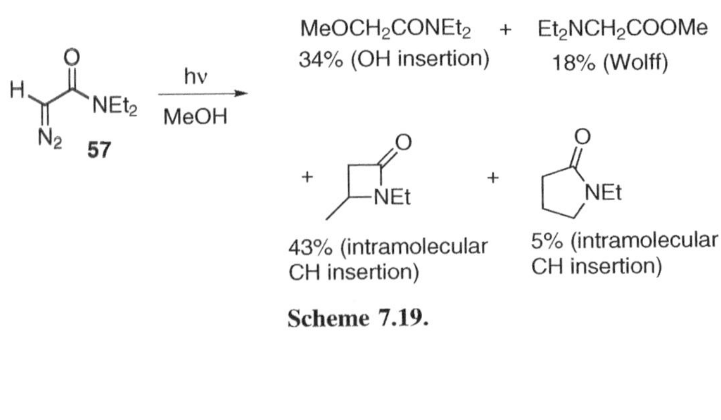

Scheme 7.19.

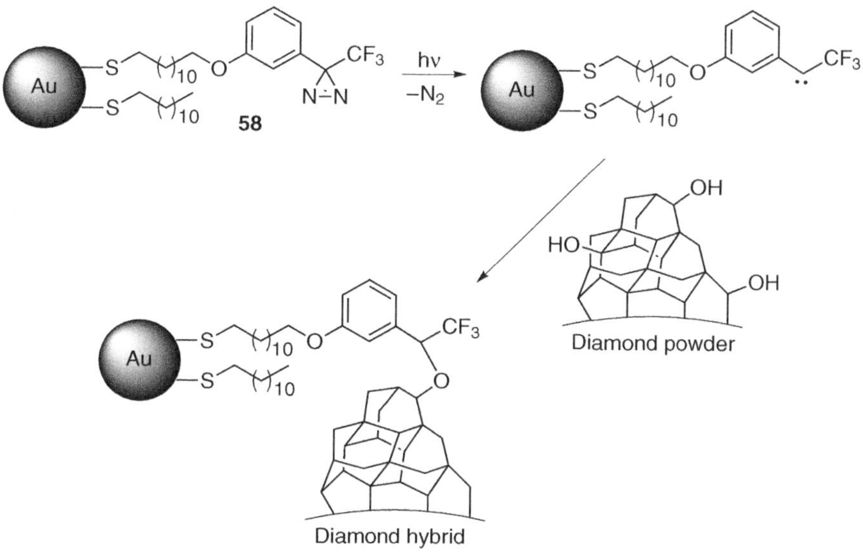

Scheme 7.20.

The OH insertion has found a variety of applications, from the glycosylation of oligosaccharides with the diazirine of gluconic acid, which occurred with some selectivity upon irradiaton at −70°C in the case of ginkgolide B (a fact that was rationalized on the basis of H-bond formation in the molecule [38]), to surface modification. An example is the formation of a covalent bond with a diamond surface upon photodecomposition of diazirine **58** (Scheme 7.20) [39].

7.1.6 Silyloxy Carbenes

Silyloxy carbenes are conveniently prepared through photochemical Brook rearrangement of acylsilanes and are useful intermediates in synthesis. These materials are colored, and a cheap 23-W compact fluorescent lamp was found to be well suited for generating the intermediate. The initial α-cleavage was followed by a 1,2-shift to give the carbene (e.g., **59**), as it was confirmed by matrix isolation [40]. These intermediates give useful reactions, both by addition and by insertion. Thus, when generated in the presence of alkynes, these species have been found to add and form silylated cyclopropenes **60** that, in turn, underwent a retro-Brook rearrangement to form silylated α,β unsaturated ketones **61** (Scheme 7.21a). This 100% atom economical reaction took place satisfactorily (except for terminal alkynes) in the intramolecular version (Scheme 7.21b). Intermolecularly the reaction also occurred, but only with electron-withdrawing substituted alkynes, and at any rate it required a long irradiation time [41]. The reaction with dimethyl fumarate gave the *trans*-dimethyl cyclopropane-1,2-dicarboxylate (with the maleate the reaction was not stereoselective, however) [42].

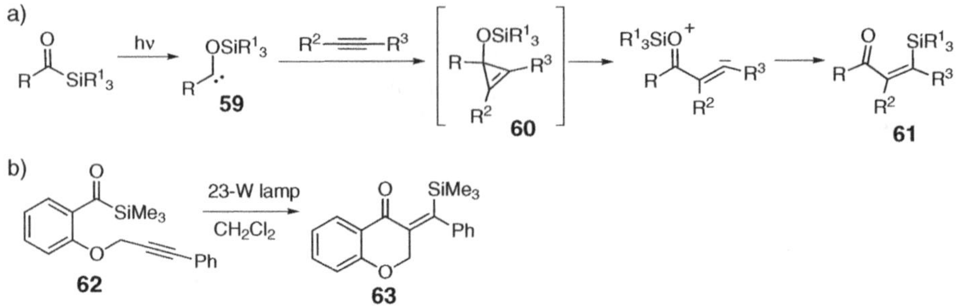

Scheme 7.21.

As for insertion reactions, important examples include O–H [43] and Si–H bonds [44]. The insertion into peracetylated free glycosides gave siloxyacetals with complete retention of the configuration. These compounds were useful pH-sensitive protecting groups (below pH 6.2) [45]. A further example involves the insertion into a B–C bond of cyclic organoboronic esters (**65**) followed by spontaneous rearrangement to a silyloxy derivative (**66**). The product obtained was converted into a ketone (**67** and **68**) by treatment with an acid (Scheme 7.22) [46].

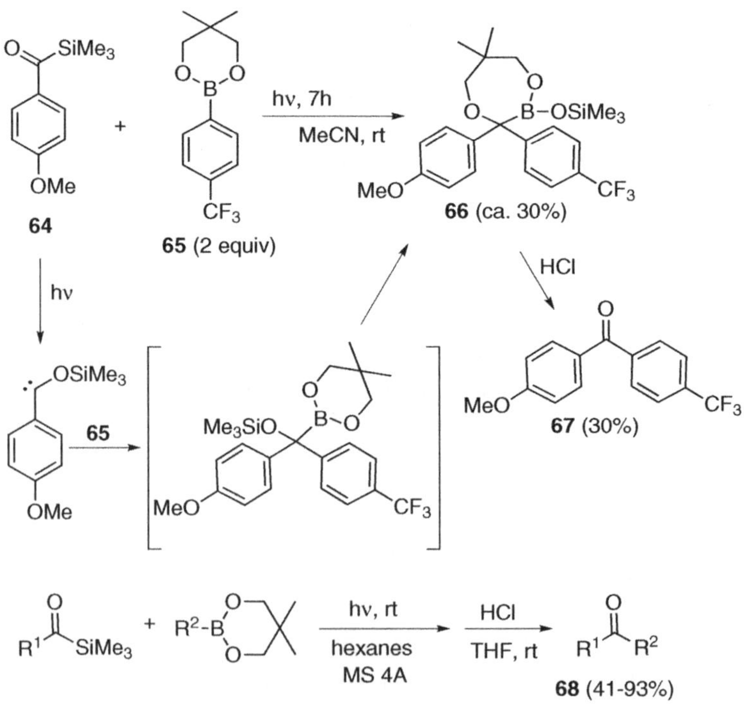

Scheme 7.22.

Under optimized conditions, which involved irradiation in hexanes in the presence of molecular sieves followed by HCl addition and refluxing, yields exceeding 90% were obtained [46].

7.2 NITRENES

7.2.1 Structure and Reactivity

Nitrenes are mainly generated by decomposition of azides [47–52]. Aromatic azides have been extensively investigated and react both thermally and photochemically. The latter way is a favored choice when mild conditions are required. Roughly, the course of the reaction can be summarized as indicated in Scheme 7.23 for the case of phenyl azide [53]. Photolysis proceeds from the singlet azide and produces the nitrene ^{1}**69** in the singlet state. This is a strong and unselective electrophile, but in aromatic derivatives intramolecular attack onto the aromatic ring prevails over any intermolecular process. This leads to a dehydroazepine (**70**)—at least potentially in equilibrium with a benzoazirine (**71**)—again an electrophile that enters in a polymerization process leading to "tars". The cumulene **70** can be conveniently trapped by nucleophiles yielding an azepine (**72**). Typical traps are amines, water [54], or (much less efficiently) alcohols. With some substituted phenylazides and with many polycyclic aryl azides, products arising from the trapping of the azirine (functionalized anilines **73**) are also formed along with or in the place of azepines (Scheme 7.23). The ground state of phenyl nitrene and of many aromatic derivatives is the triplet that is formed when either nitrogen loss occurs from the azide triplet ^{3}**69** (upon photosensitization or by direct irradiation when intersystem crossing competes with decomposition from the azide singlet) or by subsequent ISC at the nitrene level. The presence of the nitrogen atom strongly stabilizes this intermediate with respect to the corresponding carbene. Actually, at

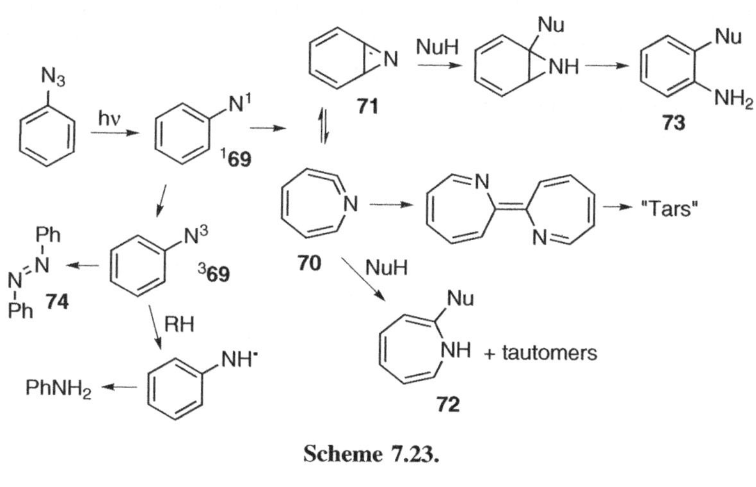

Scheme 7.23.

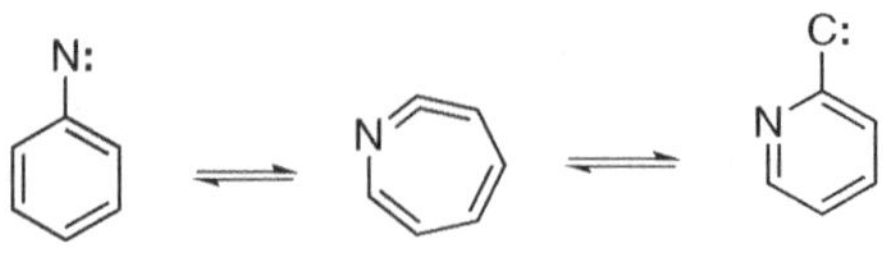

Scheme 7.24.

room temperature or below, this is a relatively persistent species. Indeed, its fate is dimerization to give the azo compound **74** (this has earned to this species the nickname of "lazy triplet"). Raising the temperature makes more apparent the radical character of triplet nitrene and typically leads to hydrogen abstraction and finally reduction to the aniline.

Summing up, the decomposition of aryl azides produces three electrophilic intermediates—namely, the singlet nitrene (difficult to trap because of the fast rearrangement), the azirine, and the dehydroazepine (which polymerizes but is longer-lived and easier to trap)—as well as the weakly radicalic triplet nitrene. The effect of substituents may thus be crucial, in particular for electron-withdrawing groups that strengthen the electrophilic character both of singlet nitrene and of the dehydroazepine. Good preparative results have been obtained in some cases, particularly with heterocyclic derivatives, but synthesis is certainly not the most valuable use of the photochemistry of aryl azides [47–52]. Rather, a variety of applications have been devised that exploit the reactivity of such intermediates for a specific action either on a molecule, as with photoaffinity labeling or the photorelease or photoactivation of drugs, or on a polymer for cross-linking or grafting, or on a surface for the immobilization of additives—for example, for adjusting wetting properties, for photolithography, or for building photoresists. The practical significance and the intellectually stimulating mechanistic issues have fostered a number of in-depth mechanistic investigations on the photochemistry of (hetero)aromatic azides by various techniques as well as complementary computational studies. This work has demonstrated the stepwise evolution of the system with the intervention of various intermediates [55–57]. Note also that matrix studies have demonstrated the equilibration of carbenes and nitrenes via ring-expanded ylides or via open-chain nitrenoid or biradicaloid intermediates (Scheme 7.24). This explains some unusual products [58].

7.2.2 Trapping of the Nitrene

As mentioned above, the first intermediate singlet aryl nitrene generally undergoes attack onto the ring rather than intermolecular trapping. However, there are many exceptions where the latter process takes place [59]. Thus, the irradiation of mesityl azide **75** in the presence of tetracyanoethylene was found to give two stable products, namely, an azomethine ylide **76** arising from trapping of the singlet nitrene (rate constant $>10^9\,M^{-1}\,s^{-1}$ and spiroazepine **77** resulting from addition to the trimethyldehydroazepine (Scheme 7.25) [60].

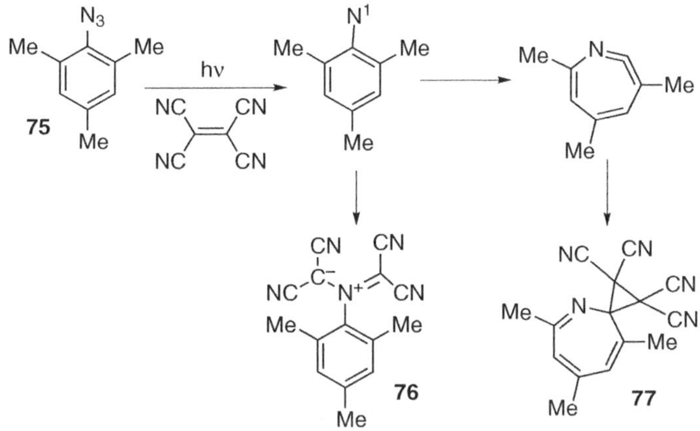

Scheme 7.25.

The presence of electron-withdrawing substituents enhanced the reactivity of the singlet. Thus, monofluorophenyl azides (**78a**) reacted via the cumulene, similarly to the parent compound, but when the 2,6-difluoro (**78b**) and the pentafluoro derivatives were irradiated in diethylamine, the singlet inserted into the N–H bond and gave the corresponding hydrazines **79** [61]. With a better nucleophile such as aniline, some trapping of the nitrene to form the asymmetric azo compound **80** occurred (by oxidation of hydrazines) also from 2-fluorophenylazide (Scheme 7.26) [62].

Likewise, some polyfluoroiodophenyl azides gave hydrazines by insertion into the N–H bond by irradiation in diethylamine–cyclohexane mixture [63].

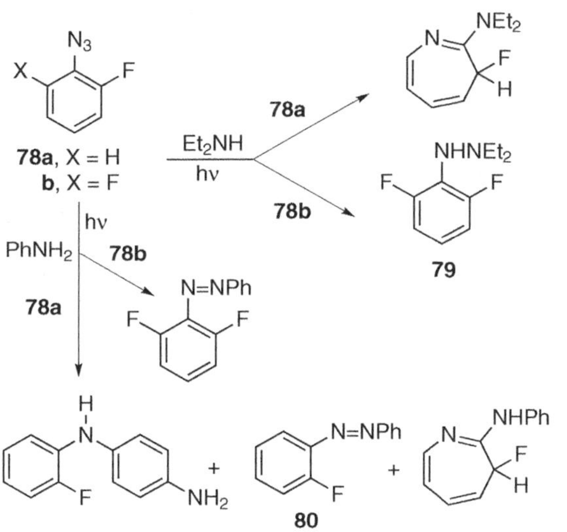

Scheme 7.26.

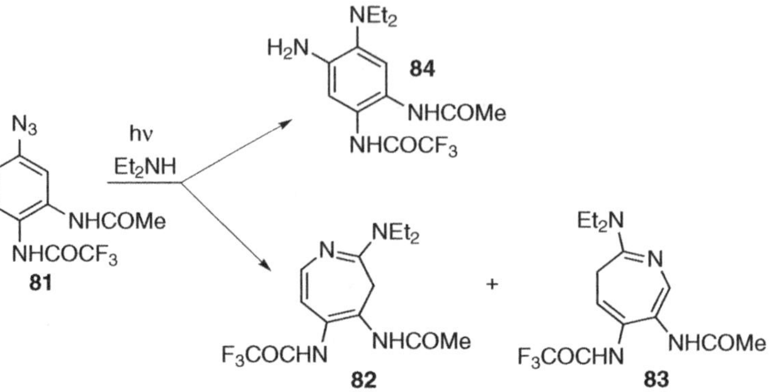

Scheme 7.27.

7.2.3 Trapping of Benzoazirine and Didehydroazepine

As mentioned, intermolecular trapping of the dehydroazepine and/or the benzoazirine are the most effective reactions from the azides in terms of yield, particularly with electron poor derivatives. As for the competition between the two above intermediates, rules have been formulated that refer to the substitution pattern, although azirines are more common with condensed aromatics, where have also been revealed spectroscopically [64]. A model studied in depth is that of 3,4-diamidophenyl azides (**81**). Apparently, in this case the azirine is formed reversibly from singlet nitrene and has an unusually long lifetime. This allows trapping by diethylamine competitively with irreversible rearrangement to the two possible dehydroazepines (**82, 83**). The last process is slow below −30°, where only the diamine **84** is formed (Scheme 7.27). On the other hand, a more hindered trap such as diisopropylamine is unable to trap the azirine [65].

The trapping by water to give azepinones, of which are known several examples [66], has been carried out also by irradiating the solution circulating in a tubing made by fluorinated polyethylene. Under these conditions secondary photolysis was minimized and the reaction was cleaner [67].

In general, polyfluorinated phenylazides (e.g., **85**), usually bearing a carboxyl function as a useful handle for attaching functions suited for the application considered (Scheme 7.28), have been studied in depth [68]. As an example,

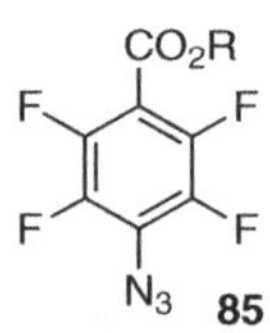

Scheme 7.28.

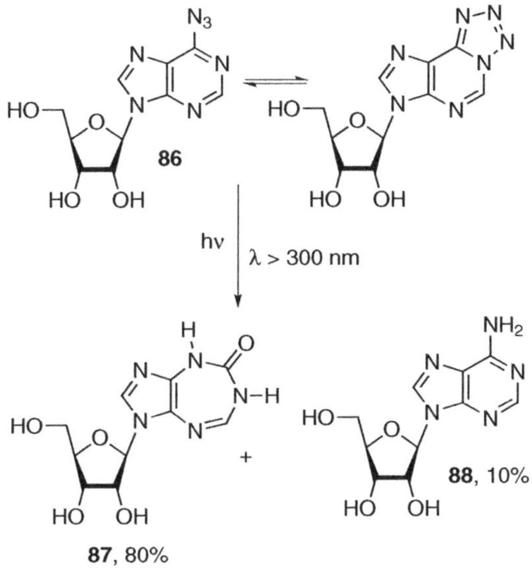

Scheme 7.29.

structures with the azide conjugated to a carbohydrate have been synthesized and applied to glycoarray fabrication [69].

2-Azidopyridines and related compounds are predominantly present in the solid state and in solution as the corresponding tetrazoles, but (acid catalyzed) equilibrium allows to observe the azide reactions. As an example, the irradiation in water of the 6-azidopurine **86** caused ring enlargement to the diazepinone **87**, with adenine **88** as a minor product (Scheme 7.29) [70].

Further transformations have been reported that can be explained with the same intermediates. As an example, 1-alkoxy-2-azidophenazine **89**-N_3 formed products that incorporated two molecules of the starting azide (**90**, **91**). The reactions were rationalized as involving addition of the nitrene or of the benzoazirine to the azide and ensuing cleavage of one of the aromatic rings. Insertion to form asymmetric azo compounds also occurred (Scheme 7.30) [71].

Ring enlargement has been observed also with benzyl azides—for example, in the formation of dihydroazepines (**93**) from the azides of podophyllotoxin and derivatives (**92**) that were obtained by photodecomposition in cyclohexene (Scheme 7.31) [72].

7.2.4 Intramolecular Trapping

Besides attack onto the ring, further intramolecular processes occur whenever a substituent is present that makes a new pathway available. Thus, *o*-tolyl nitrenes undergo a 1,4-hydrogen shift to form iminoquinone methides [73]. On the other hand, in a series of 2-(ω-phenylalkyl)phenylazides an intramolecular reaction has

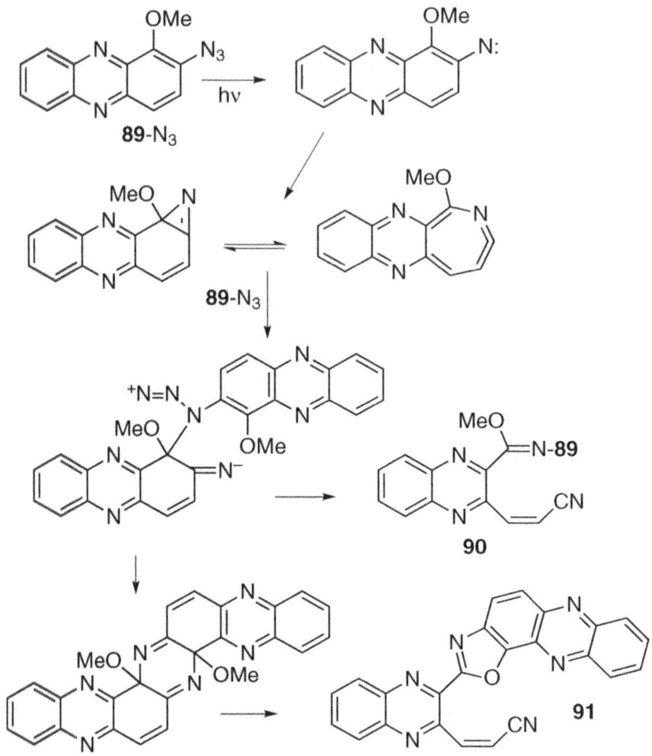

Scheme 7.30.

been found only when geometrically favored. Thus, when nitrenes **94** were generated in an apolar solvent, triplet nitrene products were obtained, mainly azo compounds **95**. With an activated (benzylic) C–H, intramolecular abstraction predominated, giving a cyclic amine (**96**), but only when the preferred conformation of the molecule had that C–H close to the nitrene ($n = 2$; with a longer chain the benzylic methylene was not favorably placed). In diethylamine, however, the dehydroazepine was

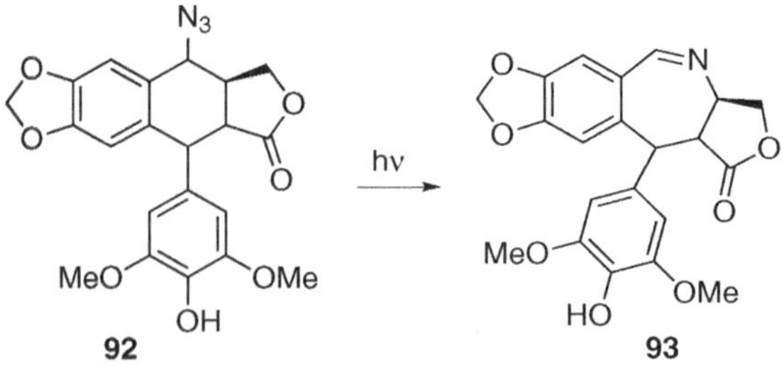

Scheme 7.31.

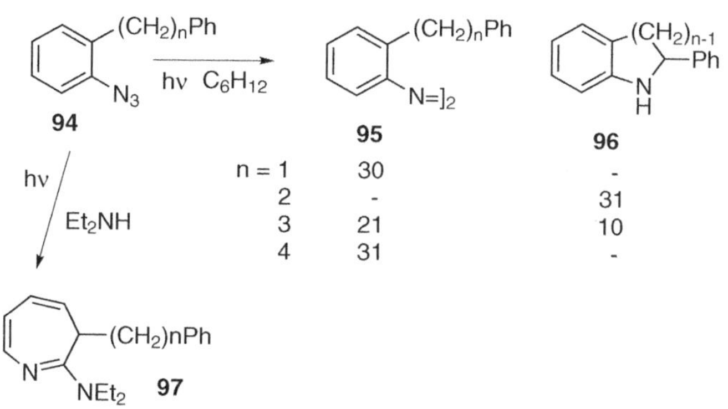

Scheme 7.32.

trapped and the ring-enlarged product **97** was obtained in every case. In ethanol a different trapping product, 2-(2-alkoxyethyl)aniline, was formed when $n=2$, presumably via a polar biradical (Scheme 7.32) [74].

The photochemistry of *o*-biphenyl azides and hetero analogues has been particularly investigated for mechanistic reasons in order to distinguish the reactions of the intermediates. As an example, the investigation on 1-(2-nitrenophenyl)dimethylpyrazole **98** (Scheme 7.33) allowed to distinguish singlet nitrene attack onto the pyrazole nitrogen, external nucleophilic trapping of the dehydroazepine, and intramolecular hydrogen abstraction from the methyl group by triplet nitrene [75, 76].

From the synthetic point of view, intramolecular attack onto a π system, whether alkene, aromatic, or heterocyclic, is a useful application of the photochemistry of aryl and vinyl azides and has been applied for annulation processes [51]. Also due to the stability of the products, these reactions can usually be carried out both thermally at a temperature in the range 80–220°C and photochemically. When the two methods have been compared, the latter one has generally been found to give a better yield [77], presumably because of the mild conditions. A concise overview is given in Scheme 7.34 and includes attack onto alkenes [78–80] and aromatics [77].

A prerogative of the photochemical method is the synthesis of small rings by addition of the nitrene. This might then expand upon temperature increase (Scheme 7.35) [81].

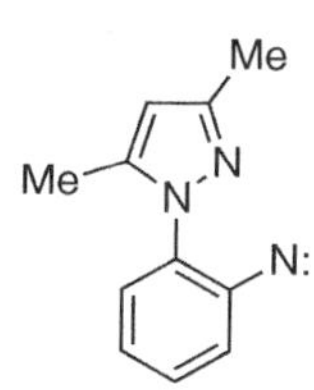

98

Scheme 7.33.

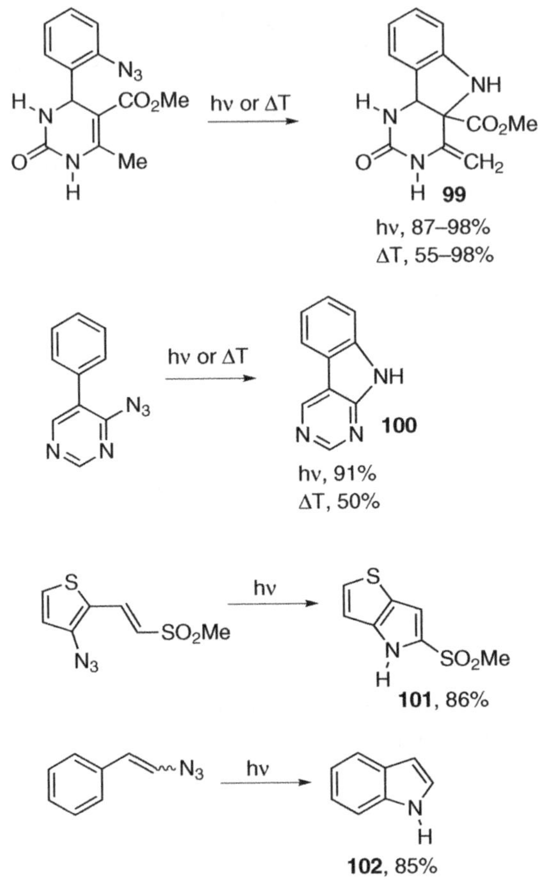

Scheme 7.34.

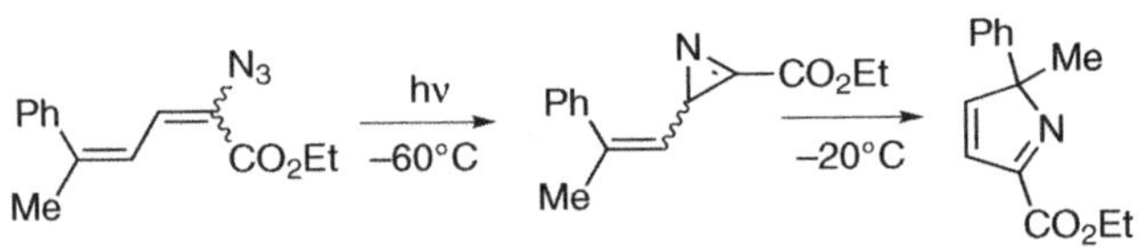

Scheme 7.35.

7.2.5 Other Effects

Under appropriate acidic conditions, singlet nitrenes are protonated and give nitrenium ions. As an example, 4-methoxyphenylazide in acid-containing alcohol yields *p*-benzoquinone by hydrolysis of the nitrenium cation [82]. These intermediates are useful for photolabeling [83]. Electron transfer may have a role. Thus, with

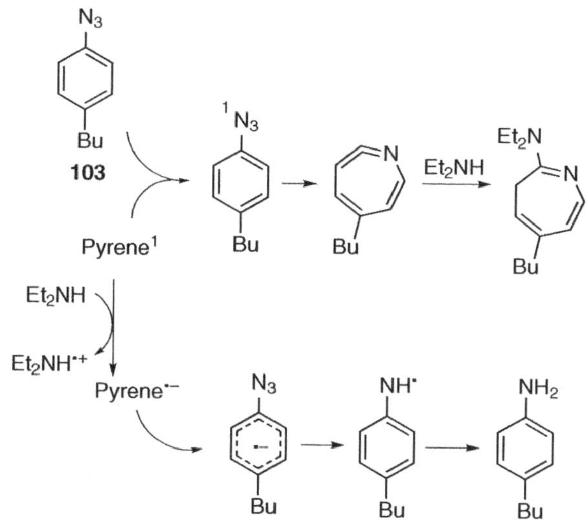

Scheme 7.36.

tertiary amines no nucleophilic trapping is possible, but when a good donor such as triethylamine is used as the solvent and an excellent acceptor such as the *p*-nitrophenylnitrene is generated, the reaction shifted to electron transfer and the amine was oxidized to the corresponding enamine, revealed as the cycloadduct with the starting azide [84].

On the other hand, a photoreductive activation seems also to be possible. Thus, pyrene-sensitized photolysis of *p*-butylphenylazide **103** in acetonitrile containing diethylamine gave a mixture of the diethylaminoazepine (via energy transfer sensitization of the azide singlet state and ring enlargement) and *p*-butylaniline (via the azide radical anion, Scheme 7.36) [85].

There is some indication that metal complexation may also guide the course of reactions via nitrenes; for example, this may explain the regioselective cyclization induced by CuI on some allenylphenyl azides [86].

7.2.6 Acylnitrenes

Acylnitrenes are also useful intermediates and excellent electrophiles. These are formed by decomposition of acylazides. This leads also to isocyanates and amines by hydrolysis. The mechanism has been studied, in order to ascertain whether the nitrene is involved as a discrete intermediate or whether N_2 elimination and rearrangement occur in a concerted way [87]. From the preparative point of view, the addition of aroylnitrenes to π bonds has been developed. This has been reported to give either aziridines or oxazolidines with olefins, oxazolines with alkynes, and dioxoles with ketones [88, 89]. The cycloaddition to C_{60} should be certainly mentioned. Thus, irradiation of aroylazides in dichloromethane in the presence of C_{60} gave stable fulleroaziridine derivatives. These could be rearranged to fullerooxazoles by boiling in tetrachloroethane [90]. Likewise, photolysis of a chlorobenzene solution of *N*-succinimidyl 4-azido-2,3,5,6-tetrafluorobenzoate in the presence of C_{60} gave a monoadduct azamethanofullerene in 10% yield (39% based on recovered C_{60}), via the nitrene. The *N*-hydroxysuccinimide (NHS)-activated ester group thus introduced in the monoadduct served as a site for the attachment of other molecules by way of an acylation reaction [91].

On the other hand, insertion into C–H bonds has also been observed. Thus, the singlet benzoyl nitrene generated photochemically from the azide inserted stereospecifically into the tertiary C–H bonds of *cis*- and *trans*-1,4-dimethylcyclohexanes. The insertion regioselectivities toward the C–H bonds were determined by use of 2-methylbutane and 2,3-dimethylbutane. This reaction did not occur from the triplet [92]. Intramolecular insertion into a C–H bond has been observed also with mesitylphosphoroyl nitrene that is similarly prepared from the azides by irradiation in MeCN–MeOH giving a dihydrobenzazaphosphole oxide, beside a Curtius-like rearrangement. With the more hindered tris(isopropyl)phenyl azide the same processes occurred along with migration.

REFERENCES

1. Candeias, N. R., Afonso, C. A. M. (2009). Developments in the photochemistry of diazo compounds. *Current Organic Chemistry*, **13**, 763–787.
2. Abdel-Wahab, A. M. A., Ahmed, S. A., Dürr, H. Carbene formation by extrusion of nitrogen. In Lenci, F., Horspool, W., Eds. *CRC Handbook of Organic Photochemistry and Photobiology*, second edition, CRC Press, Boca Raton, FL, 2004, pp. 91-1–91-37.
3. Hirai, K., Itoh, T., Tomioka, H. (2009). Persistent triplet carbenes. *Chemical Reviews*, **109**, 3275–3332.
4. Iwamoto, E., Hirai, K., Tomioka, H. (2003). A triplet carbene surviving a week in solution at room temperature. *Journal of the American Chemical Society*, **125**, 14664–14665.
5. Vignolle, J., Cattoe, X., Bourissou, D. (2009). Stable noncyclic singlet carbenes. *Chemical Reviews*, **109**, 3333–3384.
6. Bourissou, D., Guerret, O., Gabbai, F. P., Bertrand, G. (2000). Stable carbenes. *Chemical Reviews*, **100**, 39–92.
7. Arduengo, A. J., III (1999). Looking for stable carbenes. The difficulty of starting anew. *Accounts of Chemical Research*, **32**, 913–921.
8. Arduengo, A. J., III, Harlow, R. L., Kline, M. (1991). A stable crystalline carbene. *Journal of the American Chemical Society*, **113**, 361–363.
9. Myers, D. R., Senthilnathan, B. P., Platz, M. S., Jones, M., Jr., (1986). Diadamantylcarbene in solution. *Journal of the American Chemical Society*, **108**, 4232–4233.
10. Sellers, S. F., Klebach, T. C., Bollowood, F., Jones, M.Jr., von Raguè Schleyer, P. (1982). (l-Adamantyl)homoadamant-3-ene: An extraordinarily stable bridgehead olefin. *Journal of the American Chemical Society*, **104**, 5492–5493.
11. Kraska, A. R., Chang, S. T., Mosely, C. C., Shechterr, H. (1982). The effects of multiplicity and excitation on the behavior of 2-methyl-2-phenyl-propylidene. The intermolecular chemistry of 2,2-dipheylpropylidene. *Tetrahedron Letters*, **23**, 1627–1630.

12. Moss, R. A. (2006). Diaziridines. Carbene precursors by excellence. *Accounts of Chemical Research*, **39**, 267–272.
13. Moss, R. A. (2010). Dihalodiazirines and the photochemical generation of dihalocarbenes: New light on old problems. *Journal of Physical Organic Chemistry*, **23**, 293–299.
14. Moore, K. A., Vidaurri-Martinez, J. S., Thamattoor, D. M. (2012). The benzylidenecarbene-phenylacetylene rearrangement: An experimental and computational study. *Journal of the American Chemical Society*, **134**, 20037–20040.
15. Nakatani, K., Adachi, K., Tanabe, K., Saito, I. (1999). Tandem cyclizations involving carbene as an intermediate: Photochemical reactions of substituted 1,2-diketones conjugated with ene–yne. *Journal of the American Chemical Society*, **121**, 8221–8228.
16. Moss, R. A. (1989). Carbenic reactivity revised. *Accounts of Chemical Research*, **22**, 15–21.
17. Arnold, D. R., Humphreys, R. W., Leigh, W. J., Palmer, G. E. (1976). Electronic excited states of small ring compounds. 3: Cyclopropene, vinylcarbene and vinylmethylene. *Journal of the American Chemical Society*, **98**, 6225–6233.
18. Dürr, H., Scheppers, G. (1967). Photochemie kleiner Ringe, II. Reaktion des photochemisch erzeugten Carbens 1.2.3.4-Tetraphenyl-5-carbena-cyclopentadien-(1.3) mit Olefinen. *Chemische Berichte*, **100**, 3236–3245.
19. Dürr, H., Scheppers, G. (1968). Photochemische Synthese von substituierten 5H- und 7*H*-Benzocycloheptenen. *Angewandte Chemie*, **80**, 359.
20. Wang, J., Burdzinski, G., Gustavson, T. L., Platz, M. S. (2007). Ultrafast study of *p*-biphenyldiazoethane. The chemistry of the diazo excited state and the relaxed carbene. *Journal of the American Chemical Society*, **129**, 2597–2606.
21. Maas, G., Bender, S. (2000). Synthesis and ring enlargement of 2-ethoxycarbonyl-1-silacyclobutanes. *Chemical Communications*, 437–438.
22. Candeias, N. R., Gois, P. M. P., Veiro, L. F., Alfonso C. A. M. (2008). C–H carbene insertion of α-diazoacetamides by photolysis in non-conventional media. *The Journal of Organic Chemistry*, **73**, 5926–5932.
23. Baron, W. J., De Camp, M. R., Hendrick, M. E., Jones, M., Jr., Levin, H., Sohn, M. B. In Jones, M., Moss, R. A., Eds. *Carbenes I*, John Wiley & Sons, New York, 1973, Chapter 1.
24. Eaton, P. E., Hoffmann, K. L. (1987). 9-Phenyl-1(9)-homocubene, probably the most twisted olefin yet known and the carbene 1-phenyl-9-homocubylidene, its rearrangement product. *Journal of the American Chemical Society*, **109**, 5285–5286.
25. Kirmse, W. (2002). 100 Years of the Wolff rearrangement. *European Journal of Organic Chemistry*, 2193–2256.
26. Allen, A. D., Tidwell, T. T. (2012). New directions in ketene chemistry: The land of opportunity. *European Journal of Organic Chemistry*, 1081–1096.
27. Koepke, T., Pink, M., Zaleski, J. M. (2006). Photochemical preparation of pyrrole ring-contracted chlorins by the Wolff rearrangement. *Organic & Biomolecular Chemistry*, **4**, 4059–4062.
28. Rudolf, P., Buback, J., Aulbach, J., Nuernberger, P., Brixner, T. (2010). Ultrafast multi-sequential photochemistry of 5-diazo Meldrum's acid. *Journal of the American Chemical Society*, **132**, 15213–15222.
29. Yates, P., Fallis, A. G. (1968). Decomposition of 1-diazo-3-(2,2,3-trimethylcyclopent-3-enyl)propan-2-one. A novel synthesis and rearrangement or a tricyclo[3.2.1.$0^{3,6}$]octan-4-one. *Tetrahedron Letters*, **9**, 2493–2496.

30. Padwa, A., Precede, L., Semones, M. A. (1999). Model studies directed toward the total synthesis of (±)-ribasine. A tandem cyclization–cycloaddition route leading to the core skeleton. *The Journal of Organic Chemistry*, **64**, 4079–4088.

31. Danheiser, R. L., Brisbois, R. G., Kowalczyk, J. J., Miller, R. F. (1990). An annulation method for the synthesis of highly substituted polycyclic aromatic and heteroaromatic compounds. *Journal of the American Chemical Society*, **112**, 3093–3100.

32. Karady, S., Amato, J. S., Reamer, R. A., Weinstock, L. M. (1996). Annelation of aromatic oxo compounds. *Tetrahedron Letters*, **37**, 8277–8280.

33. Podlech, J., Linder, M. R. (1997). Cycloadditions of ketenes generated in the Wolff rearrangement. Stereoselective synthesis of aminoalkyl-substituted β-lactams from α-amino acids. *The Journal of Organic Chemistry*, **62**, 5873–5883.

34. Podlech, J., Steurer, S. (1999). A chiral, oxidatively cleavable auxiliary in β-lactam synthesis—Double diastereocontrol with *p*-methoxyphenethyl-substituted imines. *Synthesis*, 650–654.

35. Islami, M. R., Allen, A. D., Vukovic, S., Tidwell, T. T. (2011). *N*-Pyrrolylketene: A nonconjugated heteroarylketene. *Organic Letters*, **13**, 494–497.

36. Miller, D. J., Moody, C. J. (1995). Synthetic applications of the O–H insertion reactions of carbenes and carbenoids derived from diazocarbonyl and related diazo compounds. *Tetrahedron*, **51**, 10811–10843.

37. Tomioka, H., Kitagawa, H., Izawa, Y. (1979). Photolysis of *N,N*-diethyldiazoacetamide. Participation of a noncarbenic process in intramolecular carbon-hydrogen insertion. *The Journal of Organic Chemistry*, **44**, 3072–3075.

38. Weber, M., Vasella, A. (1997). Glycosylidene carbenes. Part 25. Glycosidation of ginkgolides B and A. *Helvetica Chimica Acta*, **80**, 2352–2367.

39. Ismaili, H., Workentin, M. S. (2011). Covalent diamond-gold nanojewel hybrids via photochemically generated carbenes. *Chemical Communications*, **47**, 7788–7790.

40. Trommer, M., Sander, W. (1996). Photochemistry and photooxidation of acylsilane. *Organometallics*, **15**, 189–193.

41. Zhang, H.-J., Becker, P., Huang, H., Pirwerdjan, R., Pan, F.-F., Bolma, C. (2012). Photochemically induced silylacylations of alkynes with acylsilanes. *Advanced Synthesis and Catalysis*, **354**, 2157–2161.

42. Dalton, J. C., Borque, R. A. (1981). Mechanistic photochemistry of acylsilanes. 2. Reaction with electron-poor olefins. *Journal of the American Chemical Society*, **103**, 699–700.

43. Bourque, R. A., Davis, P. D., Dalton, J. C. (1981). Mechanistic photochemistry of acylsilanes. 1. Reaction with alcohols. *Journal of the American Chemical Society*, **103**, 697–699.

44. Watanabe, H., Ohsawa, N., Sawai, M., Fukawa, Y., Matsumoto, H., Nagai, Y. (1975). Photochemical reaction of benzoyltrimethylsilane with substituted phenylsilanes. *Journal of Organometallic Chemistry*, **93**, 173–179.

45. Svarovsky, S. A., Taraban, M. B., Barchi, J. J., Jr. (2004). Facile photochemical synthesis of mixed siloxyacetal glycosides as potential pH-sensitized prodrugs for selective treatment of solid tumors. *Organic & Biomolecular Chemistry*, **2**, 3155–3161.

46. Ito, K., Tamashima, H., Iwasawa, N., Kusama, H. (2011). Photochemically promoted transition metal-free cross-coupling of acylsilanes with organoboronic esters. *Journal of the American Chemical Society*, **133**, 3716–3719.

47. Platz, M. S. (1995). Comparison of phenyl carbene and phenylnitrene. *Accounts of Chemical Research*, **28**, 487–492.

48. Gritsan, N., Platz, M. Photochemistry of azides: The azide/nitrene interface. In Braese, S., Banert, K., Eds. *Organic Azides Syntheses and Applications*, John Wiley & Sons, Chichester, 2009, pp. 311–372.

49. Budyka, M. F. Photochemistry of azidopyridine and related compounds. In Nylund, K., Johansson, P., Eds. *Heterocyclic Compounds*, Nova Publishers, Hauppauge, NY, 2010, pp. 225–316.

50. Bucher, G. Photochemical reactivity of azides. In Lenci, F., Horspool, W., Eds. *CRC Handbook of Organic Photochemistry and Photobiology*, second edition, CRC Press, Boca Raton, FL, 2004, pp. 44/1–44/31.

51. Soderberg, B. C. G. (2000). Synthesis of heterocycles via intramolecular annulation of nitrene intermediates. *Current Organic Chemistry*, **4**, 727–764.

52. Schuster, G. B., Platz, M. S. (1992). Photochemistry of phenyl azide. *Advances in Photochemistry*, **17**, 69–143.

53. Xue, J., Du, Y., Chuang, Y. P., Phillips, D. L., Wang, J., Luk, C., Hadad, C. M., Platz, M. S. (2008). Time-resolved resonance Raman observation of the dimerization of dehydroazepines in solution. *Journal of Physycal Chemistry A*, **112**, 1502–1510.

54. Lamara, K., Smalley, R. K. (1991). *3H*-Azepines and related systems. Part 4. Preparation of *3H*-azepin-2-ones and *6H*-azepino[*2,1-b*]quinazolin-12-ones by photo-induced ring expansions of aryl azides. *Tetrahedron*, **47**, 2277–2290.

55. Burdzinski, G. T., Gustafson, T. L., Hackett, J. C., Hadad, C. M., Platz, M. S. (2005). The direct detection of an aryl azide excited state: An ultrafast study of the photochemistry of para- and ortho-biphenyl azide. *Journal of the American Chemical Society*, **127**, 13764–13765.

56. Marcinek, A., Leyva, E., Whitt, D., Platz, M. S. (1993). Evidence for stepwise nitrogen extrusion and ring expansion upon photolysis of phenyl azide. *Journal of the American Chemical Society*, **115**, 8609–8612.

57. Sun, X.-Z., Virrels, I. G., George, M. W., Tomioka, H. (1996). Photochemistry of 2-(methoxycarbonyl)phenyl azide in solution at room temperature: a time-resolved infrared study. *Chemistry Letters*, **12**, 1089–1090.

58. Wentrup C. (2011). Nitrenes, carbenes, diradicals, and ylides. Interconversions of reactive intermediates. *Account of Chemical Research*, **44**, 393–404.

59. Abraham, H.-W. Photogenerated nitrene addition to π-bonds. In Griesbeck, A. G., Mattay, J., Eds. *Molecular and Supramolecular Photochemistry. Synthetic Organic Photochemistry*, Vol. 12, Marcel Dekker, New York, 2005, pp. 391–416.

60. Murata, S., Abe, S., Tomioka, H. (1997). Photochemical reactions of mesityl azide with tetracyanoethylene: Competitive trapping of singlet nitrene and didehydroazepine. *The Journal of Organic Chemistry*, **62**, 3055–3061.

61. Leyva, E., Sagredo, R. (1998). Photochemistry of fluorophenyl azides in diethylamine. Nitrene reaction versus ring expansion. *Tetrahedron*, **54**, 7367–7374.

62. Leyva, E., Sagredo, R., Moctezuma, E. (2004). Photochemistry of fluorophenyl azides in aniline. Asymmetric fluoroazobenzenes by N–H singlet nitrene insertion. *Journal of Fluorine Chemistry*, **125**, 741–747.

63. Cai, S. X., Keana, J. F. W. (1989). 4-Azido-2-iodo-3,5,6-trifluorophenylcarbonyl derivatives. A new class of functionalized and iodinated perfluorophenyl azide photolabels. *Tetrahedron Letters*, **30**, 5409–5412.

64. Dunkin R., Thomson, P. C. P. (1980). Infrared evidence for tricyclic azirines and didehydrobenzazepines in the matrix photolysis of azidonaphthalenes. *Journal of the Chemical Society, Chemical Communications*, 499–501.

65. Younger, C. G., Bell, R. A. (1992). Photolysis of 3,4-diamidophenyl azides: evidence for azirine intermediates. *Journal of the Chemical Society, Chemical Communications*, 1359–1361.

66. Lamara, K., Smalley, R. K. (1991). 3*H*-Azepines and related systems. Part 4. Preparation of 3*H*-azepin-2-ones and 6*H*-azepino[*2,1-b*]quinazolin-12-ones by photo-induced ring expansions of aryl azides. *Tetrahedron*, **47**, 2277–2290.

67. Bou-Hamdan, F. R., Levesque, F., O'Brien, A. G., Seeberger, P. H. (2011). Continuous flow photolysis of aryl azides: preparation of 3H-azepinones. *Beilstein Journal of Organic Chemistry*, **7**, 1124–1129.

68. Soundararajan, N., Platz, M. S. (1990). Descriptive photochemistry of polyfluorinated azide derivatives of methyl benzoate. *The Journal of Organic Chemistry*, **55**, 2034–2044.

69. Deng, L., Norberg, O., Uppalapati, S., Yan, M., Ramstroem, O. (2011). Stereoselective synthesis of light-activatable perfluorophenylazide-conjugated carbohydrates for glyco-array fabrication and evaluation of structural effects on protein binding by SPR imaging. *Organic and Biomolecular Chemistry*, **9**, 3188–3198.

70. Komodzinski, K., Nowak, J., Lepczynska, J., Milecki, J., Skalski, B. (2012). Photochemistry of 6-azidopurine ribonucleoside in aqueous solution. *Tetrahedron Letters*, **53**, 2316–2318.

71. Albini, A., Bettinetti, G., Minoli, G. (1991). Photochemical decomposition of 1-alkoxy-2-azidophenazines. Addition of nitrenes to azides. *The Journal of Organic Chemistry*, **56**, 6993–6997.

72. Laatsch, H., Ernst, B. P., Hoffmann, D. (1995). Azepines by photochemical ring enlargement of 9-azidopodophyllotoxin- and 9-azido-9′-demethylepipodophyllotoxin derivatives. *Liebigs Annalen*, **10**, 1773–1777.

73. Bucher, G. (2001). Photochemical generation of iminoquinone methides by 1,4-hydrogen migration in derivatives of o-tolylnitrene. *European Journal of Organic Chemistry*, 2447–2462.

74. Murata, S., Yoshidome, R., Satoh, Y., Kato, N., Tomioka, H. (1995). Reactivities of nitrenes generated by photolysis of 2-(ω-phenylalkyl)phenyl azides. *The Journal of Organic Chemistry*, **60**, 1428–1434.

75. Albini, A., Bettinetti, G., Minoli, G. (1999). The effect of the *p*-nitro group on the chemistry of phenylnitrene. A study via intramolecular trapping. *Journal of the Chemical Society, Perkin Transactions 2*, 2803–2807.

76. Albini, A., Bettinetti, G., Minoli, G. (1999). Photodecomposition of some para-substituted 2-pyrazolylphenyl azides. Substituents affect the phenylnitrene S-T gap more than the barrier to ring expansion. *Journal of the American Chemical Society*, **121**, 3104–3113.

77. Hyatt, J. H., Swenton, J. S. (1972). Facile synthesis of 9*H*-pyrimido[*4,5-b*]indoles from photolysis of 8-phenyltetrazolo[*1,5-c*]pyrimidines in acidic media. *The Journal of Organic Chemistry*, **37**, 3216–3220.

78. Ren, Y.-W., Wang, X.-R., Wang, W.-X., Li, B., Shi, Z.-J., Zhang, W. (2011). Photochemical and thermal cyclizations of 4-(2-azidophenyl)-3,4-dihydropyrimidin-2-ones for the synthesis of 4-methylenepyrimidino[*5,4-b*]indol-2-ones. *Tetrahedron Letters*, **52**, 192–195.

79. Gairns, R. S., Moody, C. J., Rees, C. W. (1986). Photochemical conversion of 3-azido-2-vinylthiophenes into thienopyrroles and of 2-azidostyrenes into indoles. High migratory aptitude of sulphur substituents. *Journal of the Chemical Society, Perkin Transaction* **1**, 501–506.

80. Isomura, K., Okada, M., Taniguchi, H. (1969). Azirine formation by decomposition of terminal vinyl azides. *Tetrahedron Letters*, **10**, 4072–4076.

81. Padwa, A., Smolanoff, J., Tremper, A. (1975). Intramolecular cycloaddition reactions of vinyl-substituted 2*H*-azirines. *Journal of the American Chemical Society*, **97**, 4682–4691.

82. Ramlall, P., McClelland, R. A. (1999). Photochemical generation and lifetimes in water of *p*-aryloxy- and *p*-alkoxyphenylnitrenium ions. *Journal of the Chemical Society, Perkin Transactions 2: Physical Organic Chemistry*, 225–232.

83. Voskresenska, V., Wilson, R. M., Panov, M., Tarnovsky, A. N., Krause, J. A., Vyas, S., Winter, A. H., Hadad, C. M. (2009). Photoaffinity labeling via nitrenium ion chemistry: Protonation of the nitrene derived from 4-amino-3-nitrophenyl azide to afford reactive nitrenium ion pairs. *Journal of the American Chemical Society*, **131**, 11535–11547.

84. Murata, S., Mori, Y., Satoh, Y., Yoshidome, R., Tomioka, H. (1999). New findings in photochemistry of *p*-nitrophenyl azide in the presence of an amine: Generation and trapping of enamines. *Chemistry Letters*, 597–598.

85. Murata, S., Nakatsuji, R., Tomioka, H. (1995). Mechanistic studies of pyrene-sensitized decomposition of *p*-butylphenyl azide: Generation of nitrene radical anion through a sensitizer-mediated electron transfer from amines to the azide. *Journal of the Chemical Society, Perkin Transactions 2: Physical Organic Chemistry*, 793–799.

86. Feldman, K. S., Hester D. K. II, Lopez, C. S., Faza, O. N. (2008). Allenyl azide cycloaddition chemistry. photochemical initiation and CuI mediation leads to improved regioselectivity. *Organic Letters*, **10**, 1665–1668.

87. Wentrup, C., Bornemann, H. (2005). The Curtius rearrangement of acyl azides revisited—formation of cyanate (R–O–CN). *European Journal of Organic Chemistry*, 4521–4524.

88. Buck, K., Jacobi, D., Ploegert, Y., Abraham, W. (1994). Photochemistry of acyl azides. VIII. Do acylnitrenes react as 1,3-dipoles? *Journal fuer Praktische Chemie/Chemiker-Zeitung*, **336**, 678–685.

89. Clauss, K. U., Buck, C., Abraham, W. (1995). Photochemistry of acylazides IX. Direct and sensitized photolytic generation of acylnitrenes for cycloaddition reactions. *Tetrahedron*, **51**, 7181–7192.

90. Averdung, J., Mattay, J., Jacobi, D., Abraham, W. (1995). Cycloadditions. 48. Addition of photochemically generated acylnitrenes to C_{60}. Synthesis of fulleroaziridines and thermal rearrangement to fullerooxazoles. *Tetrahedron*, **51**, 2543–2552.

91. Mingdi Yan, M., Xiong Cai, S., Keana, J. F. W. (1994). Photochemical and thermal reactions of C60 with *N*-succinimidyl 4-azido-2,3,5,6-tetrafluorobenzoate: A new method for functionalization of C_{60}. *The Journal of Organic Chemistry*, **59**, 5951–5954.

92. Inagaki, M., Shingaki, T., Nagai, T. (1991). Insertion of benzoylnitrene toward hydrocarbon carbon–hydrogen bonds. *Chemistry Letters*, **10**, 1419–1422.

8

MANIPULATING INTERMEDIATES THE PHOTOCHEMICAL WAY

Photochemical reactions are drastically different from thermal reactions. There are two key aspects in which electronically excited states differ from ground states: first, their high energy (typically $\geq$70 kcal mol^{-1} above ground states, a value of the same order of chemical bonds) and, second, their open-shell nature. Thus, their evolution occurs over a surface that has no analogy with that of thermal reactions, until the system returns to the ground-state surface (but not necessarily to a closed-shell structure; actually, open-shell intermediates such as radicals or radical ions may be formed, as seen in the previous chapters). Excitation causes a reorganization of chemical connectivity with respect to the ground state. This may be identified with a particular bond that is essentially broken, as the π bond in the $\pi\pi^*$ state of olefins, or may involve the partial weakening of several bonds, as in the $\pi\pi^*$ state of aromatics. This is reflected in the fact that reactions of excited states are often described as following a diradicalic path, in contrast with thermal reactions, most often discussed in terms of ionic intermediates or concerted paths. However, this is no absolute rule and there are thermal reactions generating diradicals, such as the Bergman [1] or the Myers–Saito [2] reactions, or at least where structures with some diradicalic character are encountered along the reaction path. As an example, a stepwise radical process has been predicted to compete with the concerted path even with some formally Woodward–Hoffmann-allowed reactions, including some cycloadditions [3] and sigmatropic rearrangements [4]. Likewise, a diradicalic path has been considered as an alternative to a concerted path for the attack by dioxiranes onto alkanes [5].

Photochemically-Generated Intermediates in Synthesis, First Edition.
Angelo Albini and Maurizio Fagnoni.

The key difference in photochemistry is that the system arrives to high-energy intermediates coming down from the vantage point of high-lying electronically excited states, whereas in thermal reactions it has to struggle up on an ascending path full of diversions. This difference should—and actually does—make it possible to generate intermediates in a smooth way and by consequence allow to control their evolution much more easily.

Furthermore, an important practical aspect is that the photon is a very active physical reagent free from many of the negative side effects "normal" chemical reagents have (actually, it is the "green" reagent by excellence). In fact, it is available when required, and shining light onto the solution is all that has to be done. Photons either pass through or are absorbed. In the latter case, these cause the reaction but, contrary to chemical reagents, they leave no residue behind that must be treated/recovered/eliminated [6]. Anyone accustomed to carrying out what is defined a green catalytic reaction that requires a small amount of catalyst (say 1% mol, but has a molecular weight >10 times that of the starting material), but actually requires also a cocatalyst (say 10% mol) and at the end requires treatment for recovering/eliminating the spent residue, often by adding a further reagent, should appreciate the difference. Take into account also that these catalysts and cocatalysts are likely to be delicate, toxic, and expensive, need to be prepared in situ, and so on.

It is hoped that the presentation in Chapters 2–7, though drastically abridged, has given a flavor of the characteristics of the photochemical access to intermediates and the (potential) advantages achievable in this way. In Chapter 8 some consideration is given to how this works, or may work.

8.1 GENERATING INTERMEDIATES

As mentioned, the issue here is the merit (if any) of photochemistry for the generation of intermediates that are—or conceivably are—accessible also through a thermal reaction—conceivably, but not in every case actually identified. In fact, photochemistry opens up the path to a variety of intermediates that have not been generated in a thermal reaction—or have been, but inefficiently, or not selectively, or in any way not conveniently. This is expected particularly when the thermal process confronts a high-energy barrier or involves an intrinsically slow process such as intersystem crossing. Therefore, photochemistry appears to have a key role in synthesis for making possible "a thermally unachievable, energetically uphill reaction" (in the previously quoted words by Noyori [7]), as well as for giving access to intermediates with a different multiplicity with respect to the reagent. In fact, the enormous amount of energy involved in photochemical reactions is generally not incorporated in the products, or only partially so. Among photochemical reactions, only those leading to severely strained structures would be markedly endothermic in the ground state. In the other cases, the photon energy is used up not for making unstable products, but for reaching smoothly high-lying intermediates that would not be easily attained thermally, and thus for introducing new chemical paths. One may notice that nature does not avail itself of light for chemical synthesis, or only rarely—not only because collecting light requires a specific organ that is exposed to solar irradiation, such as the leaf, difficult to introduce into organisms of varied architecture, but also because the presence of high-energy intermediates in the cell would require specific control and defense of the cell itself that has been elaborated for chlorophyll photosynthesis, but not for other systems. This is no problem in the much simpler situation under which chemical reactions are carried out in the laboratory.

8.1.1 Intermediates that Cannot Be Accessed Thermally

8.1.1.1 Arriving at High-Energy Intermediates via Homolysis

$$\begin{array}{l} \text{A-B}^* \longrightarrow \text{A}^\bullet + \text{B}^\bullet \\ \text{Cat}^* + \text{A-B} \longrightarrow \text{Cat-A}^\bullet + \text{B}^\bullet \end{array} \tag{8.1}$$

The key characteristics of organic chemistry, the strength of the covalent bond, limits homolysis to (relatively) weak bonds, at least under the usual conditions of a synthetic laboratory. Consider, for example, the bimolecular reduction of ketones to pinacols. On paper, a simple way to carry out the reaction would be (a) hydrogen transfer from a donor (e.g. an alcohol) and (b) coupling of the thus formed radicals. The reaction is overall exothermic, but this pathway is not accessible thermally, because the intermediates (the radicals) are too high in energy. As is well known, pinacols are prepared by using aggressive reagents such as metals or metallo-organic compounds through multistep reactions (see an example in Scheme 8.1a) [8, 9]. In contrast, electronic excitation makes indeed possible the simple method via radicals, because of the high energy and the specific radical character of the carbonyl $n\pi^*$ triplet (see Scheme 8.1b). As seen in Figure 8.1, this state abstracts a hydrogen from isopropanol. This is the only photochemical step, the system is now on the lowest surface for that configuration (the radical pair, obviously quite high in energy). Subsequent hydrogen transfer from the $Me_2C^\bullet OH$ radical to a ground state of the carbonyl leads to a second radical and the two radicals couple. The method has the added bonus that there is no spent reagent to wipe away after the reaction.

The limitation of thermal processes to the cleavage of weak bonds is general. Consider the largest field of radical chemistry, alkylation reactions. Here, radicals are generates by homolysis only when it involves atoms that are easily transferred to chain carriers, typically tin derivatives. Thus, alkyl bromides and iodides are often used as starting materials because of the easy exchange of the halide with a tin-based radical ($R_3Sn^\bullet$; see Figure 8.2 left, path *a*). Typical thermal reactions are chain processes; that is, the adduct radical R-T$^\bullet$ is able to abstract a hydrogen from tin hydride and to generate a tin radical that restarts the chain. Initiation must occur at such a rate that termination is balanced, but has no effect on the actual synthetic cycle. As shown in Figure 8.2 (left side), the steps involved in the cycle are

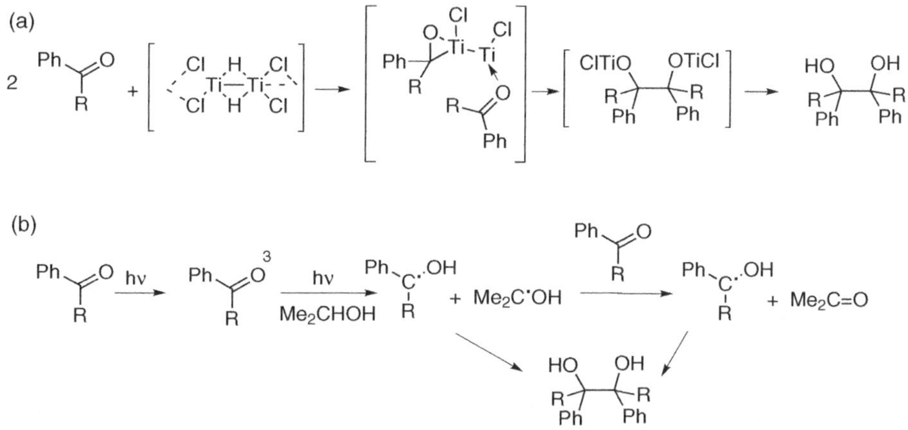

Scheme 8.1. (a) Thermal and (b) photochemical reduction of aromatic carbonyls to pinacols.

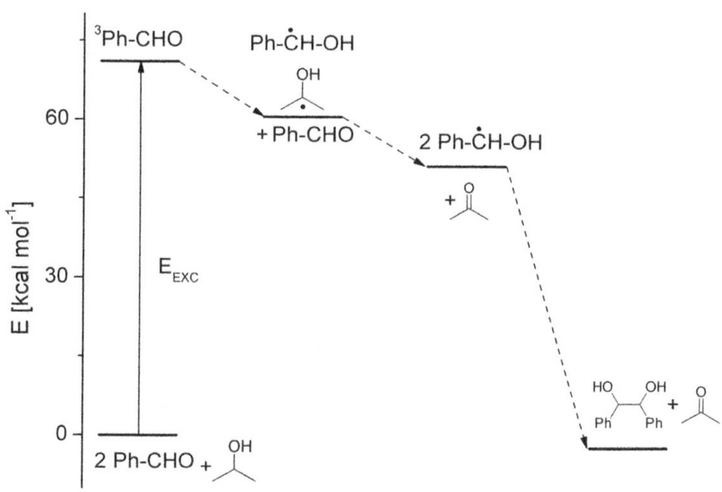

Figure 8.1. Energy diagram for the reaction in Scheme 8.1b (R = H). The $n\pi^*$ state of benzaldehyde has a radical character and sufficient energy for abstracting a hydrogen atom from isopropanol. Subsequent H donation from $Me_2C^{\cdot}OH$ to a second molecule of the aldehyde and coupling of the radicals lead to the products. The reaction would be exothermic in the ground state, but is not thermally accessible because the intermediates are too high in energy.

exothermic (the trapping step is considered thermoneutral for simplicity) and the amounts of energy involved are relatively low.

Changing from thermal to photochemical initiation in a radical chain process is possible and offers some advantage—in particular, allowing to operate at a lower temperature (see also Section 8.2). However, this may be unpractical (many common initiators do not absorb near-UV light) and at any rate is no breakthrough.

Different is the case when excited states are involved in the product formation step. On one hand, the direct excitation of various compounds causes homolysis

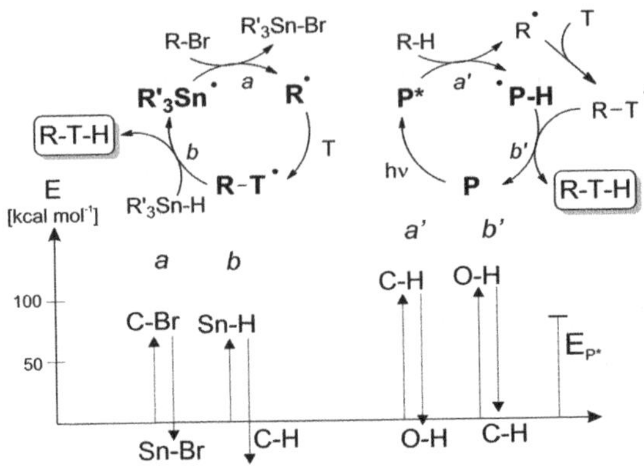

Figure 8.2. Energy diagram for radical reactions (bond broken/formed indicated by up/down arrows). **Left side:** A radical chain alkylation. The stannyl radical abstracts a bromine atom from a halide (*step a*, exothermic as a balance of the C–Br bond broken and Sn–Br formed), the alkyl radical is trapped (considered thermoneutral for simplicity), and the adduct radical abstracts a hydrogen atom from the stannane (*step b*, exothermic as a balance of Sn–H broken and C–H formed). The energy gained is sufficient and re-starts the cycle. **Right side:** A photocatalytic process. The photocatalyst P is active only in the excited state. As an example, a ketone behaves as a oxygen-centered radical. Abstraction of a hydrogen from an alkane (*step a′*, C–H cleaved, O–H formed and *b′*, C–H formed by H-donation from the reduced sensitizer occur). The steps *a′* and *b′* overall are close to thermoneutral. Furthermore, the amount of energy involved in each step is larger, suggesting a large activation barrier. Thus, a new cycle does not start unless a new photon is absorbed.

and the radicals formed add to a suitable trap, usually through a non-chain reaction. Alkyl iodides do not absorb strongly and in general undergo only a sluggish photochemical reaction, but aryl iodides (as well as arenediazonium salts upon reduction) are convenient precursors of radicals in the photo–Pschorr and related reactions, often with some advantage with respect to the thermal alternative. More importantly, many excited states have a diradical character and are able to undergo specific fragmentation reaction. In particular, the $n\pi^*$ triplet state of aldehydes and ketones has a single electron in the nonbonding orbital on the oxygen atom. This makes these states similar in structure to alkoxy radicals and thus capable to undergo more energy-demanding processes, such as the cleavage of a C–H bond with formation of a strong O–H bond (intra- or intermolecular H transfer, resulting in reduction, cleavage, or cyclization; see Chapter 4) or the addition onto a C$=$C π bond in an alkene with formation of a C–O bond, as in the $2+2$ cycloaddition (Paternò–Büchi) reaction. Hydrogen transfer leads to a radical pair or to a diradical through a process that is typical of this kind of excited state—and has very little similarity in the ground state (see Chapters 2 and 4). As shown in Figure 8.1 (left side), both the first hydrogen transfer step (from *i*-PrOH to benzaldehyde) and the latter one (from $Me_2C^{\bullet}OH$ to a second benzaldehyde molecule) confronts a high-energy barrier, and the overall process is not largely exothermic as the alkylation by alkyl bromides discussed above. Thus, a thermal chain process cannot take place.

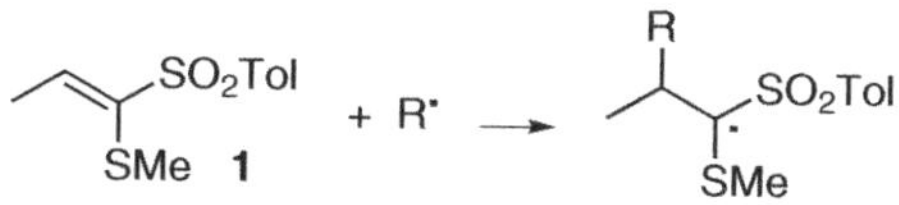

Scheme 8.2.

However, starting from the high energy level of the excited state the process is possible, but only when a photon is absorbed in every cycle.

The photochemical hydrogen abstraction has been exploited for photocatalyzed alkylation, a process where the radicals formed are trapped and the photocatalyst is regenerated by back hydrogen transfer to the radical adduct. Aromatic ketones and various other compounds—in particular, more robust species such as polytungstate anions, the uranyl cation, and other ones—are used as photocatalysts and are able to generate carbon-centered radicals from alkanes and simple aliphatic compounds. In this case, as mentioned, the aggressive hydrogen abstracting species is the excited state of the photocatalyst. The photocatalyst itself is not consumed, but has to be excited and in every cycle converts the energy of the absorbed photon into a chemical bond (light as a stoichiometric reagent) (see Figure 8.2).

The amount of energy involved is much larger than in the thermal chain reaction. However, the investment finds its return in the larger scope of this alkylation procedure and in the economy of steps it offers, since one can start directly from the alkane and make a C—H bond reactive, while the thermal reaction requires previous conversion into a halide. Several applications of photocatalysis are presented in Chapter 2, and the method appears to be undergoing a hefty development.

Photocatalysis becomes the only alternative in the cases where the radical chain does not proceed because one of the intermediates is not sufficiently reactive.

As an example, ketene thioacetal *S,S*-dioxide **1** could not be alkylated under the usual radical chain conditions with a tin hydride as the transfer agent (Scheme 8.2). This was reasonably due to the great stability of the radical adduct due to the captodative effect (thus, the energy gained in forming the final C–H bond was not sufficient for restarting the cycle—the down arrow in path *b* was too small; see Figure 8.2). However, this did not inhibit benzophenone photocatalyzed alkylation, which occurred in a high yield (89%) (see also Scheme 2.15, path *a*) [10]. Again, this is no chain process and requires a stoichiometric amount of photons, but this pays back in making the reaction viable.

8.1.1.2 Arriving at High-Energy Intermediates via Heterolysis

$$\text{A-B}^* \longrightarrow \text{A}^+ + \text{B}^- \qquad (8.2)$$

Photoheterolysis is less common, at least as a clean process, while it often competes with homolysis within several classes of benzyl and vinyl derivatives. Stabilized carbocations have been investigated mainly from the mechanistic point of view—for example, di- and tri-arylmethyl cations and fluorenyl cations from the corresponding halides and alcohols [11]. The most developed application in synthesis is for the smooth liberation of protected functions. In particular, the 9-phenylxanthyl (pixyl) group and its thia analogue have found application as protecting groups of nucleosides, since these are cleaved under neutral conditions [12].

However, the most promising item in this group is the photogeneration of the highly unstable aryl cations from halides and esters. This is a potentially large-scope synthetic method. At present, the most useful applications involve the triplet state of such intermediates, and the process is thus briefly discussed in Section 8.1.2.

8.1.1.3 Arriving at a High-Energy Intermediate via a Redox Process

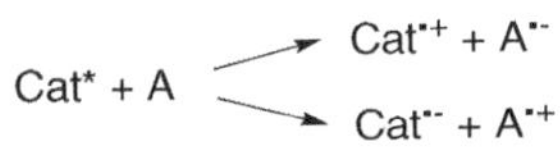

(8.3)

A general property that all excited states share is the extreme activity as oxidizing/reducing agents. This considerably extends the limits that redox reactions find in organic chemistry. In fact, redox processes involving only organic molecules are rare, because these are thermodynamically prohibited. As seen in Figure 8.3, the closed shell configuration of such compounds is strongly stabilized, while the corresponding species with one electron less or one electron more (radical ions) are

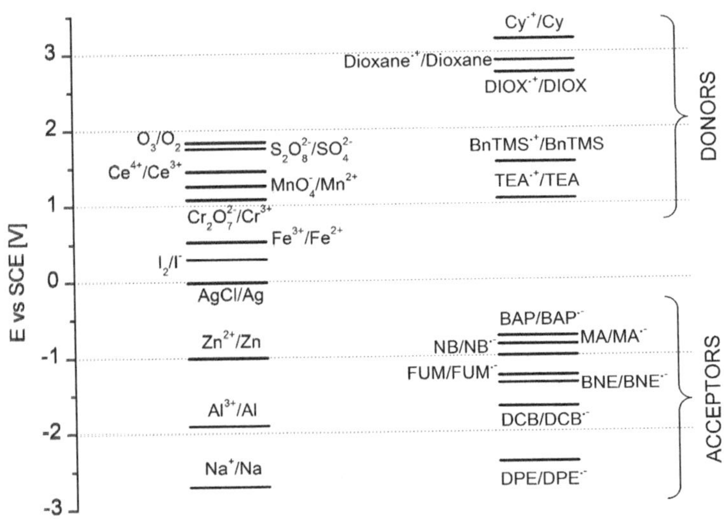

Figure 8.3. Representative redox reactions. **Left:** inorganic species. **Right:** Organic compounds DPE, 1,1-diphenylethylene; DCB, 1,4-dicyanobenzene, BNE, (*E*)-1-phenylbut-2-en-1-one; FUM, fumaronitrile; NB, nitrobenzene; MA, maleic anhydride; BAP, α-bromoacetophenone; TEA, triethylamine; BnTMS, benzyltrimethylsilane; DIOX, 1,3-dioxolane; Dioxane, 1,4-dioxane; Cy, cyclohexane. None of organic compounds are oxidized by a ground-state organic acceptor, but some are oxidized by an inorganic oxidant and all are by some excited-state oxidant (see Figures 8.4 and 8.5).

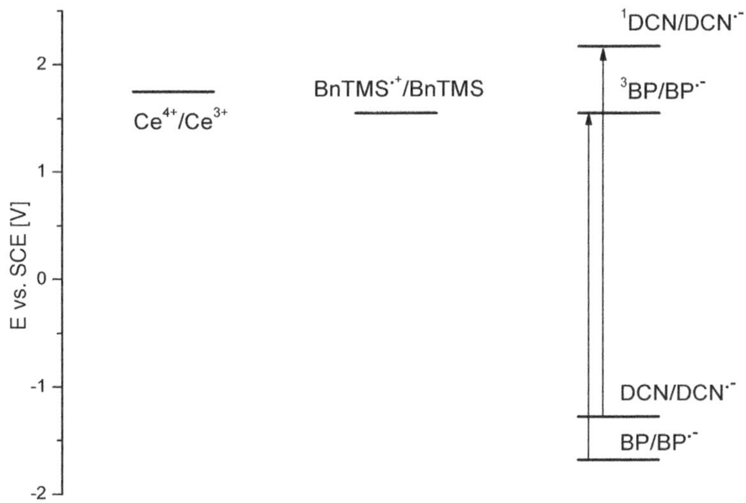

Figure 8.4. Oxidation of benzyltrimethylsilane (BnTMS). In the ground state this is possible only by using an inorganic oxidant—for example, cerium(IV). Benzophenone (BP) and 1,4-dicyanonaphthalene (DCN), modest acceptors in the ground state, become powerful oxidants in the excited state.

strongly destabilized. Therefore, if redox equilibria are ranged (see Figure 8.3), most (single electron) oxidations are grouped in a region well above that including reductions, with a gap of ~1 V between them: oxidizable compounds (donors) are grouped above 1 V vs. SCE, while oxidants (acceptors) are below 0 V vs. SCE. This contrasts with the situation in inorganic chemistry, where a thick distribution of closely lying redox processes is found at any value. Thus, it is easy to find a pair of reagents for which a redox reaction is thermodynamic accessible (e.g., permanganate oxidizes chloride to chlorine), while recourse to an inorganic reagent is required for causing electron transfer with an organic molecule (e.g., one needs Ce(IV) for oxidizing an alkene).

Electronic excitation, however, increases the redox potential by the large amount corresponding to the excited-state energy, canceling the gap and making redox processes a very common occurrence (see Figure 8.4, right side). Thus, excited (triplet) ketones are oxidants as strong as the most powerful inorganic species Ce(IV), but (singlet) excited aromatic nitriles are even more powerful. On the other hand, the reduction of a ketone requires a strong reducing agent, such as sodium, a hydride, or SmI_2, but an amine is a sufficiently good donor for reducing the corresponding triplet.

A somewhat extended view is presented in Figure 8.5, where the redox potential of a few typical acceptors and donors is indicated. As apparent, oxidation of a variety of organic molecules with excited electron-withdrawing substituted aromatics, quinones, and ketones is a viable process.

Redox processes are obviously possible also electrochemically, but photo-induced electron transfer (ET) allows to carry out such reactions under more easily

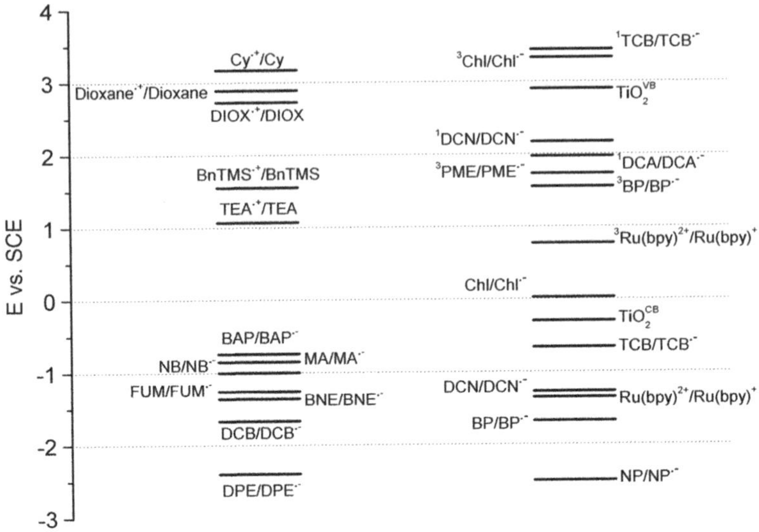

Figure 8.5. Representative photoinduced redox reactions. **Left:** Organic compounds (see Figure 8.3). **Right:** Oxidants in the ground and in the excited state. NP, naphthalene; BP, benzophenone; $Ru(bpy)^{2+}$, ruthenium(II)tris(bipyridyl) complex; DCN, 1,4-dicyanonaphthalene; TCB, 1,2,4,5-tetracyanobenzene; TiO_2, titanium dioxide; Chl, choranil; PME, tetramethyl pyromellitate. Again, none of the compound in the left column is oxidized by ground states at right, but some are by the corresponding excited states.

controlled conditions, avoiding the use of chemicals and charge-transporting salts. The potential of the approach can be appreciated from the fact that the singlet excited state of 1,2,4,5-tetracyanobenzene (TCB) has an estimated reduction potential of ~3.4 V vs. SCE. This makes possible even the oxidation of alkanes (at room temperature, in an organic solution, no additive being required, nor byproduct generated). The thus formed radical cations generally lose an electrofugal group (a proton, a carbocation, a silyl cation; at any rate the preferred path can be predicted by simple thermochemical considerations) [13]. This opens another unconventional access to radicals. Indeed, irradiation of TCB in the presence of adamantane leads cleanly to adamantyltricyanobenzene by electron transfer and proton loss from the alkane radical cation [14]. Thus, with protoadamantane (**2**), selective functionalization in position 6 (compound **3**) is obtained after electron transfer to TCB, in accord with computational predictions which evidence that the C_6–H bond is deprived of electrons and elongated in the radical cation (see Scheme 8.3) [15].

Indeed, also electron-withdrawing substituted alkanes are oxidized, as is the case for acetonitrile ($CH_3CN^{\bullet+} \rightarrow {}^{\bullet}CH_2CN$), and give trappable radicals, provided that moisture is present and avoids back electron transfer by assisting deprotonation [16]. The simple and controlled conditions of the photoinduced oxidation are useful preparatively (and the results obtained by controlling the fate of the intermediates are considered in Chapter 5), as well as a model for rationalizing all important industrial

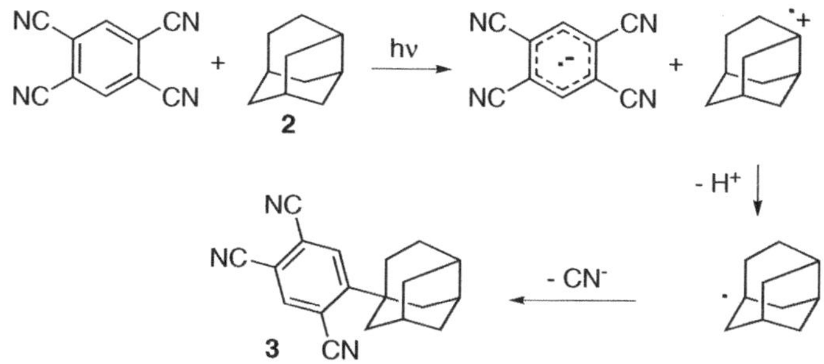

Scheme 8.3.

applications (such as alkane reforming and functionalization under heavy conditions, the behavior in superacids or with strong electrophiles [17], etc.) and the limits attainable in electrolytic processes [16].

8.1.1.4 Arriving at an Intermediate Overcoming the Poor Stability of the Precursor (A)

$$\begin{array}{ccccccc} A' & \xrightarrow{h\nu} & A'^{*} & \longrightarrow & A & \left(\underset{\Delta T}{\not\longrightarrow} \; C \right) \\ & & & & \downarrow & & \\ & & & & B & & \end{array} \qquad (8.4)$$

The thermal generation of biradicals is actually restricted to a single class of reactions, namely the cycloaromatization of enediynes and enyne-allenes in the Bergman [1] or Myers–Saito [2] reactions, yielding a *p*-benzyne or a α,3-didehydrotoluene, respectively. Not many examples are known, and a main problem lies in the fact that compounds that convert into highly reactive intermediates have by definition a poor stability and may be difficult to obtain. In fact, cyclic enyne-allenes are practically unknown, probably because the rearrangement is too fast. In this case, the photochemical generation of the precursor itself can offer a convenient entry to the reactive system. Thus, the enyne-allene moiety can be completed by deprotecting either (a) the alkyne moiety by photoelimination of CO from a cyclopropenone or (b) the allene moiety by a photo–Wolff elimination from diazoketone **4** (Scheme 8.4). The rearrangement to the diradical (or zwitterion) intermediates then proceeds directly [18, 19].

Many intermediates are much more easily generated photochemically than thermally. As an example, thermal access to highly unstable intermediates such as phenyl cations is difficult. Diazonium salts would be convenient precursors, due to the presence of such an excellent electrofugal group as the nitrogen molecule. However, this entry is troublesome because these rather unstable compounds easily undergo competing radical decomposition—a chain process. Irradiation, however, allows a much easier generation of these cations from diazonium salts, avoiding—at least in many cases—radical decomposition. This approach has the added bonus of the multiplicity option (see Section 8.1.2).

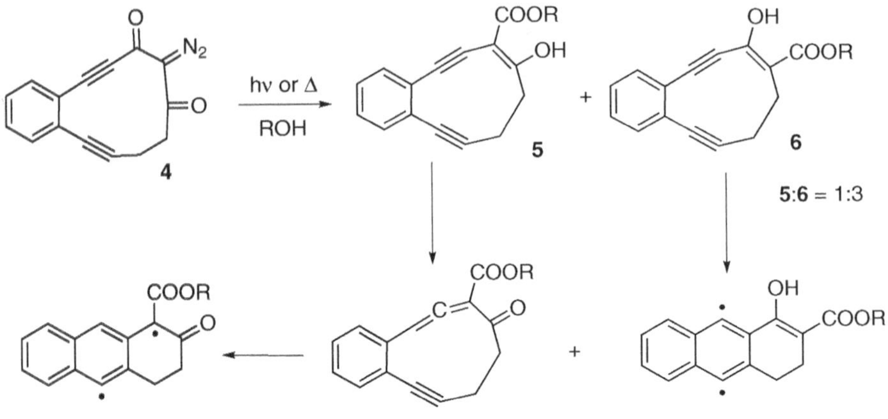

Scheme 8.4.

8.1.1.5 Avoiding the Reaction of the Precursor

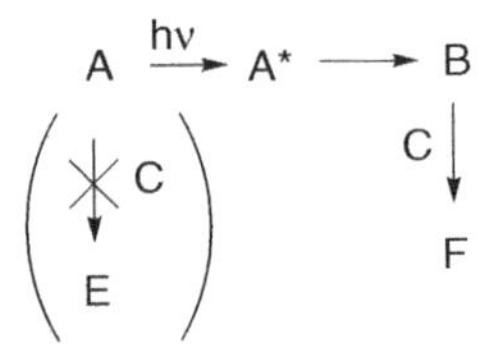

(8.5)

In some cases, the precursor has the same type of reactivity as the intermediate, but to a different degree, and competition with a thermal reaction of the precursor (A) itself may occur. There are two ways out from this problem: The first is changing to a less reactive precursor, the latter lowering the temperature. Triplet phenyl cations are synthetically useful for building aryl–carbon bonds by reaction with carbon-based nucleophiles. As mentioned above, diazonium salts are convenient precursors, but are themselves electrophiles and thus are not suitable as precursors in this case, due to the competitive thermal reaction forming azo derivatives, particularly when activated by electron-withdrawing substituents. Generation of the same intermediates from noncharged precursors such as aryl halides (or from phosphates and mesylates) is just as convenient and overcomes this limitation. For examples of the temperature effect, see Section 8.2.1.

8.1.2 Spin States that Cannot Be Accessed Thermally

$$
\begin{array}{l}
{}^{3}\text{A} \longrightarrow \text{Products} \\
\text{A}^{*} \longrightarrow {}^{3}\text{I} \longrightarrow \text{Products}
\end{array}
\qquad (8.6)
$$

Organic molecules are closed-shell species, and thus any reaction proceeding from the ground state occurs via the singlet manifold. Electronic excitation leads

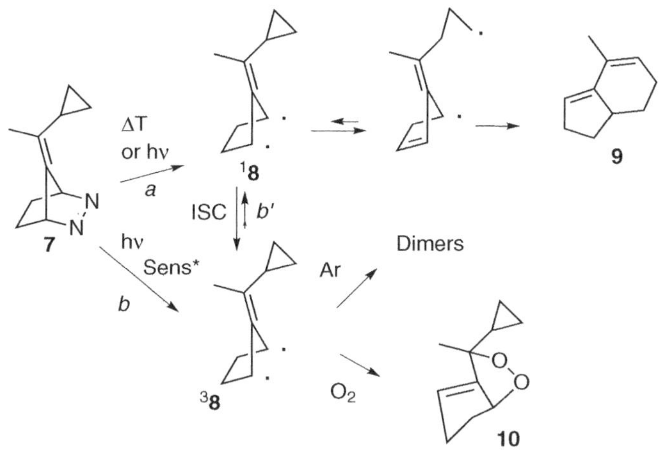

Scheme 8.5.

initially to the singlet excited state and to a process on the singlet surface; but in many classes of compounds, singlet to triplet intersystem crossing is fast enough to compete with a chemical reaction, or the latter state is formed by energy transfer photosensitization. In such cases, the triplet surface is accessed and a different chemistry occurs, not otherwise accessible. In simple systems, direct irradiation and thermolysis of a given molecule may give the same product distribution, because the same intermediates are reached at a certain point of the path followed, but triplet chemistry is generally different. A common test for a triplet mechanism in direct irradiation is in fact comparing the result with that of sensitized photochemistry. If this is the same, one may assume that ISC from the singlet to the triplet manifold is fast. An illustrative example is offered by the decomposition of a cyclic azo derivative **7** (see Scheme 8.5).

Both heating and irradiation in benzene under an inert gas of this compound led to bicyclononadiene **9** via singlet diradical ^{1}**8** (*path a*). In contrast, triplet diradical ^{3}**8** formed by benzophenone photosensitization (*path b*) or by intersystem crossing (*path b'*) gave a mixture of dimers containing the intact cyclopropyl moiety under inert gas and peroxide **10** under 5 atm O_2 pressure. Indeed, ring opening of the triplet intermediate encountered a sizeable barrier (14 kcal mol^{-1}) that hindered such a process, despite the fact that the triplet lifetime was several orders of magnitude longer than that of the corresponding singlet. This fact gave a chance to the virtually barrierless dimerization from the triplet and product distribution gave evidence about the role of ISC [20].

Consideration of the spin is important also for other reactions proceeding via biradicals, such as [2 + 2] cycloaddtions to form cyclobutanes or oxetanes, that involve 1,4-diradicals. As is well known, diradicals arising from the singlet cyclize at a rate larger than single bond rotation and thus conserve the stereochemistry; while with those formed from the triplet, recombination is slow and the stereochemistry is lost.

The spin is a key characteristic not only for diradicals, but also for even-electron intermediates such as carbenes, nitrenes, and dicoordinated carbocations, with the singlet a localized electrophile and the triplet with a diradical character. The spin-dependent chemistry occurring (for example) in carbene addition to alkenes, where a stereoselective reaction takes place with the singlet and a mixture of stereoisomers via equilibrating diradical with the triplet intermediates, is generally discussed in textbooks and requires no comment here. The presently most largely expanding approach involves the metal-catalyzed stereospecific singlet insertion into C–H bonds, which has not been considered here, as in general reactions involving metal complexes.

In the case of nitrenes, reactions via the singlet path have been advantageously carried out by irradiation in some cases, particularly for the preparation of heterocycles, probably because side paths or subsequent reactions are minimized at the low temperature of photochemistry. Not to be disregarded when considering a class of compounds that includes some explosives, photochemistry is a safer method and seems promising for scaling up.

Of more recent interest are dicoordinated carbocations and in particular phenyl or (hetero)aromatic cations, where the triplet (^{3}**11**) has a diradical character ($\pi^5\sigma^1$ structure) similar to that of triplet carbene, with one electron in the (in plane) sp^2 orbital and the other one lying out-of-plane, delocalized over the π system (Scheme 8.6). This contrasts with the localized, carbocationic character of the singlet (^{1}**11**), that has a $\pi^6\sigma^0$ structure. This makes the singlet an unselective electrophile; in practice, reactions through this path lead to solvolysis or at least to reaction with any nucleophile in a concentration high enough to trap such a short-lived species. The triplet does not react with nucleophiles acting via a nonbonding orbital. As in the example of water in the scheme (R = H), only a limited stabilization is attained in this way (a complex is formed). A π nucleophile, on the other hand, forms an adduct cation that is largely stabilized due to its distonic diradical nature and thus leads to irreversible reaction (Scheme 8.6). Indeed, irradiation in the presence of alkenes, alkynes, and (hetero)aromatics leads to arylation, and the method is beginning to show its synthetic potential, as mentioned in Section 6.3.3. With various simple aromatic derivatives, intersystem crossing is efficient and thus direct irradiation brings directly to the triplet manifold. When this is not the case, triplet sensitization can be conveniently adopted. As an example, direct irradiation of 4-cyanophenyldiazonium tetrafluoroborate in acetonitrile

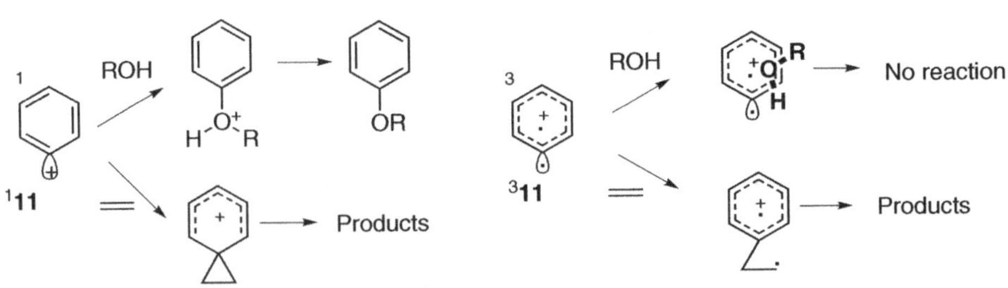

Scheme 8.6.

containing allyltrimethylsilane yielded the corresponding acetanilide via the singlet (69%), but triplet sensitization by xanthone yielded the corresponding 4-allylbenzonitrile as the only product (49%) [21].

8.2 CONTROL OF THE REACTION OF INTERMEDIATES

8.2.1 Controlling the Temperature

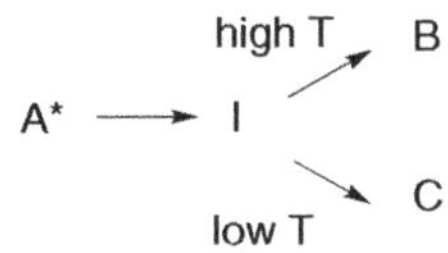

(8.7)

The scarce dependence of photochemical reactions on temperature offers a powerful synthetic advantage in three directions.

(1) Highly unstable compounds can be prepared avoiding decomposition. Compounds containing several strained bonds find in photochemistry one of the elective ways of access. Thus, hydrocarbons such as cubane, quadricyclane, dodecahydranes, and many other have been prepared through methods involving at least one photochemical step [22] and their functionalization may be advantageously carried out photochemically [23]. It has been previously noticed (see Section 8.1.1.4) that compounds that have not been isolated as yet because of their instability could be generated photochemically (or rather, the products from their own reaction have been obtained).

(2) In a second effect, a thermal reaction of the precursor of the intermediate may be avoided. Thus, the photoinduced amidation of enol ether **12** was successfully carried out at −78°C (and attack by the amidyl radical occurred predominantly at the less hindered position). At this temperature, the thermal electrophilic halogenation of the alkene by the starting chloroamide (that gave exclusively the most stabilized carbocation) was too slow to compete (see Scheme 8.7; compare Section 3.3.1.2) [24].

In a further example, the photolysis of an acylperoxide-bearing chiral substituent that is unstable at room temperature (can be stored at −20°C) was conveniently carried out at a low temperature (either −78°C or −196°C) and gave the product of decarboxylation with no loss of stereochemistry, avoiding side thermal reactions (compare Scheme 4.42 in Section 4.3.2) [25].

(3) In the third way, the reaction of the intermediates is controlled. A better yield from a photochemical reaction carried out at a lower temperature with respect to a thermal reaction is a common occurrence and several examples are documented, in particular in radical chemistry. As an example, a 14-membered ring was built via a radical–radical coupling after expulsion of four CO_2 molecules. Photolysis gave a 70% yield, larger than the 40% obtained thermally (compare Scheme 4.33 in Section 4.2.5) [26].

Of particular interest is controlling the competition between different paths. Thus, acyl radicals **13•** are conveniently photogenerated directly from aldehyes, both

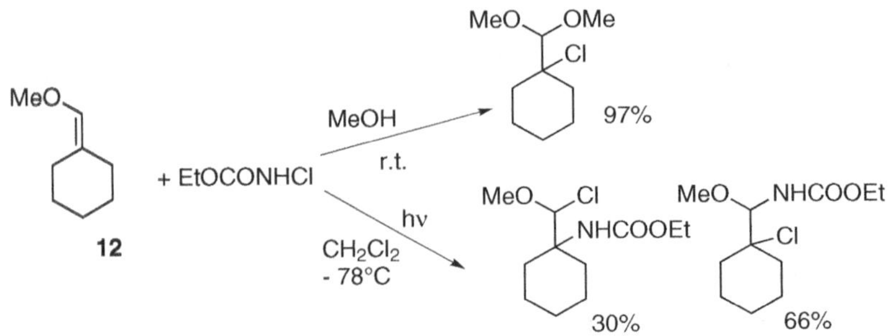

Scheme 8.7.

aliphatic [27] and aromatic [28] by photocatalytic hydrogen abstraction, and are trapped by electron-poor olefins, thus offering a convenient path to unsymmetrical ketones (see Scheme 8.8). However, the reaction is limited by the competitive decarbonylation of radicals **13·** to **14·** that is significant or predominating in the case of secondary and tertiary aldehydes. Indeed, by lowering the temperature to −20°C the proportion of acylation was much increased (compare Section 2.5.6) [27].

The temperature effect on the competition between different paths is quite common in radical chain processes. Thus, allyl α-bromophenylacetate underwent practically only reductive dehalogenation in the presence of tributyl tin hydride under the usual thermal conditions (AIBN, 80°C). However, changing the initiation process by using hexamethyldistannane photosensitized by a ketone allowed to carry out the reaction at a lower temperature and to control the chemistry of the α-acetyl radical [29]. Thus, the dimer was obtained in a non-hydrogen-donating solvent at 75°C, and the product from 5-endo cyclization onto the aromatic ring and subsequent reduction was obtained in a moderately hydrogen-donating medium, such as *i*-PrOH at 54°C (compare Section 2.2.1) [29]. Better yields of cyclization profiting of the temperature lowering permitted by photochemical initiation were obtained also in other cases. As an example, this was reported for the reaction of 2-iodo-*N*-(2-arylethyl)imidazole under irradiation with respect to thermal tin hydride-mediated reaction, where reduction generally predominated (compare Scheme 2.29 in Section 2.4) [30].

Possibly, the relatively limited application of photochemical initiation in radical chain processes is due to the low absorbance at convenient wavelengths of traditional initiators. As mentioned in Chapter 2, however, sensitization can overcome this

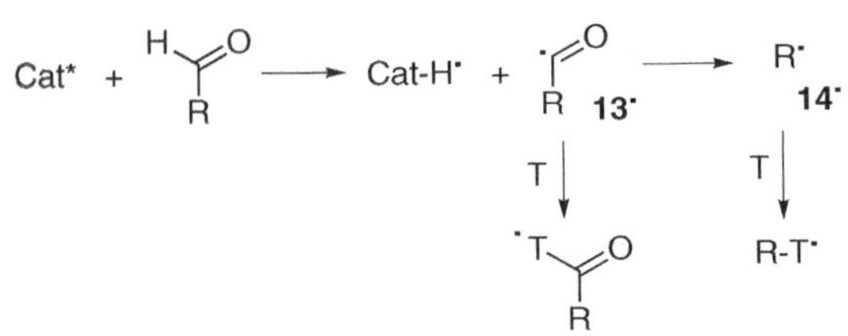

Scheme 8.8.

limitation and makes it possible to exploit the increased versatility that the control of temperature certainly involves in this kind of reactions.

8.2.2 Controlling Electron Transfer

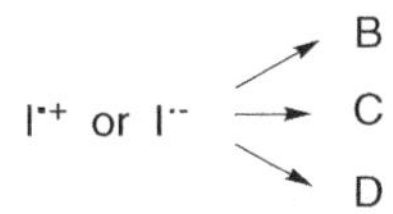

(8.8)

As shown in Chapter 5, many photochemical processes involve electron transfer. The situation of the photochemical experiment is radically different from that encountered in electrochemistry or when chemical reagents are used. Thus, reduction at the cathode surface occurs where the local concentration of electrons is much higher than in solution, and conversely of holes in anodic oxidation. The electron transfer step thus occurs under particular conditions, although a few angstroms away from the electrode surface the potential difference drops to negligible and the environment does not differ from that of homogeneous solutions and leads to a local high concentration of the intermediates. Likewise, treating with a reducing or oxidizing chemical reagent, that has to be used in a rather high concentration to be effective, makes the environment unambiguously reducing or respectively oxidizing. Apart from the exact mechanism of the first electron transfer step (outer sphere, inner sphere, catalyzed), the different situation affects the ensuing course of the reaction, as shown in Scheme 8.9, for the case of the oxidation of an organic compound D–X by photoexcited electron acceptor A–Y*. Single electron transfer generates the corresponding radical cation D–X$^{\bullet+}$. A chemical reaction with a nucleophile may occur at this stage (*path a*); for example, an alkene radical cation reacts with a diene in a [4 + 2] cycloaddition. Oxidation by a chemical reagent (e.g., $Ar_3N^{\bullet+}SbF_6^-$, an aminyl radical cation) or at the cathode brings about some risk that a second electron transfer step gives the dication D–X^{2+} and thus a different

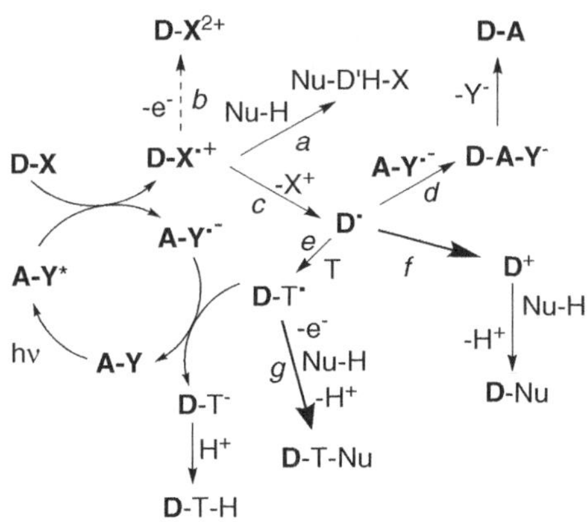

Scheme 8.9.

chemistry (*path b*). On the contrary, this is kinetically forbidden under photochemical conditions due to the minimal steady-state concentration of the only oxidant, the acceptor excited state. Furthermore, when the radical cation has available a rapid monomolecular reaction (e.g., deprotonation to give an alkyl radical D$^{\cdot}$, *path c*), the available paths under irradiation involve interception of the radical by the acceptor radical anion (*path d*) or by a radicophile trap T (*path e*). With a chemical oxidant or at the cathode, however, a further oxidation step is likely, involving either the initial radical or the adduct radical (*paths f* and *g*, respectively, as highlighted in the scheme by using heavy arrows). These steps lead to the corresponding cations D$^+$ and are followed by nucleophilic trapping.

Thus, even if single electron oxidation is the common first step, the course of a photochemical ET process tends to be different from that of chemical or electrochemical reactions. In the last two cases, some steps occur at the cation level (marked by heavy arrows) and give oxidized products (D–Nu, D–T–Nu; overall, two ET steps in the same direction are involved). A fitting example is the generation of radicals from enolizable dicarboxyls **15** by cerium or thallium salts. The radicals add to nucleophilic olefins, and the radical adducts are further oxidized, adding a nucleophile and losing a proton to give the final products **16** (Scheme 8.10a).

Under photochemical conditions, on the contrary, ET and deprotonation lead to a radical and, by attack to olefin (electrophilic in this case), lead to radical adducts, with no risk of further oxidation. Indeed, these steps are followed by reduction. Photoinduced ET generates *both* a reduced and an oxidized species. Thus, while as mentioned the minimal steady-state concentration of the excited acceptor A–Y* makes a second ET step in the *same* direction unlikely, the parallel occurring accumulation of the reduced species A–Y$^{\cdot-}$ makes *back* electron transfer efficient, as indicated in Scheme 8.9 by formation of products involving *no net* ET (e.g., D–T–H) with respect to the starting reagents. In the latter case the initial oxidation is compensated for by final back electron transfer regenerating the acceptor, unless this is incorporated in a product (e.g., D–A). This is exemplified by alkyl radical addition to α,β-unsaturated esters (Scheme 8.10b; compare Section 5.3).

As for multiple electron transfer, a representative example, though of limited synthetic value, involves the comparison between the photocatalyzed (by 1,4-dicyanobenzene, DCB) oxidation and the anodic oxidation of 1,4-bis(methylene) cyclohexane **17** in MeOH/MeCN 1 to 3 mixed solvent (Scheme 8.11).

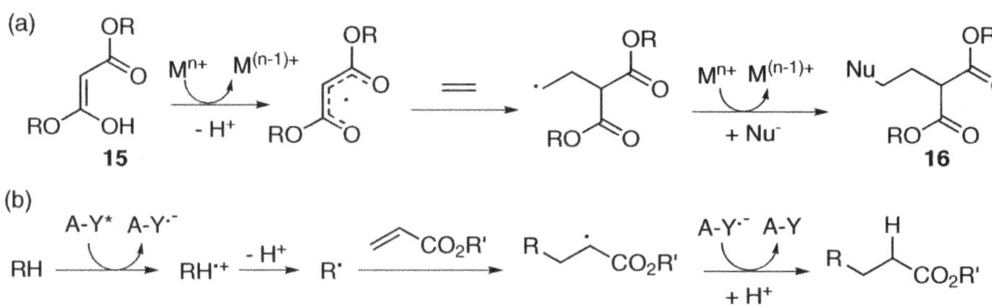

Scheme 8.10.

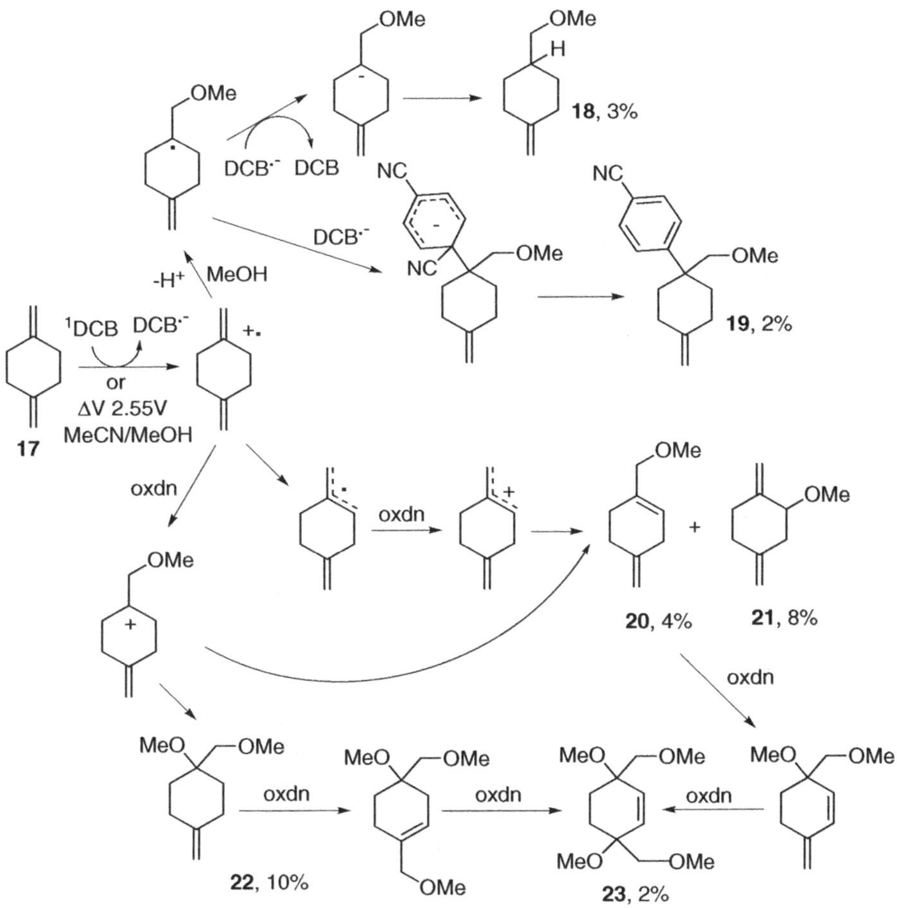

Scheme 8.11.

Under photocatalytic conditions (upper part), only a small amount of the alkene was converted. The products arose from methanol addition to the radical cation. The resulting neutral radical was either reduced by $DCB^{\bullet -}$, giving the anion that was then protonated to give ether **18**, or added to it, giving the substituted nitrile **19** (compare products DA and Nu–D'H–X in Scheme 8.9). Thus, no net oxidation took place.

As for the controlled electrochemistry (lower part), this involved the same radical cation that likewise deprotonated yielding an allyl radical, but this was further oxidized to the cation that in turn added methanol, giving products **20** and **21**. Alternatively, methanol addition preceded oxidation, giving products **20** and **22**. These ethers were more oxidized than the starting material (compare DNu in Scheme 8.9). Furthermore, consecutive processes easily took place, as indicated by the products of double and triple oxidation observed (e.g., **23**) [31].

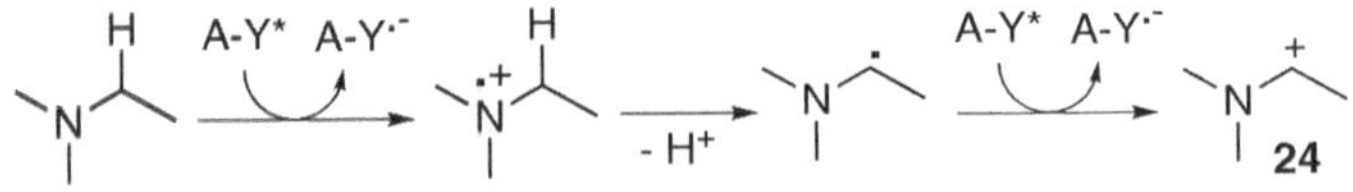

24

Scheme 8.12.

For the same reasons, the occurring of consecutive reactions is less often encountered in photochemical reactions, a fact that has been noted in several cases and results in a cleaner process.

Thus, the absence of an intermediate able to stop the process by back electron transfer is probably the cause of polymerization of 1-methoxy-1,3-cyclohexadiene taking place in the thermal $Ar_3N^{\bullet+}SbF_6^-$-catalyzed ET reaction, where the oxidant is a relatively abundant ground-state species. On the contrary, this reagent underwent a cycloaddition reaction (70% yield) when irradiated in MeCN in the presence of 25 mol% DCB [32], where the oxidant is the DCB excited state.

Net oxidation is possible also photochemically only when the first intermediate is further oxidized by the *ground-state* photocatalyst also in this case, namely when this has a low E_{ox}. As reported in Section 5.3.1.2, oxidation and deprotonation of amines gives α-amino radicals. The low oxidation potential of these intermediates makes a further oxidation step likely. In such a case, iminium cations **24**, formed via an overall two-electron process, are formed (see Scheme 8.12; compare D^+ in Scheme 8.9).

In another example involving the photocatalytic intramolecular [2 + 2] cycloaddition of a bis-enone to give the *cis*-diacylcyclobutane (see Scheme 5.80 in Section 5.5.1), it was noted that this occurred with both a lower yield and less diastereoselectivity under electrochemical conditions than under photocatalysis. The reaction involved reduction of the enone to the radical anion followed by intramolecular attack to the other enone moiety functioning as Michael acceptor. The isolated primary product gave the secondary products both at the cathode and upon prolonged photocatalysis. Thus, these were due to the occurring of a further electron transfer step that was likely at the high local concentration of electrons at the cathode surface. This made likely over-reduction before diffusion, while such a process was slow for the free, solvated compounds formed under photochemical conditions [33].

Certainly, the photocatalytic method suffers from the handicap of easy back electron transfer within the original radical ion pair that may hinder separation and drastically reduce the efficiency of radical ions reaction (Scheme 8.13a). However, in many cases it has been proven that adding a cocatalyst—that is, a secondary donor or acceptor that has no reaction of its own, but facilitates ions diffusion in the solution and thus lengthens their lifetime—makes the reaction more effective, as illustrated in Scheme 8.13b. Cocatalysts (CoCat) are usually aromatic hydrocarbons. A typical application is the alkylation of alkenes photocatalyzed by aromatic nitriles. In most of the reported examples, the high concentration at which these additives have been used in order to be effective (0.1 M) is detrimental for an actual synthetic application, however.

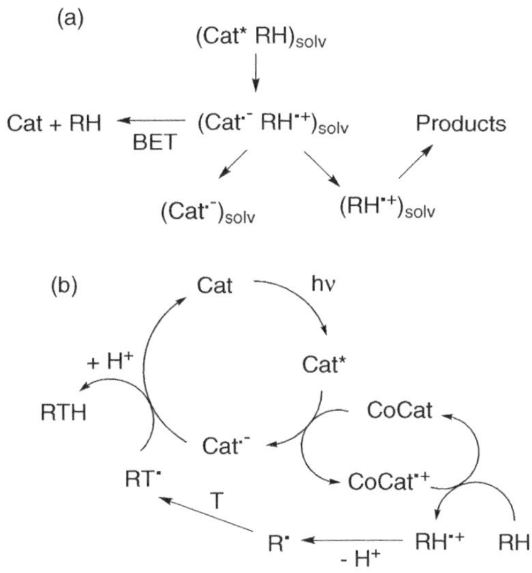

Scheme 8.13.

More elaborated models—for example, solid heterogeneous photocatalysts—should be explored and may offer a more convenient solution to the problem. A further approach for fighting back electron transfer is the addition of ions that may increase the ionic strength of the medium and thus stabilize ions, or more specifically cations that act as Lewis acids and complex the radical anions (see Scheme 5.83 in Section 5.5.2). Alternatively, one may shift to the triplet multiplicity. Using a catalyst that acts through the much longer-lived triplet state has been proven advantageous in some cases. Thus, the alkylation of electrophilic alkenes with stannanes as precursors of the active radicals was efficient by using TCB as the photocatalyst, but the reaction with alkynes occurred only sluggishly under these conditions. The efficiency was recovered by substituting an aromatic ester (e.g., tetramethylpiromellitate, TMPM, acting via the triplet) for the nitrile as the photocatalyst because BET was slower in this case, leaving more time to the radical for adding to the alkyne. The strategy, on the other hand, could not be extended to the alkylation of unsaturated ketones,

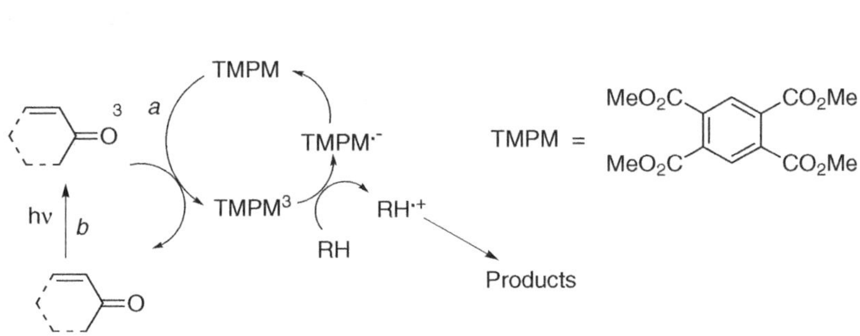

Scheme 8.14.

because the strong absorption by such substrates prevented light absorption by either the nitrile or the ester (see Scheme 8.14, path *a*). However, the ester was still a valid choice, because the ketones intersystem crossed to the triplet state and were able to sensitize by energy transfer the aromatic ester, which then acted as an ET photocatalyst (path *b*) reentering the usual path [34].

Control by using a cocatalyst was effective also in the reaction [35] of either dicyano- or tetracyanobenzene with malonodinitrile and dimethylhexadiene to give the corresponding NOCAS products (see also Section 5.2.2.3). With the latter acceptor, a monosubstitued product was obtained in the absence (a disubstituted in the presence) of phenanthrene. This demonstrated the potential of a secondary donor for accelerating the process by preventing BET and thus prolonging the lifetime of the active radical cation.

8.2.3 Control by the Choice of the Physical State

$$(A^*)_C \longrightarrow (I)_C \longrightarrow \begin{cases} B \\ C \\ D \end{cases} \qquad (8.9)$$

This is a very rich field that cannot be discussed easily in this presentation, and the interested (and presumably advanced level) reader is referred to the specialized literature that is summarily quoted where appropriate.

8.2.3.1 Reactions in Alternative Media The high rate and the fact that no heating is required make photochemical reactions excellent candidates for inducing selectivity by some external agent, since molecules will scarcely move before reacting and supramolecular architectures built for preordering the reagent will not be affected by heating. Interesting effects have been obtained in "alternative" media, such as supercritical fluids, voids in polymeric films, ionic liquids, gels, microemulsions, micelles, vesicles, and passive voids in dendrimers [36].

As an example, the photodecomposition of diazonium salts **25** in an ionic liquid, [bmin][Tf$_2$N], has been found to lead to O-attack by the anion (to give compound **26** as the main product), although this is a poor nucleophile, rather than to the fluoride expected from the Schiemann reaction (see Scheme 8.15) [37]. Further cases of

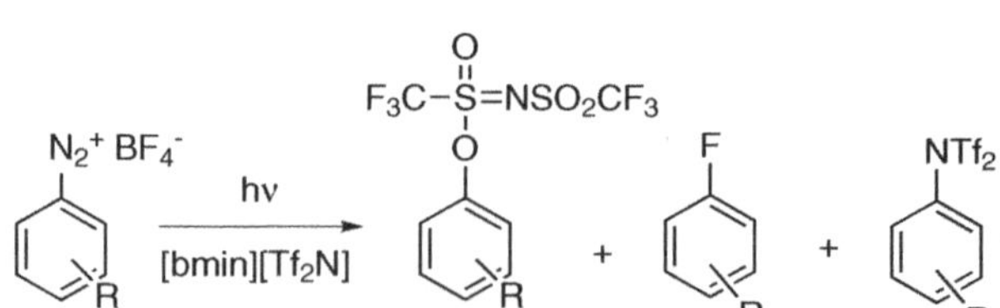

25

26, main product

Scheme 8.15.

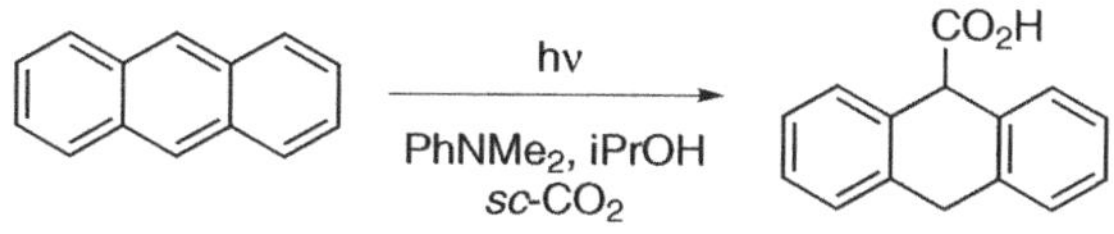

Scheme 8.16.

medium incorporation have been found in supercritical CO_2. As an example, irradiation of anthracene in the presence of *N,N*-dimethylaniline as electron donor and isopropanol as proton donor has been found to lead to carboxylation in a good yield (see Scheme 8.16) [38].

Of particular interest is asymmetric photochemistry. Several examples of the general way of carrying out diastereoselective reactions by using chiral auxiliaries are reported in Chapter 4. A more "typically photochemical" approach is either that of using chiral photosensitizers or photocatalysts that show a preferential interaction (physical or chemical, respectively) with one of the enantiomers or that of inducing enantioselectivity by using a suitable template. These methods would be an ideal way for chiral synthesis, but are of difficult application because the short lifetime of excited state hinders an effective interaction with the reagent. To meet with some success, the interaction must occur at the ground state and excitation must involve a preformed complex. As an example, template **27**—which bears both a chiral moiety derived from Kemp's acid and the photocatalyst, benzophenone—has been used with success for chiral synthesis. Thus, the photochemical reaction occurred in the complex and involved electron and then proton transfer from the pyrroline moiety of quinolone **28** to the ketone excited moiety. This gave a radical that, in turn, added to the ipso position in the quinolone with preference for the less hindered face (product **29,** Scheme 8.17). This resulted in up to 70% ee [39].

Several examples where the chiral moiety is part of the photocatalyst molecules have been reported (see Chapter 4). When ET is involved, an intrinsic problem is generated. Thus, a ionic medium is desired for efficient ET, but at the same time favors separation of the ions because of the effective solvation and thus prevents an intimate contact between the chiral catalyst and the reagent. An improvement in the enantioselectivity has been obtained for the anti-Markovnikov addition of alcohols

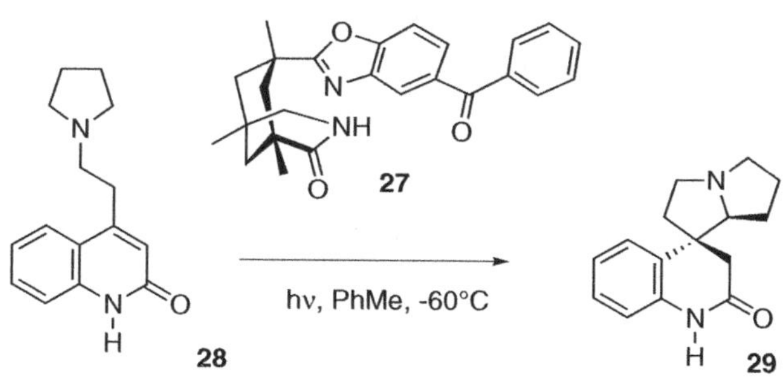

Scheme 8.17.

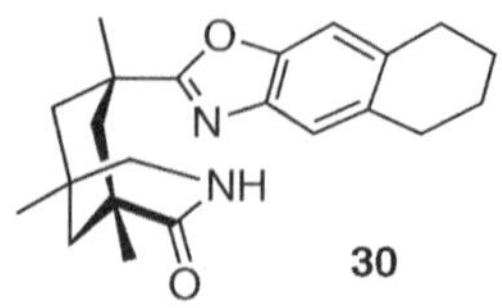

30

Figure 8.6. Oxazole incorporating Kemp's acid scaffold.

onto 1,1-diphenylethylene by using a photocatalyst bearing a chiral moiety of ionic character (e.g., a sugar derivative), but carrying out the reaction in supercritical CO_2. In this way the local ionic microenvironment facilitates ET, but separation of the radical ions in the bulk is hindered by the use of a nonpolar medium [40].

Alternatively, a chiral template can be used. Compounds based on Kemp's acid (e.g., **30**, Figure 8.6) have been used in solution [41] and have been immobilized by binding to Wang resin [42] for enantioselective cycloadditions (intramolecular of ω-alkenylquinolones). Good results were obtained with the intramolecular [2 + 2] cycloaddition of alkenylquinolones and the [4 + 4] cycloaddition of pyridine and dienes [41–43].

Cyclodextrins have also been used. As an example, decarboxylation of mesityl methylbutanoate (**31**) appears to occur concertedly, with complete retention in solution (see Scheme 8.18). When the racemic mixture was irradiated in the presence of β-CD, one of the enantiomers was destroyed with some preference [44].

Photochemical reactions of molecules in inclusion complexes and zeolites have also been explored [45–47]. Zeolites are not *per se* chiral, but can be modified by appending a chiral group. Thus, the α-cleavage of racemic benzoin methyl ether (**32**), when carried out in solution, gave a mixture of products, mainly the symmetric dimers **33** and **34**, from the recombination of the radicals (see Scheme 8.19). However, in the presence of chiral modified zeolite, nonsymmetric recombination to give the starting material occurred with a small enantioselectivity (see formulae **35**), and the main product **36** resulted from a different mode of bonding (Scheme 8.19) [48].

Irradiation in the solid state has given important results (as mentioned in Chapter 4, conditions suited for this case must be adopted; e.g., a suspension of nanocrystals may be convenient, as discussed in Chapter 1) [49–54]. Among the type of chemistry presented in previous chapters, typical applications in the solid state include homolysis and radical recombination, as in the case of the decabonylation of ketones having substituents stabilized as radicals, such as benzyl or adamantyl [55].

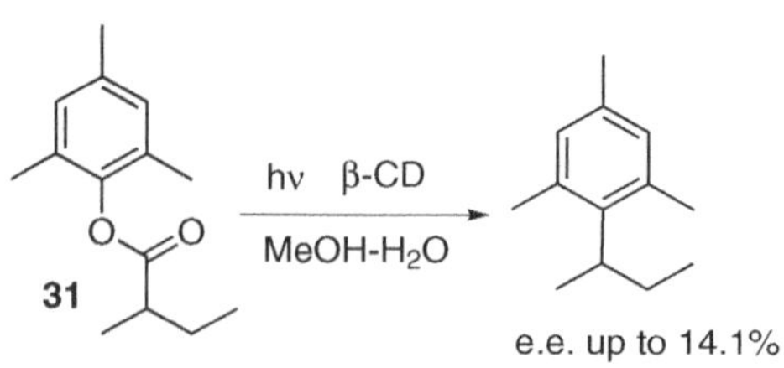

Scheme 8.18.

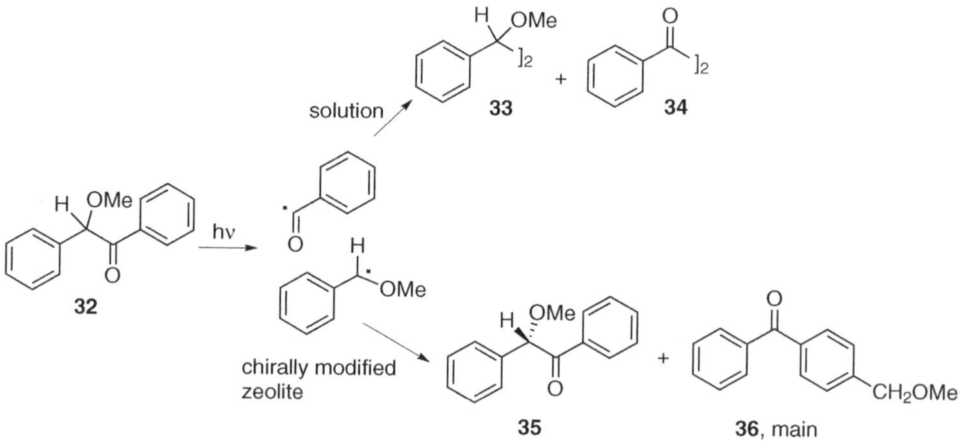

Scheme 8.19.

Asymmetric synthesis in single crystal offers specific possibilities through the irradiation of co-crystals of an achiral substrate with a chiral inductor or of an achiral molecule that forms chiral crystals, a not very common occurrence [56].

8.2.3.2 ***Homogeneous versus Heterogenesous Photocatalysis*** Heterogeneous catalysts have a large place in photochemistry. Examples of molecules that are active as homogeneous photocatalysts and have been in some way immobilized or bonded to a surface or incorporated in the hole of a material have been explored to some degree, and at least in some cases are advantageous both for the expeditious recovery and for an increased robustness and efficiency [57, 58]. However, the term heterogeneous photocatalysis is mainly referred to the use of semiconductors powders, oxides, and sulfides, a discipline that has evolved for the main part separately from the rest of photochemistry and is largely used for the degradation of pollutants, but rarely for chemical synthesis. However, this offers a useful way for the generation of intermediates for synthetic purpose [59]. As an example, photocatalytic reactions of cyclic alkenes and enol ethers mediated by ZnS or CdS in the presence of a sacrificial donor led to the oxidative dimerization. In the presence of aromatic azo compounds or of aromatic imines along with alkenes, reductive dimerization of the acceptor, oxidative dimerization of the donor, and coupling took place in a proportion that depended on structure, catalyst, and conditions [59]. As mentioned in Chapter 5, an application is the photoinduced single-electron oxidation of compounds containing a benzyl group followed by α-fragmentation to give radicals. This is possible also in a homogeneous medium, but under these conditions offers a further possibility of control. Thus, TiO_2-photocatalyzed reactions of compounds of formula 4-$YC_6H_4CH_2X$ in the presence of maleic acid or anhydride or of fumaric acid led to the corresponding benzylated or dibenzylated products. In every case, the reaction was initiated by hole/electron transfer at the semiconductor surface. This led to benzyl radicals on one hand and the

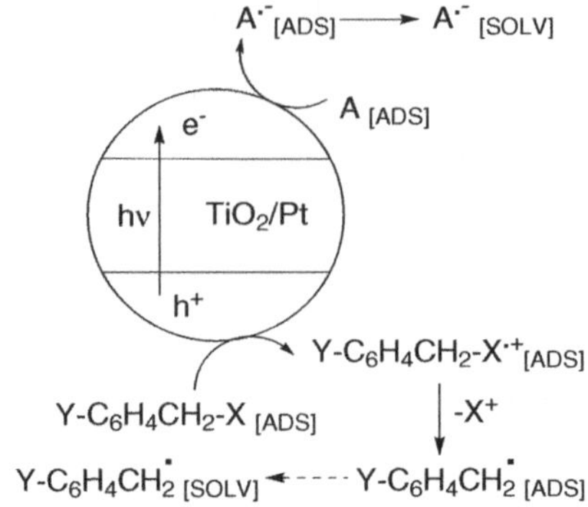

Scheme 8.20.

radical anions of the electron-withdrawing substituted alkene on the other one. These intermediates migrate in the solution and couple. With good donors (Y = OMe), electron transfer is fast and the rate-determining step is the cleavage of the radical cation (fast for benzyltrimethylsilane, less so for phenylacetic acid, too slow to have any role for the methyl derivatives). With poor donors (Y = H), however, the electron transfer step becomes rate-determining, and this requires that both hole and electron transfer occur, since otherwise reverse electron transfer will reduce back the radical cation before it cleaves. This depends on the structure of the alkene, which must be present in a high enough concentration at the catalyst surface. Platinizing the surface (0.3 or 0.75% load) made the reaction faster and led to some methylbenzene and an increased proportion of dibenzylated derivatives. Apparently, the metal has a multi-faceted effect, by both facilitating electron transfer and enhancing the local concentration of benzyl radicals, and further promotes hydrogen transfer (see Scheme 8.20) [60].

8.2.4 Control of the Process Through the Irradiation Conditions

8.2.4.1 Optimizing Light Absorption As mentioned in Chapter 1, it is essential that light be absorbed by the intended reagent. Apart from the quite demanding case of solids mentioned above, also in solution an evaluation of this factor is advised, because no absorbance means no reaction, but too high an absorbance often means that a mixture (or tars) will be obtained. Dilution generally brings better results, although this choice is not generally viable for preparative purpose. Consider the best possible case, namely an aliphatic ketone that undergoes α-fragmentation. The absorption is moderate (ε_{max} ~30 L mol^{-1} cm^{-1}) and the product does not absorb (see next section). Still, in a synthetic procedure the α-cleavage of a ketone was reported to be clean in dilute solution (0.015 M; see Chapter 4), but less so at a larger concentration; at any rate, impurities developed upon prolonged irradiation [61]. This said, it is probably fair to state that the situation is not uniformly negative. Several successful preparations with a substrate concentration 0.1 to 0.5 M have been published (see Table 8.1), and the fact that most reported photochemical procedures

Table 8.1. Exemplificative Photochemical Reactions Developed in a Scale from ~10 to > 100 g

Reaction	Light source, year of publication
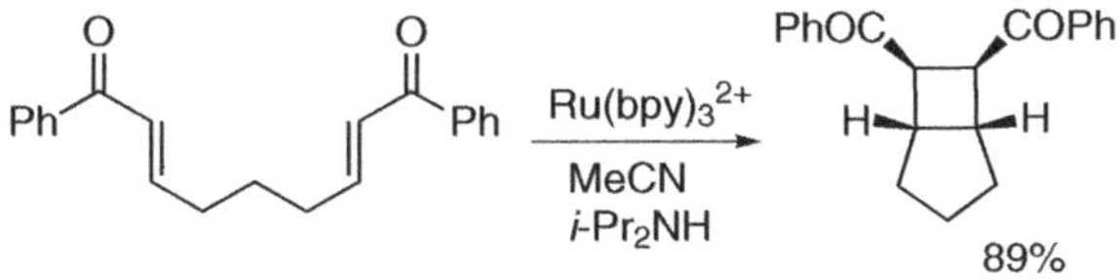	23-W fluorescent bulb or laboratory lighting, 2008 [68]
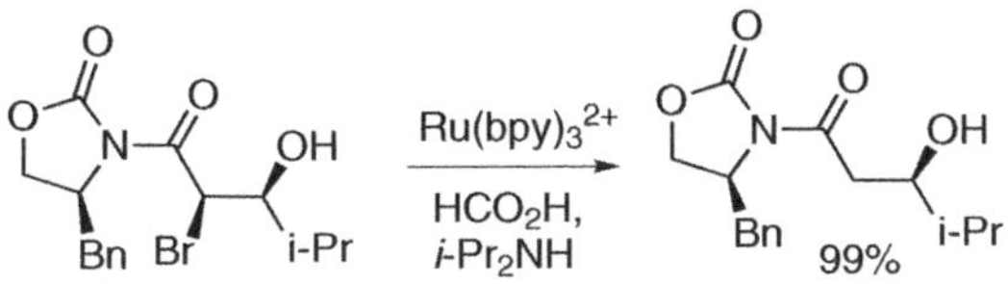	14-W fluorescent bulb, 2009 [69]
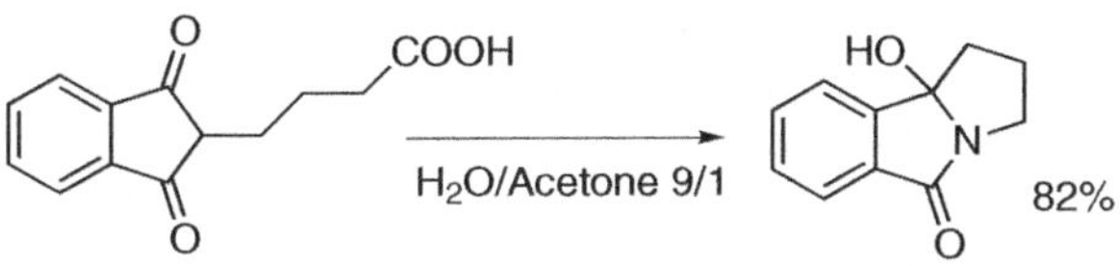	XeCl excimer lamp, 2003 [70]
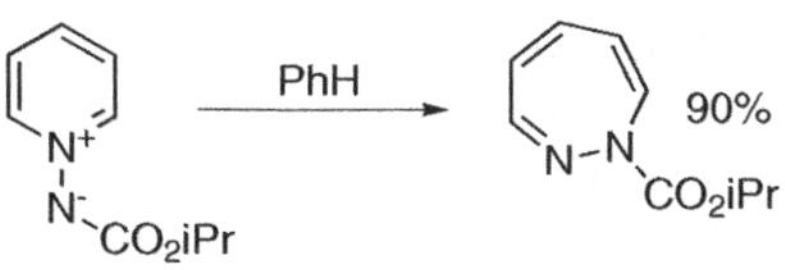	2000-W Hg arc, medium pressure, 2003 [71]
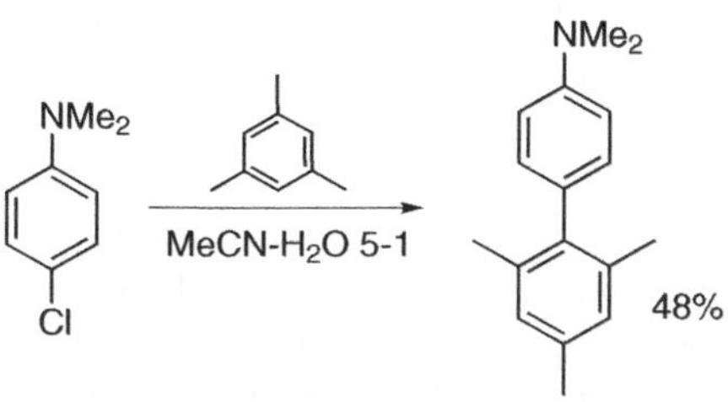	Natural sunlight, 2010 [72]
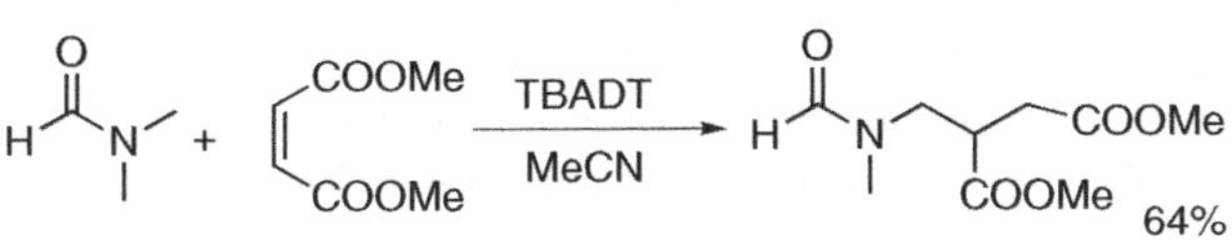	Natural sunlight, 2009 [73]
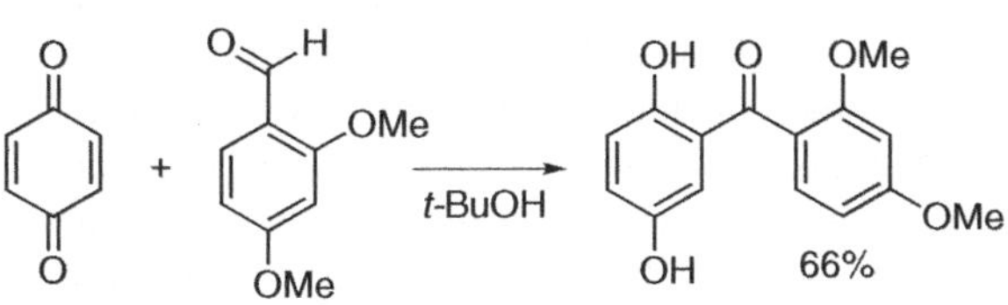	Medium concentrated sunlight, 2006 [74]
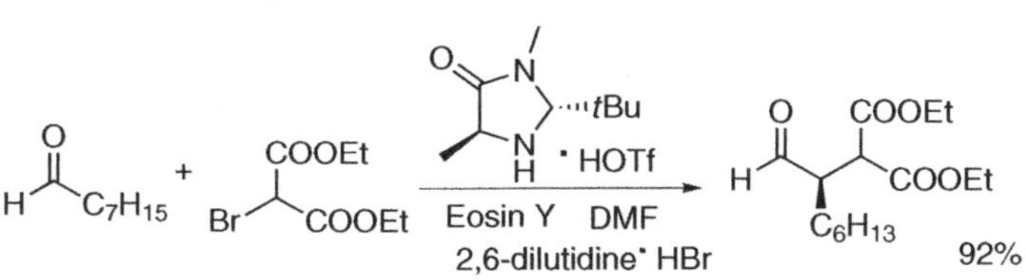	Flow reactor, 23-W fluorescent bulb, 2012 [75]

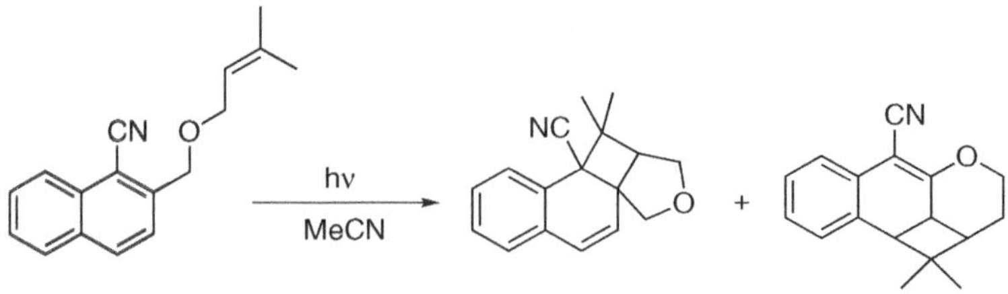

37

batch (1 hour) **38**, 72% **39**, 28%
microreactor (3 min) 90% 10%

Scheme 8.21.

involve dilute solutions (typically 1 to 5×10^{-3} M) is probably because such studies are mechanistically oriented and there has been less interest for the synthetic application. The choice of more synthetically oriented irradiation conditions is—and will become more and more—quite important, in particular with the adoption of low optical path reactors (microreactors). As an example, the dimethoxybenzophenone-photocatalyzed isopropanol addition to furanone in a LED-driven microchip reactor has been reported [62]. The reaction reached completion in only 5 min, a much better performance in comparison to the usual batch reaction. The improvement was ascribed both to the larger surface-to-volume ratio of the reaction in the microreactor and to the more uniform absorption of light within the reactor microchannels (see Chapter 1). Likewise, the intramolecular cycloaddition of the allylnaphthyl ether **37** in MeCN occurred in 3 minutes, while in a batch reactor this required about 1 hour and gave a better regioselectivity in favor of the adduct **38** (Scheme 8.21) [63].

8.2.4.2 Selective Irradiation Synthetically useful photochemical reactions give products that absorb less than the reagent. No wonder, also among the thermal counterpart clean reactions are those giving a product less reactive than the reagent (well-known negative examples are the alkylation of amines and of aromatics). Thus, the reaction of the primary product must be avoided. In photochemistry, this rule translates into the requirement that the product must not absorb in the wavelength window used for the excitation of the reagent. If it does, it may either slow down the reaction (the so-called "inner filter" effect) or introduce secondary processes. Indeed, most of the best-known photochemical reactions lead to a product that absorbs at lower wavelength or less intensively than the reagent—for example, oxetanes from ketones where the $n\pi^*$ band around 280 nm is lost or reactions such as the formation of acylcyclobutanes from α,β-unsaturated ketones where the conjugation is lost and so is the strong absorbance around 300 nm. In such cases, the effects mentioned above are minimized by using the common low-pressure mercury arcs (phosphor coated in the latter case). This holds for simple derivatives. The presence of substituents or of other absorbing moieties complicates the issue; and being able to choose a narrow wavelength range for the irradiation may offer a clear advantage, particularly when cheap light sources such as LEDs will be available over all of the

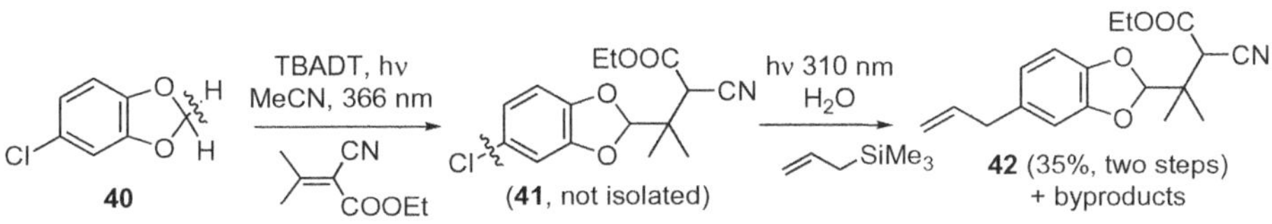

Scheme 8.22.

UV (at the moment, this is possible for the visible spectrum and—less efficiently—for part of the UV). In principle, the selective irradiation of a single compound in a mixture is possible, although this may be not as simple, given the breadth of absorption bands of molecules and the fact that sensitization may lead to excitation also of compounds not directly absorbing.

On the other hand, two subsequent photochemical steps have been carried out in a single operation in other cases by simply prolonging the irradiation. An example of a clean procedure involving two photochemical steps is shown in Scheme 4.19 in Section 4.2.2. Thus, cyclohexane-1,2-dicarboxyl imide suffered α-cleavage in the first step to give an α,β-unsaturated imide that was converted upon further irradiation into a azetidine-2,4-dione [64]. The primary product absorbed three times as much as the reagent at 254 nm, thus conditions were optimal for the occurring of the second step.

Two processes under different conditions have also been carried out in a one-pot procedure. Thus, decatungstate (TBADT) photocatalysis of a chlorobenzodioxole (**40**) caused hydrogen abstraction and alkylation of a cyanoacrylate. The product (**41**) was not isolated, but water and allyltrimethylsilane were added and the lamp changed to one emitting at shorter wavelength. Under these conditions, the chlorine was activated and allylation to give compound **42** ensued (see Scheme 8.22) [65].

It must be stressed that excitation in any part of the absorption spectrum of a compound gives the same result, since even if high-lying singlets are initially formed, these convert to the lowest-lying state of each multiplicity, S_1 and T_1 at a rate larger than any chemical processes. Exceptions are only a few—in particular, thiones and azulene derivatives. In such cases, S_2 and S_1 are so far away from each other that internal conversion is slow and the first formed state has a chemistry of its own.

Obviously, different results are obtained also from compounds containing two separate chromophores. In such a case, different results are achieved by irradiation in different regions of the spectrum. As an example, compounds **43** contain two separate chromophores, the cross-conjugated ketone and the nonconjugated ketone in position 17. Irradiation with a 310-nm centered lamp excites primarily the latter, while irradiation at 254 or at 366 nm excites the former [66]. The product distribution in fact changes accordingly, giving either **44** or **45** as the main product (see Scheme 8.23). Such a clear-cut difference is not always noted because energy transfer may follow and lead also to the state not directly excited.

In bimolecular reactions, an important case is the formation of complexes. These are different ground-state species and thus have a different absorption spectrum, contrary to the case of complexes formed by the excited state (exciplexes; see Scheme 8.24). A new well distinguished band may emerge and selective irradiation into it may result in a cleaner process. Note that the species formed by excitation of

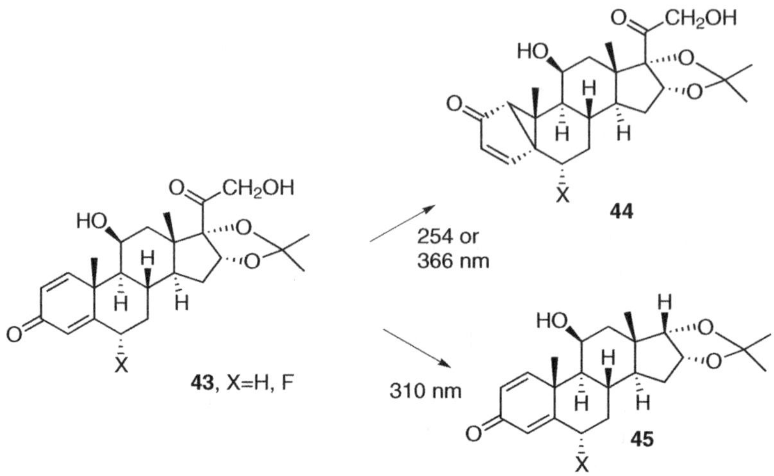

Scheme 8.23.

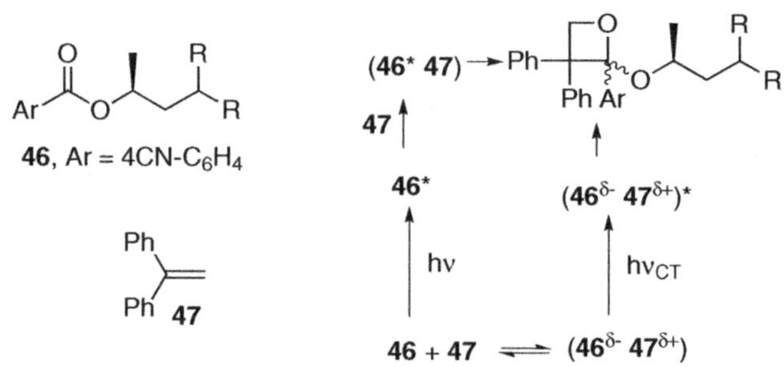

Scheme 8.24.

the complex does not necessarily coincide with the exciplex. As an example, irradiation of chiral cyanobenzoates **46** in the presence of alkenes **47** yielded two oxetanes in a different diastereoisomeric ratio when irradiated in the specific CT band ($h\nu_{CT}$) appearing when the two reagents were mixed (the ground-state complex) or in the region where the noncomplexed ester absorbed (and it was the excited state that was complexed) [67].

8.3 CONCLUSION AND OUTLOOK

$$A \xrightarrow{h\nu} B \qquad A \xrightarrow[\Delta T \text{ or reagent}]{} B \text{ or } C \tag{8.10}$$

As indicated in Chapter 1, provided that some typical pitfalls are avoided, carrying out a photochemical reaction is really an easy and inexpensive way to make chemistry. The only general precaution is the preference for deaerated conditions (except of course for oxygenations), similar to the case of radicals but less dramatic, because excited states are shorter-lived than radicals and are less easily intercepted. Otherwise, photochemistry seems to be completely unaware of the techniques introduced in the last decades for "advanced" synthesis, mainly based on organometallics. Thus, exhaustive dehydration/purification of the solvents, low temperature (unless by choice), the use of easily flammable or other types of dangerous solvent/reagents, and so on, are rare in photochemistry. The recent trend toward visible light-induced reactions, which are expeditiously carried out by using domestic lamps or solar light, further increases the appeal of the method. Certainly there is no problem in carrying out a reaction on the 10-mg to 10-g scale. Although the largest part of photochemistry literature is still centered on mechanistic and applicative aspects, the synthetic facet is increasingly being considered, and the application on a gram scale or above seems to gain more and more attention. In particular, reactions occur satisfactorily by using either fluorescent bulbs or simply sunlight as the light source (see Section 1.3.1.3). Certainly a casual look at photochemical literature may lead to some disconcert. Leaving mechanistic studies alone, and limiting consideration to preparations reported in the 10- to 100-g scale, still an exceedingly variety of methods appear to have been used. An idea is given in Table 8.1, where lamps vary from high-power arcs to solar light and it is not always easy to predict whether changing the source for another one will lead to an important change in the yield or in the time required for conversion. Some general guidelines have been given in Chapter 1 and may give some help. At any rate, this should not discourage beginning to explore a new photochemical reaction by using the less expensive homemade arrangements. In >90% of the cases this approach will guide us toward the best solution.

The peculiar advantage of photochemical reactions, namely their ability to occur relatively independently on conditions, results in a large versatility by controlling the further evolution of the primarily formed intermediates. This fact is far from being realized by the synthetic community, but should be an emerging field in the near future. Reactions have been rapidly reviewed in Chapter 2–7 according to the type of intermediate formed. It appears that photochemical methods have a different role in the various cases. Photochemical initiation has at the moment a limited role in radical chain alkylation procedures, but the possibility of carrying out the process at a lower temperature in such highly temperature-dependent reactions should improve the application (provided that an appropriate sensitizer is used, otherwise initiation is inefficient). With precursors such as Barton esters and similar derivatives that absorb UV light more efficiently, photoinitiation has a larger scope. Photocatalysis widens by far the field by introducing alkanes and simple aliphatic derivatives as radicals precursors. Heteroatom-centered radicals are likewise accessible, arising from precursors D—X—Y with a weak X—Y bond, with varied preference for thermal or photochemical generation.

Carbanions and enolates have rarely been generated photochemically. On the contrary, photoheterolysis is effective as a route to carbocations, particularly for vinyl cations, where in some cases it competes with homolysis [76]. Significant is the case of aryl cations, that are practically accessible only through the photochemical path [77]. Carbenes and nitrenes have a long established tradition for photochemical generation, and diradicals are a photochemical province. Radical ions are a quite active field of research, where the different results between the photochemical approach and other methods such as electrochemistry or chemical redox processes have been evidenced many times (compare Chapter 5 and Section 8.2.2) and the versatility of the former has been exploited to some degree.

Not surprisingly, the role of photochemistry is more developed and promising for open-shell structures, such as radicals or radical ions (and, characteristically, of triplet cations). In these fields it would be advisable to test every new reaction by a photochemical method.

REFERENCES

1. Alabugin, I. V., Yang, W.-Y., Pal, R. Photochemical Bergman cyclization and related photoreactions of enediynes. In Griesbeck, A., Oelgemoeller, M., Ghetti, F., Eds. *CRC Handbook of Organic Photochemistry and Photobiology*, third Edition, CRC Press, Boca Raton, FL, 2012, pp. 549–592.
2. Polukhtine, A., Karpov, G., Pandithavidana, D. R., Kuzmin, A., Popik, V. V. (2010). Photochemical triggering of the Bergman and Myers–Saito cyclizations. *Australian Journal of Chemistry*, **63**, 1099–1107.
3. Lan, Y., Houk, K. N. (2010). Mechanism and stereoselectivity of the stepwise 1 3-dipolar cycloadditions between a thiocarbonyl ylide and electron-deficient dipolarophiles: A computational investigation. *Journal of the American Chemical Society*, **132**, 17921–17927.
4. Beno, B. R., Fennen, J., Houk, K. N., Lindner, H. J., Hafner, K. (1998). [5, 5] Sigmatropic rearrangement. DFT Prediction of a diradical mechanism for a Woodward–Hoffmann "allowed" thermal pericyclic reaction. *Journal of the American Chemical Society*, **120**, 10490–10493.
5. Freccero, M., Gandolfi, R., Sarzi-Amadé, M., Rastelli, A. (2001). Transition structures for the stepwise insertion of oxygen into alkane tertiary C—H bonds by dimethyldioxirane. *Tetrahedron Letters*, **42**, 2739–2742.
6. Albini, A., Fagnoni. M. Photochemistry as a green synthetic method. In Mordini., A, Feigl, F., Eds. *New Methodologies and Techniques for a Sustainable Organic Chemistry. NATO Science Series II: Mathematics, Physics and Chemistry*, Vol. **246**, 2008, pp. 279–293.
7. Noyori, P. (2010). Insight: Green chemistry: The key to our future. *Tetrahedron*, **66**, 1028.
8. Kim, Y., Do, Y., Park, S. (2011). Mechanistic study of half-titanocene-based reductive pinacol coupling reaction. *Bulletin of the Korean Chemical Society*, **32**, 3973–3978.
9. Bogdanovic, B., Bolte, A. (1995). A comparative study of the McMurry reaction utilizing [$HTiCl(THF)_{\approx 0.5}$], $TiCl_3(DME)_{1.5}$–Zn(Cu) and $TiCl_2{\cdot}LiCl$ as coupling reagents. *Journal of Organometallic Chemistry*, **502**, 109–121.
10. Gonzalez-Cameno, A. M., Mella, M., Fagnoni, M., Albini, A. (2000). Photochemical alkylation of ketene dithioacetal *S*, S-dioxides. An example of captodative olefin functionalization. *The Journal of Organic Chemistry*, **65**, 297–303.

11. Abraham, H.-W. Stabilized carbocations generated by photoheterolysis. In Griesbeck, A., Oelgemoeller, M., Ghetti, F., Eds. *CRC Handbook of Organic Photochemistry and Photobiology*, third edition, CRC Press, Boca Raton, FL, 2012, pp. 419–443.
12. Coleman, M. P., Boyd, M. K. (1999). The *S*-pixyl group: An efficient photocleavable protecting group for the 5′ hydroxy function of deoxyribonucleosides. *Tetrahedron Letters*, **40**, 7911–7915.
13. Shubina T. E., Fokin A. A. (2011). Hydrocarbon σ radical cations. *WIREs Computational Molecular Science*, **1**, 661–679.
14. Mella, M., Freccero, M., Albini, A. (1995). Photoinduced SET for the functionalization of alkanes. *Journal of the Chemical Society, Chemical Communication*, 41–42.
15. Fokin, A. A. Tkachenko, A. B., Shubina, T. E. Gunchenko, P. A. Gusev, D. V., Vohs, J. K., Robinson, G. H., Yurchenko, A. G., Schreiner, P. R. (2002). The protoadamantane radical cation. *European Journal of Organic Chemistry*, 3844–3849.
16. Fagnoni, M., Vanossi, M., Mella, M., Albini, A. (1996). Limits to the oxidation of organic substrates. SET oxidative processes of commonly used solvents as revealed through the photochemical reaction with 1,2,4,5-benzenetetracarbonitrile. *Tetrahedron*, **52**, 1785–1796.
17. Schreiner, P. R., Fokin A. A. (2004). Selective alkane C—H bond functionalizations utilizing oxidative single electron transfer and organocatalysis. *Chemistry Records*, **3**, 247–257.
18. Zhegalova N. G., Popik, V. V. (2011). Wolff rearrangement of β-alkynyl-α-diazo-β-ketoesters: Light-induced acetylene–allene isomerization and its use for activation of enediynes. *Journal of Physical Organic Chemistry*, **24**, 969–975.
19. Nakatani, K., Maekawa, S., Tanabe, K., Saito, I. (1995). α-Diazo ketones as photochemical DNA cleavers: A mimic for the radical generating system of neocarzinostatin chromophore. *Journal of the American Chemical Society*, **117**, 10635–10644.
20. Adam, W., Finzel, R. (1992). Cyclopropylcarbinyl ring opening as mechanistic probe for differentiation between the singlet and triplet spin states of trimethylenemethane diradicals (non-Kekule species). *Journal of the American Chemical Society*, **114**, 4563–4568.
21. Milanesi, S., Fagnoni, M., Albini, A. (2005). (Sensitized) photolysis of diazonium salts as a mild general method for the phenyl cations. Chemoselectivity of the singlet and the triplet 4-substituted phenyl cations. *Journal of Organic Chemistry*, **70**, 603–610.
22. Bethke, S., Brand, S., Treptow, B., Gleiter, R. (2002). Strained hydrocarbons from cyclic diynes- preparation and reactivity. *Journal of Physical Organic Chemistry*, **15**, 484–489.
23. Eaton, P. E., Gilardi, R. L., Zhang, M. X. (2000). Polynitrocubanes: Advanced high-density, high-energy materials. *Advanced Materials*, **12**, 1143–1148.
24. Caron, G., Lessard, J. (1993). The regiochemistry of the radical addition of *N*-chloroamides to enol ethers. *Tetrahedron*, **49**, 8039–8058.
25. Spantulescu, M. D., Jain, R. P., Derksen, D. J., Vederas, J. C. (2003). Photolysis of diacyl peroxides: A radical-based approach for the synthesis of functionalized amino acids. *Organic Letters*, **5**, 2963–2965.
26. Feldhues, M., Schäfer, H. J. (1986). Selective mixed coupling of carboxylic acids—III: Synthesis of cyclopentadecanone from cyclic tetraacyl diperoxides. *Tetrahedron*, **42**, 1285–1290.

27. Esposti, S., Dondi, D., Fagnoni, M., Albini, A. (2007). Acylation of electrophilic olefins through decatungstate-photocatalyzed activation of aldehydes. *Angewandte Chemie International Edition*, **46**, 2531–2534.
28. Ravelli, D., Zema, M., Mella, M., Fagnoni, M., Albini, A. (2010). Benzoyl radicals from (hetero)aromatic aldehydes. Decatungstate photocatalyzed synthesis of substituted aromatic ketones. *Organic and Biomolecular Chemistry*, **8**, 4158–4164.
29. Harendza, M., Junggebauer, J., Lessmann, K., Neumann, W. P., Tews, H. (1993). Mild photochemical generation of stannyl radicals $Bu_3Sn^\bullet$ from hexabutyl distannane for organic synthesis. *Synlett*, 286–288.
30. Clyne, M. A., Aldabbagh, F. (2006). Photochemical intramolecular aromatic substitutions of the imidazol-2-yl radical are superior to those mediated by Bu_3SnH. *Organic and Biomolecular Chemistry*, **4**, 268–277.
31. de Lijser, H. J. P., Arnold, D. R. (1977). Interconversion and rearrangement of radical cations. Part 2.1. Photoinduced electron transfer and electrochemical oxidation of 1,4-bis(methylene)cyclohexane. *Journal of the Chemical Society, Perkin Transaction 2*, 1369–1380.
32. Pabon, R. A., Bellville, D. J., Bauld, N. L. (1983). Cation radical Diels–Alder reactions of electron-rich dienophiles. *Journal of the American Chemical Society*, **105**, 5158–5159.
33. Shen, M., Du, J., Yoon T. P. quoted in Ischay, M., Yoon, T. P. (2012). Accessing the synthetic chemistry of radical ions. *European Journal of Organic Chemistry*, 3359–3372.
34. Ohashi, M., Nakatani, K., Maeda, H., Mizuno, K. (2008). Photoinduced tandem three-component coupling of propanedinitrile, 2, 5-dimethylhexa-2,4-diene, and cyanoarenes. *The Journal of Organic Chemistry*, **73**, 8348–8351.
35. Fagnoni, M., Mella, M., Albini, A. (1997). Tandem energy transfer–electron transfer in the photosensitized alkylation of α, β-unsatured ketones. *Journal of Physical Organic Chemistry*, **10**, 777–780.
36. Mori, T., Inoue, Y. Photochemistry in alternative media. In Griesbeck, A., Oelgemoeller, M., Ghetti, F., Eds. *CRC Handbook of Organic Photochemistry and Photobiology*, third edition, CRC Press, Boca Raton, FL, 2012, pp. 249–276.
37. Laali, K. K., Okazaki, T., Bunge, S. D. (2007). *N*-(Trifluoromethylsulfonyl)aryloxytrifluoromethylsulfoximines [ArO-SO(CF_3)=NTf] and *N*-aryltriflimides Ar-N$(Tf)_2$ by thermal and photolytic dediazoniation of [ArN_2][BF_4] in [BMIM][Tf_2N] ionic liquid: Exploiting the ambident nucleophilic character of a "nonnucleophilic" anion. *The Journal of Organic Chemistry*, **72**, 6758–6762.
38. Chateauneuf, J. E., Zhang, J., Foote, J., Brink, J., Perkovic, M. W. (2002). Photochemical fixation of supercritical carbon dioxide: The production of a carboxylic acid from a polyaromatic hydrocarbon. *Advanced Environmental Research*, **6**, 487–493.
39. Bauer, A., Westkamper, F., Grimme, S., Bach, T. (2005). Catalytic enantioselective reactions driven by photoinduced electron transfer. *Nature*, **436**, 1139–1140.
40. Asaoka, S., Wada, T., Inoue, Y. (2003). Microenvironmental polarity control of electron-transfer photochirogenesis. Enantiodifferentiating polar addition of 1,1-diphenyl-1-alkenes photosensitized by saccharide naphthalenecarboxylates. *Journal of the American Chemical Society*, **125**, 3008–3027.
41. Breitenlechner, S., Selig, P., Bach, T. Chiral organocatalysts for enantioselective photochemical reactions. In Reetz, M. T., List, B., Jaroch, S., Weinmann, H., Eds. *Organocatalysis*, Springer-Verlag, Berlin, 2008, pp. 255–280.

42. Breitenlechner, S., Bach, T. (2008). A polymer-bound chiral template for enantioselective photochemical reactions. *Angewandte Chemie International Edition*, **47**, 7957–7959.

43. Bach, T., Bergmann, H., Harms, K. (2001). Enantioselective photochemical reactions of 2-pyridones in solution. *Organic Letters*, **3**, 601–603.

44. Mori, T., Weiss, R. G., Inoue, Y. (2004). Mediation of conformationally controlled photodecarboxylations of chiral and cyclic aryl esters by substrate structure, temperature, pressure, and medium constraints. *Journal of the American Chemical Society*, **126**, 8961–8975.

45. Pitchumani, K., Madhavan, D. Induced diastereoselectivity in photodecarboxylation reactions. In Lenci, F., Horspool, W., Eds. *CRC Handbook of Organic Photochemistry and Photobiology*, second edition, CRC Press, Boca Raton, FL, 2004, 65-1–14-14.

46. Sivaguru, J., Shailaja, J., Uppili, S., Ponchot, K., Joy, A., Arunkumar, N., Ramamurthy, V. Achieving enantio- and diastereoselectivities in photoreactions through the use of a confined space. In Toda, F., Ed. *Organic Solid-State Reactions*, Kluwer Academic Publishers, Dordrecht, The Netherlands, 2002, pp. 159–188.

47. Turro, N. J. (1995). Supramolecular organic and inorganic photochemistry: Radical pair recombination in micelles, electron transfer on starburst dendrimers, and the use of DNA as a molecular wire. *Pure and Applied Chemistry*, **67**, 199–208.

48. Kaprinidis, N. A., Landis, M. S., Turro, N. J. (1997). Supramolecular control of photochemical enantiomeric induction and radical pair recombination in zeolites. *Tetrahedron Letters*, **38**, 2609–2612.

49. Scheffer, J. R. (2004). The solid-state ionic chiral auxiliary approach to asymmetric induction in photochemical reactions. *Molecular and Supramolecular Photochemistry*, **11**, 463–483.

50. Kaftory, M. Structural aspects and templation of photochemistry in solid-state supramolecular systems. In Ramamurthy, V., Inoue, Y., Eds. *Supramolecular Photochemistry*, John Wiley & Sons, Hoboken, NJ, 2011, pp. 229–266.

51. Sonoda, Y. (2011). Solid-state [2 + 2] photodimerization and photopolymerization of α, ω-diarylpolyene monomers: Effective utilization of noncovalent intermolecular interactions in crystals. *Molecules*, **16**, 119–148.

52. Yang, C. Xia, W. (2009) Solid-state asymmetric photochemical studies using the ionic chiral auxiliary approach. *Chemistry—An Asian Journal*, **4**, 1774–1784.

53. Nagarathinam, M., Peedikakkal, A., Malik P., Vittal, J. J. (2008). Stacking of double bonds for photochemical [2 + 2] cycloaddition reactions in the solid state. *Chemical Communications*, **42**, 5277–5288.

54. Cheung E. Y., Harris K. D. M., Kang T., Scheffer J. R. Trotter, J. (2006). Structure–reactivity correlations for solid-state enantioselective photochemical reactions established directly from powder X-ray diffraction. *Journal of the American Chemical Society*, **128**, 15554–15555.

55. Garcia-Garibay, M. A., Campos, L. M. Photochemical decarbonylation of ketones: Recent advances and reactions in crystalline solids. In Lenci, F., Horspool, W., Eds. *CRC Handbook of Organic Photochemistry and Photobiology*, second edition, CRC Press, Boca Raton, FL, 2004, pp. 48-1–48-41.

56. Sakamoto, M. (2004). Absolute asymmetric photochemistry using spontaneous chiral crystallization. *Molecular and Supramolecular Photochemistry*, **11**, 415–461.

57. Bernardi Bayarri, B., Carbonell, E., Gimenez, J., Esplugas, S., Garcia, H. (2002). Higher intrinsic photocatalytic efficiency of 2,4,6-triphenylpyrylium-based photocatalysts compared to TiO_2 P-25 for the degradation of 2,4-dichlorophenol using solar simulated light. *Chemosphere*, **72**, 67–74.
58. Bonesi, S. M., Carbonell, E., Garcia, H., Fagnoni, M., Albini, A. (2008). Photocatalytic oxidation of aliphatic and aromatic sulfides in the presence of silica adsorbed or zeolite-encapsulated 2,4,6-triphenyl(thia)pyrylium. *Applied Catalysis B—Environmental*, **79**, 368–375.
59. Kisch H. (2013). Semiconductor photocatalysis. Mechanistic and synthetic aspects. *Angewandte Chemie International Edition*, **52**, 812–847.
60. Cermenati L., Albini A. (2001). Titanium dioxide photocatalysis for radical alkylation. *Journal of Advanced Oxidation Technology*, **5**, 58–66.
61. Dake, G. R., Fenster, E. E., Patrick, B. O. (2008). A synthetic approach to the fusicoccane A–B ring fragment based on a Pauson–Khand cycloaddition/Norrish Type 1 fragmentation. *The Journal of Organic Chemistry*, **73**, 6711–6715.
62. Shvydkiv, O., Yavorskyy, A., Nolan, K., Youssef, A., Riguet, E., Hoffmann N., Oelgemöller, M. (2010). Photosensitized addition of isopropanol to furanones in a 365 nm UV-LED microchip. *Photochemical Photobiological Sciences*, **9**, 1601–1603.
63. Mukae, H., Maeda, H., Nashihara, S., Mizuno, K. (2007). Intramolecular photocycloaddition of 2-(2-alkenyloxymethyl)naphthalene-1-carbonitriles using glass-made microreactors. *Bulletin of the Chemical Society of Japan*, **80**, 1157–1161.
64. Maruyama, K., Ishitoku, T., Kubo, Y. (1981). Photochemistry of aliphatic imides. Synthesis of azetidine-2,4-diones via photochemical isomerization of succinimides and *N*-formyl-*N*-methyl α,β-unsaturated amides. *The Journal of Organic Chemistry*, **46**, 27–34.
65. Ravelli, D., Albini, A., Fagnoni, M. (2011). Smooth photocatalytic preparation of 2-substituted-1, 3-benzodioxoles. *Chemistry—A European Journal*, **17**, 572–579.
66. Ricci, A., Fasani, E., Mella, M., Albini, A. (2001). Noncommunicating photoreaction paths in some pregna-1,4-diene-3,20-diones. *The Journal of Organic Chemistry*, **66**, 8086–8093.
67. Matsumura, K., Mori, T., Inoue, Y. (2010). Solvent and temperature effects on diastereodifferentiating Paternò–Büchi reaction of chiral alkyl cyanobenzoates with diphenylethene upon direct versus charge-transfer excitation. *The Journal of Organic Chemistry*, **75**, 5461–5469.
68. Ischay, M. A., Anzovino, M. E., Du, J., Yoon, T. P. (2008). Efficient visible light photocatalysis of [2 + 2] enone cycloadditions. *Journal of the American Chemical Society* **130**, 12886–12887.
69. Narayanam, J. M. R., Tucker, J. W., Stephenson, C. R. J. (2009). Electron-transfer photoredox catalysis: Development of a tin-free reductive dehalogenation reaction. *Journal of the American Chemical Society*, **131**, 8756–8757.
70. Griesbeck, A. G., Maptue, N., Bondock, S., Oelgemöller, M. (2003). The excimer radiation system: A powerful tool for preparative organic photochemistry. A technical note. *Photochemical and Photobiological Sciences*, **2**, 450–451.
71. Streith, J., Leuttringe, P., Astasi, M. (1971). Photochemical synthesis of 1, 2-diazepines. V. Synthesis and rearrangements of 1,2-diazepines, *The Journal of Organic Chemistry*, **36**, 2962–2967.

72. Dichiarante, V., Fagnoni, M., Albini, A. (2010). Solarylations via 4-aminophenyl cations. *The Journal of Organic Chemistry*, **75**, 1271–1276.

73. Protti, S., Ravelli, D., Fagnoni, M., Albini, A. (2009). Solar light-driven photocatalyzed alkylations. Chemistry on the window ledge. *Chemical Communications*, 7351–7353.

74. Schiel, C., Oelgemöller, M., Ortner, J., Mattay, J. (2001). Green photochemistry: The solar-chemical Photo–Friedel–Crafts acylation of quinones. *Green Chemistry*, **3**, 224–228.

75. Neumann, M., Zeitler, K. (2012). Application of microflow conditions to visible light photoredox catalysis. *Organic Letters*, **14**, 2658–2661.

76. Kitamura, T. C—X bond fission in alkene system. In Lenci, F., Horspool, W., Eds. *CRC Handbook of Organic Photochemistry and Photobiology*, second edition, CRC Press, Boca Raton, FL, 2004, pp. 11-1–11-10.

77. Dichiarante, V., Fagnoni, M. (2008). Aryl cation chemistry as an emerging versatile tool for metal-free arylations. *Synlett*, 787–800.

INDEX

Photochemically-Generated Intermediates in Synthesis, First Edition.
Angelo Albini and Maurizio Fagnoni.

Printed and bound by CPI Group (UK) Ltd, Croydon, CR0 4YY
16/09/2026
14963417-0004